# DATSUN
## F10 & 310
### 1976-1981
#### SHOP MANUAL

By
**ALAN AHLSTRAND**

**ERIC JORGENSEN**
Editor

**JEFF ROBINSON**
Publisher

**CLYMER PUBLICATIONS**

World's largest publisher of books
devoted exclusively to automobiles and motorcycles

**12860 MUSCATINE STREET · P.O. BOX 20 · ARLETA, CALIFORNIA 91331**

Copyright © 1980, 1981 Clymer Publications

All rights reserved. No part of this publication may be reproduced, stored in a retrieval system or transmitted, in any form or by any means, electronic, mechanical, photocopying, recording or otherwise, without the written permission of Clymer Publications.

FIRST EDITION
First Printing July, 1980

SECOND EDITION
Revised by Alan Ahlstrand to include 1980-1981 models
First Printing August, 1981
Second Printing June, 1983

Printed in U.S.A.

ISBN: 0-89287-318-3

Production Coordinators, Mary Jo Yeager and Mark Whalen

COVER
Photographed by Michael Brown Photographic Productions, Los Angeles, California. Assisted by Chris Ryfle. Car courtesy of Nissan Motors, U.S.A.

**Chapter One**
General Information

**Chapter Two**
Troubleshooting

**Chapter Three**
Lubrication, Maintenance, and Tune-up

**Chapter Four**
Engine

**Chapter Five**
Fuel and Exhaust Systems

**Chapter Six**
Cooling System and Heater

**Chapter Seven**
Electrical System

**Chapter Eight**
Clutch and Transmission

**Chapter Nine**
Brakes

**Chapter Ten**
Rear Suspension

**Chapter Eleven**
Front Suspension and Steering

**Chapter Twelve**
Body

**Chapter Thirteen**
Air Conditioning

**Supplement**
1980 and Later Service Information

Index

Wiring Diagrams

# CONTENTS

QUICK REFERENCE DATA ................................................................................................. IX

## CHAPTER ONE
### GENERAL INFORMATION ................................................................................................. 1
Manual organization
Service hints
Safety first
Expendable supplies
Shop tools
Emergency tool kit
Troubleshooting and tune-up equipment

## CHAPTER TWO
### TROUBLESHOOTING ................................................................................................. 9
Starting system
Charging system
Engine performance
Engine oil pressure light
Fuel system (carburetted)
Fuel system (fuel injected)
Fuel pump test (mechanical and electric)
Emission control systems
Engine noises
Electrical accessories
Cooling system
Clutch
Manual transmission/transaxle
Automatic transmission
Brakes
Steering and suspension
Tire wear analysis
Wheel balancing

## CHAPTER THREE
### LUBRICATION, MAINTENANCE AND TUNE-UP ................................................................. 33
Fuel stop checks
Scheduled maintenance
Tune-up
Spark plugs
Distributor

## CHAPTER FOUR
### ENGINE ................................................................................................. 54
Engine removal (F10)
Engine installation (F10)
Engine removal (310)
Engine installation (310)
Disassembly sequences
Rocker assembly
Cylinder head
Valves/valve seats
Front cover, timing chain and sprockets
Camshaft
Connecting rods and pistons
Crankshaft
Cylinder block inspection
Flywheel

## CHAPTER FIVE
## FUEL AND EXHAUST SYSTEMS .................................................................................................. 83

- Air cleaner
- Carburetor
- Automatic choke circuit
- Carburetor cooling fan
- Fuel pump
- Intake and exhaust manifolds
- Throttle opener
- Exhaust gas recirculation system
- Early fuel evaporative system
- Air injection system
- Air induction system
- Evaporative control system
- Exhaust system
- Throttle linkage
- Fuel tank and lines

## CHAPTER SIX
## COOLING AND HEATING SYSTEM .............................................................................................. 123

- Cooling system flushing
- Thermostat
- Radiator
- Fan
- Water pump
- Heater

## CHAPTER SEVEN
## ELECTRICAL SYSTEM .................................................................................................................. 130

- Battery
- Alternator
- Starter
- Lighting system
- Switches
- Instruments
- Horn
- Windshield wipers/blades
- Fuses and fusible links
- Ignition system
- Spark timing control system

## CHAPTER EIGHT
## CLUTCH AND TRANSMISSION ................................................................................................... 169

- Clutch
- Clutch bleeding
- Clutch removal
- Clutch inspection
- Clutch installation
- Release bearing replacement
- Transmission
- Shift linkage

## CHAPTER NINE
## BRAKES .......................................................................................................................................... 201

- Front brakes
- Rear brakes
- Master cylinder
- Brake booster
- Proportioning valve
- Brake bleeding
- Adjustments

## CHAPTER TEN
## REAR SUSPENSION ...................................................................................................................... 221

- Rear suspension (coil springs)
- Rear suspension (leaf springs)
- Rear wheel bearings

## CHAPTER ELEVEN
### FRONT SUSPENSION AND STEERING .................................................................................. 230
Wheel alignment
Shock absorber replacement
Spring replacement
Sway bar
Transverse links
Ball-joints
Wheel bearings
Steering

## CHAPTER TWELVE
### BODY ........................................................................................................................................ 244
Body
Bumpers
Grille
Front fenders
Hood
Doors
Rear quarter windows
Trunk lid (F10 sedan)
Rear door (F10 hatchback)
Tailgate (F10 wagon)
Rear door (310)
Instrument panel
Seats

## CHAPTER THIRTEEN
### AIR CONDITIONING ............................................................................................................... 268
System operation
Get to know your vehicle's system
Routine maintenance
Refrigerant
Troubleshooting

## SUPPLEMENT
### 1980 AND LATER SERVICE INFORMATION ....................................................................... 277
Scheduled maintenance
Tune-up
Engine removal
Valves and valve seats
Air cleaner
Carburetor
Automatic choke circuit
Exhaust gas recirculation system
Air induction system
Air injection system
Fuel shutoff system
Exhaust system
Charging system
Lighting system
Ignition system
Spark timing control system
Master cylinder
Wheel alignment
Power steering

## INDEX ........................................................................................................................................ 309

## WIRING DIAGRAMS ............................................................................................... END OF BOOK

# DATSUN
## F10 & 310
### 1976-1981
SHOP MANUAL

# QUICK REFERENCE DATA

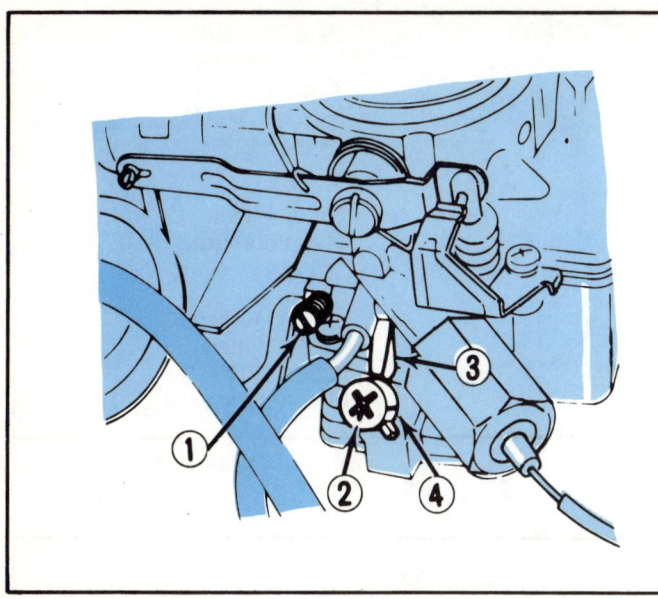

## CARBURETOR ADJUSTING SCREWS

1. Idle speed screw
2. Idle mixture screw (except 1981 California)
3. Stopper
4. Idle limiter cap

## F10 TIMING MARKS

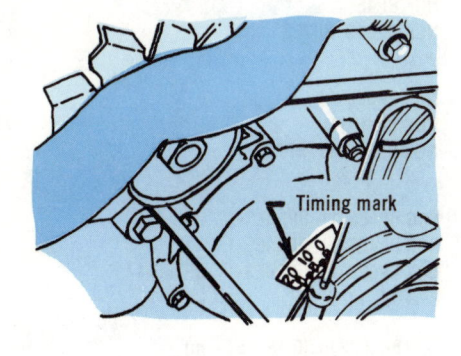

Timing mark

## 310 TIMING MARKS

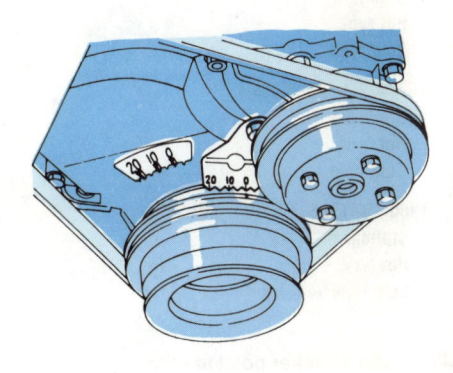

### RECOMMENDED LUBRICANTS

| | |
|---|---|
| Engine | API Service SE |
| Manual transmission | API GL-4 |
| Power steering | Dexron |
| Differential | API GL-5 |
| Steering gear | API GL-4 |
| Brake and clutch fluid | DOT 3 |
| Antifreeze | Ethylene glycol base with corrosion inhibitor |
| Windshield washer fluid | Windshield washer solvent/antifreeze |

## APPROXIMATE REFILL CAPACITIES

| | |
|---|---|
| Engine oil | |
|   1976-1977 F10 | 3⅝ qt. (3.4 liters) |
|   1978 F10 | 3½ qt. (3.3 liters) |
|   310 | 3⅜ qt. (3.2 liters) |
| Transmission oil | 4⅞ pt. (2.3 liters) |
| Cooling system | |
|   F10 | 7 qt. (6.6 liters) |
|   310 | 6 1/4 qt. (5.9 liters) |
| Fuel tank | |
|   1976 F10 sedan and hatchback | 10⅝ gal. (40 liters) |
|   1976 F10 wagon | 9¼ gal. (30 liters) |
|   1977-1978 F10 (all models) | 10⅝ gal. (40 liters) |
|   310 | 13¼ gal. (50 liters) |

## TUNE-UP SPECIFICATIONS

| | |
|---|---|
| Valve clearance (hot engine) | 0.014 in. (0.35 mm) |
| Spark plug gap | |
|   All 1976 models, 1977 and later Canadian models | 0.031-0.035 in. (0.8-0.9 mm) |
|   1977 and later U.S. models | 0.039-0.043 in. (1.0-1.1 mm) |
| Spark plug type | |
|   All 1976 models | |
|     Standard | NGK BP5ES |
|     Hot type | NGK BP4E |
|     Cold type | NGK BP6ES, BP7ES |
|   1977-1979 U.S. models | |
|     Standard | NGK BP5ES-11 |
|     Hot type | NGK BP4E-11 |
|     Cold type | NGK BP6ES-11, BP7ES-11 |
|   1980-1981 U.S. models | |
|     Standard | NKG BP5ES-11, BPR5ES-11 |
|     Hot type | NGK BP4ES-11, BPR4ES-11 |
|     Cold type | NGK BP6ES-11, BPR6ES-11, BP7ES-11, BPR7ES-11 |
| Point gap (breaker point ignition) | 0.018-0.022 in. (0.45-0.55 mm) |
| Dwell angle (breaker point ignition) | 49-55° |
| Ignition timing | |
|   All F10's | 10° BTDC at 700 rpm |
|   1979 California 310's | 5° BTDC at 700 rpm |
|   1979 non-California 310's | 10° BTDC at 700 rpm |
|   1980 U.S. | 8 +/- 2° BTDC at 750 +/- 50 rpm |
|   1980 Canada | 10 +/- 2° BTDC at 750 +/- 50 rpm |
|   1981 | 5 +/- 2° BTDC at 750 +/- 50 rpm |
| Idle speed | |
|   1976-1979 | 700 rpm |
|   1980-on | 750 +/- 50 rpm |
| Firing order | 1-3-4-2 |

## PLUG WIRE NUMBERS

FRONT →

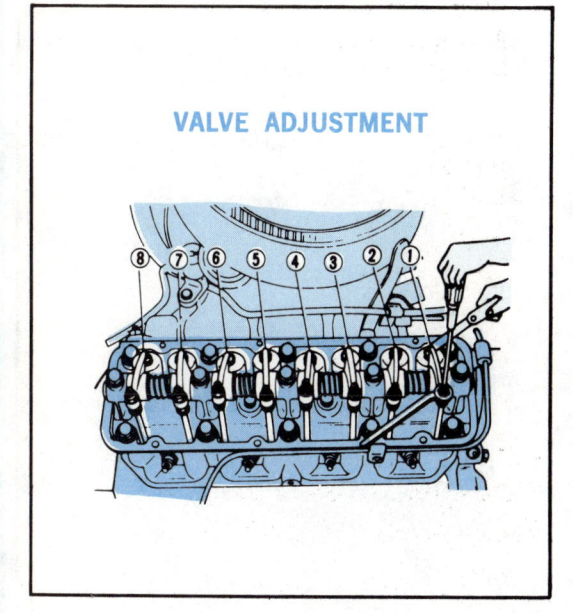

## LUBRICANT VISCOSITY

Engine Oil

- 20W-20, 20W-40, 20W-50 →
- 10W-30, 10W-40 →
- 10W
- ← 5W-30
- ← *5W-20

* Not recommended for sustained high-speed driving.

Gear Oil

- ← 75W
- ← 80W
- 85W
- 90
- 140 →

| °F | —20 | 0 | 20 | 40 | 60 | 80 | 100 |
|---|---|---|---|---|---|---|---|
| (°C) | (—29) | (—18) | (—7) | (4) | (15) | (27) | (38) |

**Temperature Range Anticipated Before Next Oil Change**

## VALVE ADJUSTMENT

## TIRE PRESSURES

| Size | Pressure (psi) |
|---|---|
| 155SR-13 | 24* |
| 165/70SR-13 | 24* |
| 155-13/6.15-13-4PR | |
|   Front | 28 |
|   Rear | 26 |

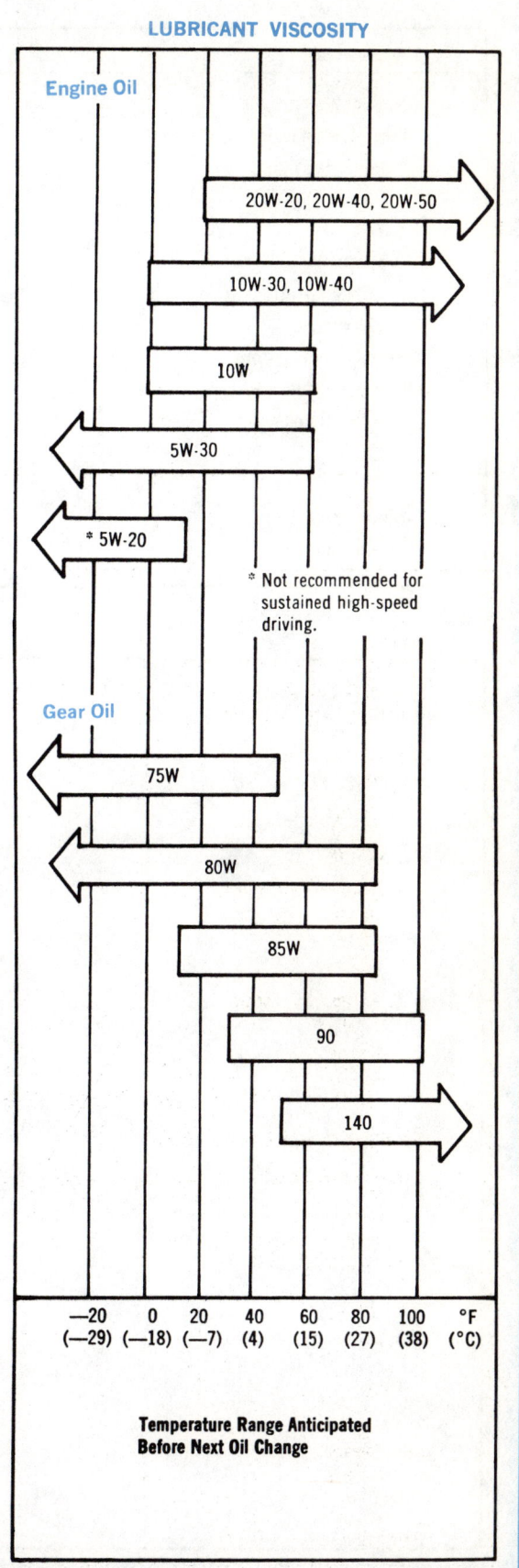

*On 310's, add 4 psi for speeds over 60 mph (100 kph).

## BULB CHART

| Application | Wattage | Trade number |
|---|---|---|
| Headlights | 50/40 | 6012 |
| Front turn/parking light | 23/8 | 1034 |
| Side marker lights | | |
|   F10 | 8 | 67 |
|   310 | 5 | 158 |
| Brake/taillights | | |
|   F10 sedan and hatchback, all 310 | 23/8 | 1034 |
|   F10 wagon | 23/5 | — |
| Rear turn indicators | 23 | 1078 |
| Back-up lights | 23 | 1078 |
| License plate light | | |
|   F10 | 7.5 | 89 |
|   310 | 8 | — |
| Dash lights | 3.4 | 194 |
| Interior light | 10 | — |

—NOTES—

# INTRODUCTION

This detailed, comprehensive manual covers 1976-1981 Datsun F10 and 310 models. The expert text gives complete information on maintenance, repair, and overhaul. Hundreds of photos and drawings guide you through every step. The book includes all you need to know to keep your car running right.

Where repairs are practical for the owner/mechanic, complete procedures are given. Equally important, difficult jobs are pointed out. Such operations are usually more economically performed by a dealer or independent garage.

A shop manual is a reference. You want to be able to find information fast. As in all Clymer books, this one is designed with this in mind. All chapters are *thumb* tabbed. Important items are indexed at the rear of the book. Finally, all the most frequently used specifications and capacities are summarized on the *Quick Reference* pages at the front of the book.

Keep the book handy. Carry it in your glove box. It will help you to better understand your car, lower repair and maintenance costs, and generally improve your satisfaction with your vehicle.

# CHAPTER ONE

# GENERAL INFORMATION

The troubleshooting, tune-up, maintenance, and step-by-step repair procedures in this book are written for the owner and home mechanic. The text is accompanied by useful photos and diagrams to make the job as clear and correct as possible.

Troubleshooting, tune-up, maintenance, and repair are not difficult if you know what tools and equipment to use and what to do. Anyone not afraid to get their hands dirty, of average intelligence, and with some mechanical ability can perform most of the procedures in this book.

In some cases, a repair job may require tools or skills not reasonably expected of the home mechanic. These procedures are noted in each chapter and it is recommended that you take the job to your dealer, a competent mechanic, or machine shop.

## MANUAL ORGANIZATION

This chapter provides general information and safety and service hints. Also included are lists of recommended shop and emergency tools as well as a brief description of troubleshooting and tune-up equipment.

Chapter Two provides methods and suggestions for quick and accurate diagnosis and repair of problems. Troubleshooting procedures discuss typical symptoms and logical methods to pinpoint the trouble.

Chapter Three explains all periodic lubrication and routine maintenance necessary to keep your vehicle running well. Chapter Three also includes recommended tune-up procedures, eliminating the need to constantly consult chapters on the various subassemblies.

Subsequent chapters cover specific systems such as the engine, transmission, and electrical systems. Each of these chapters provides disassembly, repair, and assembly procedures in a simple step-by-step format. If a repair requires special skills or tools, or is otherwise impractical for the home mechanic, it is so indicated. In these cases it is usually faster and less expensive to have the repairs made by a dealer or competent repair shop. Necessary specifications concerning a particular system are included at the end of the appropriate chapter.

When special tools are required to perform a procedure included in this manual, the tool is illustrated either in actual use or alone. It may be possible to rent or borrow these tools. The inventive mechanic may also be able to find a suitable substitute in his tool box, or to fabricate one.

The terms NOTE, CAUTION, and WARNING have specific meanings in this manual. A NOTE provides additional or explanatory information. A CAUTION is used to emphasize areas where equipment damage could result if proper precautions are not taken. A WARNING is used to stress those areas where personal injury or death could result from negligence, in addition to possible mechanical damage.

## SERVICE HINTS

Observing the following practices will save time, effort, and frustration, as well as prevent possible injury.

Throughout this manual keep in mind two conventions. "Front" refers to the front of the vehicle. The front of any component, such as the transmission, is that end which faces toward the front of the vehicle. The "left" and "right" sides of the vehicle refer to the orientation of a person sitting in the vehicle facing forward. For example, the steering wheel is on the left side. These rules are simple, but even experienced mechanics occasionally become disoriented.

Most of the service procedures covered are straightforward and can be performed by anyone reasonably handy with tools. It is suggested, however, that you consider your own capabilities carefully before attempting any operation involving major disassembly of the engine.

Some operations, for example, require the use of a press. It would be wiser to have these performed by a shop equipped for such work, rather than to try to do the job yourself with makeshift equipment. Other procedures require precision measurements. Unless you have the skills and equipment required, it would be better to have a qualified repair shop make the measurements for you.

Repairs go much faster and easier if the parts that will be worked on are clean before you begin. There are special cleaners for washing the engine and related parts. Brush or spray on the cleaning solution, let it stand, then rinse it away with a garden hose. Clean all oily or greasy parts with cleaning solvent as you remove them.

### WARNING
*Never use gasoline as a cleaning agent. It presents an extreme fire hazard. Be sure to work in a well-ventilated area when using cleaning solvent. Keep a fire extinguisher, rated for gasoline fires, handy in any case.*

Much of the labor charge for repairs made by dealers is for the removal and disassembly of other parts to reach the defective unit. It is frequently possible to perform the preliminary operations yourself and then take the defective unit in to the dealer for repair, at considerable savings.

Once you have decided to tackle the job yourself, make sure you locate the appropriate section in this manual, and read it entirely. Study the illustrations and text until you have a good idea of what is involved in completing the job satisfactorily. If special tools are required, make arrangements to get them before you start. Also, purchase any known defective parts prior to starting on the procedure. It is frustrating and time-consuming to get partially into a job and then be unable to complete it.

Simple wiring checks can be easily made at home, but knowledge of electronics is almost a necessity for performing tests with complicated electronic testing gear.

During disassembly of parts keep a few general cautions in mind. Force is rarely needed to get things apart. If parts are a tight fit, like a bearing in a case, there is usually a tool designed to separate them. Never use a screwdriver to pry apart parts with machined surfaces such as cylinder head and valve cover. You will mar the surfaces and end up with leaks.

Make diagrams wherever similar-appearing parts are found. You may think you can remember where everything came from — but mistakes are costly. There is also the possibility you may get sidetracked and not return to work for days or even weeks — in which interval, carefully laid out parts may have become disturbed.

Tag all similar internal parts for location, and mark all mating parts for position. Record number and thickness of any shims as they are removed. Small parts such as bolts can be iden-

# GENERAL INFORMATION

tified by placing them in plastic sandwich bags that are sealed and labeled with masking tape.

Wiring should be tagged with masking tape and marked as each wire is removed. Again, do not rely on memory alone.

When working under the vehicle, do not trust a hydraulic or mechanical jack to hold the vehicle up by itself. Always use jackstands. See **Figure 1**.

Disconnect battery ground cable before working near electrical connections and before disconnecting wires. Never run the engine with the battery disconnected; the alternator could be seriously damaged.

Protect finished surfaces from physical damage or corrosion. Keep gasoline and brake fluid off painted surfaces.

Frozen or very tight bolts and screws can often be loosened by soaking with penetrating oil like Liquid Wrench or WD-40, then sharply striking the bolt head a few times with a hammer and punch (or screwdriver for screws). Avoid heat unless absolutely necessary, since it may melt, warp, or remove the temper from many parts.

Avoid flames or sparks when working near a charging battery or flammable liquids, such as brake fluid or gasoline.

No parts, except those assembled with a press fit, require unusual force during assembly. If a part is hard to remove or install, find out why before proceeding.

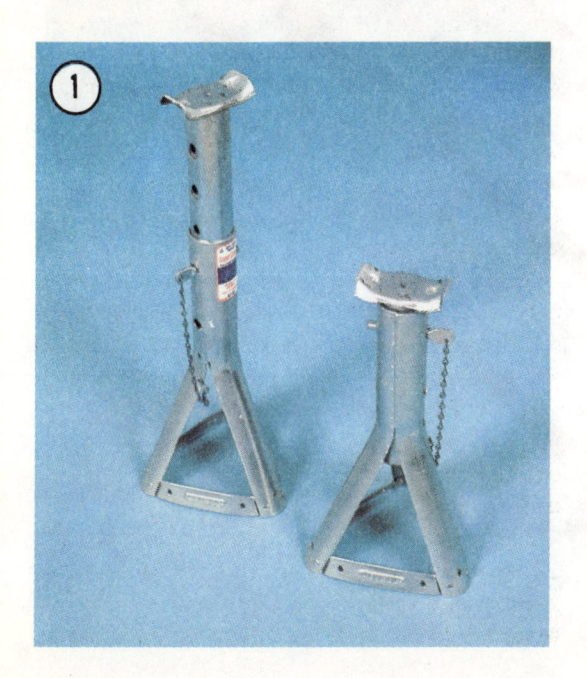

Cover all openings after removing parts to keep dirt, small tools, etc., from falling in.

When assembling two parts, start all fasteners, then tighten evenly.

The clutch plate, wiring connections, brake shoes, drums, pads, and discs should be kept clean and free of grease and oil.

When assembling parts, be sure all shims and washers are replaced exactly as they came out.

Whenever a rotating part butts against a stationary part, look for a shim or washer. Use new gaskets if there is any doubt about the condition of old ones. Generally, you should apply gasket cement to one mating surface only, so the parts may be easily disassembled in the future. A thin coat of oil on gaskets helps them seal effectively.

Heavy grease can be used to hold small parts in place if they tend to fall out during assembly. However, keep grease and oil away from electrical, clutch, and brake components.

High spots may be sanded off a piston with sandpaper, but emery cloth and oil do a much more professional job.

Carburetors are best cleaned by disassembling them and soaking the parts in a commercial carburetor cleaner. Never soak gaskets and rubber parts in these cleaners. Never use wire to clean out jets and air passages; they are easily damaged. Use compressed air to blow out the carburetor, but only if the float has been removed first.

Take your time and do the job right. Do not forget that a newly rebuilt engine must be broken in the same as a new one. Refer to your owner's manual for the proper break-in procedures.

## SAFETY FIRST

Professional mechanics can work for years and never sustain a serious injury. If you observe a few rules of common sense and safety, you can enjoy many safe hours servicing your vehicle. You could hurt yourself or damage the vehicle if you ignore these rules.

1. Never use gasoline as a cleaning solvent.

2. Never smoke or use a torch in the vicinity of flammable liquids such as cleaning solvent in open containers.

3. Never smoke or use a torch in an area where batteries are being charged. Highly explosive hydrogen gas is formed during the charging process.

4. Use the proper sized wrenches to avoid damage to nuts and injury to yourself.

5. When loosening a tight or stuck nut, be guided by what would happen if the wrench should slip. Protect yourself accordingly.

6. Keep your work area clean and uncluttered.

7. Wear safety goggles during all operations involving drilling, grinding, or use of a cold chisel.

8. Never use worn tools.

9. Keep a fire extinguisher handy and be sure it is rated for gasoline (Class B) and electrical (Class C) fires.

## EXPENDABLE SUPPLIES

Certain expendable supplies are necessary. These include grease, oil, gasket cement, wiping rags, cleaning solvent, and distilled water. Also, special locking compounds, silicone lubricants, and engine cleaners may be useful. Cleaning solvent is available at most service stations and distilled water for the battery is available at most supermarkets.

## SHOP TOOLS

For proper servicing, you will need an assortment of ordinary hand tools (**Figure 2**).

As a minimum, these include:

a. Combination wrenches
b. Sockets
c. Plastic mallet
d. Small hammer
e. Snap ring pliers
f. Gas pliers
g. Phillips screwdrivers
h. Slot (common) screwdrivers
i. Feeler gauges
j. Spark plug gauge
k. Spark plug wrench

Special tools necessary are shown in the chapters covering the particular repair in which they are used.

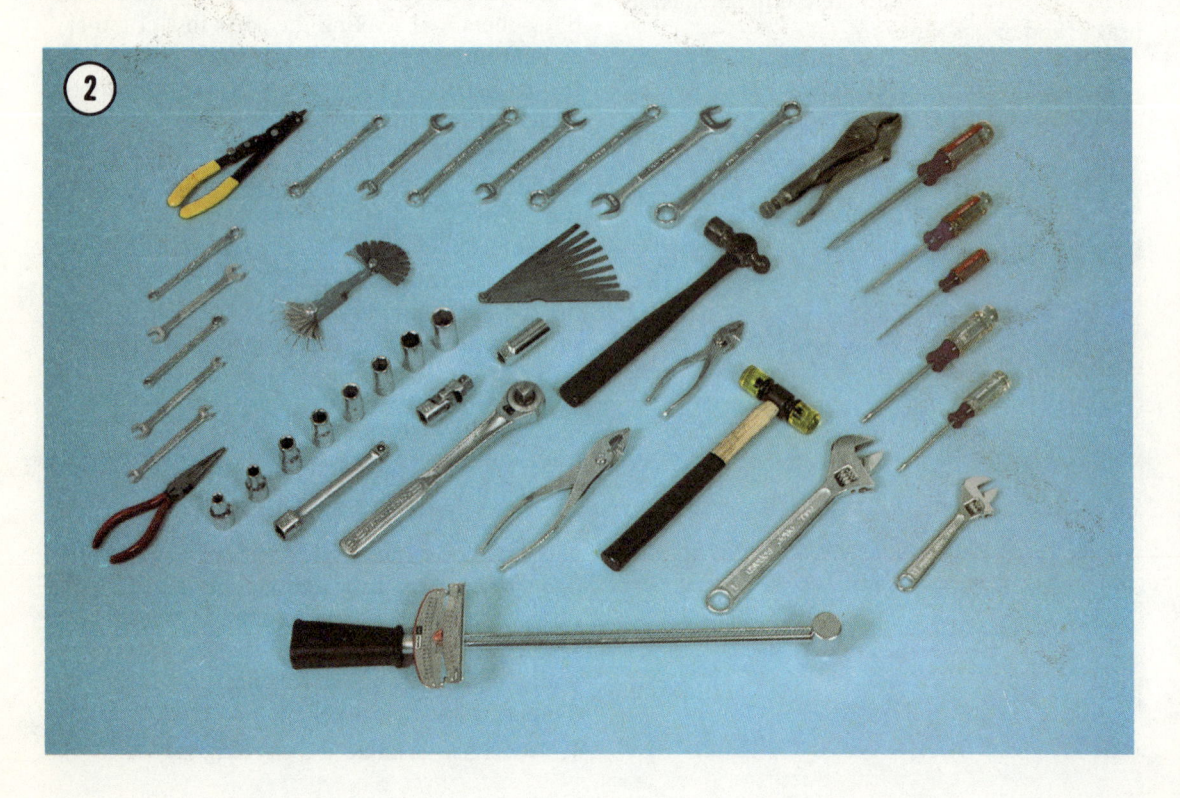

# GENERAL INFORMATION

Engine tune-up and troubleshooting procedures require other special tools and equipment. These are described in detail in the following sections.

## EMERGENCY TOOL KIT

A small emergency tool kit kept in the trunk is handy for road emergencies which otherwise could leave you stranded. The tools listed below and shown in **Figure 3** will let you handle most roadside repairs.

a. Combination wrenches
b. Crescent (adjustable) wrench
c. Screwdrivers — common and Phillips
d. Pliers — conventional (gas) and needle nose
e. Vise Grips
f. Hammer — plastic and metal
g. Small container of waterless hand cleaner
h. Rags for clean up
i. Silver waterproof sealing tape (duct tape)
j. Flashlight
k. Emergency road flares — at least four
l. Spare drive belts (water pump, alternator, etc.)

## TROUBLESHOOTING AND TUNE-UP EQUIPMENT

**Voltmeter, Ohmmeter, and Ammeter**

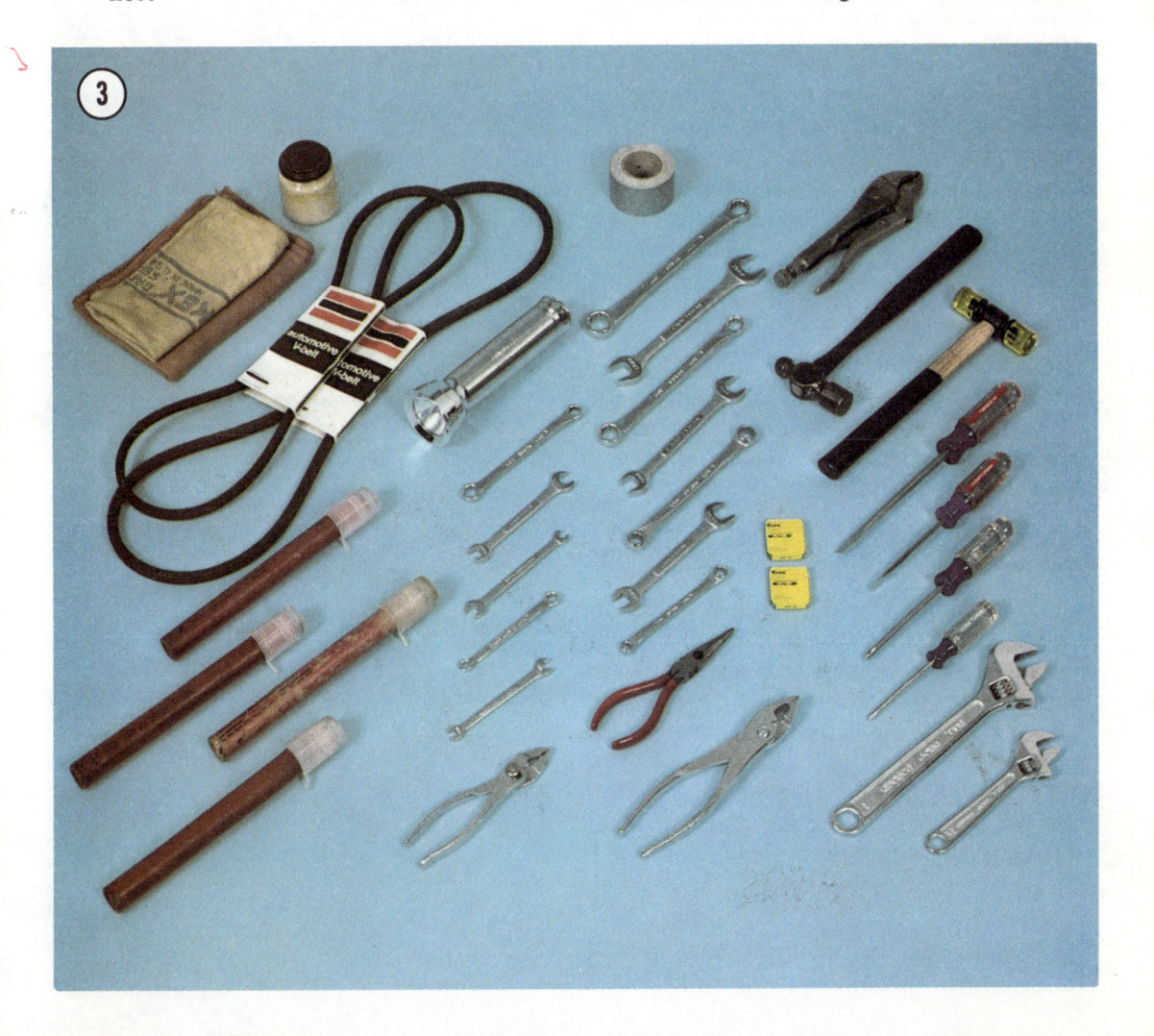

For testing the ignition or electrical system, a good voltmeter is required. For automotive use, an instrument covering 0-20 volts is satisfactory. One which also has a 0-2 volt scale is necessary for testing relays, points, or individual contacts where voltage drops are much smaller. Accuracy should be ± ½ volt.

An ohmmeter measures electrical resistance. This instrument is useful for checking continuity (open and short circuits), and testing fuses and lights.

The ammeter measures electrical current. Ammeters for automotive use should cover 0-50 amperes and 0-250 amperes. These are useful for checking battery charging and starting current.

Several inexpensive VOM's (volt-ohm-milliammeter) combine all three instruments into one which fits easily in any tool box. See **Figure 4**. However, the ammeter ranges are usually too small for automotive work.

## Hydrometer

The hydrometer gives a useful indication of battery condition and charge by measuring the specific gravity of the electrolyte in each cell. See **Figure 5**. Complete details on use and interpretation of readings are provided in the electrical chapter.

## Compression Tester

The compression tester measures the compression pressure built up in each cylinder. The results, when properly interpreted, can indicate general cylinder and valve condition. See **Figure 6**.

## Vacuum Gauge

The vacuum gauge (**Figure 7**) is one of the easiest instruments to use, but one of the most difficult for the inexperienced mechanic to interpret. The results, when interpreted with other findings, can provide valuable clues to possible trouble.

To use the vacuum gauge, connect it to a vacuum hose that goes to the intake manifold. Attach it either directly to the hose or to a T-fitting installed into the hose.

NOTE: *Subtract one inch from the reading for every 1,000 ft. elevation.*

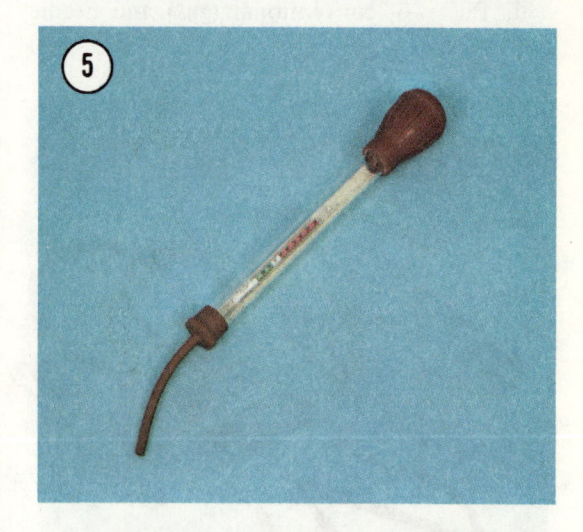

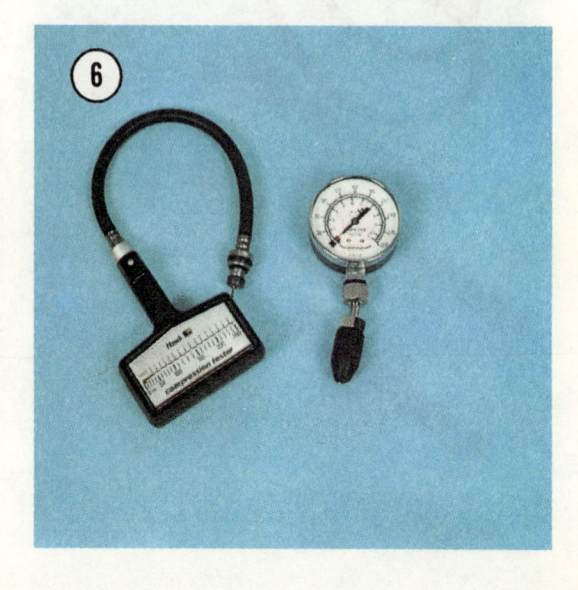

# GENERAL INFORMATION

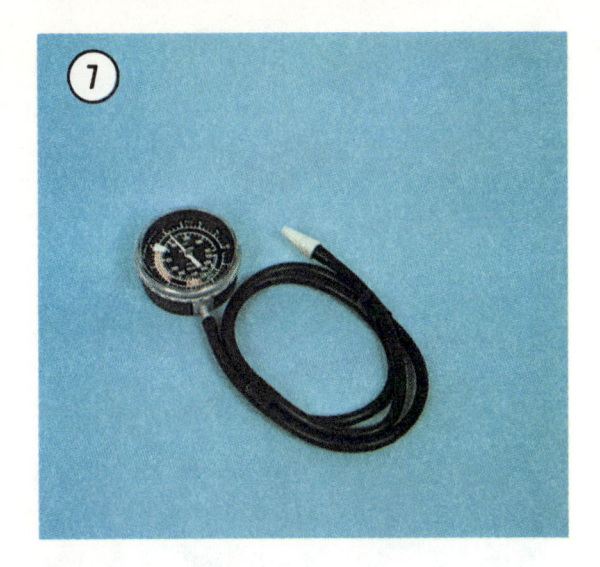

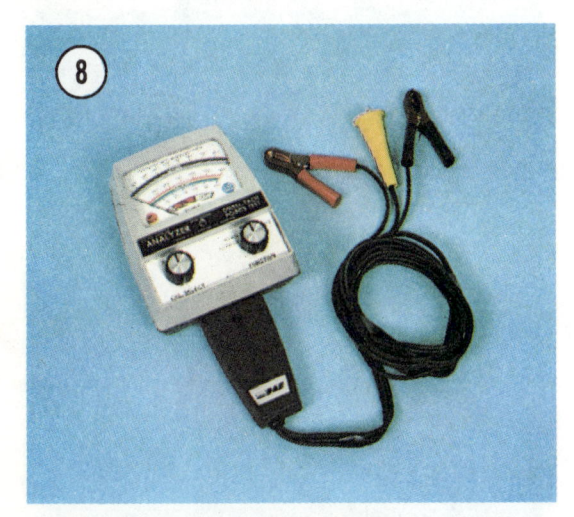

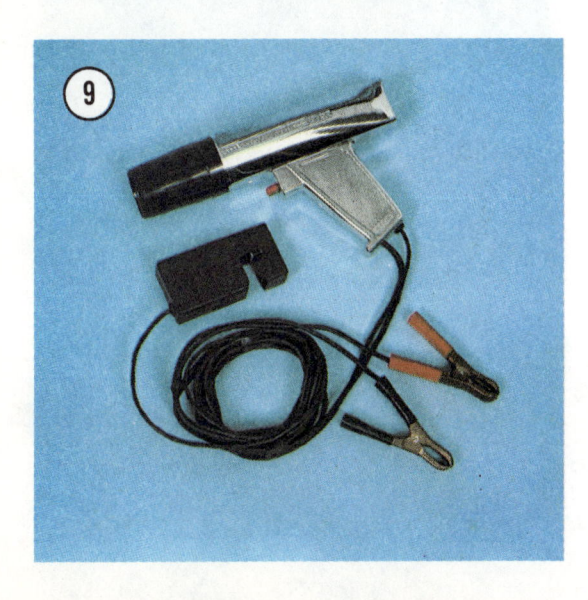

## Fuel Pressure Gauge

This instrument is invaluable for evaluating fuel pump performance. Fuel system troubleshooting procedures in this manual use a fuel pressure gauge. Usually a vacuum gauge and fuel pressure gauge are combined.

## Dwell Meter (Contact Breaker Point Ignition Only)

A dwell meter measures the distance in degrees of cam rotation that the breaker points remain closed while the engine is running. Since this angle is determined by breaker point gap, dwell angle is an accurate indication of breaker point gap.

Many tachometers intended for tuning and testing incorporate a dwell meter as well. See **Figure 8**. Follow the manufacturer's instructions to measure dwell.

## Tachometer

A tachometer is necessary for tuning. See **Figure 8**. Ignition timing and carburetor adjustments must be performed at the specified idle speed. The best instrument for this purpose is one with a low range of 0-1,000 or 0-2,000 rpm for setting idle, and a high range of 0-4,000 or more for setting ignition timing at 3,000 rpm. Extended range (0-6,000 or 0-8,000 rpm) instruments lack accuracy at lower speeds. The instrument should be capable of detecting changes of 25 rpm on the low range.

## Strobe Timing Light

This instrument is necessary for tuning, as it permits very accurate ignition timing. The light flashes at precisely the same instant that No. 1 cylinder fires, at which time the timing marks on the engine should align. Refer to Chapter Three for exact location of the timing marks for your engine.

Suitable lights range from inexpensive neon bulb types ($2-3) to powerful xenon strobe lights ($20-40). See **Figure 9**. Neon timing lights are difficult to see and must be used in dimly lit areas. Xenon strobe timing lights can be used outside in bright sunlight. Both types work on this vehicle; use according to the manufacturer's instructions.

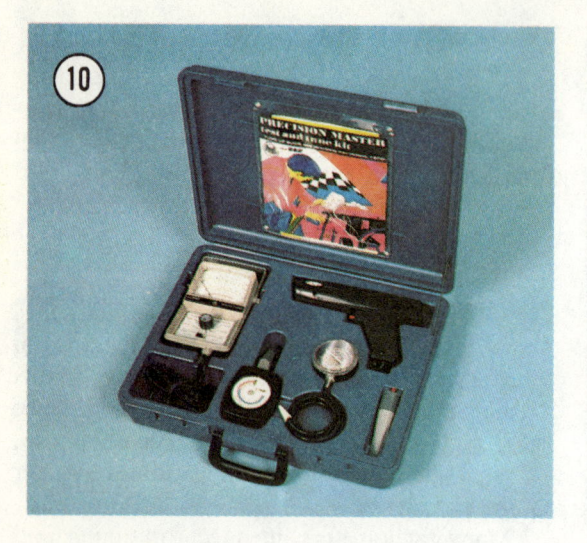

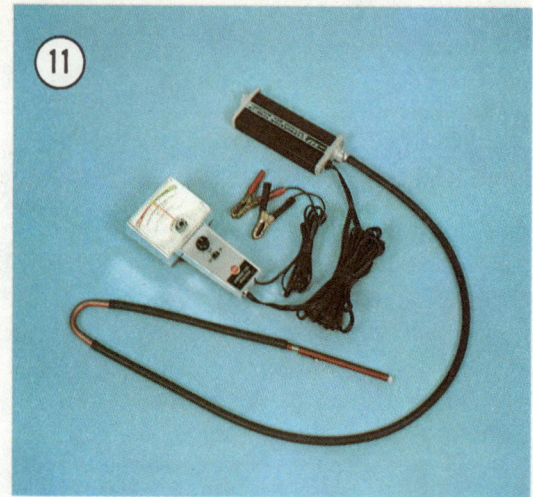

### Tune-up Kits

Many manufacturers offer kits that combine several useful instruments. Some come in a convenient carry case and are usually less expensive than purchasing one instrument at a time. **Figure 10** shows one of the kits that is available. The prices vary with the number of instruments included in the kit.

### Exhaust Gas Analyzer

Of all instruments described here, this is the least likely to be owned by a home mechanic. This instrument samples the exhaust gases from the tailpipe and measures the thermal conductivity of the exhaust gas. Since different gases conduct heat at varying rates, thermal conductivity of the exhaust is a good indication of gases present.

An exhaust gas analyzer is vital for accurately checking the effectiveness of exhaust emission control adjustments. They are relatively expensive to buy ($70 and up), but must be considered essential for the owner/mechanic to comply with today's emission laws. See **Figure 11**.

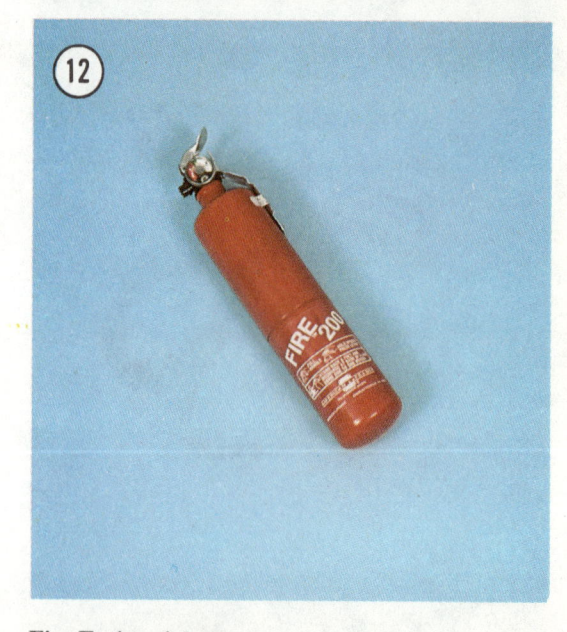

### Fire Extinguisher

A fire extinguisher is a necessity when working on a vehicle. It should be rated for both *Class B* (flammable liquids — gasoline, oil, paint, etc.) and *Class C* (electrical — wiring, etc.) type fires. It should always be kept within reach. See **Figure 12**.

# CHAPTER TWO

# TROUBLESHOOTING

Troubleshooting can be a relatively simple matter if it is done logically. The first step in any troubleshooting procedure must be defining the symptoms as closely as possible. Subsequent steps involve testing and analyzing areas which could cause the symptoms. A haphazard approach may eventually find the trouble, but in terms of wasted time and unnecessary parts replacement, it can be very costly.

The troubleshooting procedures in this chapter analyze typical symptoms and show logical methods of isolation. These are not the only methods. There may be several approaches to a problem, but all methods must have one thing in common — a logical, systematic approach.

## STARTING SYSTEM

The starting system consists of the starter motor and the starter solenoid. The ignition key controls the starter solenoid, which mechanically engages the starter with the engine flywheel, and supplies electrical current to turn the starter motor.

Starting system troubles are relatively easy to find. In most cases, the trouble is a loose or dirty electrical connection. **Figures 1 and 2** provide routines for finding the trouble.

## CHARGING SYSTEM

The charging system consists of the alternator (or generator on older vehicles), voltage regulator, and battery. A drive belt driven by the engine crankshaft turns the alternator which produces electrical energy to charge the battery. As engine speed varies, the voltage from the alternator varies. A voltage regulator controls the charging current to the battery and maintains the voltage to the vehicle's electrical system at safe levels. A warning light or gauge on the instrument panel signals the driver when charging is not taking place. Refer to **Figure 3** for a typical charging system.

Complete troubleshooting of the charging system requires test equipment and skills which the average home mechanic does not possess. However, there are a few tests which can be done to pinpoint most troubles.

Charging system trouble may stem from a defective alternator (or generator), voltage regulator, battery, or drive belt. It may also be caused by something as simple as incorrect drive belt tension. The following are symptoms of typical problems you may encounter.

1. *Battery dies frequently, even though the warning lamp indicates no discharge* — This can be caused by a drive belt that is slightly too

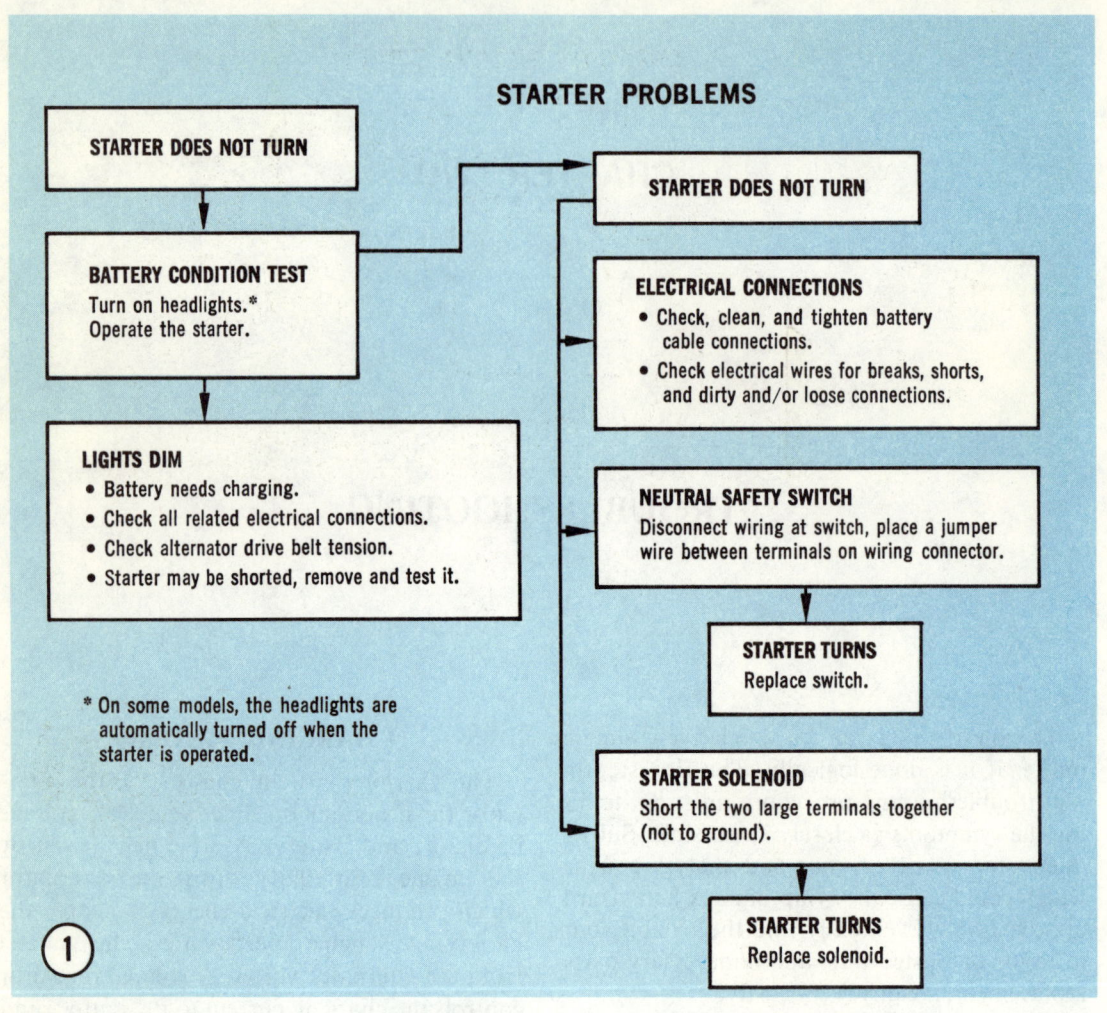

# CHAPTER TWO

## STARTER PROBLEMS

**STARTER DOES NOT TURN**
↓
**BATTERY CONDITION TEST**
Turn on headlights.*
Operate the starter.
↓
**LIGHTS DIM**
- Battery needs charging.
- Check all related electrical connections.
- Check alternator drive belt tension.
- Starter may be shorted, remove and test it.

\* On some models, the headlights are automatically turned off when the starter is operated.

**STARTER DOES NOT TURN**
↓
**ELECTRICAL CONNECTIONS**
- Check, clean, and tighten battery cable connections.
- Check electrical wires for breaks, shorts, and dirty and/or loose connections.

**NEUTRAL SAFETY SWITCH**
Disconnect wiring at switch, place a jumper wire between terminals on wiring connector.
↓
**STARTER TURNS**
Replace switch.

**STARTER SOLENOID**
Short the two large terminals together (not to ground).
↓
**STARTER TURNS**
Replace solenoid.

① 

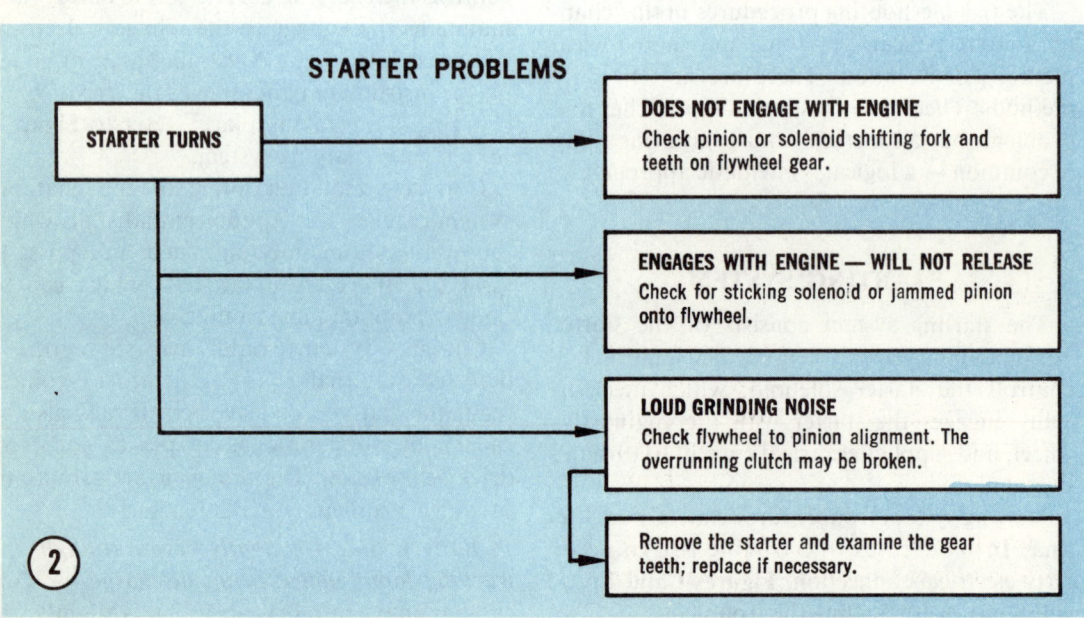

## STARTER PROBLEMS

**STARTER TURNS**

**DOES NOT ENGAGE WITH ENGINE**
Check pinion or solenoid shifting fork and teeth on flywheel gear.

**ENGAGES WITH ENGINE — WILL NOT RELEASE**
Check for sticking solenoid or jammed pinion onto flywheel.

**LOUD GRINDING NOISE**
Check flywheel to pinion alignment. The overrunning clutch may be broken.

Remove the starter and examine the gear teeth; replace if necessary.

②

# TROUBLESHOOTING

**CHARGING SYSTEM CIRCUIT**

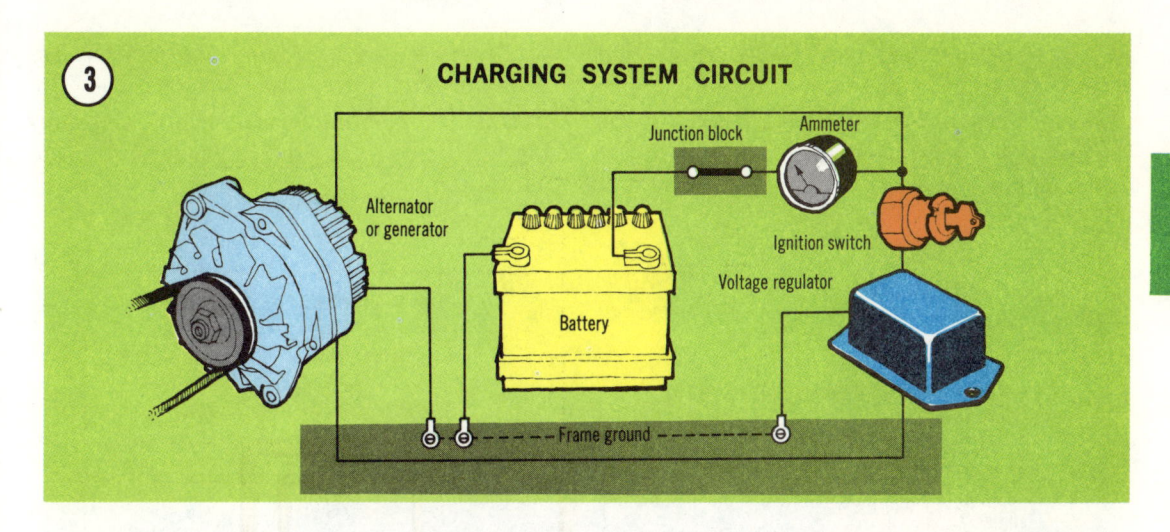

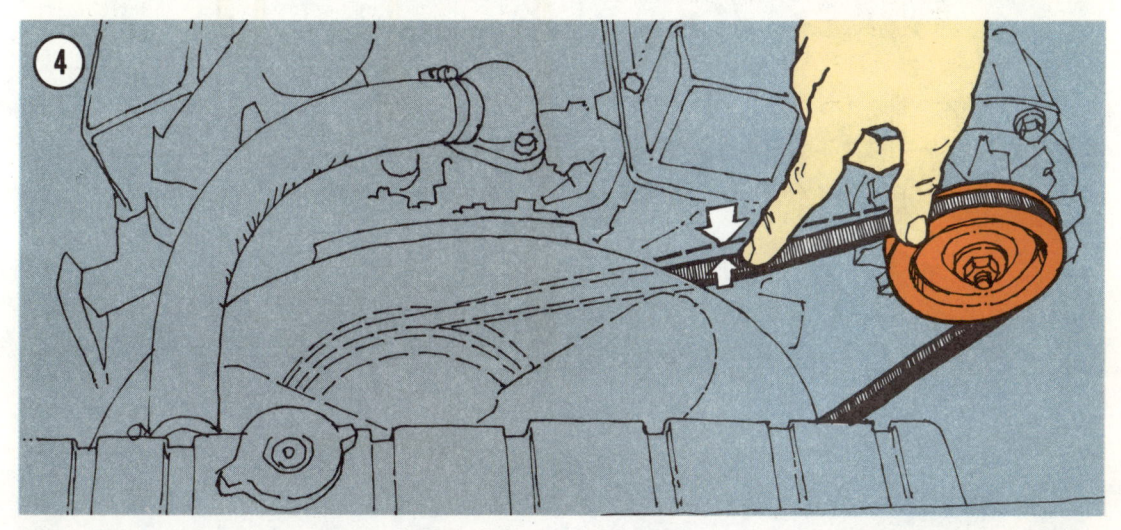

loose. Grasp the alternator (or generator) pulley and try to turn it. If the pulley can be turned without moving the belt, the drive belt is too loose. As a rule, keep the belt tight enough that it can be deflected about ½ in. under moderate thumb pressure between the pulleys (**Figure 4**). The battery may also be at fault; test the battery condition.

2. *Charging system warning lamp does not come on when ignition switch is turned on* — This may indicate a defective ignition switch, battery, voltage regulator, or lamp. First try to start the vehicle. If it doesn't start, check the ignition switch and battery. If the car starts, remove the warning lamp; test it for continuity with an ohmmeter or substitute a new lamp. If the lamp is good, locate the voltage regulator and make sure it is properly grounded (try tightening the mounting screws). Also the alternator (or generator) brushes may not be making contact. Test the alternator (or generator) and voltage regulator.

3. *Alternator (or generator) warning lamp comes on and stays on* — This usually indicates that no charging is taking place. First check drive belt tension (**Figure 4**). Then check battery condition, and check all wiring connections in the charging system. If this does not locate the trouble, check the alternator (or generator) and voltage regulator.

4. *Charging system warning lamp flashes on and off intermittently* — This usually indicates the charging system is working intermittently.

## CHAPTER TWO

Check the drive belt tension (**Figure 4**), and check all electrical connections in the charging system. Check the alternator (or generator). *On generators only*, check the condition of the commutator.

5. *Battery requires frequent additions of water, or lamps require frequent replacement* — The alternator (or generator) is probably overcharging the battery. The voltage regulator is probably at fault.

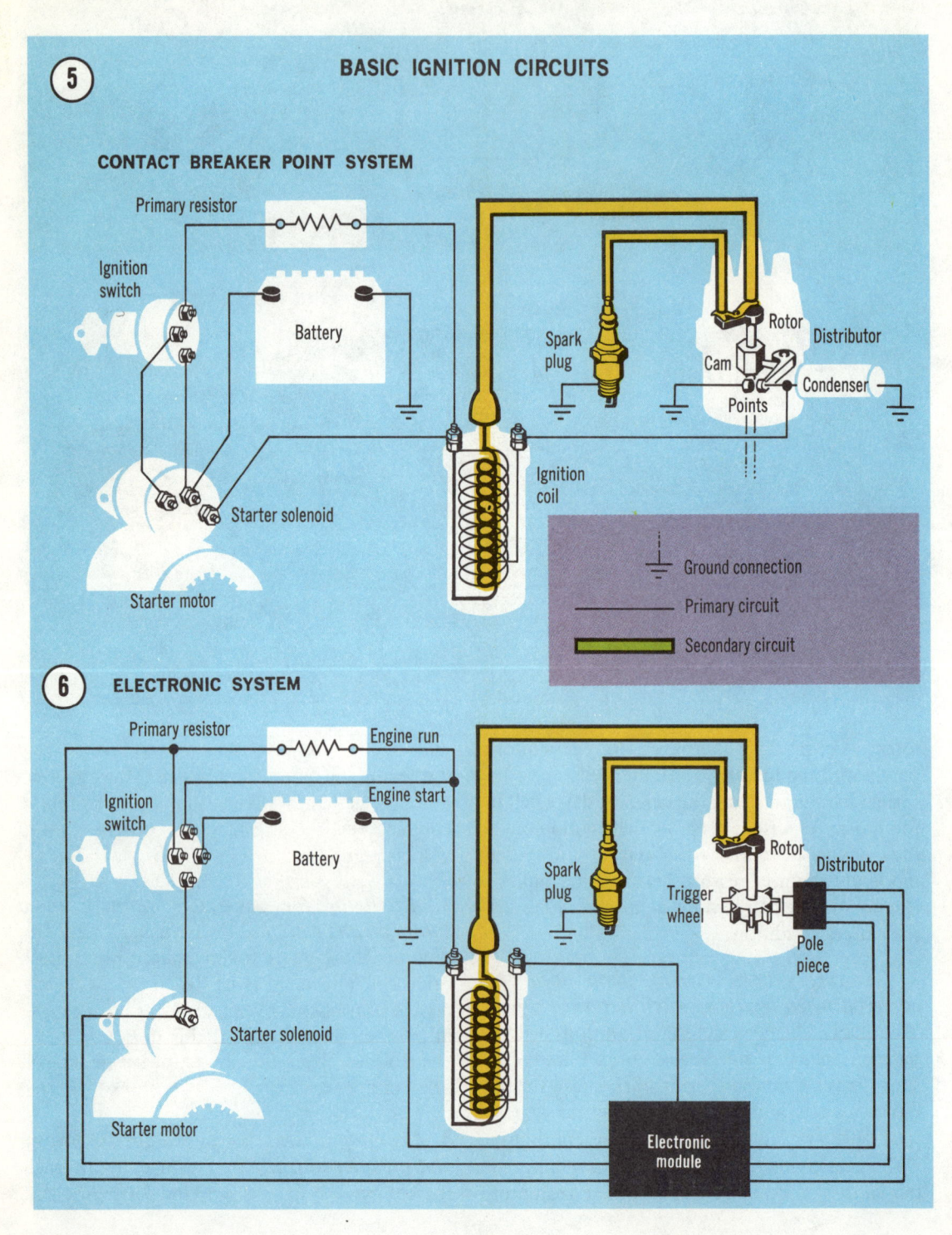

**BASIC IGNITION CIRCUITS**

**⑤ CONTACT BREAKER POINT SYSTEM**

**⑥ ELECTRONIC SYSTEM**

# TROUBLESHOOTING

6. *Excessive noise from the alternator (or generator)* — Check for loose mounting brackets and bolts. The problem may also be worn bearings or the need of lubrication in some cases. If an alternator whines, a shorted diode may be indicated.

## IGNITION SYSTEM

The ignition system may be either a conventional contact breaker type or an electronic ignition. See electrical chapter to determine which type you have. **Figures 5 and 6** show simplified diagrams of each type.

Most problems involving failure to start, poor performance, or rough running stem from trouble in the ignition system, particularly in contact breaker systems. Many novice troubleshooters get into trouble when they assume that these symptoms point to the fuel system instead of the ignition system.

Ignition system troubles may be roughly divided between those affecting only one cylinder and those affecting all cylinders. If the trouble affects only one cylinder, it can only be in the spark plug, spark plug wire, or portion of the distributor associated with that cylinder. If the trouble affects all cylinders (weak spark or no spark), then the trouble is in the ignition coil, rotor, distributor, or associated wiring.

The troubleshooting procedures outlined in **Figure 7** (breaker point ignition) or **Figure 8**

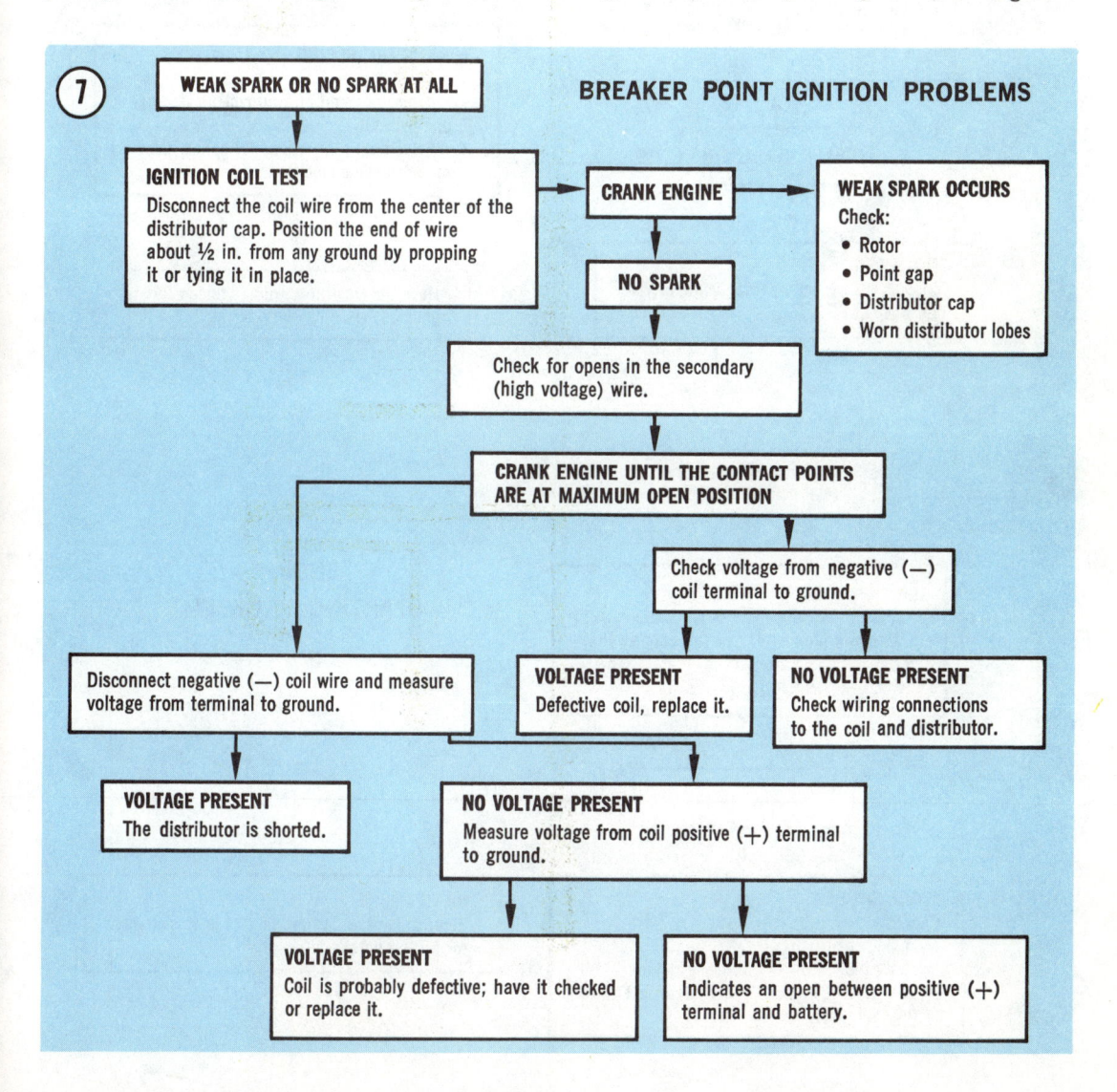

---

**⑦ BREAKER POINT IGNITION PROBLEMS**

**WEAK SPARK OR NO SPARK AT ALL**

**IGNITION COIL TEST**
Disconnect the coil wire from the center of the distributor cap. Position the end of wire about ½ in. from any ground by propping it or tying it in place.

→ **CRANK ENGINE**

→ **WEAK SPARK OCCURS**
Check:
- Rotor
- Point gap
- Distributor cap
- Worn distributor lobes

**NO SPARK**

Check for opens in the secondary (high voltage) wire.

**CRANK ENGINE UNTIL THE CONTACT POINTS ARE AT MAXIMUM OPEN POSITION**

Check voltage from negative (—) coil terminal to ground.

**VOLTAGE PRESENT**
Defective coil, replace it.

**NO VOLTAGE PRESENT**
Check wiring connections to the coil and distributor.

Disconnect negative (—) coil wire and measure voltage from terminal to ground.

**VOLTAGE PRESENT**
The distributor is shorted.

**NO VOLTAGE PRESENT**
Measure voltage from coil positive (+) terminal to ground.

**VOLTAGE PRESENT**
Coil is probably defective; have it checked or replace it.

**NO VOLTAGE PRESENT**
Indicates an open between positive (+) terminal and battery.

(electronic ignition) will help you isolate ignition problems fast. Of course, they assume that the battery is in good enough condition to crank the engine over at its normal rate.

## ENGINE PERFORMANCE

A number of factors can make the engine difficult or impossible to start, or cause rough running, poor performance and so on. The majority of novice troubleshooters immediately suspect the carburetor or fuel injection system. In the majority of cases, though, the trouble exists in the ignition system.

The troubleshooting procedures outlined in **Figures 9 through 14** will help you solve the majority of engine starting troubles in a systematic manner.

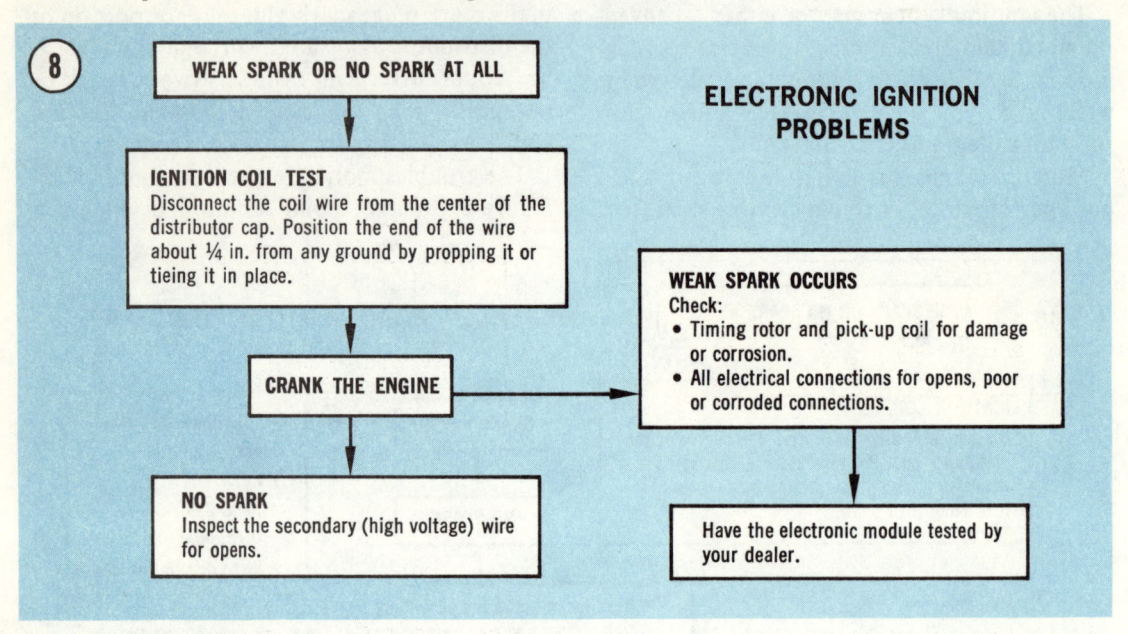

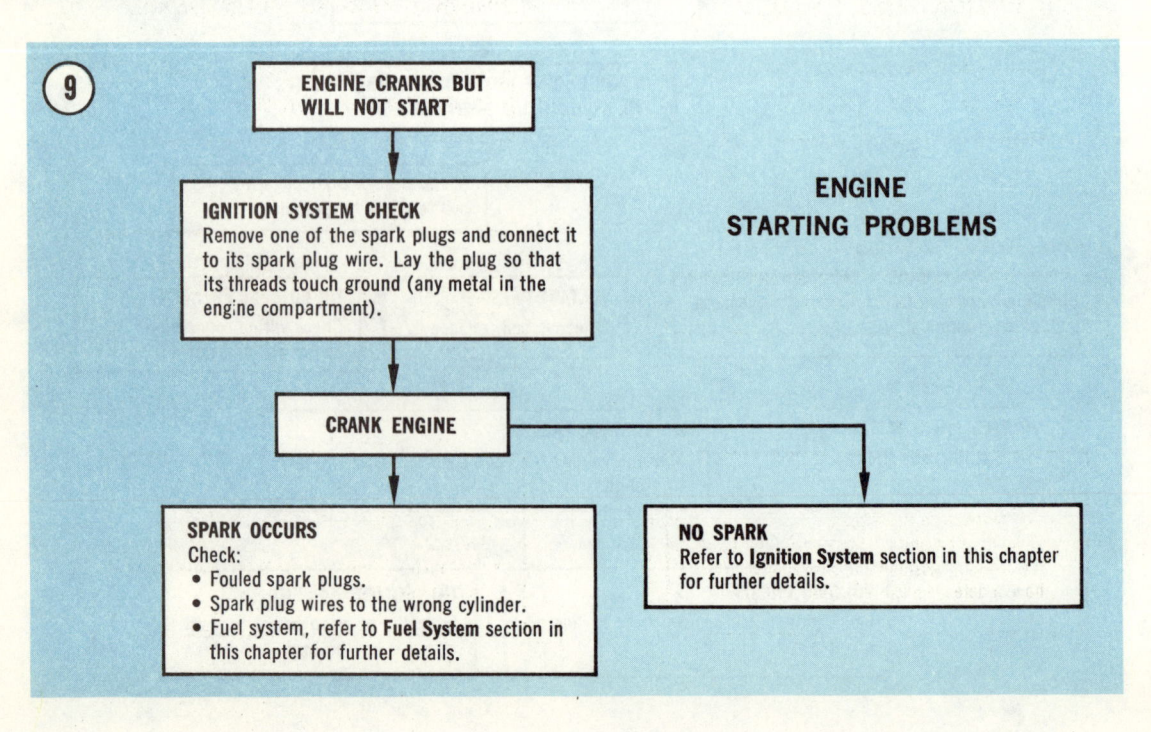

**⑧ ELECTRONIC IGNITION PROBLEMS**

WEAK SPARK OR NO SPARK AT ALL
↓
**IGNITION COIL TEST**
Disconnect the coil wire from the center of the distributor cap. Position the end of the wire about ¼ in. from any ground by propping it or tieing it in place.
↓
**CRANK THE ENGINE** →
↓
**NO SPARK**
Inspect the secondary (high voltage) wire for opens.

**WEAK SPARK OCCURS**
Check:
- Timing rotor and pick-up coil for damage or corrosion.
- All electrical connections for opens, poor or corroded connections.
↓
Have the electronic module tested by your dealer.

**⑨ ENGINE STARTING PROBLEMS**

ENGINE CRANKS BUT WILL NOT START
↓
**IGNITION SYSTEM CHECK**
Remove one of the spark plugs and connect it to its spark plug wire. Lay the plug so that its threads touch ground (any metal in the engine compartment).
↓
**CRANK ENGINE** →
↓
**SPARK OCCURS**
Check:
- Fouled spark plugs.
- Spark plug wires to the wrong cylinder.
- Fuel system, refer to **Fuel System** section in this chapter for further details.

**NO SPARK**
Refer to **Ignition System** section in this chapter for further details.

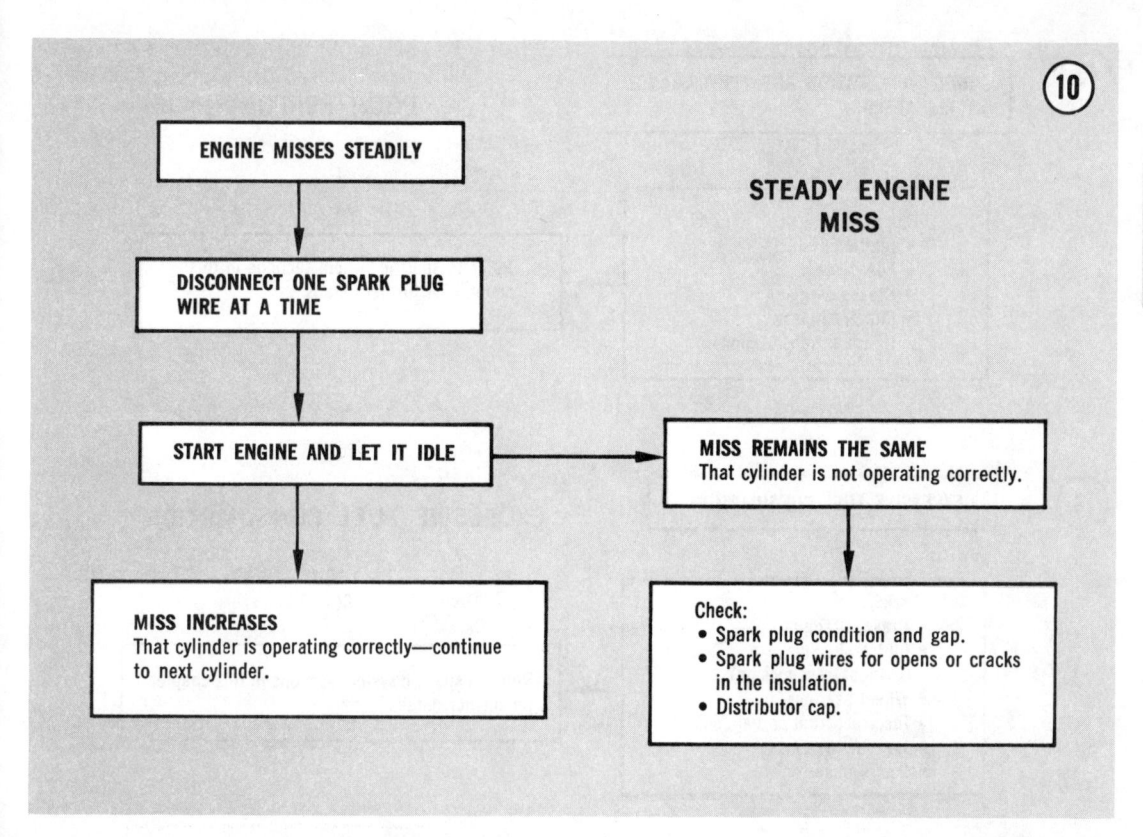

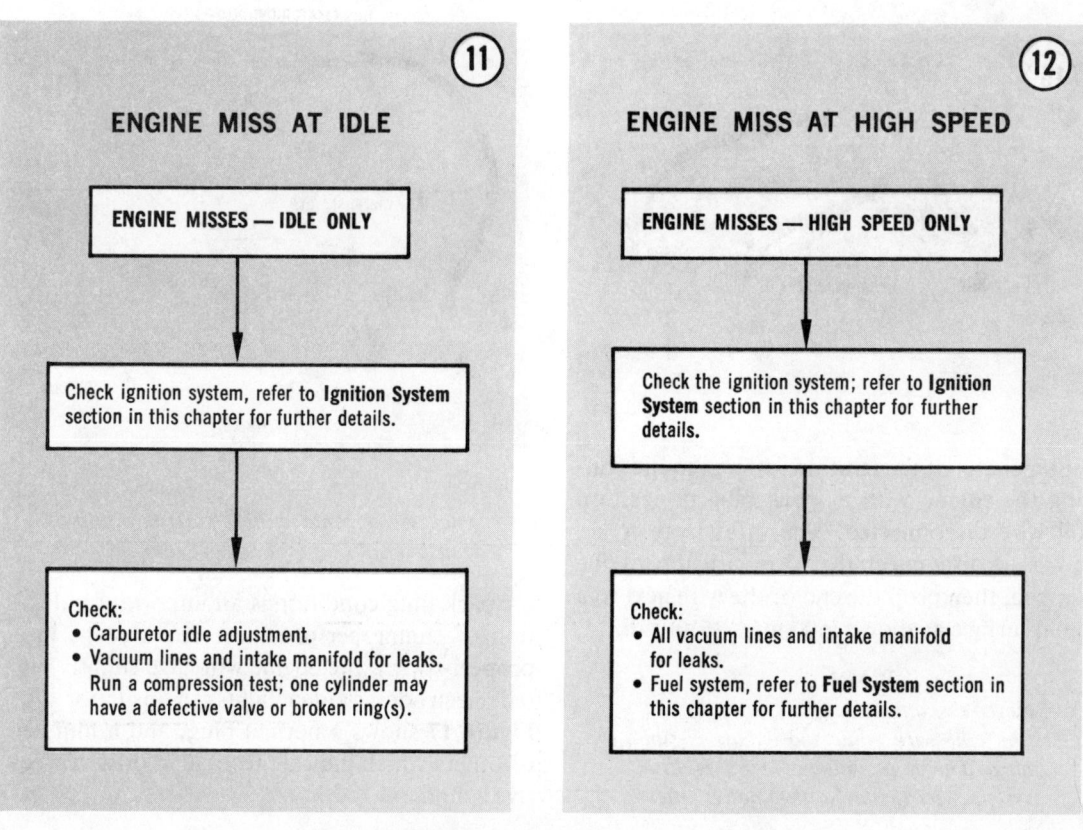

# TROUBLESHOOTING

⑩

**STEADY ENGINE MISS**

```
ENGINE MISSES STEADILY
          ↓
DISCONNECT ONE SPARK PLUG
     WIRE AT A TIME
          ↓
START ENGINE AND LET IT IDLE  →  MISS REMAINS THE SAME
          ↓                       That cylinder is not operating correctly.
                                          ↓
MISS INCREASES                    Check:
That cylinder is operating         • Spark plug condition and gap.
correctly—continue                 • Spark plug wires for opens or cracks
to next cylinder.                    in the insulation.
                                   • Distributor cap.
```

⑪ **ENGINE MISS AT IDLE**

ENGINE MISSES — IDLE ONLY
↓
Check ignition system, refer to **Ignition System** section in this chapter for further details.
↓
Check:
- Carburetor idle adjustment.
- Vacuum lines and intake manifold for leaks. Run a compression test; one cylinder may have a defective valve or broken ring(s).

⑫ **ENGINE MISS AT HIGH SPEED**

ENGINE MISSES — HIGH SPEED ONLY
↓
Check the ignition system; refer to **Ignition System** section in this chapter for further details.
↓
Check:
- All vacuum lines and intake manifold for leaks.
- Fuel system, refer to **Fuel System** section in this chapter for further details.

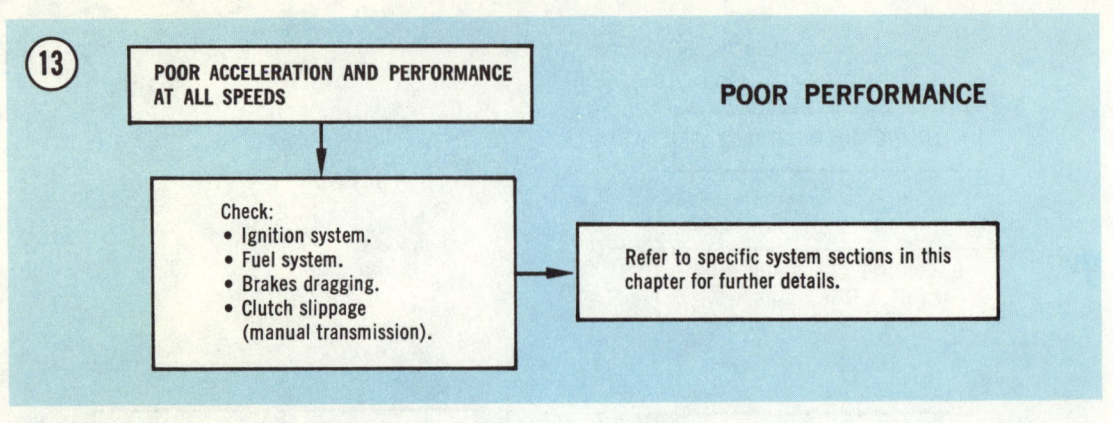

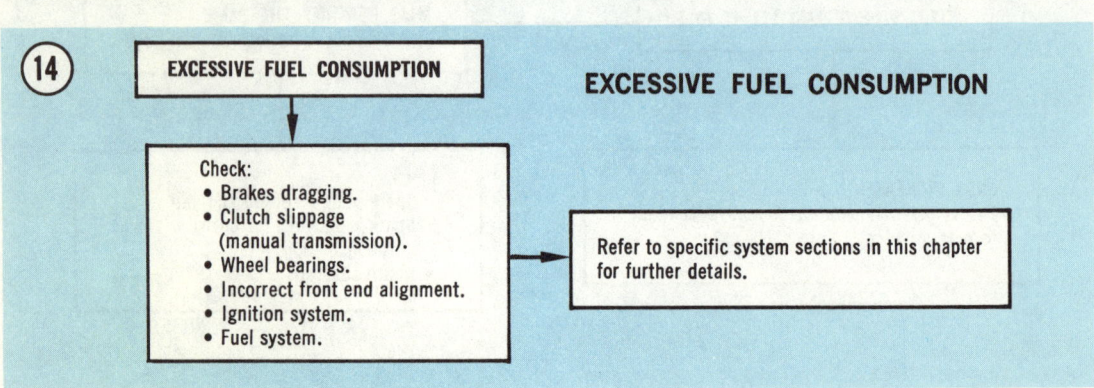

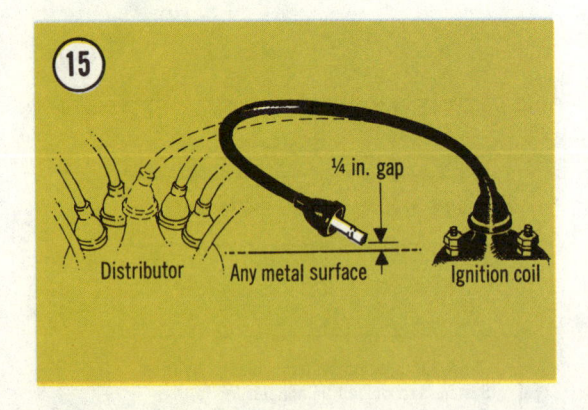

## CHAPTER TWO

**(13)** POOR ACCELERATION AND PERFORMANCE AT ALL SPEEDS

**POOR PERFORMANCE**

Check:
- Ignition system.
- Fuel system.
- Brakes dragging.
- Clutch slippage (manual transmission).

→ Refer to specific system sections in this chapter for further details.

**(14)** EXCESSIVE FUEL CONSUMPTION

**EXCESSIVE FUEL CONSUMPTION**

Check:
- Brakes dragging.
- Clutch slippage (manual transmission).
- Wheel bearings.
- Incorrect front end alignment.
- Ignition system.
- Fuel system.

→ Refer to specific system sections in this chapter for further details.

**(15)** Distributor — ¼ in. gap — Any metal surface — Ignition coil

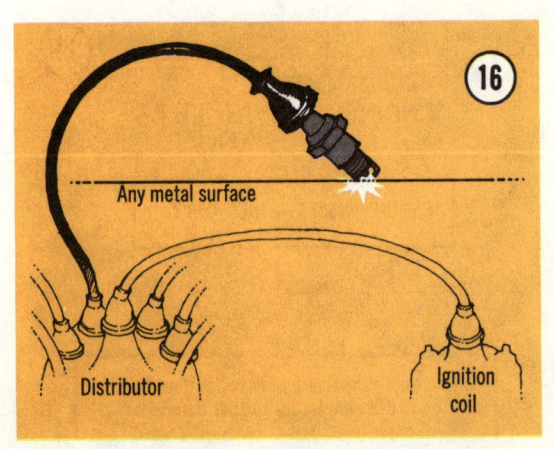

**(16)** Any metal surface — Distributor — Ignition coil

Some tests of the ignition system require running the engine with a spark plug or ignition coil wire disconnected. The safest way to do this is to disconnect the wire with the engine stopped, then prop the end of the wire next to a metal surface as shown in **Figures 15 and 16**.

### WARNING
*Never disconnect a spark plug or ignition coil wire while the engine is running. The high voltage in an ignition system, particularly the newer high-energy electronic ignition systems could cause serious injury or even death.*

Spark plug condition is an important indication of engine performance. Spark plugs in a properly operating engine will have slightly pitted electrodes, and a light tan insulator tip. **Figure 17** shows a normal plug, and a number of others which indicate trouble in their respective cylinders.

# TROUBLESHOOTING

⑰

### NORMAL
- Appearance—Firing tip has deposits of light gray to light tan.
- Can be cleaned, regapped and reused.

### CARBON FOULED
- Appearance—Dull, dry black with fluffy carbon deposits on the insulator tip, electrode and exposed shell.
- Caused by—Fuel/air mixture too rich, plug heat range too cold, weak ignition system, dirty air cleaner, faulty automatic choke or excessive idling.
- Can be cleaned, regapped and reused.

### OIL FOULED
- Appearance—Wet black deposits on insulator and exposed shell.
- Caused by—Excessive oil entering the combustion chamber through worn rings, pistons, valve guides or bearings.
- Replace with new plugs (use a hotter plug if engine is not repaired).

### LEAD FOULED
- Appearance — Yellow insulator deposits (may sometimes be dark gray, black or tan in color) on the insulator tip.
- Caused by—Highly leaded gasoline.
- Replace with new plugs.

### LEAD FOULED
- Appearance—Yellow glazed deposits indicating melted lead deposits due to hard acceleration.
- Caused by—Highly leaded gasoline.
- Replace with new plugs.

### OIL AND LEAD FOULED
- Appearance—Glazed yellow deposits with a slight brownish tint on the insulator tip and ground electrode.
- Replace with new plugs.

### FUEL ADDITIVE RESIDUE
- Appearance — Brown colored hardened ash deposits on the insulator tip and ground electrode.
- Caused by—Fuel and/or oil additives.
- Replace with new plugs.

### WORN
- Appearance — Severely worn or eroded electrodes.
- Caused by—Normal wear or unusual oil and/or fuel additives.
- Replace with new plugs.

### PREIGNITION
- Appearance — Melted ground electrode.
- Caused by—Overadvanced ignition timing, inoperative ignition advance mechanism, too low of a fuel octane rating, lean fuel/air mixture or carbon deposits in combustion chamber.

### PREIGNITION
- Appearance—Melted center electrode.
- Caused by—Abnormal combustion due to overadvanced ignition timing or incorrect advance, too low of a fuel octane rating, lean fuel/air mixture, or carbon deposits in combustion chamber.
- Correct engine problem and replace with new plugs.

### INCORRECT HEAT RANGE
- Appearance—Melted center electrode and white blistered insulator tip.
- Caused by—Incorrect plug heat range selection.
- Replace with new plugs.

**2**

**18** **OIL INDICATOR**

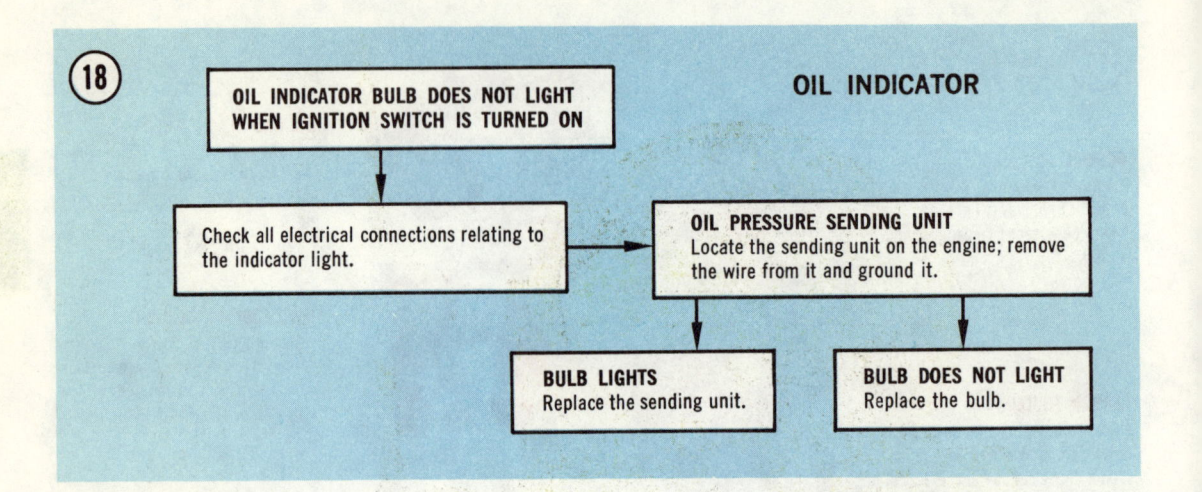

```
OIL INDICATOR BULB DOES NOT LIGHT
WHEN IGNITION SWITCH IS TURNED ON
            │
            ▼
Check all electrical connections relating to  ──▶  OIL PRESSURE SENDING UNIT
the indicator light.                                Locate the sending unit on the engine; remove
                                                    the wire from it and ground it.
            │                                               │
            ▼                                               ▼
   BULB LIGHTS                                      BULB DOES NOT LIGHT
   Replace the sending unit.                        Replace the bulb.
```

## ENGINE OIL PRESSURE LIGHT

Proper oil pressure to the engine is vital. If oil pressure is insufficient, the engine can destroy itself in a comparatively short time.

The oil pressure warning circuit monitors oil pressure constantly. If pressure drops below a predetermined level, the light comes on.

Obviously, it is vital for the warning circuit to be working to signal low oil pressure. Each time you turn on the ignition, but before you start the car, the warning light should come on. If it doesn't, there is trouble in the warning circuit, not the oil pressure system. See **Figure 18** to troubleshoot the warning circuit.

Once the engine is running, the warning light should stay off. If the warning light comes on or acts erratically while the engine is running there is trouble with the engine oil pressure system. *Stop the engine immediately*. Refer to **Figure 19** for possible causes of the problem.

## FUEL SYSTEM (CARBURETTED)

Fuel system problems must be isolated to the fuel pump (mechanical or electric), fuel lines, fuel filter, or carburetor. These procedures assume the ignition system is working properly and is correctly adjusted.

1. *Engine will not start* — First make sure that fuel is being delivered to the carburetor. Remove the air cleaner, look into the carburetor throat, and operate the accelerator

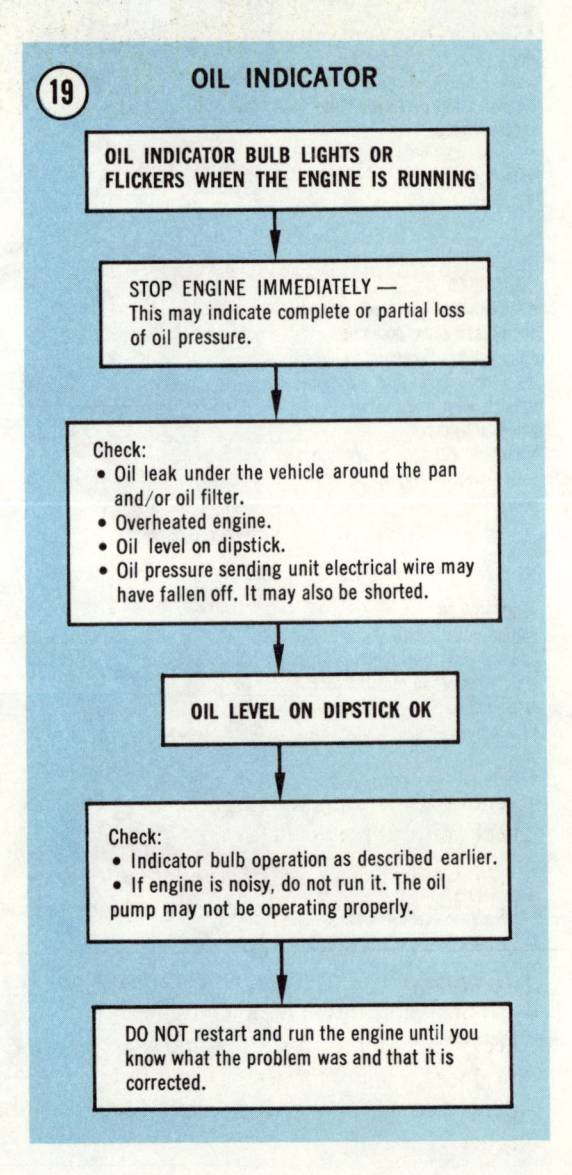

**19** **OIL INDICATOR**

```
OIL INDICATOR BULB LIGHTS OR
FLICKERS WHEN THE ENGINE IS RUNNING
            │
            ▼
STOP ENGINE IMMEDIATELY —
This may indicate complete or partial loss
of oil pressure.
            │
            ▼
Check:
• Oil leak under the vehicle around the pan
  and/or oil filter.
• Overheated engine.
• Oil level on dipstick.
• Oil pressure sending unit electrical wire may
  have fallen off. It may also be shorted.
            │
            ▼
OIL LEVEL ON DIPSTICK OK
            │
            ▼
Check:
• Indicator bulb operation as described earlier.
• If engine is noisy, do not run it. The oil
  pump may not be operating properly.
            │
            ▼
DO NOT restart and run the engine until you
know what the problem was and that it is
corrected.
```

# TROUBLESHOOTING

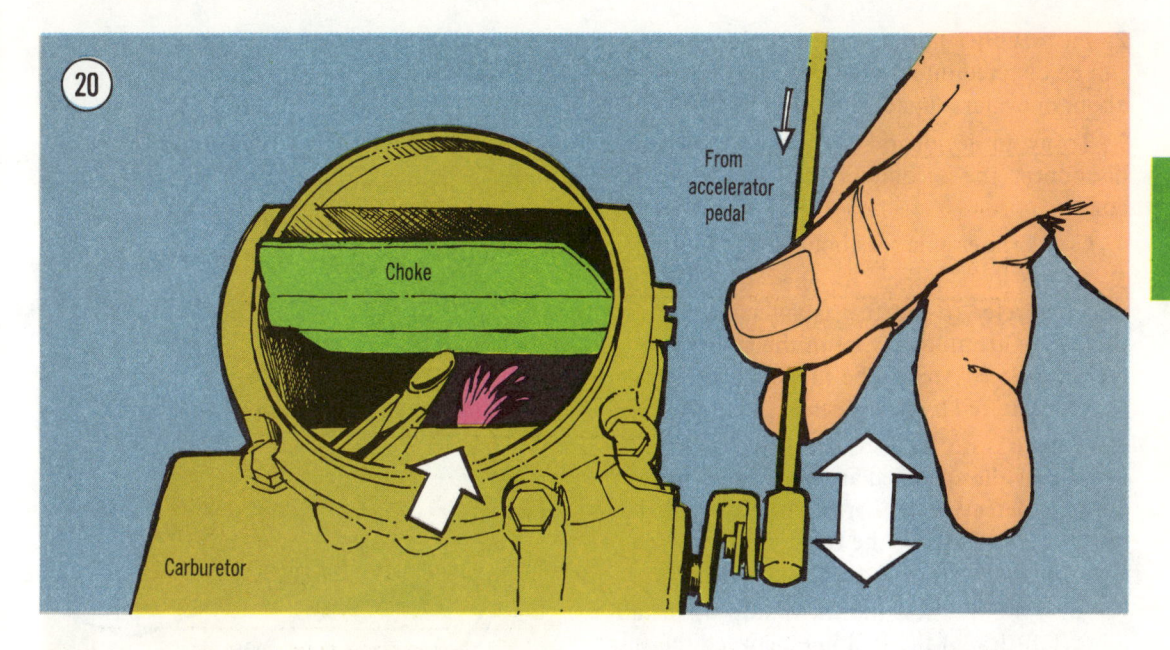

**Figure 20** (Choke, Carburetor, From accelerator pedal)

linkage several times. There should be a stream of fuel from the accelerator pump discharge tube each time the accelerator linkage is depressed (**Figure 20**). If not, check fuel pump delivery (described later), float valve, and float adjustment. If the engine will not start, check the automatic choke parts for sticking or damage. If necessary, rebuild or replace the carburetor.

2. *Engine runs at fast idle* — Check the choke setting. Check the idle speed, idle mixture, and decel valve (if equipped) adjustment.

3. *Rough idle or engine miss with frequent stalling* — Check idle mixture and idle speed adjustments.

4. *Engine "diesels" (continues to run) when ignition is switched off* — Check idle mixture (probably too rich), ignition timing, and idle speed (probably too fast). Check the throttle solenoid (if equipped) for proper operation. Check for overheated engine.

5. *Stumbling when accelerating from idle* — Check the idle speed and mixture adjustments. Check the accelerator pump.

6. *Engine misses at high speed or lacks power* — This indicates possible fuel starvation. Check fuel pump pressure and capacity as described in this chapter. Check float needle valves. Check for a clogged fuel filter or air cleaner.

7. *Black exhaust smoke* — This indicates a badly overrich mixture. Check idle mixture and idle speed adjustment. Check choke setting. Check for excessive fuel pump pressure, leaky floats, or worn needle valves.

8. *Excessive fuel consumption* — Check for overrich mixture. Make sure choke mechanism works properly. Check idle mixture and idle speed. Check for excessive fuel pump pressure, leaky floats, or worn float needle valves.

## FUEL SYSTEM (FUEL INJECTED)

Troubleshooting a fuel injection system requires more thought, experience, and know-how than any other part of the vehicle. A logical approach and proper test equipment are essential in order to successfully find and fix these troubles.

It is best to leave fuel injection troubles to your dealer. In order to isolate a problem to the injection system make sure that the fuel pump is operating properly. Check its performance as described later in this section. Also make sure that fuel filter and air cleaner are not clogged.

## FUEL PUMP TEST (MECHANICAL AND ELECTRIC)

1. Disconnect the fuel inlet line where it enters the carburetor or fuel injection system.

2. Fit a rubber hose over the fuel line so fuel can be directed into a graduated container with about one quart capacity. See **Figure 21**.

3. To avoid accidental starting of the engine, disconnect the secondary coil wire from the coil.

4. Crank the engine for about 30 seconds.

5. If the fuel pump supplies the specified amount (refer to the fuel chapter later in this book), the trouble may be in the carburetor or fuel injection system. The fuel injection system should be tested by your dealer.

6. If there is no fuel present or the pump cannot supply the specified amount, either the fuel pump is defective or there is an obstruction in the fuel line. Replace the fuel pump and/or inspect the fuel lines for air leaks or obstructions.

7. Also pressure test the fuel pump by installing a T-fitting in the fuel line between the fuel pump and the carburetor. Connect a fuel pressure gauge to the fitting with a short tube (**Figure 22**).

8. Reconnect the primary coil wire, start the engine, and record the pressure. Refer to the fuel chapter later in this book for the correct pressure. If the pressure varies from that specified, the pump should be replaced.

9. Stop the engine. The pressure should drop off very slowly. If it drops off rapidly, the outlet valve in the pump is leaking and the pump should be replaced.

## EMISSION CONTROL SYSTEMS

Major emission control systems used on nearly all U.S. models include the following:

a. Positive crankcase ventilation (PCV)
b. Thermostatic air cleaner
c. Air injection reaction (AIR)
d. Fuel evaporation control
e. Exhaust gas recirculation (EGR)

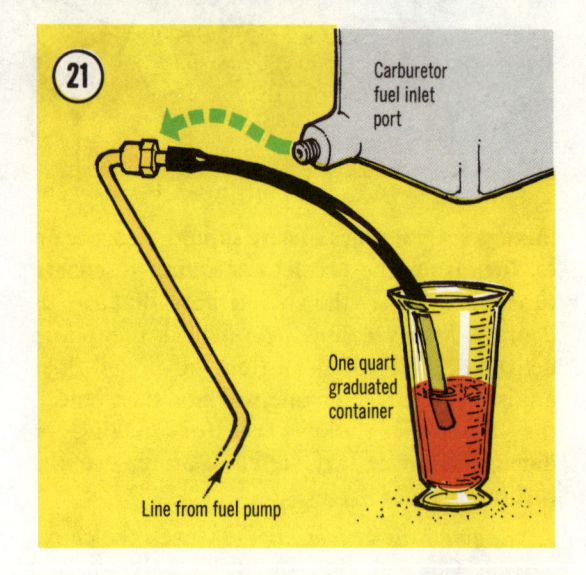

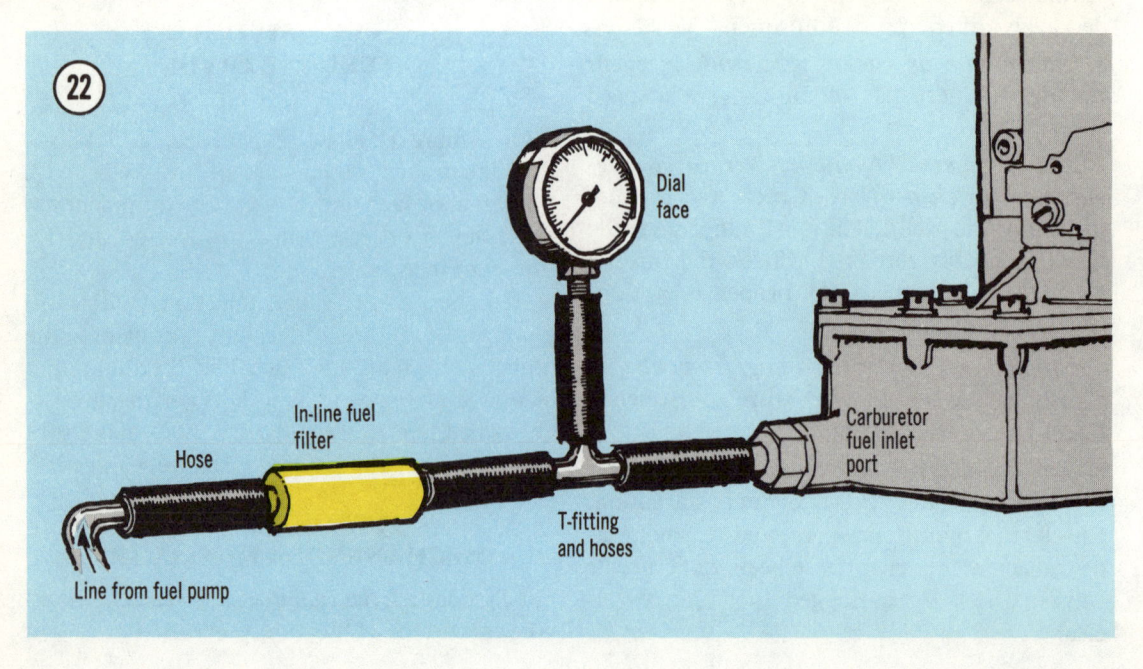

# TROUBLESHOOTING

Emission control systems vary considerably from model to model. Individual models contain variations of the four systems described here. In addition, they may include other special systems. Use the index to find specific emission control components in other chapters.

Many of the systems and components are factory set and sealed. Without special expensive test equipment, it is impossible to adjust the systems to meet state and federal requirements.

Troubleshooting can also be difficult without special equipment. The procedures described below will help you find emission control parts which have failed, but repairs may have to be entrusted to a dealer or other properly equipped repair shop.

With the proper equipment, you can test the carbon monoxide and hydrocarbon levels. **Figure 23** provides some sources of trouble if the readings are not correct.

### Positive Crankcase Ventilation

Fresh air drawn from the air cleaner housing scavenges emissions (e.g., piston blow-by) from the crankcase, then the intake manifold vacuum draws emissions into the intake manifold. They can then be reburned in the normal combustion process. **Figure 24** shows a typical system. **Figure 25** provides a testing procedure.

### Thermostatic Air Cleaner

The thermostatically controlled air cleaner maintains incoming air to the engine at a predetermined level, usually about 100°F or higher. It mixes cold air with heated air from the exhaust manifold region. The air cleaner in-

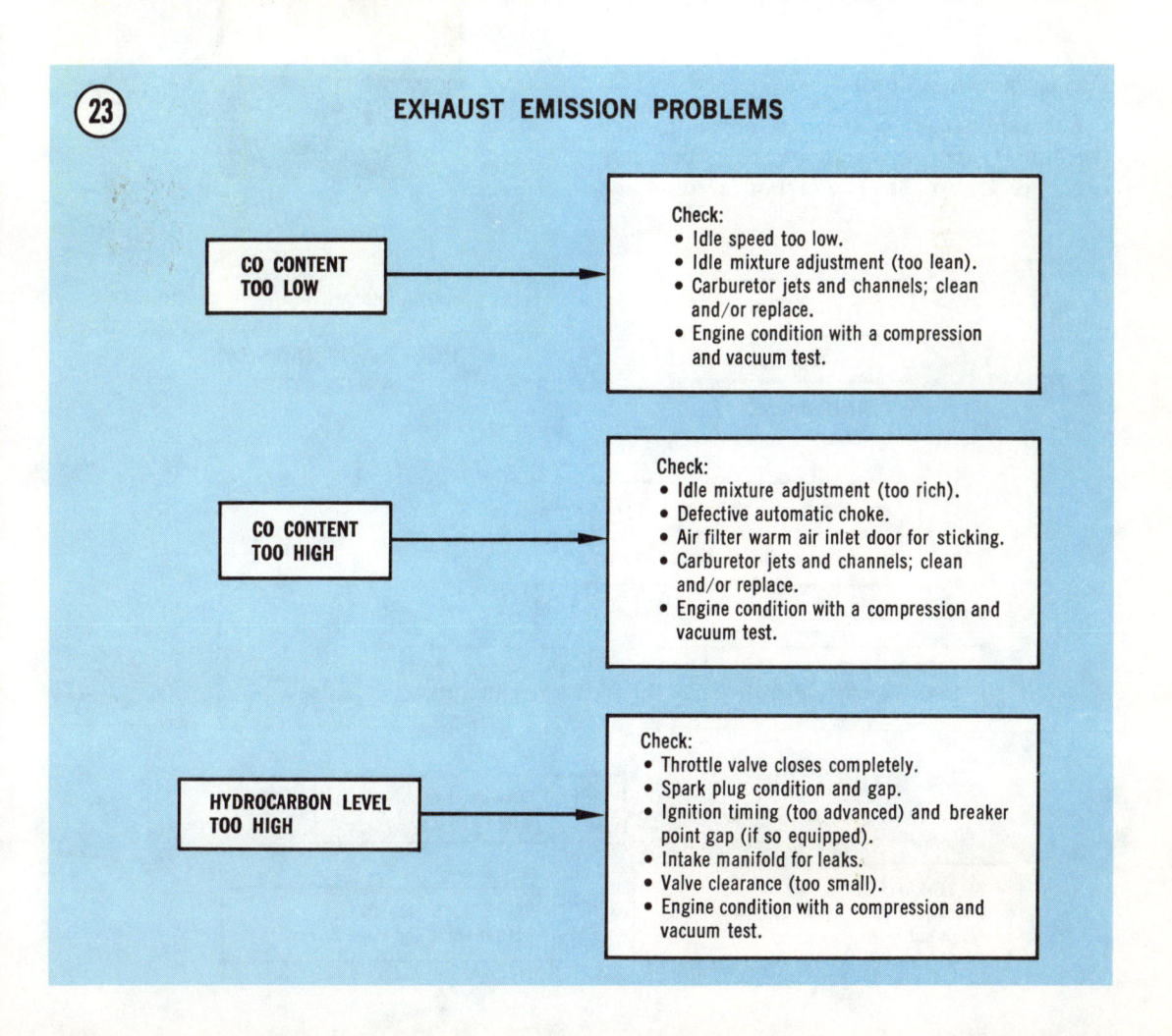

**(23) EXHAUST EMISSION PROBLEMS**

**CO CONTENT TOO LOW** →
Check:
- Idle speed too low.
- Idle mixture adjustment (too lean).
- Carburetor jets and channels; clean and/or replace.
- Engine condition with a compression and vacuum test.

**CO CONTENT TOO HIGH** →
Check:
- Idle mixture adjustment (too rich).
- Defective automatic choke.
- Air filter warm air inlet door for sticking.
- Carburetor jets and channels; clean and/or replace.
- Engine condition with a compression and vacuum test.

**HYDROCARBON LEVEL TOO HIGH** →
Check:
- Throttle valve closes completely.
- Spark plug condition and gap.
- Ignition timing (too advanced) and breaker point gap (if so equipped).
- Intake manifold for leaks.
- Valve clearance (too small).
- Engine condition with a compression and vacuum test.

cludes a temperature sensor, vacuum motor, and a hinged door. See **Figure 26**.

The system is comparatively easy to test. See **Figure 27** for the procedure.

**Air Injection Reaction System**

The air injection reaction system reduces air pollution by oxidizing hydrocarbons and carbon monoxide as they leave the combustion chamber. See **Figure 28**.

The air injection pump, driven by the engine, compresses filtered air and injects it at the exhaust port of each cylinder. The fresh air mixes with the unburned gases in the exhaust and promotes further burning. A check valve prevents exhaust gases from entering and damaging the air pump if the pump becomes inoperative, e.g., from a fan belt failure.

**Figure 29** explains the testing procedure for this system.

**Fuel Evaporation Control**

Fuel vapor from the fuel tank passes through the liquid/vapor separator to the carbon canister. See **Figure 30**. The carbon absorbs and

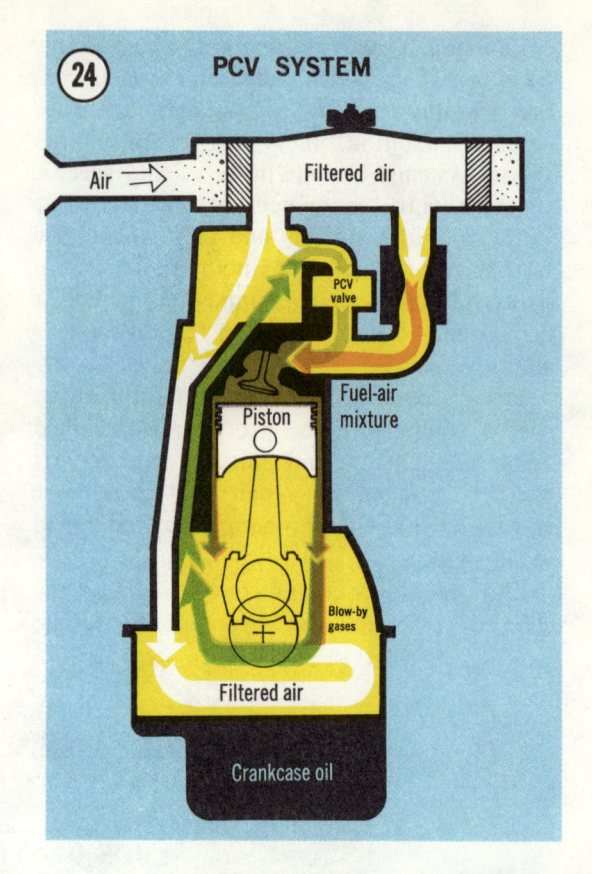

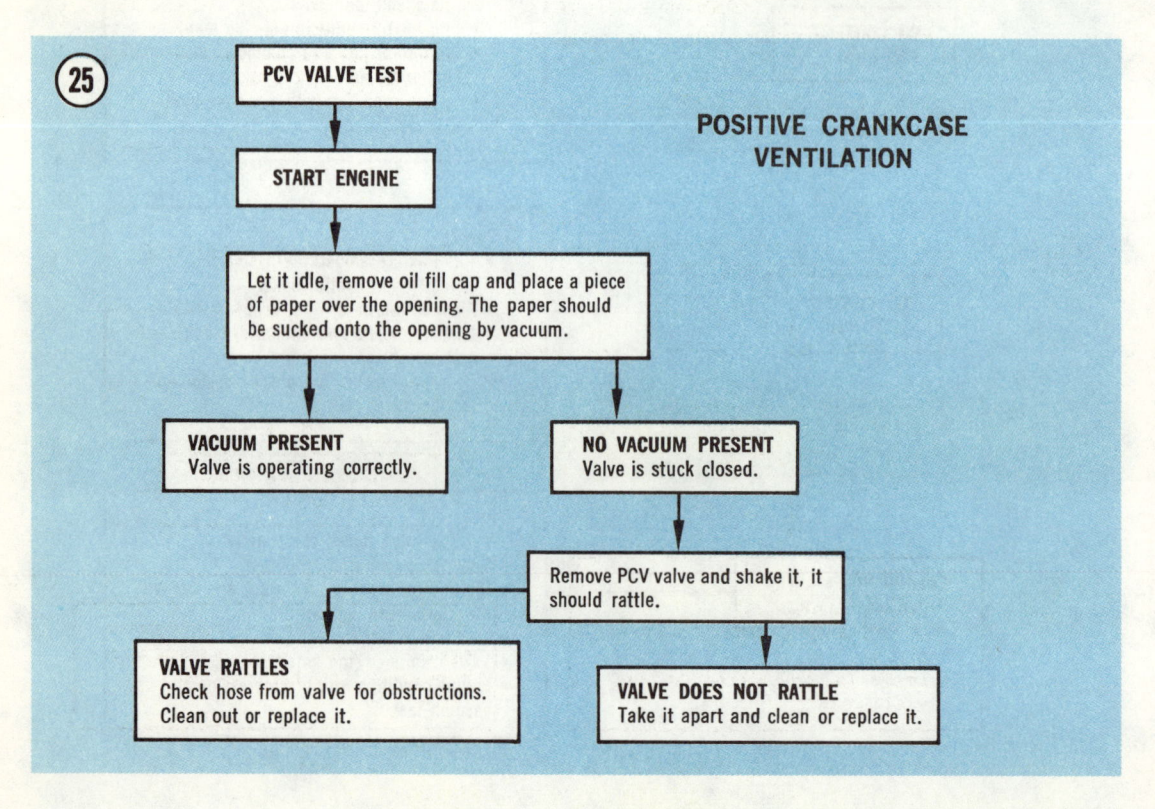

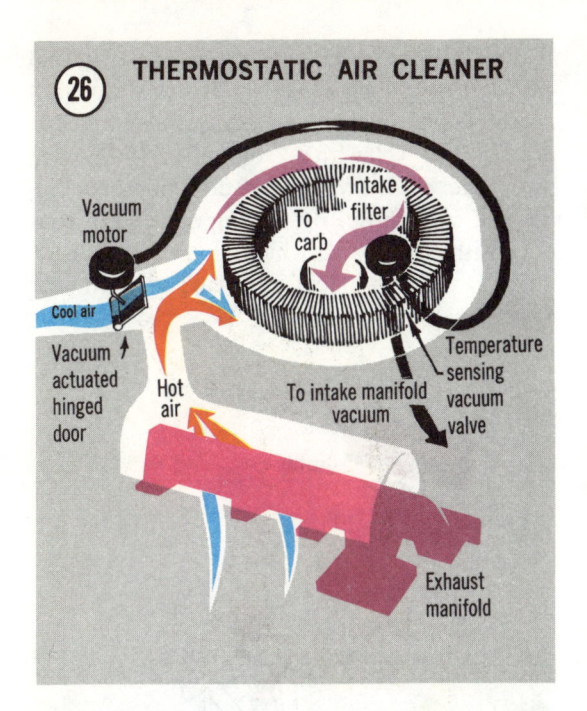

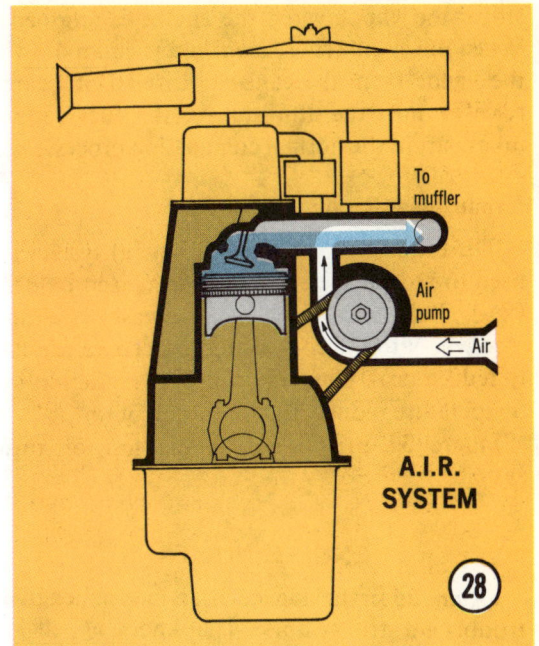

# TROUBLESHOOTING 23

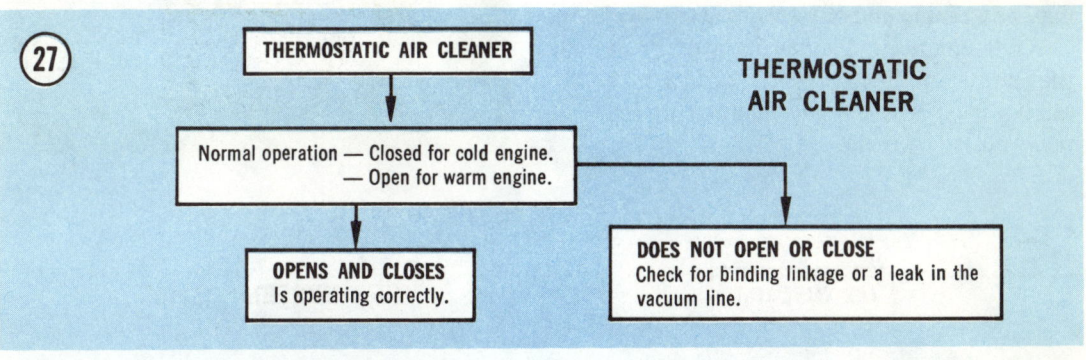

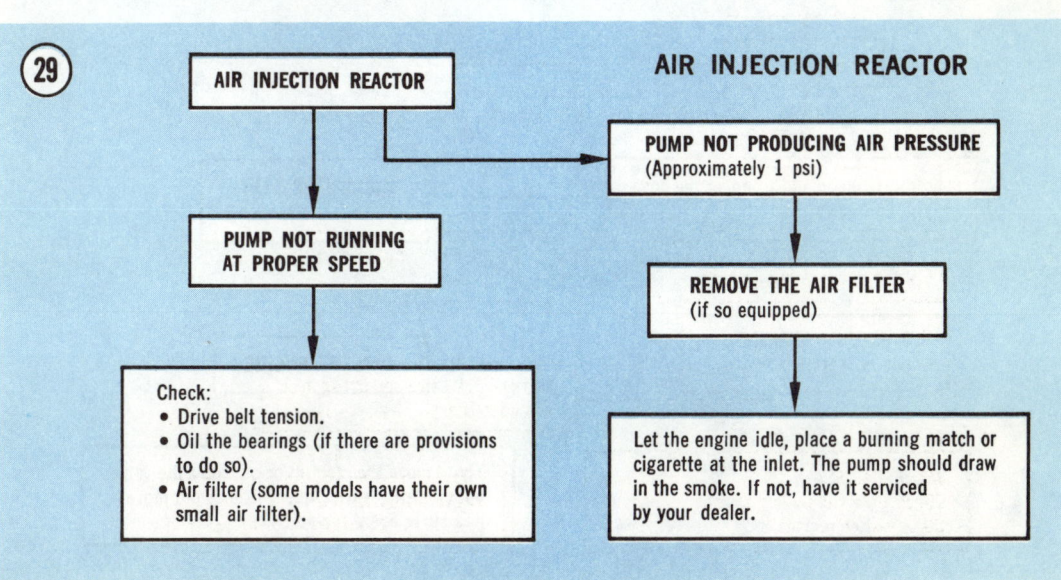

**㉖ THERMOSTATIC AIR CLEANER**

Vacuum motor
Cool air
Vacuum actuated hinged door
Hot air
To carb
Intake filter
Temperature sensing vacuum valve
To intake manifold vacuum
Exhaust manifold

**A.I.R. SYSTEM** ㉘

To muffler
Air pump
Air

---

㉗ **THERMOSTATIC AIR CLEANER**

THERMOSTATIC AIR CLEANER
↓
Normal operation — Closed for cold engine.
— Open for warm engine.

**OPENS AND CLOSES**
Is operating correctly.

**DOES NOT OPEN OR CLOSE**
Check for binding linkage or a leak in the vacuum line.

---

㉙ **AIR INJECTION REACTOR**

AIR INJECTION REACTOR

**PUMP NOT RUNNING AT PROPER SPEED**

Check:
• Drive belt tension.
• Oil the bearings (if there are provisions to do so).
• Air filter (some models have their own small air filter).

**PUMP NOT PRODUCING AIR PRESSURE**
(Approximately 1 psi)

**REMOVE THE AIR FILTER**
(if so equipped)

Let the engine idle, place a burning match or cigarette at the inlet. The pump should draw in the smoke. If not, have it serviced by your dealer.

stores the vapor when the engine is stopped. When the engine runs, manifold vacuum draws the vapor from the canister. Instead of being released into the atmosphere, the fuel vapor takes part in the normal combustion process.

### Exhaust Gas Recirculation

The exhaust gas recirculation (EGR) system is used to reduce the emission of nitrogen oxides (NOx). Relatively inert exhaust gases are introduced into the combustion process to slightly reduce peak temperatures. This reduction in temperature reduces the formation of NOx.

**Figure 31** provides a simple test of this system.

### ENGINE NOISES

Often the first evidence of an internal engine trouble is a strange noise. That knocking, clicking, or tapping which you never heard before may be warning you of impending trouble.

While engine noises can indicate problems, they are sometimes difficult to interpret correctly; inexperienced mechanics can be seriously misled by them.

Professional mechanics often use a special stethoscope which looks similar to a doctor's stethoscope for isolating engine noises. You can do nearly as well with a "sounding stick" which can be an ordinary piece of doweling or a section of small hose. By placing one end in contact with the area to which you want to listen and the other end near your ear, you can hear

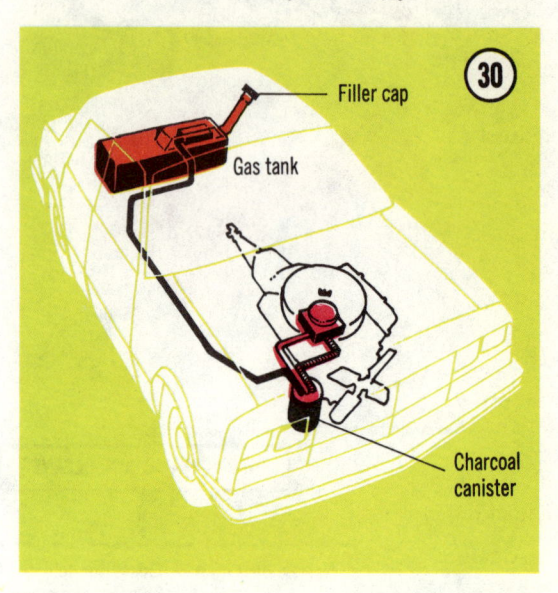

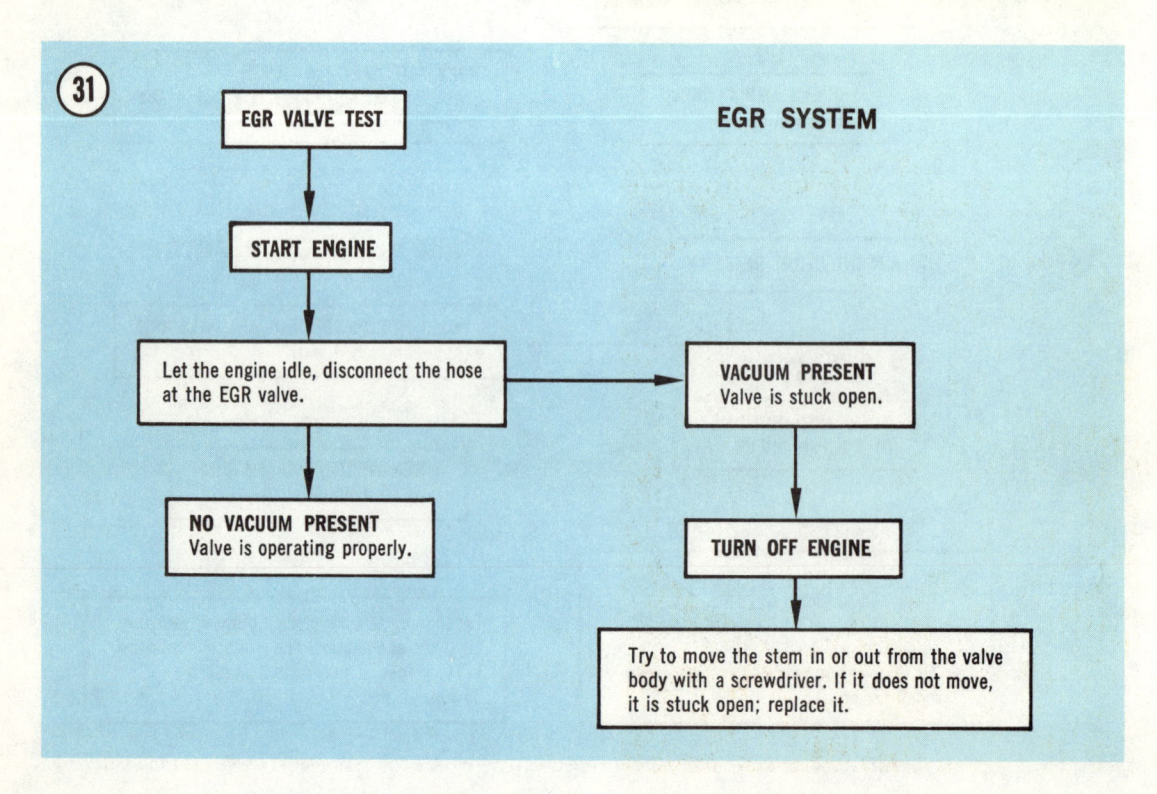

# TROUBLESHOOTING

sounds emanating from that area. The first time you do this, you may be horrified at the strange noises coming from even a normal engine. If you can, have an experienced friend or mechanic help you sort the noises out.

### Clicking or Tapping Noises

Clicking or tapping noises usually come from the valve train, and indicate excessive valve clearance.

If your vehicle has adjustable valves, the procedure for adjusting the valve clearance is explained in Chapter Three. If your vehicle has hydraulic lifters, the clearance may not be adjustable. The noise may be coming from a collapsed lifter. These may be cleaned or replaced as described in the engine chapter.

A sticking valve may also sound like a valve with excessive clearance. In addition, excessive wear in valve train components can cause similar engine noises.

### Knocking Noises

A heavy, dull knocking is usually caused by a worn main bearing. The noise is loudest when the engine is working hard, i.e., accelerating hard at low speed. You may be able to isolate the trouble to a single bearing by disconnecting the spark plugs one at a time. When you reach the spark plug nearest the bearing, the knock will be reduced or disappear.

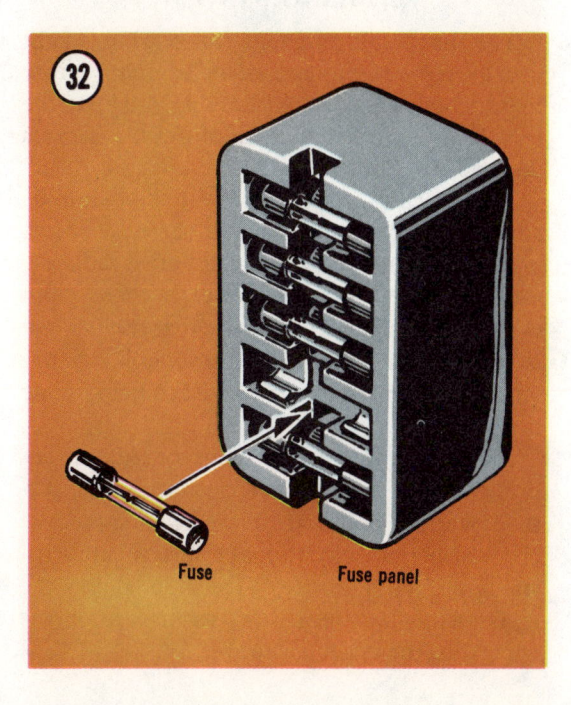

Worn connecting rod bearings may also produce a knock, but the sound is usually more "metallic." As with a main bearing, the noise is worse when accelerating. It may even increase further just as you go from accelerating to coasting. Disconnecting spark plugs will help isolate this knock as well.

A double knock or clicking usually indicates a worn piston pin. Disconnecting spark plugs will isolate this to a particular piston, however, the noise will *increase* when you reach the affected piston.

A loose flywheel and excessive crankshaft end play also produce knocking noises. While similar to main bearing noises, these are usually intermittent, not constant, and they do not change when spark plugs are disconnected.

Some mechanics confuse piston pin noise with piston slap. The double knock will distinguish the piston pin noise. Piston slap is identified by the fact that it is always louder when the engine is cold.

## ELECTRICAL ACCESSORIES

### Lights and Switches (Interior and Exterior)

1. *Bulb does not light* — Remove the bulb and check for a broken element. Also check the inside of the socket; make sure the contacts are clean and free of corrosion. If the bulb and socket are OK, check to see if a fuse has blown or a circuit breaker has tripped. The fuse panel **(Figure 32)** is usually located under the instrument panel. Replace the blown fuse or reset the circuit breaker. If the fuse blows or the breaker trips again, there is a short in that circuit. Check that circuit all the way to the battery. Look for worn wire insulation or burned wires.

If all the above are all right, check the switch controlling the bulb for continuity with an ohmmeter at the switch terminals. Check the switch contact terminals for loose or dirty electrical connections.

2. *Headlights work but will not switch from either high or low beam* — Check the beam selector switch for continuity with an ohmmeter

at the switch terminals. Check the switch contact terminals for loose or dirty electrical connections.

3. *Brake light switch inoperative* — On mechanically operated switches, usually mounted near the brake pedal arm, adjust the switch to achieve correct mechanical operation. Check the switch for continuity with an ohmmeter at the switch terminals. Check the switch contact terminals for loose or dirty electrical connections.

4. *Back-up lights do not operate* — Check light bulb as described earlier. Locate the switch, normally located near the shift lever. Adjust switch to achieve correct mechanical operation. Check the switch for continuity with an ohmmeter at the switch terminals. Bypass the switch with a jumper wire; if the lights work, replace the switch.

**Directional Signals**

1. *Directional signals do not operate* — If the indicator light on the instrument panel burns steadily instead of flashing, this usually indicates that one of the exterior lights is burned out. Check all lamps that normally flash. If all are all right, the flasher unit may be defective. Replace it with a good one.

2. *Directional signal indicator light on instrument panel does not light up* — Check the light bulbs as described earlier. Check all electrical connections and check the flasher unit.

3. *Directional signals will not self-cancel* — Check the self-cancelling mechanism located inside the steering column.

4. *Directional signals flash slowly* — Check the condition of the battery and the alternator (or generator) drive belt tension (**Figure 4**). Check the flasher unit and all related electrical connections.

**Windshield Wipers**

1. *Wipers do not operate* — Check for a blown fuse or circuit breaker that has tripped; replace or reset. Check all related terminals for loose or dirty electrical connections. Check continuity of the control switch with an ohmmeter at the switch terminals. Check the linkage and arms for loose, broken, or binding parts. Straighten out or replace where necessary.

2. *Wiper motor hums but will not operate* — The motor may be shorted out internally; check and/or replace the motor. Also check for broken or binding linkage and arms.

3. *Wiper arms will not return to the stowed position when turned off* — The motor has a special internal switch for this purpose. Have it inspected by your dealer. Do not attempt this yourself.

**Interior Heater**

1. *Heater fan does not operate* — Check for a blown fuse or circuit breaker that has tripped. Check the switch for continuity with an ohmmeter at the switch terminals. Check the switch contact terminals for loose or dirty electrical connections.

2. *Heat output is insufficient* — Check the heater hose/engine coolant control valve usually located in the engine compartment; make sure it is in the open position. Ensure that the heater door(s) and cable(s) are operating correctly and are in the open position. Inspect the heat ducts; make sure that they are not crimped or blocked.

## COOLING SYSTEM

The temperature gauge or warning light usually signals cooling system troubles before there is any damage. As long as you stop the vehicle at the first indication of trouble, serious damage is unlikely.

In most cases, the trouble will be obvious as soon as you open the hood. If there is coolant or steam leaking, look for a defective radiator, radiator hose, or heater hose. If there is no evidence of leakage, make sure that the fan belt is in good condition. If the trouble is not obvious, refer to **Figures 33 and 34** to help isolate the trouble.

Automotive cooling systems operate under pressure to permit higher operating temperatures without boil-over. The system should be checked periodically to make sure it can withstand normal pressure. **Figure 35** shows the equipment which nearly any service station has for testing the system pressure.

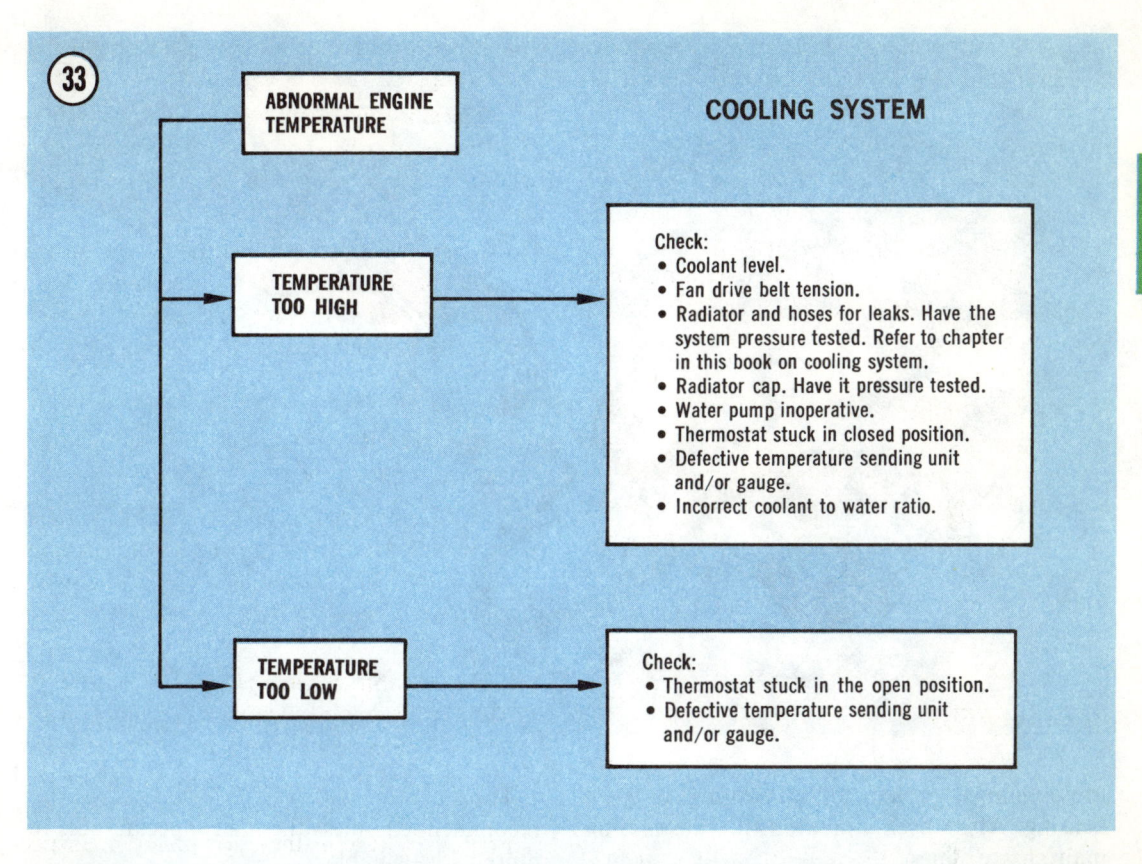

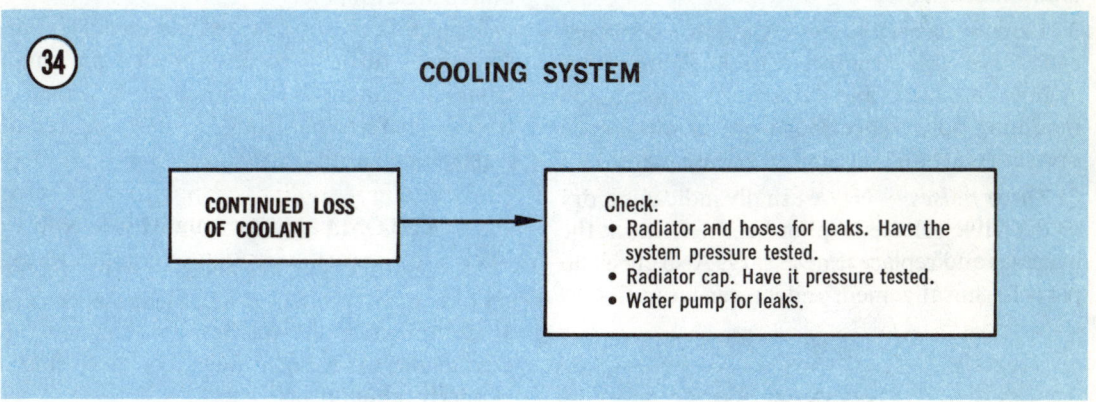

# TROUBLESHOOTING

㉝ **COOLING SYSTEM**

ABNORMAL ENGINE TEMPERATURE

TEMPERATURE TOO HIGH →
Check:
- Coolant level.
- Fan drive belt tension.
- Radiator and hoses for leaks. Have the system pressure tested. Refer to chapter in this book on cooling system.
- Radiator cap. Have it pressure tested.
- Water pump inoperative.
- Thermostat stuck in closed position.
- Defective temperature sending unit and/or gauge.
- Incorrect coolant to water ratio.

TEMPERATURE TOO LOW →
Check:
- Thermostat stuck in the open position.
- Defective temperature sending unit and/or gauge.

㉞ **COOLING SYSTEM**

CONTINUED LOSS OF COOLANT →
Check:
- Radiator and hoses for leaks. Have the system pressure tested.
- Radiator cap. Have it pressure tested.
- Water pump for leaks.

## CLUTCH

All clutch troubles except adjustments require transmission removal to identify and cure the problem.

1. *Slippage* — This is most noticeable when accelerating in a high gear at relatively low speed. To check slippage, park the vehicle on a level surface with the handbrake set. Shift to 2nd gear and release the clutch as if driving off. If the clutch is good, the engine will slow and stall. If the clutch slips, continued engine speed will give it away.

Slippage results from insufficient clutch pedal free play, oil or grease on the clutch disc, worn pressure plate, or weak springs.

2. *Drag or failure to release* — This trouble usually causes difficult shifting and gear clash, especially when downshifting. The cause may be excessive clutch pedal free play, warped or bent pressure plate or clutch disc, broken or

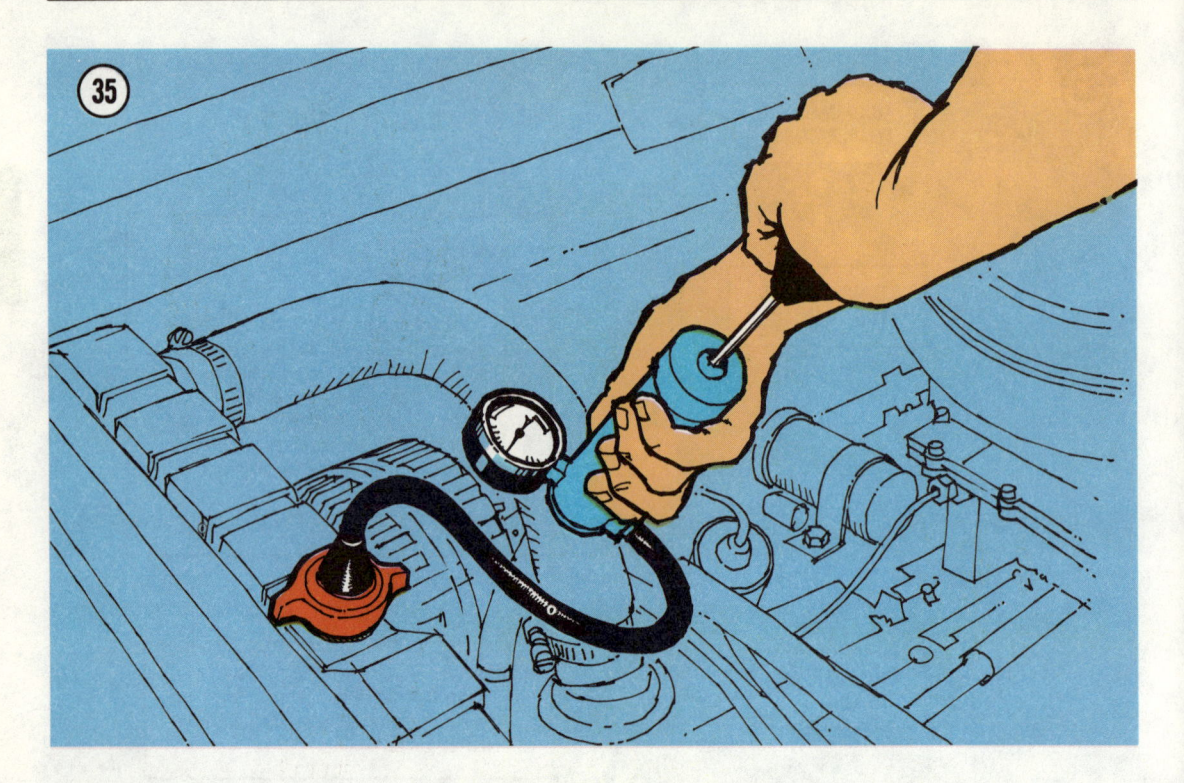

loose linings, or lack of lubrication in pilot bearing. Also check condition of transmission main shaft splines.

3. *Chatter or grabbing* — A number of things can cause this trouble. Check tightness of engine mounts and engine-to-transmission mounting bolts. Check for worn or misaligned pressure plate and misaligned release plate.

4. *Other noises* — Noise usually indicates a dry or defective release or pilot bearing. Check the bearings and replace if necessary. Also check all parts for misalignment and uneven wear.

## MANUAL TRANSMISSION/TRANSAXLE

Transmission and transaxle troubles are evident when one or more of the following symptoms appear:

a. Difficulty changing gears
b. Gears clash when downshifting
c. Slipping out of gear
d. Excessive noise in NEUTRAL
e. Excessive noise in gear
f. Oil leaks

Transmission and transaxle repairs are not recommended unless the many special tools required are available.

Transmission and transaxle troubles are sometimes difficult to distinguish from clutch troubles. Eliminate the clutch as a source of trouble before installing a new or rebuilt transmission or transaxle.

## AUTOMATIC TRANSMISSION

Most automatic transmission repairs require considerable specialized knowledge and tools. It is impractical for the home mechanic to invest in the tools, since they cost more than a properly rebuilt transmission.

Check fluid level and condition frequently to help prevent future problems. If the fluid is orange or black in color or smells like varnish, it is an indication of some type of damage or failure within the transmission. Have the transmission serviced by your dealer or competent automatic transmission service facility.

## BRAKES

Good brakes are vital to the safe operation of the vehicle. Performing the maintenance speci-

# TROUBLESHOOTING

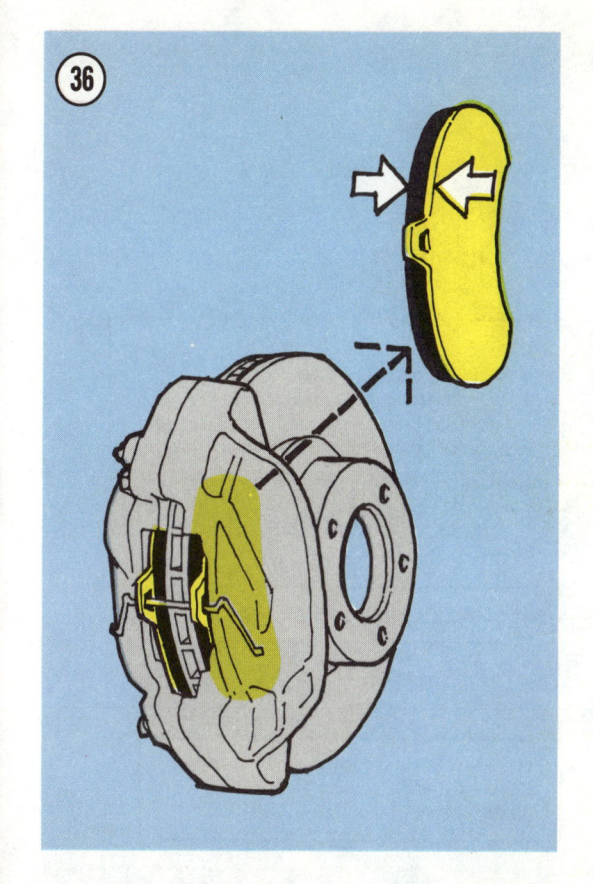

fied in Chapter Three will minimize problems with the brakes. Most importantly, check and maintain the level of fluid in the master cylinder, and check the thickness of the linings on the disc brake pads (**Figure 36**) or drum brake shoes (**Figure 37**).

If trouble develops, **Figures 38 through 40** will help you locate the problem. Refer to the brake chapter for actual repair procedures.

## STEERING AND SUSPENSION

Trouble in the suspension or steering is evident when the following occur:

a. Steering is hard
b. Car pulls to one side
c. Car wanders or front wheels wobble
d. Steering has excessive play
e. Tire wear is abnormal

Unusual steering, pulling, or wandering is usually caused by bent or otherwise misaligned suspension parts. This is difficult to check without proper alignment equipment. Refer to the suspension chapter in this book for repairs that you can perform and those that must be left to a dealer or suspension specialist.

If your trouble seems to be excessive play, check wheel bearing adjustment first. This is the most frequent cause. Then check ball-joints as described below. Finally, check tie rod end ball-joints by shaking each tie rod. Also check steering gear, or rack-and-pinion assembly to see that it is securely bolted down.

## TIRE WEAR ANALYSIS

Abnormal tire wear should be analyzed to determine its causes. The most common causes are the following:

a. Incorrect tire pressure
b. Improper driving
c. Overloading
d. Bad road surfaces
e. Incorrect wheel alignment

## CHAPTER TWO

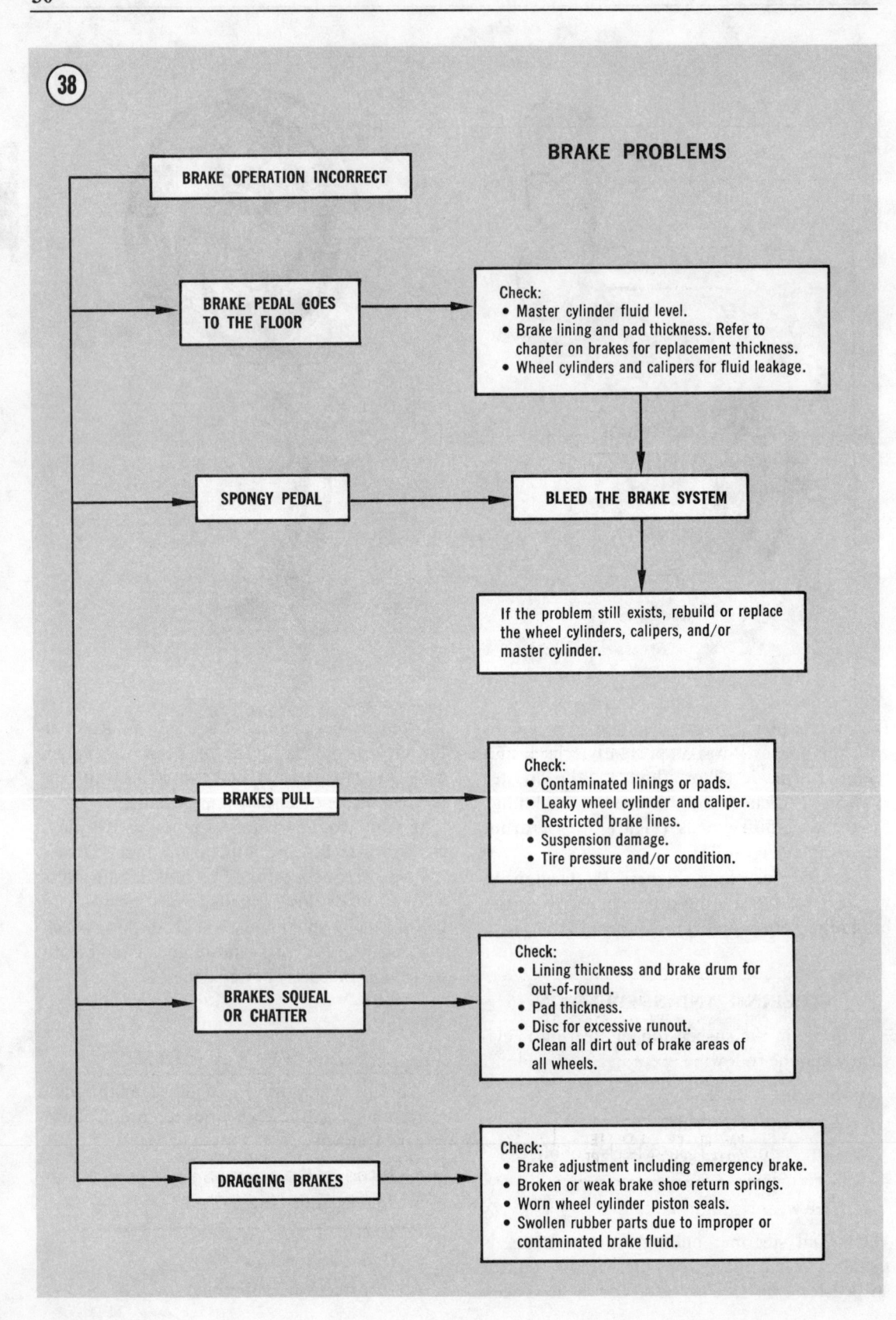

**(38)**

## BRAKE PROBLEMS

**BRAKE OPERATION INCORRECT**

**BRAKE PEDAL GOES TO THE FLOOR** →

Check:
- Master cylinder fluid level.
- Brake lining and pad thickness. Refer to chapter on brakes for replacement thickness.
- Wheel cylinders and calipers for fluid leakage.

**SPONGY PEDAL** → **BLEED THE BRAKE SYSTEM**

If the problem still exists, rebuild or replace the wheel cylinders, calipers, and/or master cylinder.

**BRAKES PULL** →

Check:
- Contaminated linings or pads.
- Leaky wheel cylinder and caliper.
- Restricted brake lines.
- Suspension damage.
- Tire pressure and/or condition.

**BRAKES SQUEAL OR CHATTER** →

Check:
- Lining thickness and brake drum for out-of-round.
- Pad thickness.
- Disc for excessive runout.
- Clean all dirt out of brake areas of all wheels.

**DRAGGING BRAKES** →

Check:
- Brake adjustment including emergency brake.
- Broken or weak brake shoe return springs.
- Worn wheel cylinder piston seals.
- Swollen rubber parts due to improper or contaminated brake fluid.

# TROUBLESHOOTING

### ⑨ BRAKE PROBLEMS

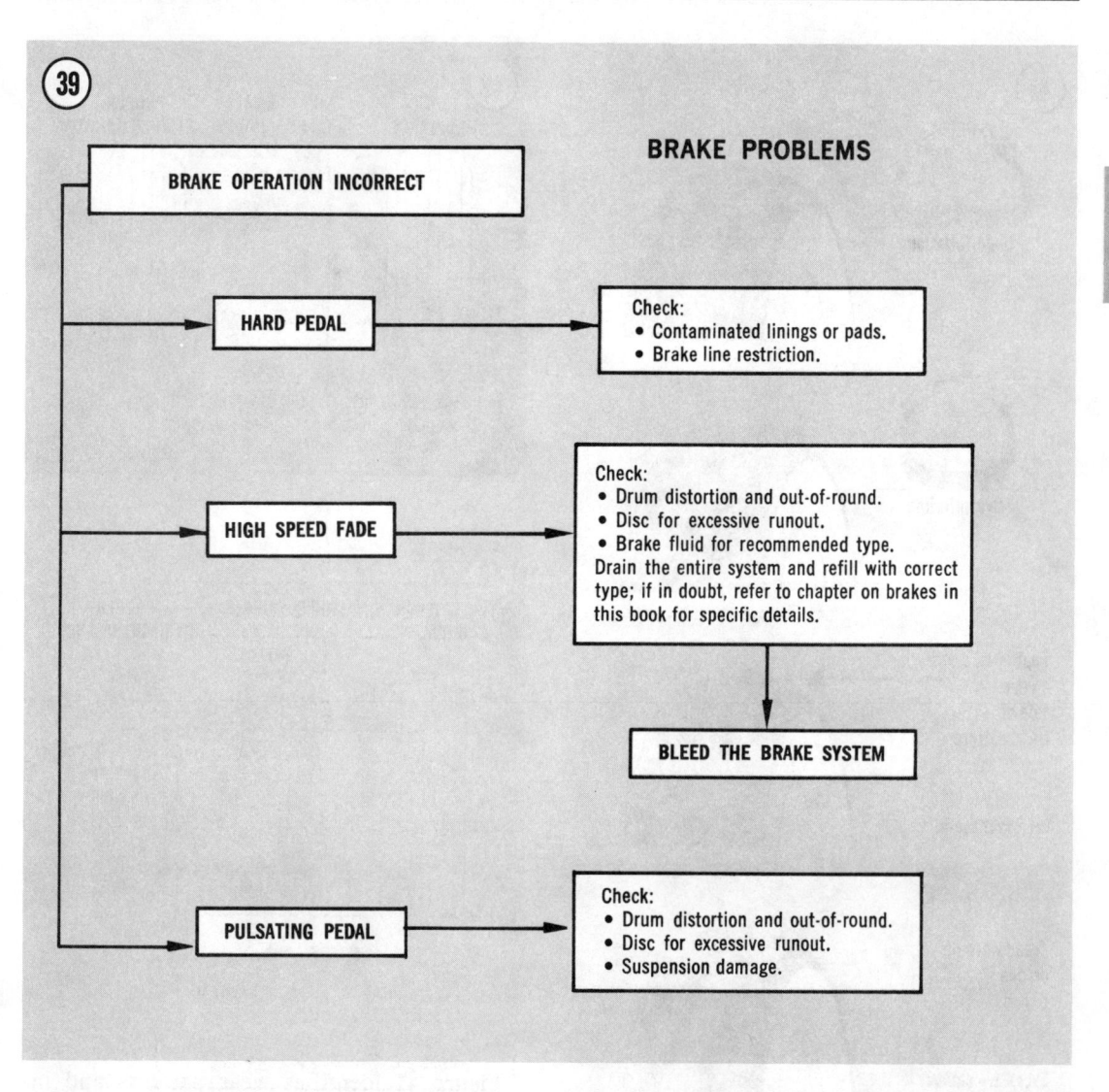

**BRAKE OPERATION INCORRECT**

→ **HARD PEDAL** → Check:
- Contaminated linings or pads.
- Brake line restriction.

→ **HIGH SPEED FADE** → Check:
- Drum distortion and out-of-round.
- Disc for excessive runout.
- Brake fluid for recommended type.
Drain the entire system and refill with correct type; if in doubt, refer to chapter on brakes in this book for specific details.

↓ **BLEED THE BRAKE SYSTEM**

→ **PULSATING PEDAL** → Check:
- Drum distortion and out-of-round.
- Disc for excessive runout.
- Suspension damage.

---

### ⑩ BRAKE PROBLEMS

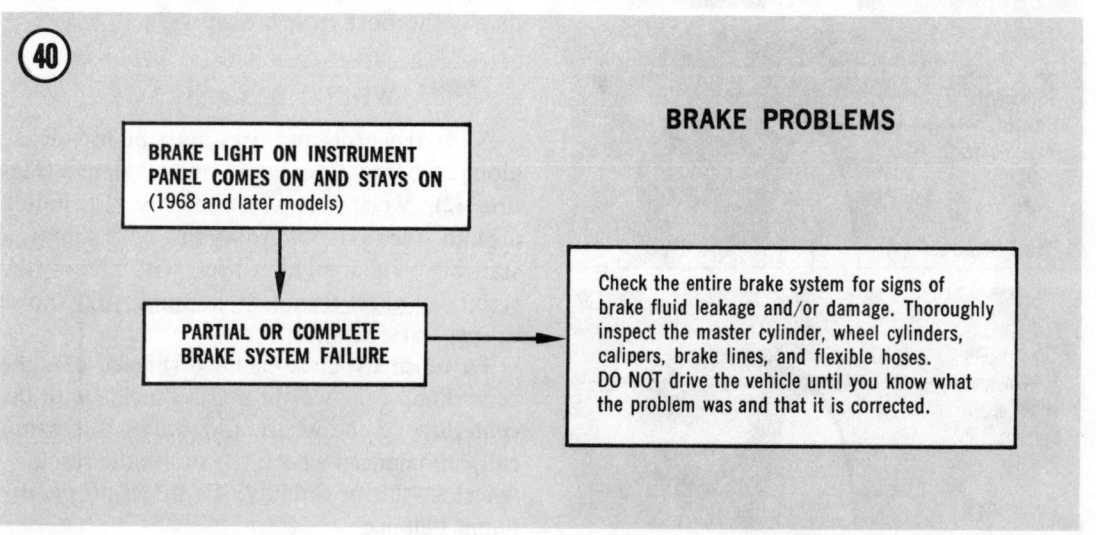

**BRAKE LIGHT ON INSTRUMENT PANEL COMES ON AND STAYS ON**
(1968 and later models)

↓

**PARTIAL OR COMPLETE BRAKE SYSTEM FAILURE** → Check the entire brake system for signs of brake fluid leakage and/or damage. Thoroughly inspect the master cylinder, wheel cylinders, calipers, brake lines, and flexible hoses. DO NOT drive the vehicle until you know what the problem was and that it is corrected.

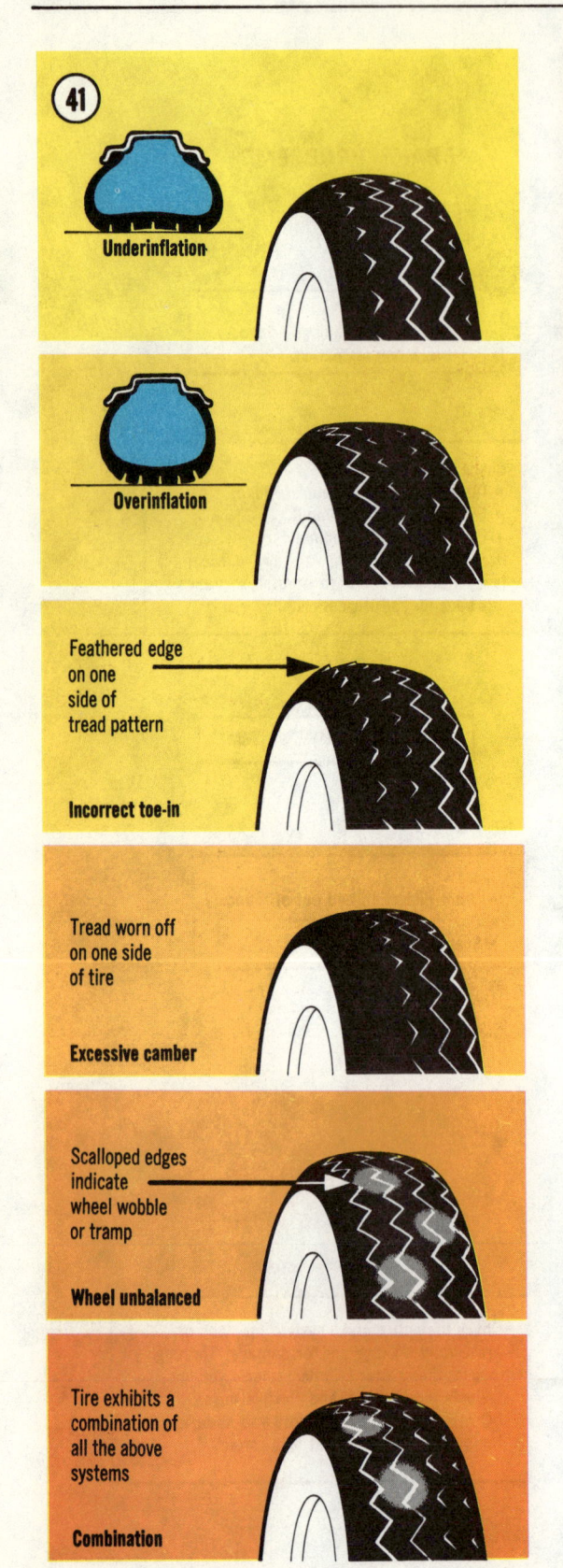

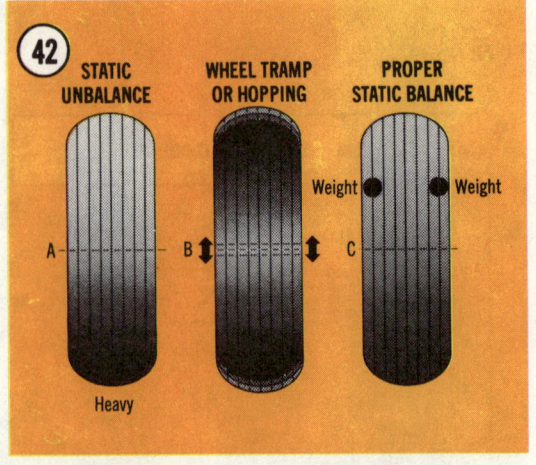

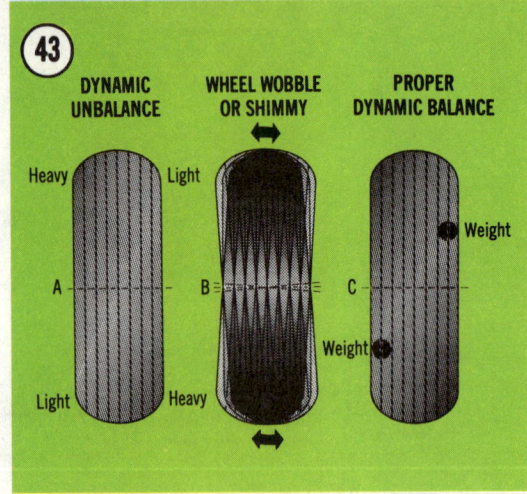

**Figure 41** identifies wear patterns and indicates the most probable causes.

## WHEEL BALANCING

All four wheels and tires must be in balance along two axes. To be in static balance (**Figure 42**), weight must be evenly distributed around the axis of rotation. (A) shows a statically unbalanced wheel; (B) shows the result — wheel tramp or hopping; (C) shows proper static balance.

To be in dynamic balance (**Figure 43**), the centerline of the weight must coincide with the centerline of the wheel. (A) shows a dynamically unbalanced wheel; (B) shows the result — wheel wobble or shimmy; (C) shows proper dynamic balance.

NOTE: If you own a 1980 or 1981 model, first check the Supplement at the back of the book for any new service information.

# CHAPTER THREE

# LUBRICATION, MAINTENANCE AND TUNE-UP

This chapter deals with the maintenance necessary to keep your car running properly. **Tables 1, 2, and 3** are service schedules. Some procedures are done at fuel stops, while others are done at specified intervals of miles or time.

The service schedules are intended for cars given normal use. More frequent service is required under the following conditions:

a. Stop-and-go driving
b. Constant high-speed driving
c. Severe dust
d. Rough or salted roads
e. Very hot, very cold, or rainy weather

Some maintenance procedures are included in the *Tune-Up* section at the end of the chapter, and detailed instructions will be found there. Other steps are included in various chapters. Chapter references are included with these steps.

## FUEL STOP CHECKS

1. Check engine oil level on the dipstick (**Figures 1 and 2**). Top up to the "H" mark on the dipstick, if necessary, using a grade recommended in **Tables 4 and 5**. Add oil through the filler hole on top of the engine (**Figure 3**).

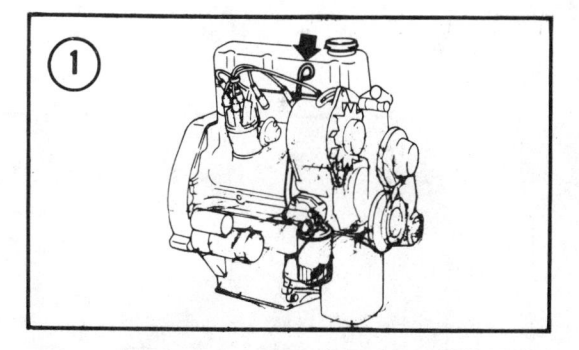

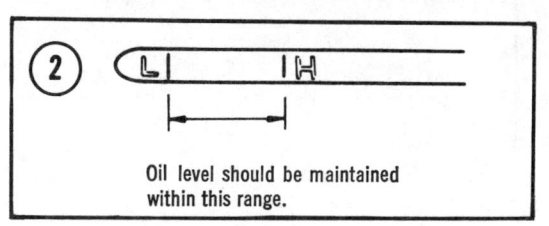

Oil level should be maintained within this range.

**Table 1   FUEL STOP CHECKS**

| Item | Procedure |
|---|---|
| Engine oil | Check level |
| Coolant | Check level |
| Battery electrolyte | Check level |
| Windshield washers | Check container level |
| Brake fluid | Check level |
| Clutch fluid | Check level |
| Tire pressures | Check |

## CHAPTER THREE

**Table 2    SCHEDULED MAINTENANCE, 1976-1977**

| | |
|---|---|
| Every 6,250 miles<br>(6 months) | Engine oil and filter<br>Transmission oil check<br>Brakes<br>Hinges, latches, locks<br>Leak inspection |
| Every 12,500 miles<br>(12 months) | Drive belts<br>Coolant hoses and connections<br>Vacuum lines<br>Air cleaner<br>Choke plate and linkage<br>Brake fluid<br>Steering and suspension<br>Wheels and tires<br>Brakes<br>Evaporative emission control system<br>Tune-up |
| Every 25,000 miles<br>(24 months) | Coolant<br>Air filters<br>Fuel filter<br>PCV system<br>Ball-joints<br>Front and rear wheel bearings<br>Transmission oil change |

**Table 3    SCHEDULED MAINTENANCE, 1978-ON**

| | |
|---|---|
| Every 7,500 miles<br>(6 months) | Engine oil and filter<br>Transmission oil check<br>Brakes<br>Hinges, latches, locks<br>Leak inspection |
| Every 15,000 miles<br>(12 months) | Drive belts<br>Coolant hoses and connections<br>Vacuum lines<br>Air cleaner<br>Choke plate and linkage<br>Brake fluid<br>Steering and suspension<br>Wheels and tires<br>Brakes<br>Evaporative emission control system<br>Tune-up |
| Every 30,000 miles<br>(24 months) | Coolant<br>Air filters<br>Air induction valve filter<br>Fuel filter<br>PCV system<br>Ball-joints<br>Front and rear wheel bearings<br>Transmission oil change |

# LUBRICATION, MAINTENANCE, AND TUNE-UP

### Table 4  RECOMMENDED LUBRICANTS

| | |
|---|---|
| Engine | API Service SE |
| Manual transmission | API GL-4 |
| Automatic transmission | Dexron |
| Differential | API GL-5 |
| Steering gear | API GL-4 |
| Brake and clutch fluid | DOT 3 |
| Antifreeze | Ethylene glycol base with corrosion inhibitor |
| Windshield washer fluid | Windshield washer solvent/antifreeze |

2. Check coolant level (**Figure 4**). It should be ¾ to 1½ in. (20-30mm) below the base of the filler neck.

> **WARNING**
> *Do not remove the radiator cap when the engine is hot. The cap could fly off, followed by a fountain of hot coolant.*

3. Check battery electrolyte level. On translucent batteries, it should be between the marks on the battery case (**Figure 5**). On black batteries, it should be even with the bottom of the filler wells. See **Figure 6**.

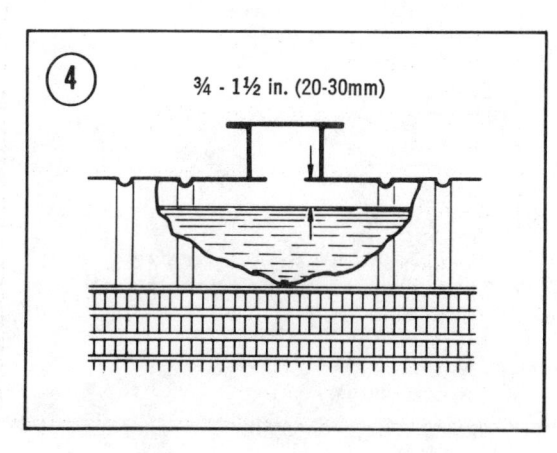

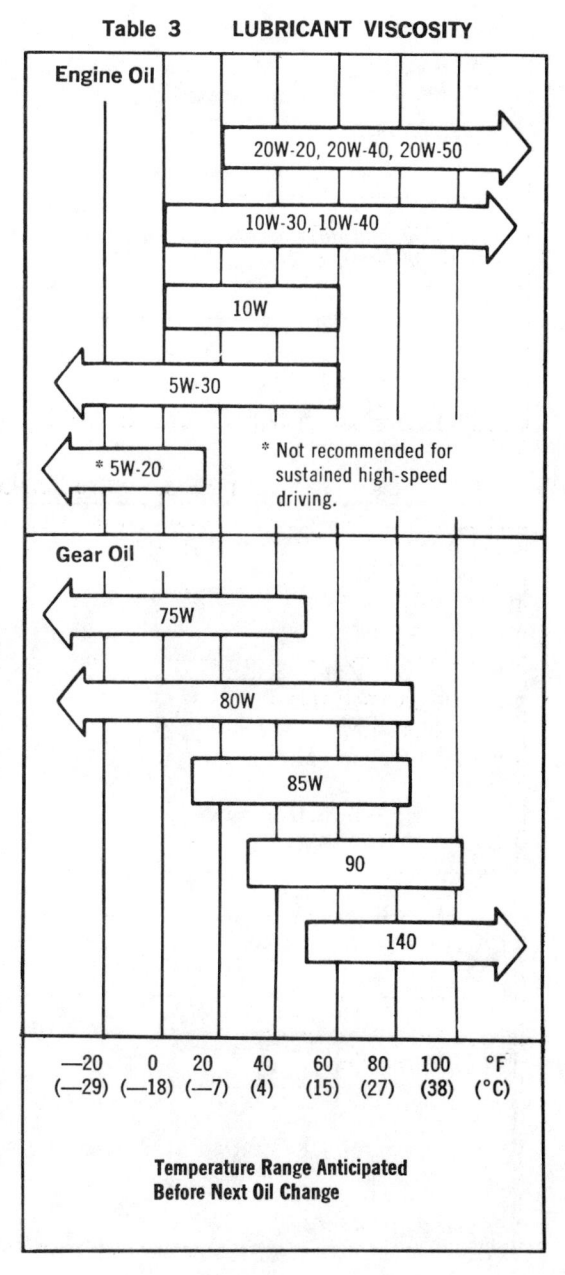

### Table 3  LUBRICANT VISCOSITY

**Engine Oil**
- 20W-20, 20W-40, 20W-50
- 10W-30, 10W-40
- 10W
- 5W-30
- *5W-20  *Not recommended for sustained high-speed driving.

**Gear Oil**
- 75W
- 80W
- 85W
- 90
- 140

Temperature range: —20 (—29), 0 (—18), 20 (—7), 40 (4), 60 (15), 80 (27), 100 (38) °F (°C)

**Temperature Range Anticipated Before Next Oil Change**

③

④ ¾ - 1½ in. (20-30mm)

4. Check fluid level in the windshield washer tank (**Figure 7**). It should be kept full. Use windshield washer solvent, following manufacturer's instructions.

> **CAUTION**
> *Do not use radiator antifreeze in the washer tank. The runoff may damage the car's paint.*

5. Check fluid level in the brake and clutch master cylinders (**Figure 8**). Since the reservoirs are translucent, this can be done at a glance. Fluid should be between the lines on the reservoirs. If low, top up with brake fluid marked DOT 3. The same fluid is used for clutch and brakes.

> **CAUTION**
> *Do not remove reservoir caps unless topping up fluid. Clean the area around the caps before removal.*

6. Check tire pressures (**Table 6**). This should be done when the tires are cold (after driving less than one mile). When the tires heat up from driving, the air in them expands and gives false high pressure readings.

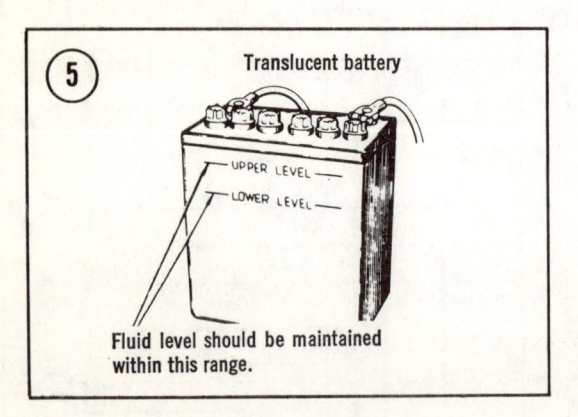

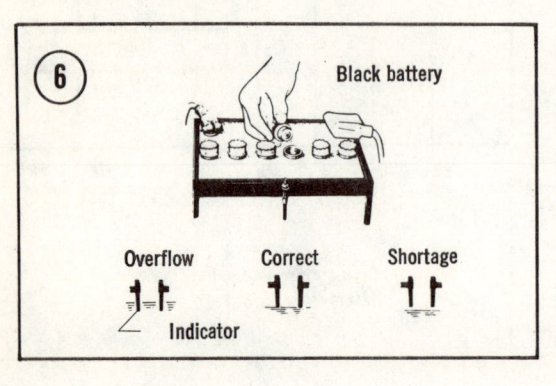

## SCHEDULED MAINTENANCE

### Engine Oil and Filter

If the car is given normal use, change the oil when recommended in **Table 2 or 3**. If it is used for stop-and-go driving, in dusty areas, or left idling for long periods, change the oil every 3,000 miles or 3 months.

Use an oil recommended in **Tables 4 and 5**. The rating (SE) is usually printed on top of the can (**Figure 9**).

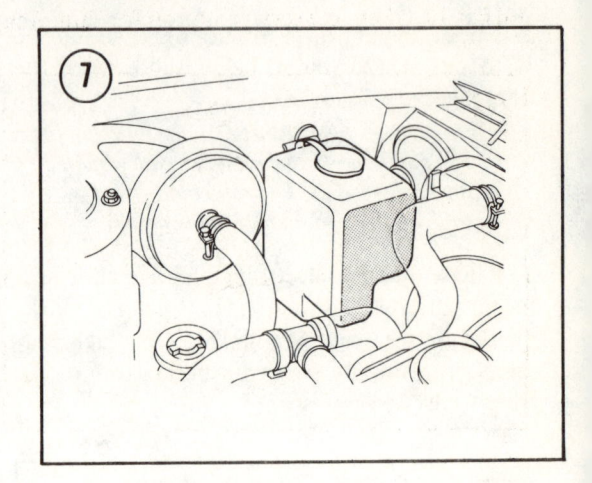

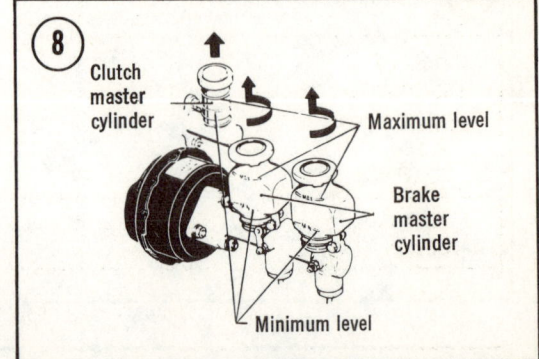

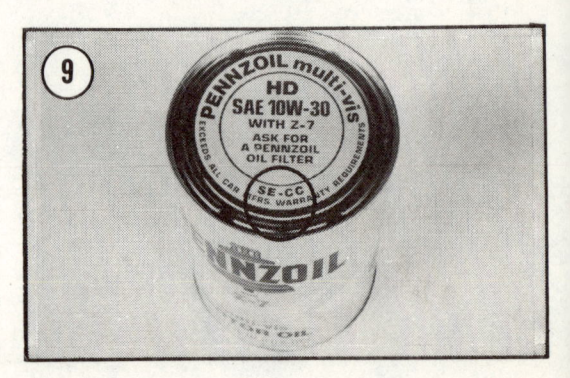

# LUBRICATION, MAINTENANCE, AND TUNE-UP

To drain the oil and change the filter, you will need:

a. Drain pan
b. Can opener and funnel
c. Filter wrench
d. Oil (4 quarts)
e. Oil filter

There are a number of ways of discarding the old oil safely. The easiest way is to pour it from the drain pan into a gallon bleach bottle. The oil can be taken to a service station for dumping or, where permitted, thrown in your household trash.

1. Warm engine to operating temperature, then shut it off.

2. Put the drain pan under the drain plug (**Figure 10**). Remove the plug and let the oil drain for at least 10 minutes.

3. Unscrew the oil filter (**Figure 10**) counterclockwise. Use a filter wrench if the filter is too tight to remove by hand.

4. Wipe the gasket surface on the engine block clean with a lint-free cloth.

5. Coat the neoprene gasket on the new filter with clean engine oil.

6. Screw the filter onto the engine *by hand* until the gasket just touches the engine block. At this point, there will be a very slight resistance when turning the filter.

7. Tighten the filter $\frac{2}{3}$ turn more *by hand*. If the filter wrench is used, the filter will probably be overtightened. This will cause a leak.

8. Install the oil pan drain plug. Tighten it securely.

9. Remove the oil filler cap (**Figure 10**).

10. Pour oil into the engine. Capacity is listed in **Table 7**.

11. Start the engine and let it idle. The instrument panel oil pressure light will remain on for 15-30 seconds, then go out.

#### CAUTION
*Do not rev the engine to make the oil pressure light go out. It takes time for the oil to reach all areas of the engine, and revving it could damage dry parts.*

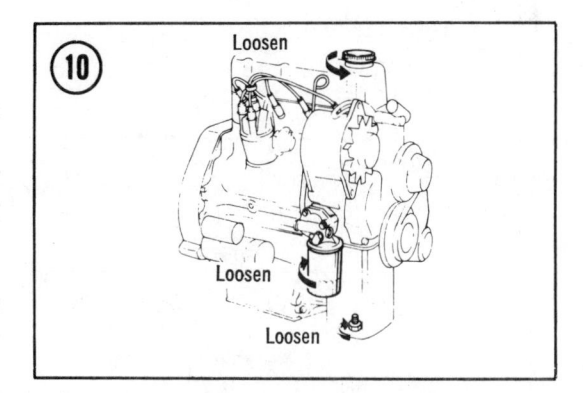

### Table 6  TIRE PRESSURES

| Size | Pressure (psi) |
| --- | --- |
| 155SR-13 | 24* |
| 165/70SR-13 | 24* |
| 155-13/6.15-13-4PR | |
| Front | 28 |
| Rear | 26 |

*On 310's, add 4 psi for speeds over 60 mph (100 kph).

### Table 7  APPROXIMATE REFILL CAPACITIES

| | |
| --- | --- |
| Engine oil (including filter) | |
| 1976-1977 F10 | $3\frac{5}{8}$ qt. (3.4 liters) |
| 1978 F10 | $3\frac{1}{2}$ qt. (3.3 liters) |
| 310 | $3\frac{3}{8}$ qt. (3.2 liters) |
| Transmission oil | $4\frac{7}{8}$ pt. (2.3 liters) |
| Cooling system | |
| F10 | 7 qt. (6.6 liters) |
| 310 | 6-1/4 qt. (5.9 liters) |
| Fuel tank | |
| 1976 F10 sedan and hatchback | $10\frac{5}{8}$ gal. (40 liters) |
| 1976 F10 wagon | $9\frac{1}{4}$ gal. (30 liters) |
| 1977-1978 F10 (all models) | $10\frac{5}{8}$ gal. (40 liters) |
| 310 | $13\frac{1}{4}$ gal. (50 liters) |

12. While the engine is running, check the drain plug and oil filter for leaks.

13. Turn the engine off. Let the oil settle for several minutes, then check level on the dipstick (**Figures 1 and 2**). Add oil if necessary to bring the level up to the "H" mark, but *do not* overfill.

### Manual Transmission Oil

To check, remove the filler plug from the front of the transmission (**Figure 11**). Make sure oil is up to the bottom of the filler plug threads. If necessary, top up with an oil recommended in **Tables 4 and 5**.

### Brakes

Check front brake pads for wear. Check pedal free play. Test the brake booster and proportioning valve. See Chapter Nine.

At alternate inspections, check rear brake linings for wear, and check wheel cylinders for fluid leaks. See *Rear Brakes*, Chapter Ten.

> NOTE: *If salt is used on the roads, inspect front and rear brakes every 3,000 miles or 3 months.*

### Hinges, Latches, Locks

Lightly grease the hood latch and trunk or tailgate lock with molybdenum disulphide grease. Apply 1-2 drops of oil to hinges on doors, hood, and tailgate or trunk lid. Lubricate striker plates with a non-staining stick lube such as Door Ease. Lubricate lock tumblers by applying a thin coat of Lubriplate, lock oil, or graphite to the key. Insert the key and work the lock several times. Wipe the key clean.

> NOTE: *If salt is used on the roads, do this service every 3,000 miles or 3 months.*

### Leak Inspection

The engine should be checked visually for leaks. Check the oil pan drain plug, oil pan gasket, oil filter, engine front cover, and oil pump. Greasy looking dirt at these points may indicate an oil leak. Inspect the radiator and hose connections for coolant residue or rust.

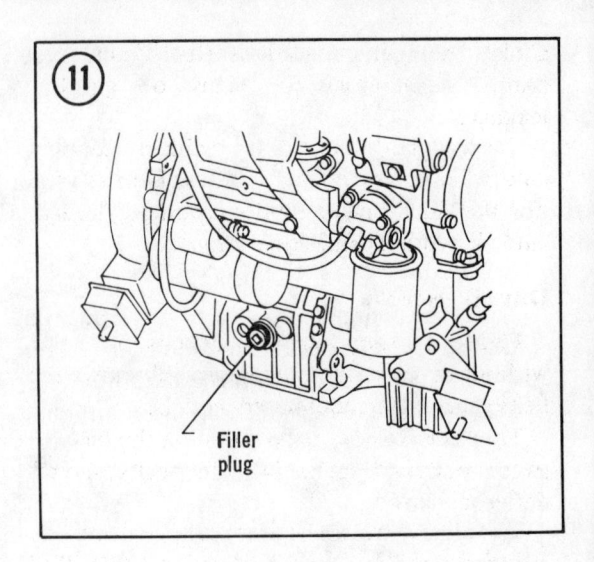

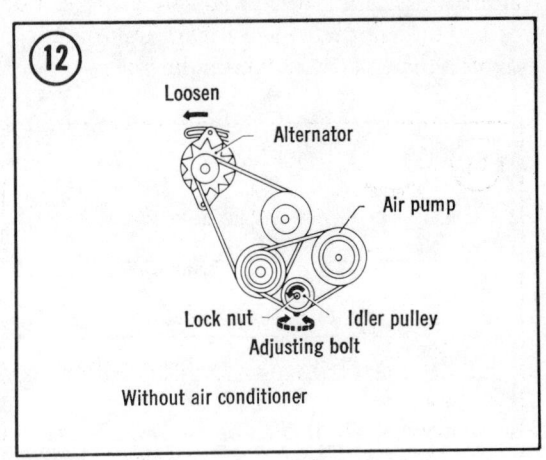

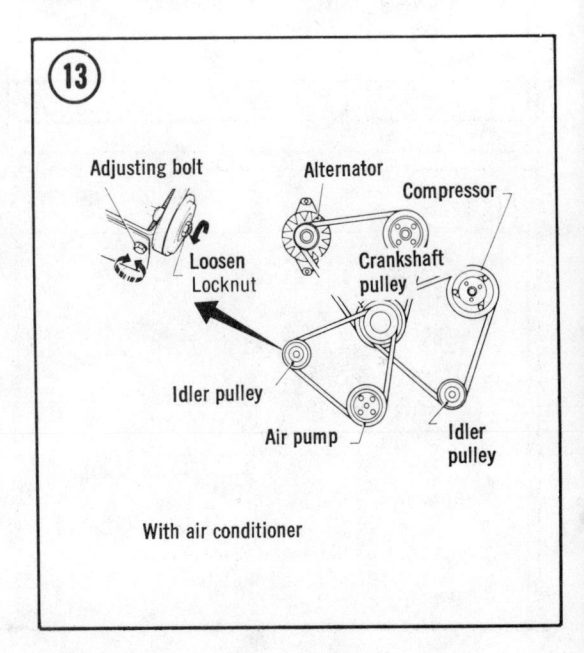

# LUBRICATION, MAINTENANCE, AND TUNE-UP

Check the fuel connections (fuel filter, fuel pump, carburetor) for signs of gasoline leakage.

Inspect the brake master cylinder, calipers, and wheel cylinders for wetness. Do the same for the clutch master and operating cylinders, and all hydraulic line connections.

## Drive Belts (F10)

**Figure 12** shows the drive belts on F10's without air conditioning; **Figure 13** shows the belts used on air conditioned models.

To check tension, press down on the belt between pulleys (**Figure 14**). The belts should deflect as shown.

To adjust the alternator belt, loosen the alternator mounting and adjusting bolts. Pull or pry the alternator away from the engine to tighten the belt, then tighten the bolts.

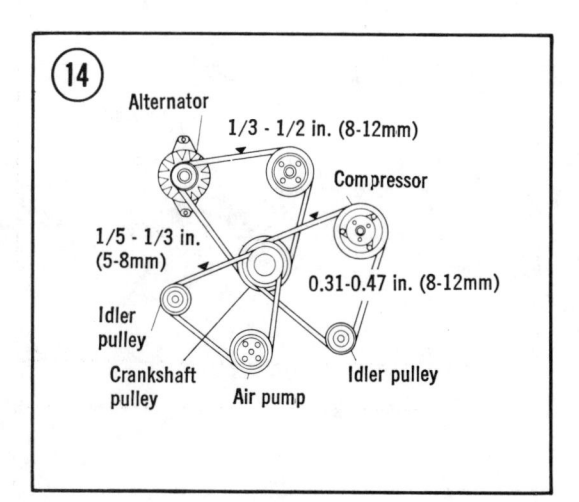

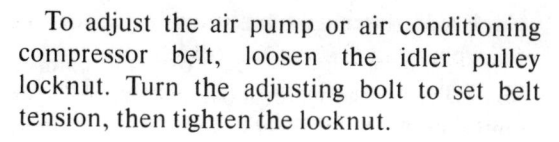

To adjust the air pump or air conditioning compressor belt, loosen the idler pulley locknut. Turn the adjusting bolt to set belt tension, then tighten the locknut.

## Drive Belts (310)

To check tension, press on the belts midway between pulleys (**Figure 15**). All belts should deflect 1/3-1/2 in. (8-12 mm).

To adjust the alternator belt, loosen the alternator mounting and adjusting bolts (**Figure 16**). Pull or pry the alternator away from the engine to tighten the belt, then tighten the bolts.

To adjust the air pump or air conditioning compressor belt, loosen the idler pulley locknut (**Figure 17**). Turn the adjusting bolt to change belt tension, then tighten the locknut.

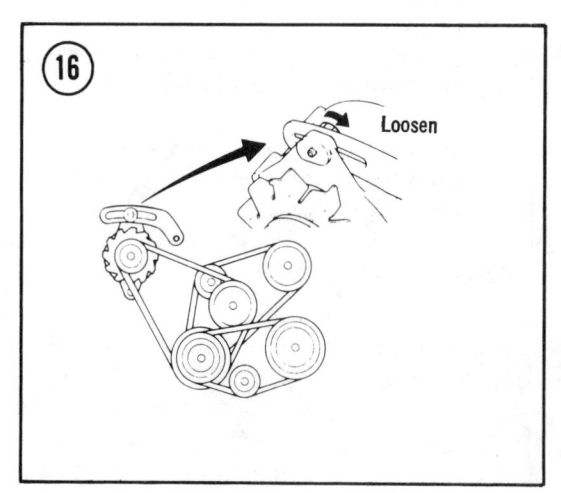

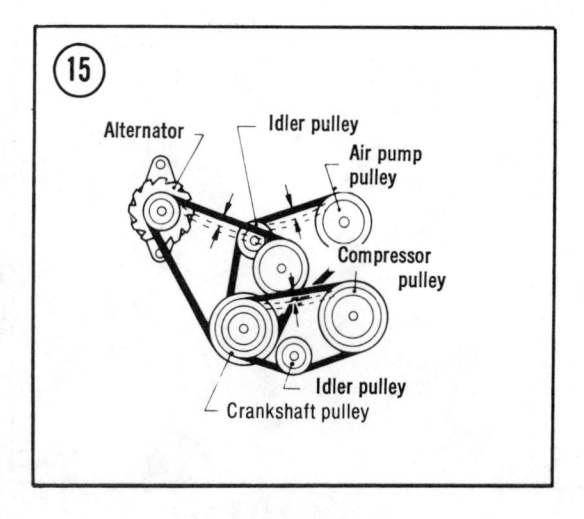

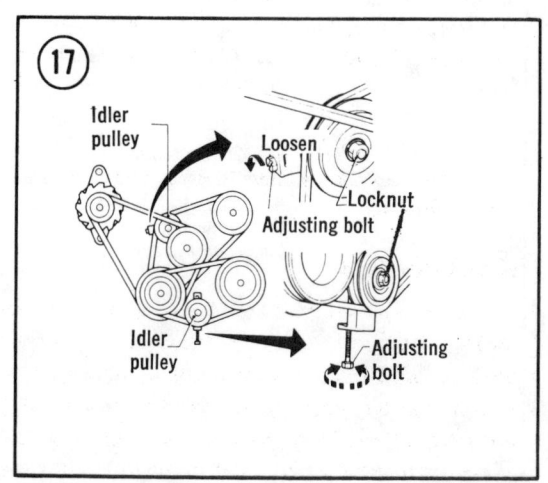

**CHAPTER THREE**

**Coolant Hoses and Connections**

Inspect all cooling system hoses and connections, including heater hoses. Replace hoses that are cracked, deteriorated, or extremely soft. Make sure all clamps are tight.

**Vacuum Lines**

Check vacuum lines for cracks or deterioration. Refer to the following illustrations:

a. **Figure 18** — 1976 California
b. **Figure 19** — 1976 non-California, 1977 Canada
c. **Figure 20** — 1977 U.S. models
d. **Figure 21** — 1978 U.S. models
e. **Figure 22** — 1978 Canada
f. **Figure 23** — 1979 U.S. models
g. **Figure 24** — 1979 Canada

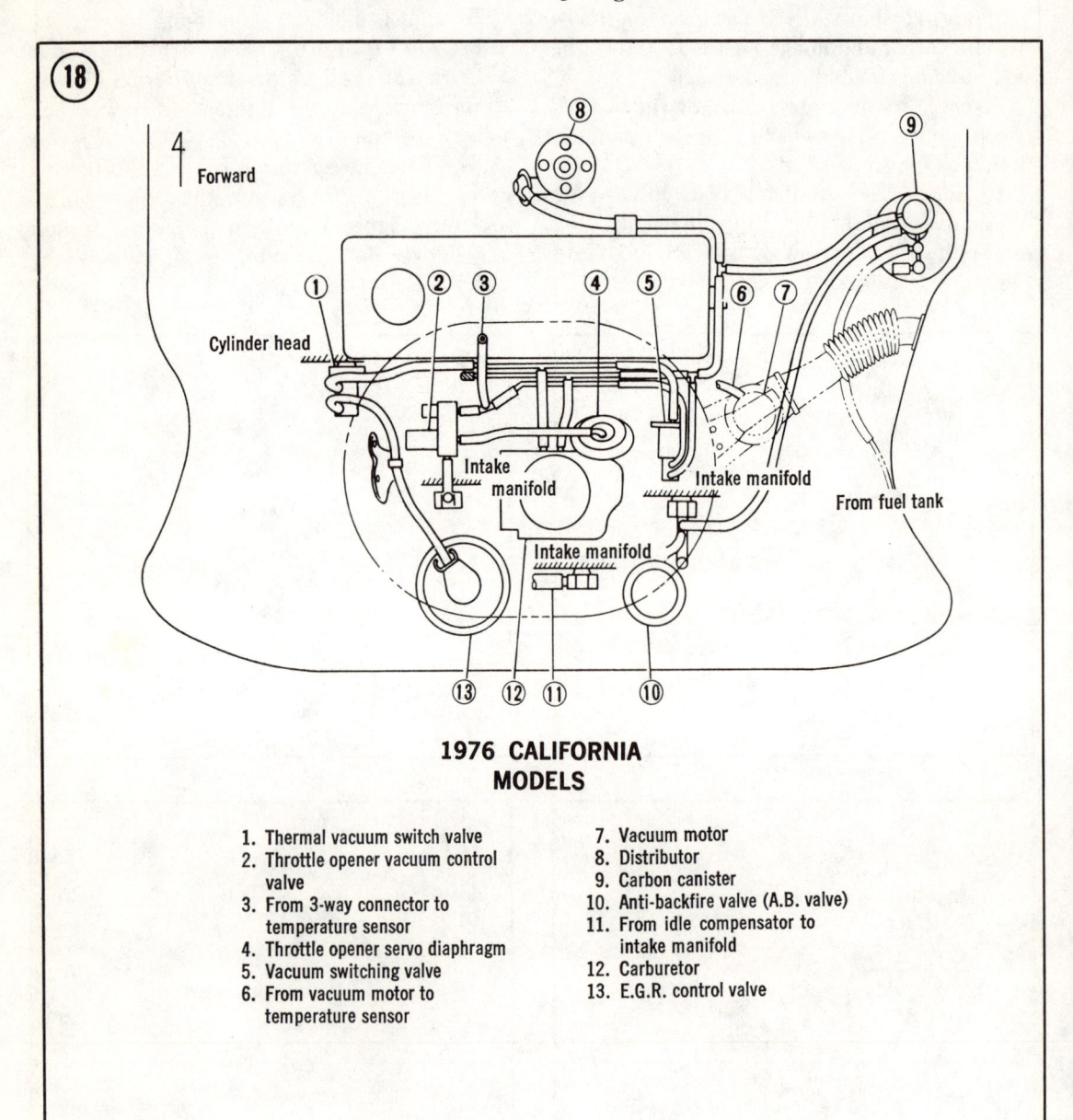

**1976 CALIFORNIA MODELS**

1. Thermal vacuum switch valve
2. Throttle opener vacuum control valve
3. From 3-way connector to temperature sensor
4. Throttle opener servo diaphragm
5. Vacuum switching valve
6. From vacuum motor to temperature sensor
7. Vacuum motor
8. Distributor
9. Carbon canister
10. Anti-backfire valve (A.B. valve)
11. From idle compensator to intake manifold
12. Carburetor
13. E.G.R. control valve

# LUBRICATION, MAINTENANCE, AND TUNE-UP

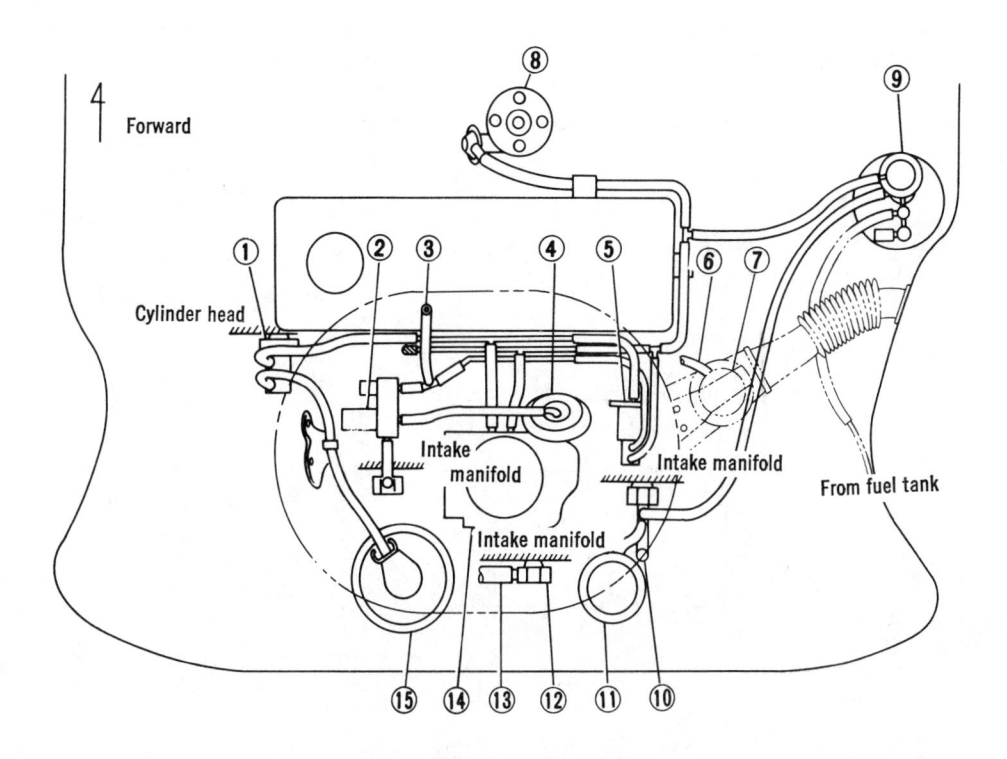

**⑲**

**1976 NON-CALIFORNIA MODELS**
**1977 CANADIAN MODELS**

1. Thermal vacuum switch valve
2. Throttle opener vacuum control valve
3. From 3-way connector to temperature sensor
4. Throttle opener servo diaphragm
5. Vacuum switching valve
6. From vacuum motor to temperature sensor
7. Vacuum motor
8. Distributor
9. Carbon canister
10. From Master-Vac
11. Anti-backfire valve (A.B. valve)
12. From A.B. valve
13. From idle compensator to intake manifold
14. Carburetor
15. E.G.R. control valve

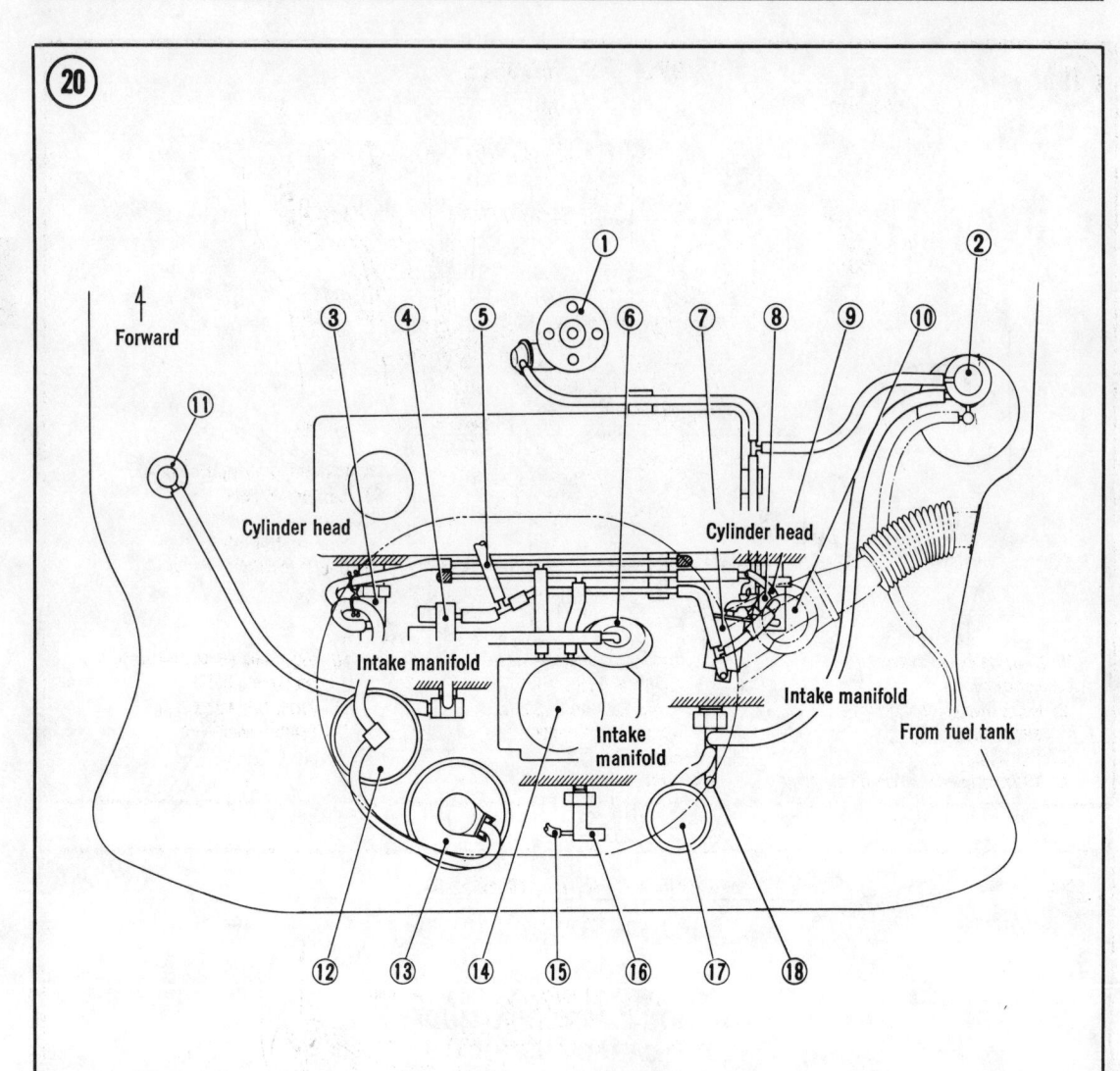

**1977 U.S. MODELS**

1. Distributor
2. Carbon canister
3. Thermal vacuum valve-E.G.R.
4. Throttle opener vacuum control valve
5. From 3-way connector to temperature sensor
6. Throttle opener servo diaphragm
7. Vacuum switching valve
8. Thermal vacuum valve-T.C.S.
9. From vacuum motor to temperature sensor
10. Vacuum motor
*11. Combined air control valve (C.A.C. valve California models)
12. Back pressure transducer valve (B.P.T. valve)
13. E.G.R. control valve
14. Carburetor
15. From idle compensator to intake manifold
16. From anti-backfire valve (A.B. valve)
17. A.B. valve
18. From Master-Vac

# LUBRICATION, MAINTENANCE, AND TUNE-UP

**㉑**  **1978 U.S. MODELS**

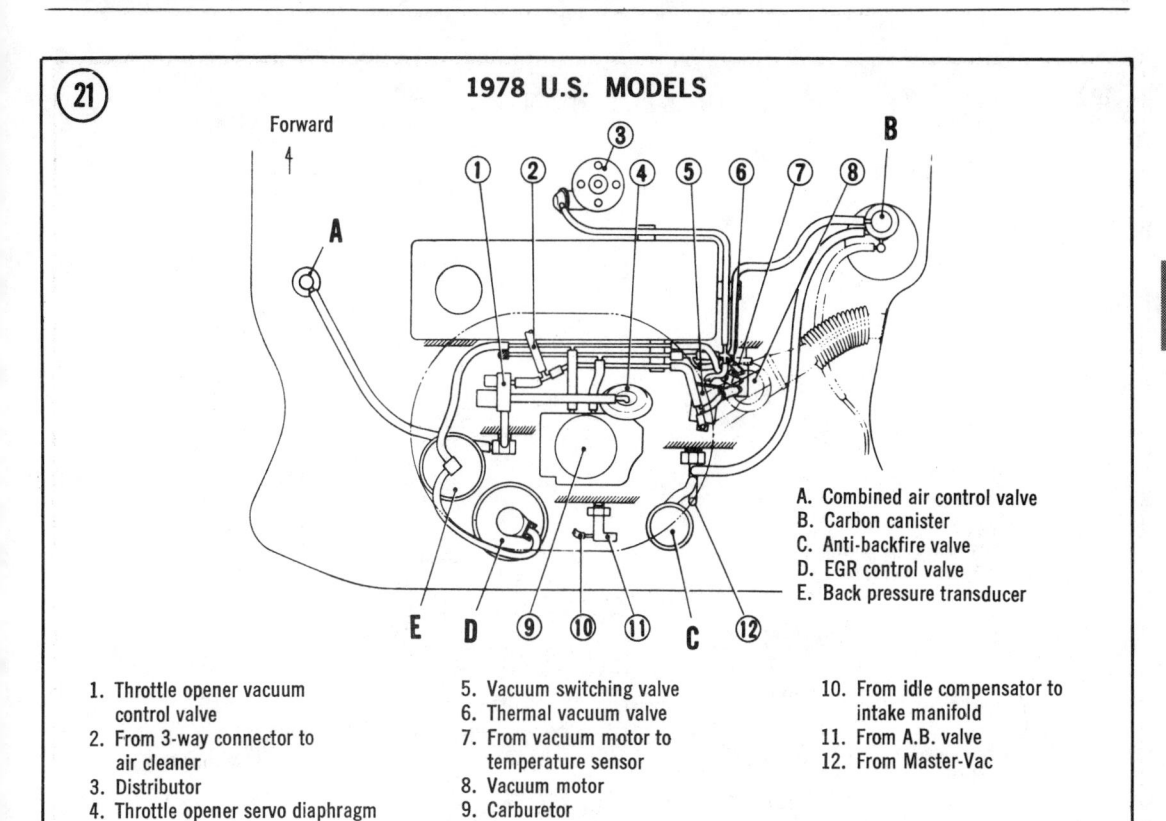

A. Combined air control valve
B. Carbon canister
C. Anti-backfire valve
D. EGR control valve
E. Back pressure transducer

1. Throttle opener vacuum control valve
2. From 3-way connector to air cleaner
3. Distributor
4. Throttle opener servo diaphragm
5. Vacuum switching valve
6. Thermal vacuum valve
7. From vacuum motor to temperature sensor
8. Vacuum motor
9. Carburetor
10. From idle compensator to intake manifold
11. From A.B. valve
12. From Master-Vac

**㉒**  **1978 CANADA MODELS**

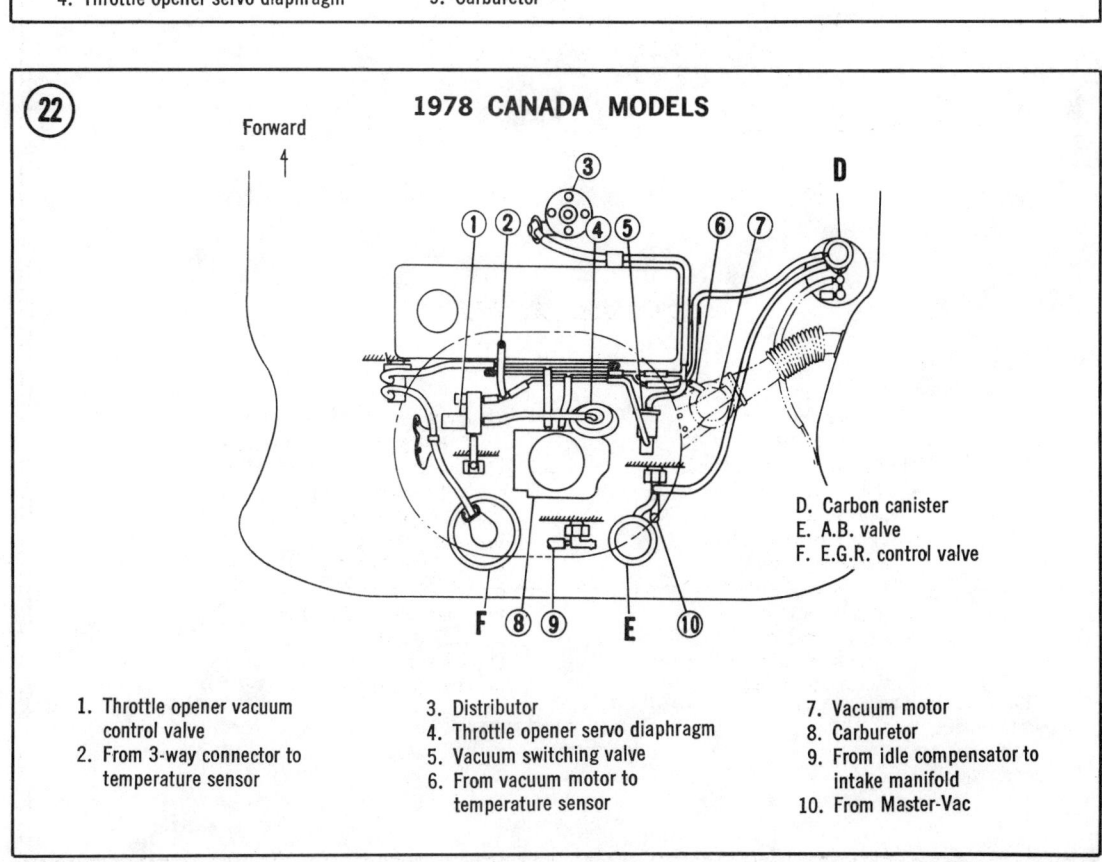

D. Carbon canister
E. A.B. valve
F. E.G.R. control valve

1. Throttle opener vacuum control valve
2. From 3-way connector to temperature sensor
3. Distributor
4. Throttle opener servo diaphragm
5. Vacuum switching valve
6. From vacuum motor to temperature sensor
7. Vacuum motor
8. Carburetor
9. From idle compensator to intake manifold
10. From Master-Vac

CHAPTER THREE

## 1979 U.S. MODELS

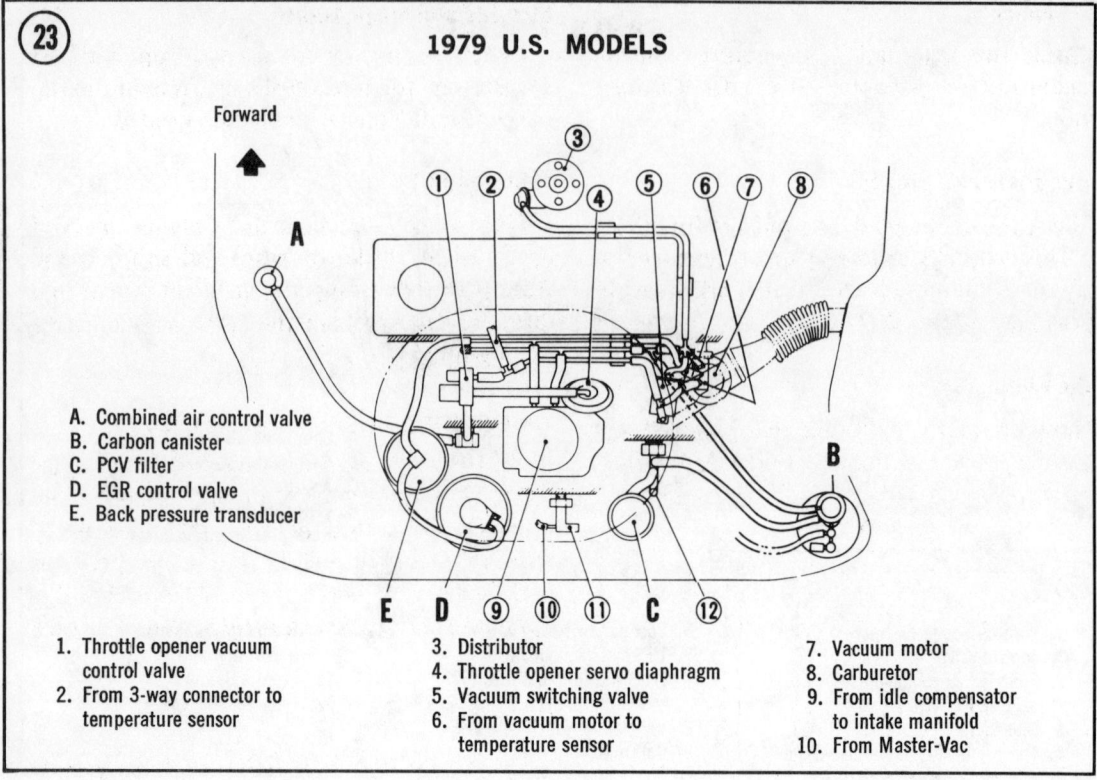

A. Combined air control valve
B. Carbon canister
C. PCV filter
D. EGR control valve
E. Back pressure transducer

1. Throttle opener vacuum control valve
2. From 3-way connector to temperature sensor
3. Distributor
4. Throttle opener servo diaphragm
5. Vacuum switching valve
6. From vacuum motor to temperature sensor
7. Vacuum motor
8. Carburetor
9. From idle compensator to intake manifold
10. From Master-Vac

## 1979 CANADA MODELS

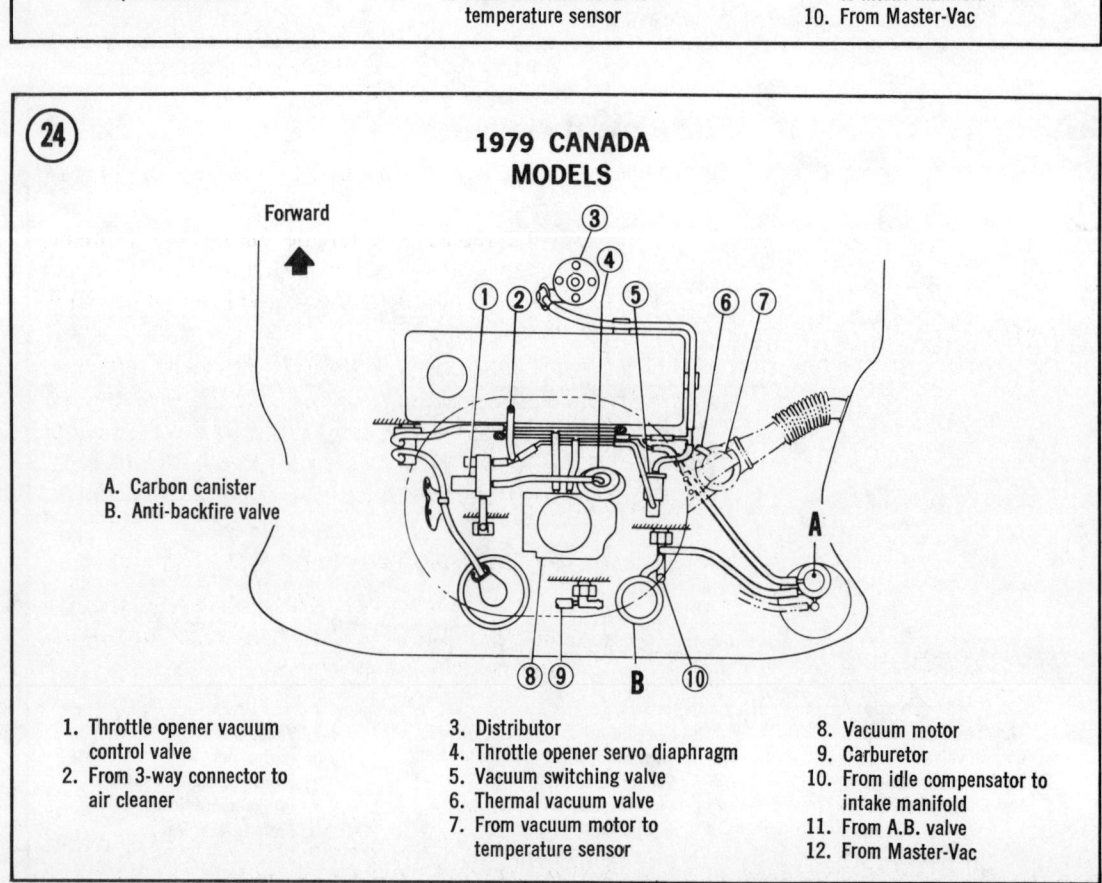

A. Carbon canister
B. Anti-backfire valve

1. Throttle opener vacuum control valve
2. From 3-way connector to air cleaner
3. Distributor
4. Throttle opener servo diaphragm
5. Vacuum switching valve
6. Thermal vacuum valve
7. From vacuum motor to temperature sensor
8. Vacuum motor
9. Carburetor
10. From idle compensator to intake manifold
11. From A.B. valve
12. From Master-Vac

# LUBRICATION, MAINTENANCE, AND TUNE-UP

## Air Cleaner

Check the automatic temperature control mechanism. See Chapter Five, *Air Cleaner* section.

## Choke Plate and Linkage

Check the choke plate and linkage for sticking. This is usually caused by dirt. Clean with a spray-on carburetor cleaner, then lubricate with a spray lube such as WD-40.

## Brake Fluid

Pump out all the old brake fluid, and fill the hydraulic system with new fluid. See *Brake Bleeding*, Chapter Ten.

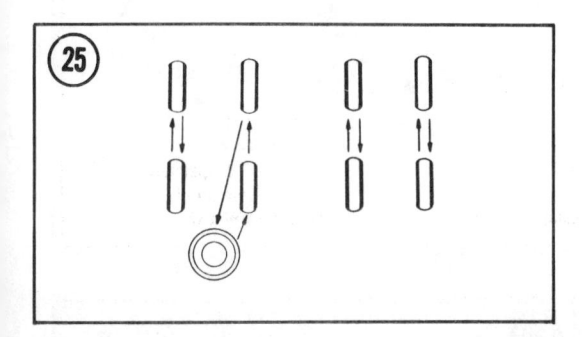

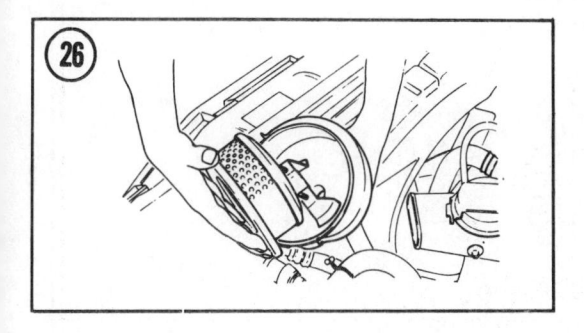

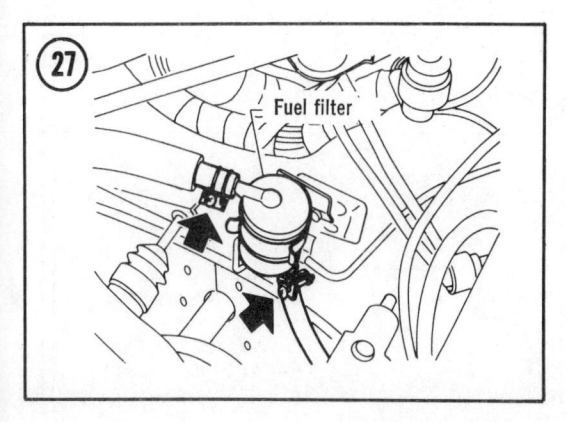

## Steering and Suspension

Check the steering linkage and front and rear suspensions for loose fasteners, missing parts, or damage. Tighten or replace as needed.

## Wheels and Tires

Have wheel alignment and balance checked by a Datsun dealer or front end shop. If the front tires are wearing at a different rate or in a different manner than the rears, rotate them. See **Figure 25**.

## Evaporative Emission Control System

On 1976-1977 models, inspect the vapor lines every 12,500 miles or 12 months. Replace the carbon canister filter every 25,000 miles or 24 months. On 1978 and later models, perform both steps at 30,000 miles or 24 months. See *Evaporative Emission Control System*, Chapter Five.

## Coolant

Drain, flush, and refill the cooling system. See *Cooling System Flushing*, Chapter Six.

## Air Filters

Replace filter elements in the engine air cleaner and air pump air cleaner.

To replace the engine air cleaner element, remove the cover nut and clips. Lift the cover off, take out the element, and install a new one. Reinstall the cover.

To replace the air pump air cleaner element, disconnect the air hose and remove the air cleaner from the hood ledge. Remove the cover and take out the element (**Figure 26**). Installation is the reverse of removal.

## Air Induction Valve Filter

This is used on 1978 and later Canadian models. Replace as described under *Air Induction System*, Chapter Five.

## Fuel Filter

Replace the fuel filter (**Figure 27**). Disconnect the inlet and outlet lines, take the filter out of its clip, and install a new one.

## PCV System

1. Replace the air cleaner PCV filter (**Figure 28**).
2. Disconnect the hose from the PCV valve. Unscrew the valve (**Figure 29**) and screw in a new one.
3. Blow out the PCV hose with compressed air (**Figure 30**). If the hose is cracked or brittle, replace it.

## Ball-joints

Lubricate the front suspension ball-joints with multipurpose grease. Remove the plug (**Figure 31**) and install a grease nipple. Inject grease with a grease gun (**Figure 32**) until all old grease is forced out.

## Front and Rear Wheel Bearings

Remove, clean, repack, and adjust the wheel bearings. See *Wheel Bearings*, Chapter Ten or Eleven.

## Manual Transmission Oil

Drive the car to warm the oil, then place a container beneath the transmission. Remove the drain and filler plugs (**Figure 33** and **Figure 34**) and let the oil drain completely. Reinstall the drain plug and fill the transmission with an oil recommended in **Table 4** or **Table 5**. Capacity is listed in **Table 7**. Oil level should be up to the bottom of the filler plug threads.

## TUNE-UP

Under normal conditions, a tune-up should be done every 12,500 miles (through 1977), 15,000 miles (1978-on), or once a year. More frequent tune-ups may be needed if the car is used primarily for stop-and-go driving, or left idling for long periods.

Since different engine systems interact, a tune-up should be done in the following order:
  a. Compression check
  b. Valve adjustment
  c. Spark plug replacement
  d. Distributor inspection
  e. Ignition timing
  f. Carburetor adjustment

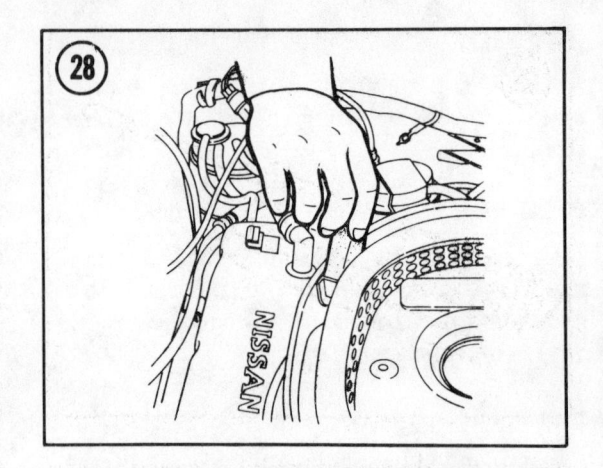

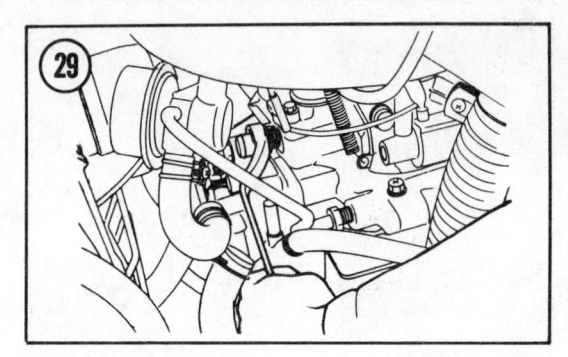

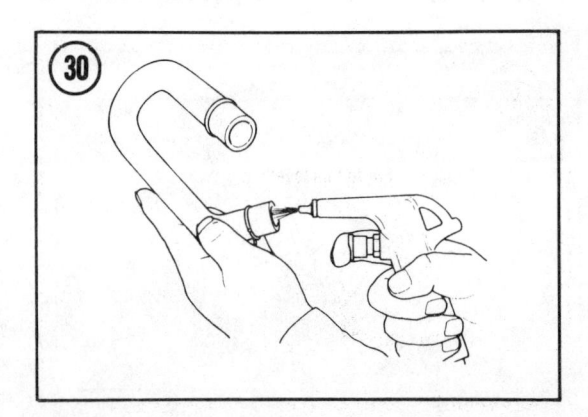

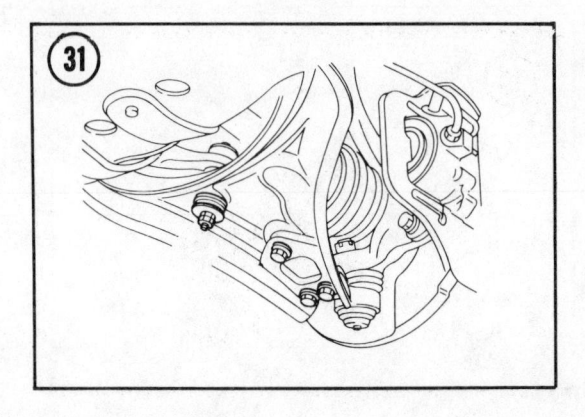

# LUBRICATION, MAINTENANCE, AND TUNE-UP

**(32)**

**(33)** Drain plug

**(34)** Fill to this level / Filler plug

**(35)**

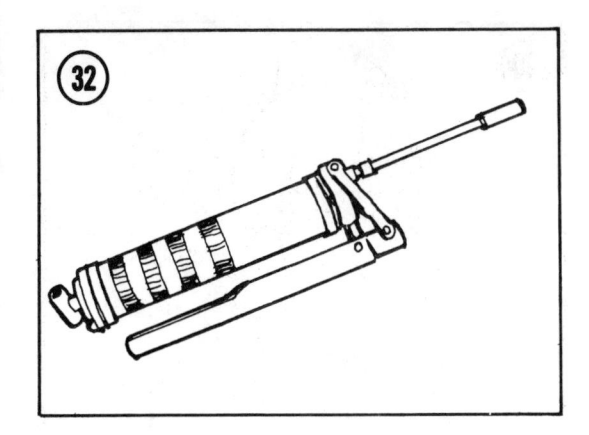

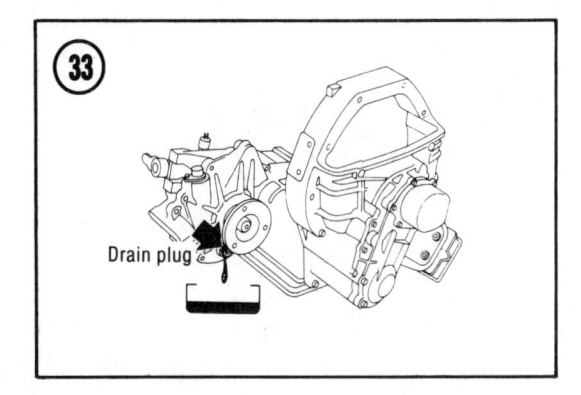

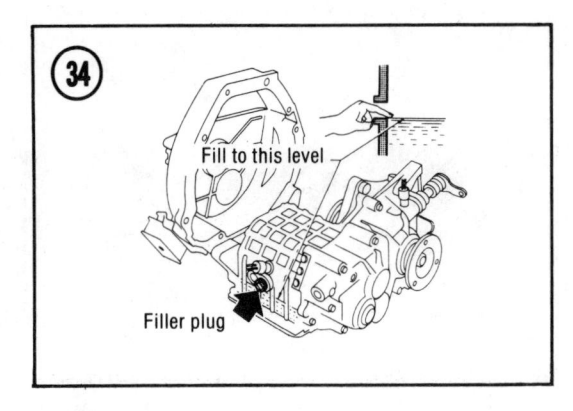

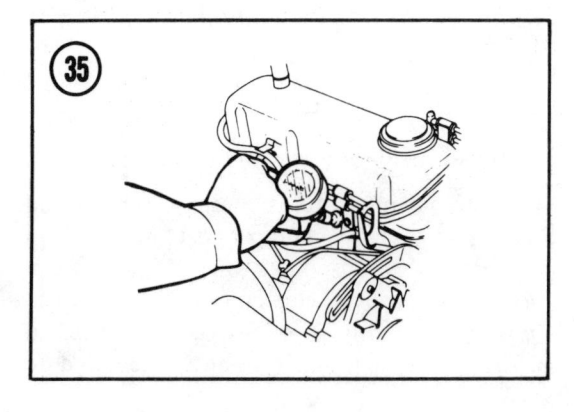

## COMPRESSION TEST

There are 2 types of compression test: "wet" and "dry." These tests are interpreted together to isolate problems in cylinders and valves. The dry compression test is done first. Test as follows.

1. Warm the engine to normal operating temperature.
2. Remove the air cleaner as described under *Air Cleaner*, Chapter Five. Make sure the choke is completely open. If it isn't, inspect the choke linkage and electrical circuit. See *Carburetor* and *Automatic Choke Circuit*, Chapter Five.
3. Remove the spark plugs. See *Spark Plugs* in this chapter for correct procedures.
4. Connect the compression tester to one cylinder following manufacturer's instructions. **Figure 35** shows a hand-held compression tester in use. You can also use the screw-in type described in Chapter One.

*NOTE*
*Hand-held compression testers require 2 people, one to hold the compression tester and one to crank the engine. Screw-in compression testers only require one person.*

5. Crank the engine over until there is no further increase in compression reading.
6. Remove the tester and write down the reading.
7. Repeat Steps 4-6 for each cylinder. Compare results with **Table 8** in this chapter.

When interpreting the results, actual readings are not as important as the differences in readings. Low readings, although they may be even, are a sign of wear. Low readings in 2 adjacent cylinders may indicate a defective head gasket. No cylinder should test at less than 80 per cent of the highest cylinder. A greater difference indicates worn or broken rings, leaky or sticking valves, a defective head gasket, or a combination of all.

If the dry compression test indicates a problem, isolate the cause with a wet compression test. This is done in the same way as the dry compression test, except that about one tablespoon of oil is poured down the spark plug holes before performing Steps 4-6. If the wet compression readings are much greater than the dry readings, the trouble is probably due to worn or broken rings. If there is little difference between wet and dry readings, the trouble is probably due to leaky or sticking valves. If 2 adjacent cylinders are low and the wet and dry readings are close, the head gasket may be damaged.

## VALVE ADJUSTMENT

1. Warm the engine to normal operating temperature.
2. Remove the spark plugs. This makes it easier to turn the engine.

*NOTE*
*See **Spark Plugs** later in this chapter for correct removal procedures.*

3. Remove the rocker arm cover.
4. Turn the engine until No. 1 piston is at top dead center on its compression stroke. When this occurs, the 0 degree mark on the timing scale will align with the crankshaft pulley notch. See **Figure 36** (F10) or **Figure 37** (310). The distributor rotor will also point to No. 1 terminal in the distributor cap.

*NOTE*
*Be sure to check rotor position as well as the timing marks. The 0 degree mark also lines up when No. 4 cylinder is at TDC on its compression stroke.*

5. With No. 1 cylinder at TDC, adjust valves 1, 2, 3, and 5. See **Figure 38**. To adjust a valve, insert a 0.014 in. (0.35mm) feeler gauge between the rocker arm and valve stem as shown. It should just slip in. If it is hard to insert or fits loosely, loosen the locknut. Turn the adjusting screw to change clearance, then tighten the locknut.
6. After tightening, recheck the clearance.
7. Turn the engine in its normal direction (clockwise, viewed from the radiator end) one full turn. The 0 degree mark should align with the pulley notch, and the rotor should point exactly opposite the direction it pointed in Steps 4-6.

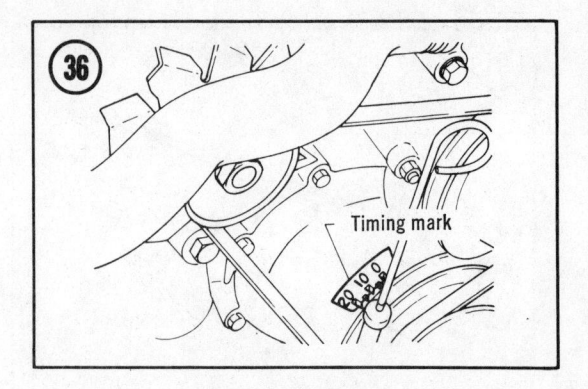

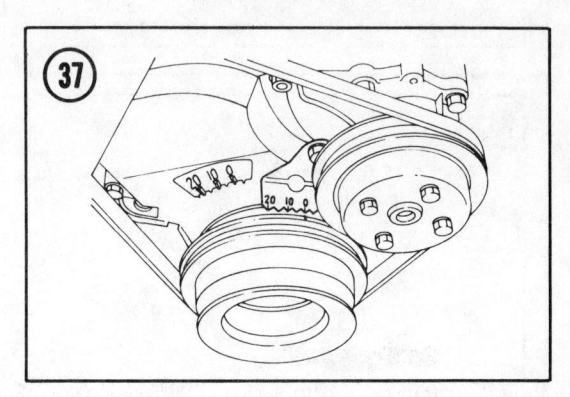

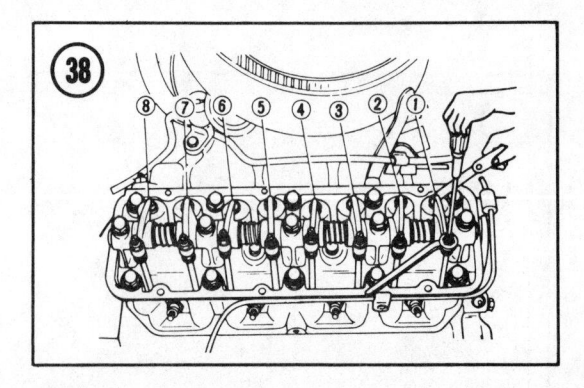

8. Adjust valves 4, 6, 7, and 8 (**Figure 38**). These valves are adjusted in the same manner as valves 1, 2, 3, and 5.
9. Install the rocker arm cover.

## SPARK PLUGS

### Removal

1. Blow out any foreign matter from around spark plugs with compressed air. Use a compressor if you have one. If not, most household vacuum cleaners can be set up to blow air. When most of the outlet is blocked with fingers, enough pressure is created to

# LUBRICATION, MAINTENANCE, AND TUNE-UP

2. Mark spark plug wires with the cylinder numbers so you can reconnect them properly. **Figure 39** shows cylinder numbers and corresponding terminals in the distributor cap.

*NOTE*
*To make labels, wrap a small strip of masking tape around each wire.*

3. Disconnect spark plug wires. Pull off by grasping the connector, *not* the wire. See **Figure 40**. Pulling on the wire may break it.

*NOTE*
*If the boots seem to be stuck, twist them 1/2 turn to break the seal. Do not pull on boots with pliers. The pliers could cut the insulation, causing an electrical short.*

4. Remove spark plugs with a 13/16 in. spark plug socket. Keep the plugs in order so you know which cylinder they came from.
5. Examine each spark plug. Compare its condition to the illustrations in Chapter Two. Spark plug condition indicates engine condition, and can warn of developing trouble.
6. Discard the plugs. Although they could be cleaned, gapped, and reused if in good condition, they rarely last very long; new plugs are only $4-5 a set and will be far more reliable.

## Gapping and Installing the Plugs

New plugs should be carefully gapped to insure a reliable, consistent spark. Use a special spark plug gapping tool with a wire gauge. See **Figure 41** or **Figure 42**.
1. Remove the new plugs from the boxes. See if the small end pieces (**Figure 43**) are screwed on or loose in the box. If loose, screw them on.
2. Find the correct spark plug gap for your car in **Table 8**. Insert the correct diameter gauge wire between the plug electrodes as shown in **Figure 42**. If the gap is correct, you will feel a slight drag as you pull the wire through. If there is no drag, or if the gauge won't fit through, bend the side electrode with the gapping tool (**Figure 44**) to change the gap.
3. Put a *small* drop of oil on the threads of each spark plug.
4. Crank the starter for about 5 seconds to blow any dirt away from the spark plug holes.

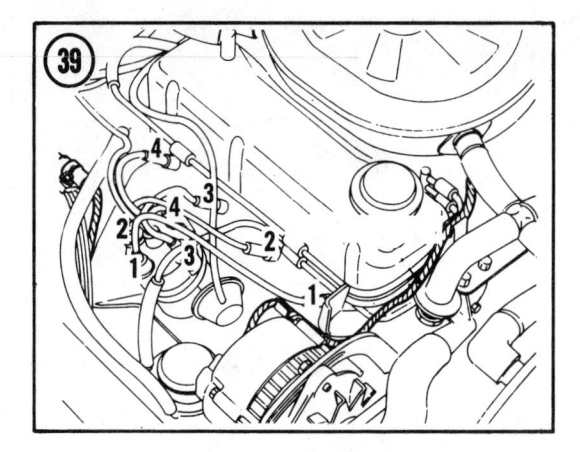

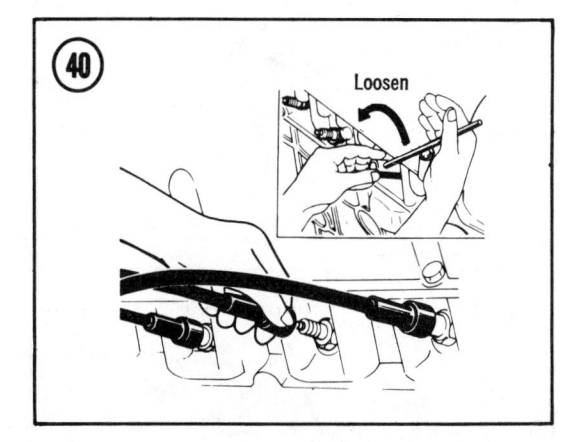

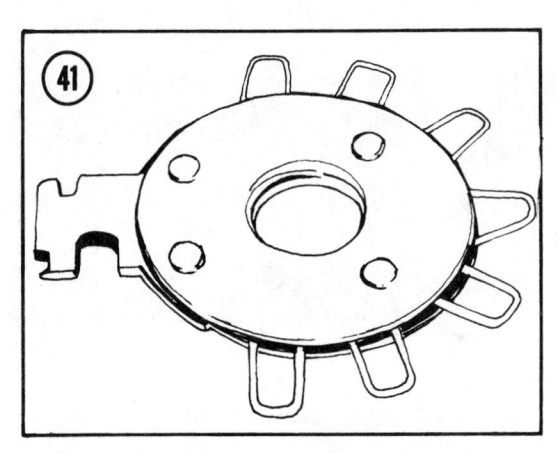

blow dirt away. Another method is to use a can of compressed inert gas, available from photo stores.

*CAUTION*
*When spark plugs are removed, dirt around the plugs can fall into the spark plug holes. This can cause expensive engine damage.*

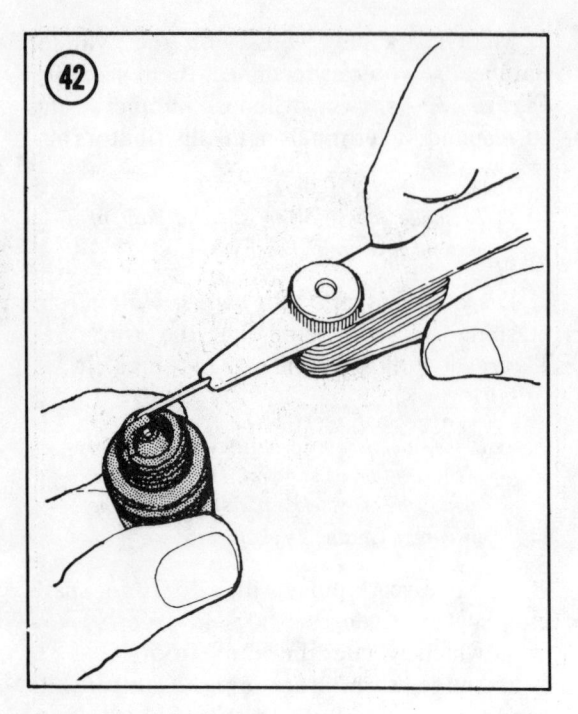

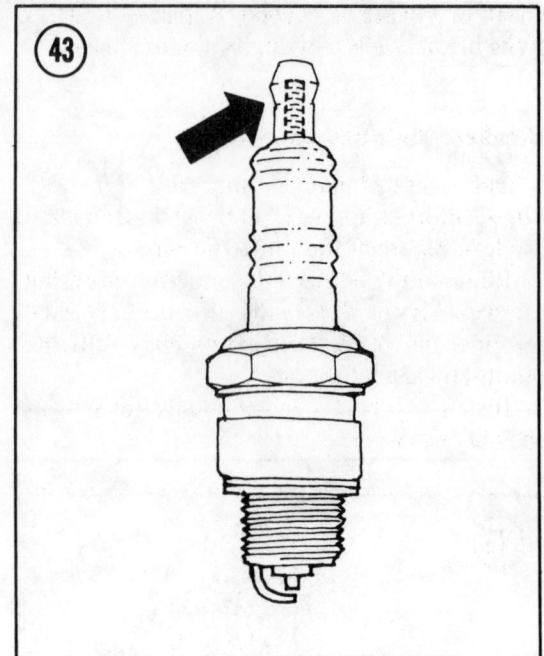

5. Screw each plug in by hand until it seats. Very little effort is required. If force is necessary, the plug is cross-threaded. Unscrew it and try again.

6. Tighten the spark plugs. If you have a torque wrench, tighten to 11-14 ft.-lb. (1.5-2.0 mkg). If not, tighten the plugs with fingers, then tighten an additional 1/4-1/2 turn with the plug wrench.

*NOTE*
*Do not overtighten. This prevents the plugs from sealing.*

## DISTRIBUTOR

The distributor controls the ignition coil, causing it to produce high-voltage current. The distributor then routes the current to the right spark plug at the right time for good engine operation.

On 1976-1977 non-California models, and on 1978 Canadian models, the distributor uses breaker points to control the ignition coil. On all other models, a magnetic pulse system replaces the points.

### Distributor Cap, Wires, and Rotor

1. Pry back the distributor cap clips and remove the cap.

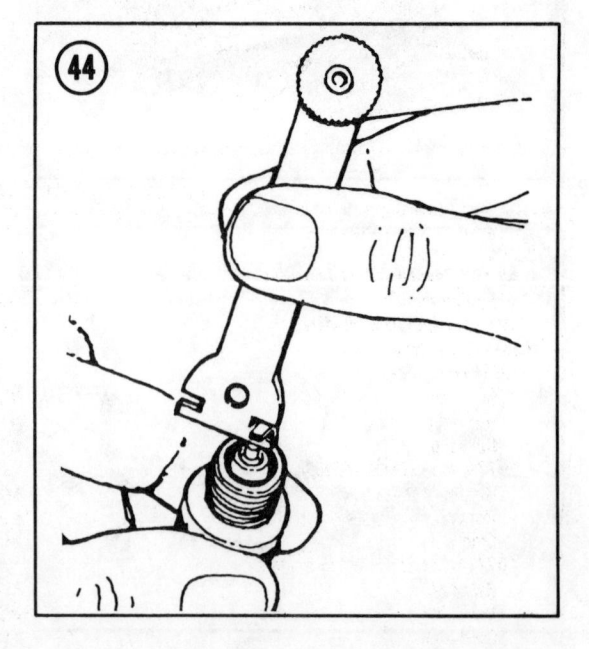

2. If you have an ohmmeter, connect it between each wire end and distributor cap terminal (**Figure 45**). Resistance should be less than 30,000 ohms. If it is higher, remove the wire and test it separately. If resistance is still too high, replace the wire. If not, replace the distributor cap.

If you don't have an ohmmeter, check the distributor cap terminals for dirt or corrosion.

# LUBRICATION, MAINTENANCE, AND TUNE-UP

Clean or replace as needed. Replace the wires if the insulation is melted, brittle, or cracked.

## Breaker Points and Condenser

This section applies only to 1976-1977 non-California models, and 1978 Canadian models. **Figure 46** identifies the parts.

1. Remove the points and condenser securing screws (**Figure 47**). Slide the primary lead terminal out of the distributor body. Lift the points and condenser out.
2. Install new points and condenser exactly as the old ones were.

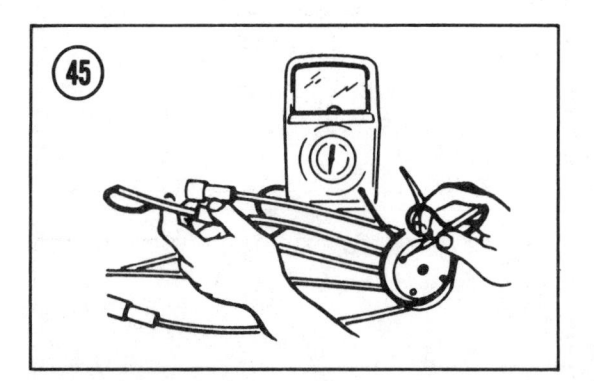

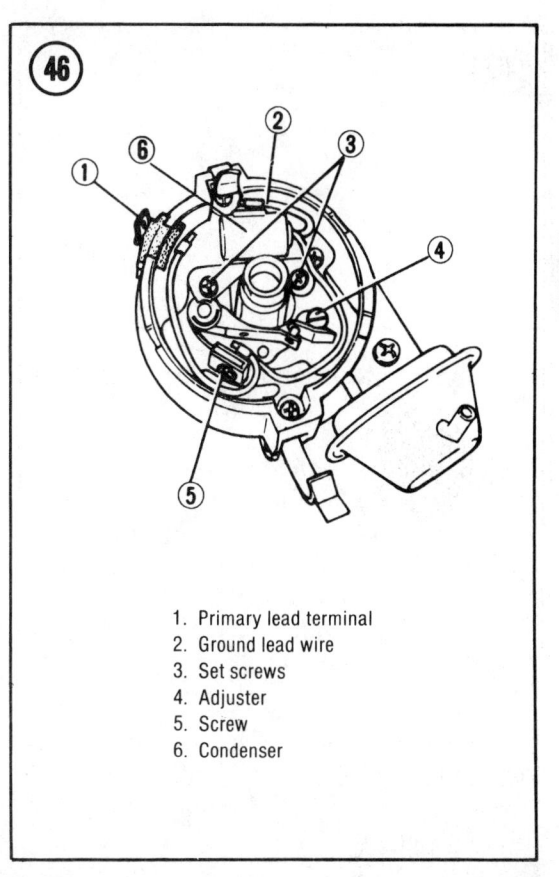

1. Primary lead terminal
2. Ground lead wire
3. Set screws
4. Adjuster
5. Screw
6. Condenser

Table 8  TUNE-UP SPECIFICATIONS

| | |
|---|---|
| Valve clearance (hot engine) | 0.014 in. (0.35mm) |
| Spark plug gap | |
|   All 1976 models, 1977 and later Canadian models | 0.031-0.035 in. (0.8-0.9mm) |
|   1977 and later U.S. models | 0.039-0.043 in. (1.0-1.1mm) |
| Spark plug type | |
|   All 1976 models | |
|     Standard | NGK BP5ES |
|     Hot type | NGK BP4E |
|     Cold type | NGK BP6ES, BP7ES |
|   1977 and later U.S. models | |
|     Standard | NGK BP5ES-11 |
|     Hot type | NGK BP4E-11 |
|     Cold type | NGK BP6ES-11, BP7ES-11 |
|   1977 and later Canadian models | |
|     Standard | NKG BPR5ES |
|     Hot type | NGK BPR4ES |
|     Cold type | NGK BPR6ES |
| Point gap (breaker point ignition) | 0.018-0.022 in. (0.45-0.55 mm) |
| Dwell angle (breaker point ignition) | 49-55° |
| Ignition timing | |
|   All F10's, non-California 310's | 10° BTDC at 700 rpm |
|   California 310's | 5° BTDC at 700 rpm |
| Idle speed | |
|   1976-1979 | 700 rpm |
| Firing order | 1-3-4-2 |

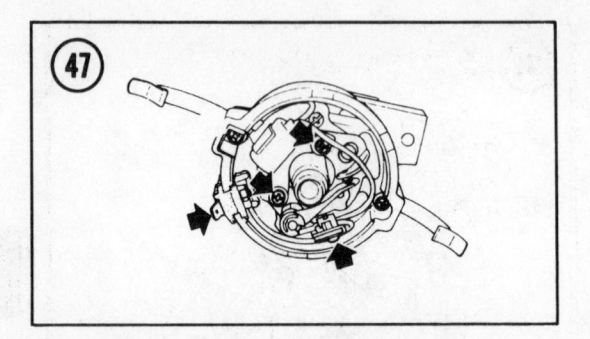

3. Apply a *small* amount of distributor cam lubricant to the cam lobes. Any excess may be thrown off and foul the points.

4. Using a crayon or felt pen, make alignment marks on the distributor body and engine. Loosen the lockscrew (**Figure 48**, F10; **Figure 49**, 310). Turn the distributor until the cam lobes open the points to the maximum gap. Measure the gap with a feeler gauge (**Figure 50**). It should be 0.018-0.022 in. (0.45-0.55mm). If incorrect, loosen the points securing screws. Turn the eccentric adjusting screw as shown to change the gap, then tighten the securing screws.

5. Install the distributor cap and rotor. Connect the spark plug wires to the plugs Refer to the labels made when the wires were disconnected, or see **Figure 39**. Realign match marks and tighten distributor lockscrew.

**Ignition Timing**

Ignition timing requires a stroboscopic timing light of the type described in Chapter One. Connect the light according to manufacturer's instructions.

1. Clean the crankshaft pulley and timing marks. See **Figure 51** (F10) or **Figure 52** (310).
2. Find your engine's timing mark in **Table 8**. Apply white paint to the pulley notch and the appropriate point on the timing scale.
3. Warm up the engine. Connect the timing light and an accurate tune-up tachometer.
4. Start the engine and check idle speed. It should be 700 rpm. If necessary, adjust as described under *Idle Speed and Mixture Adjustment* in this chapter.
5. Point the timing light at the timing marks. If timing is incorrect, loosen the distributor lockscrew (**Figure 48**, F10; **Figure 49**, 310).

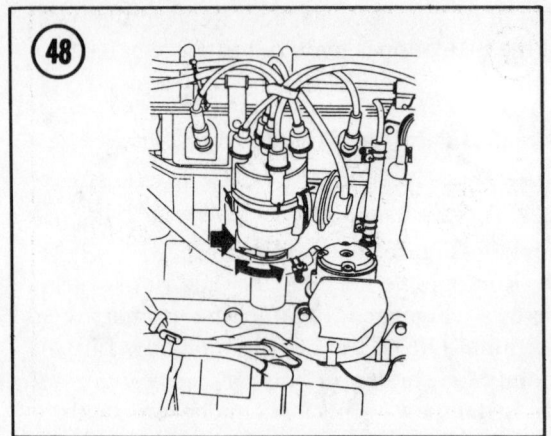

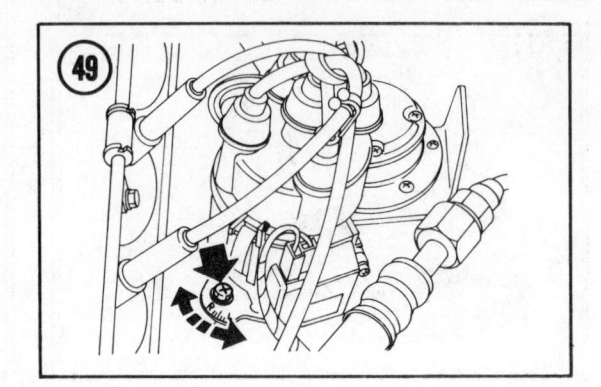

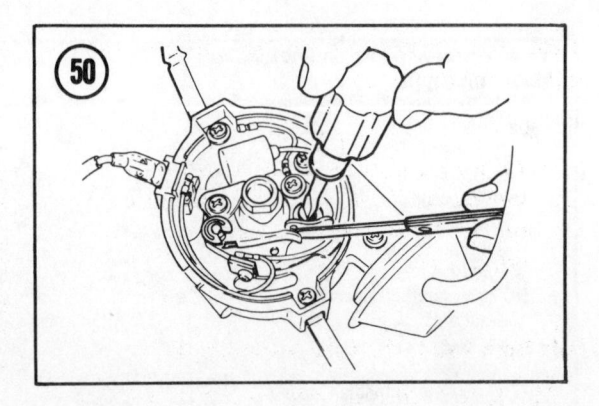

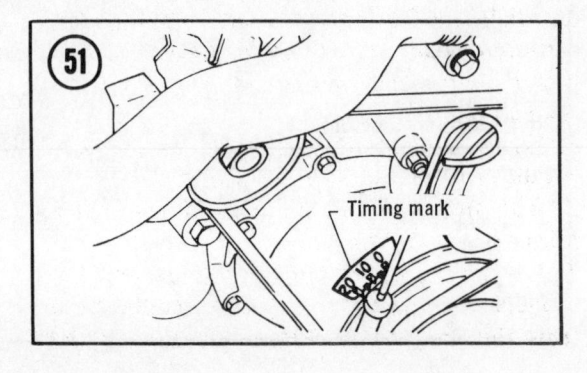

# LUBRICATION, MAINTENANCE, AND TUNE-UP

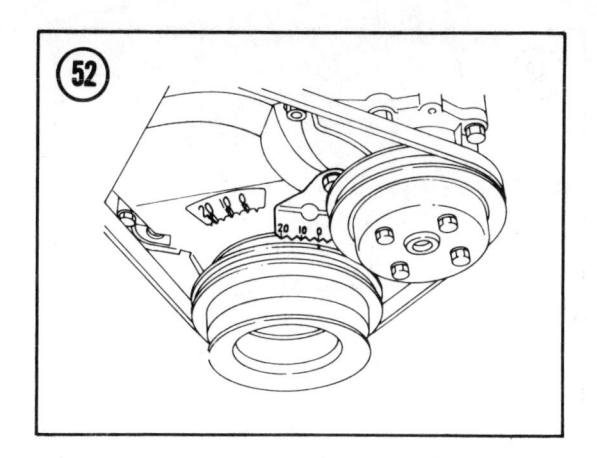

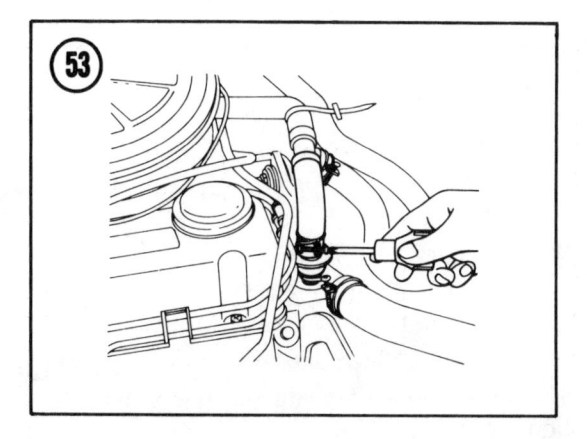

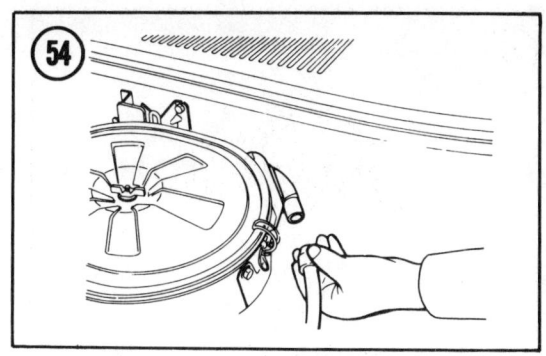

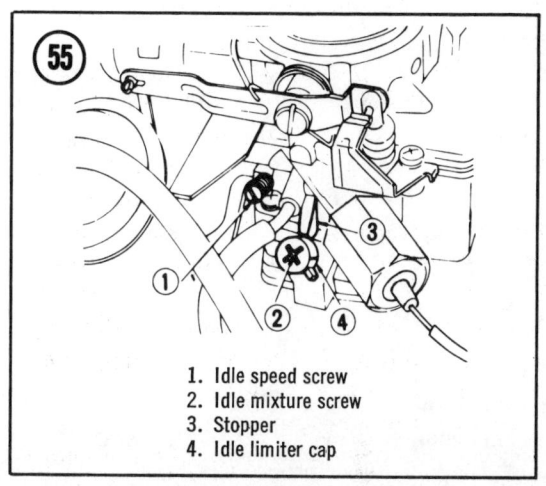

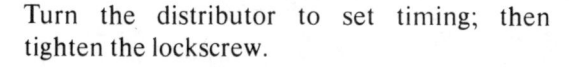

1. Idle speed screw
2. Idle mixture screw
3. Stopper
4. Idle limiter cap

Turn the distributor to set timing; then tighten the lockscrew.

*WARNING*
*Do not touch the spark plug or coil wires with the engine running. This can cause a painful shock, even if the insulation is in perfect condition.*

### Idle Speed and Mixture Adjustment

The factory recommends using a CO meter, especially for California models. Although the carburetor can be adjusted without a CO meter, the car may not comply with emission regulations.

1. Warm the engine until the temperature needle is in the middle of the gauge.
2. Run the engine at about 2,000 rpm for 5 minutes with the hood open.
3. Let the engine idle for 10 minutes. On U.S. models, use the time to disconnect the hose from the check valve (**Figure 53**) and cap the valve. On Canadian models, disconnect and cap the air induction hose (**Figure 54**).
4. Rev the engine to 1,500-2,000 rpm 2 or 3 times, then let it idle for one minute.
5. Check idle speed. It should be 700 rpm. If necessary, adjust by turning the idle speed screw (**Figure 55**).
6. If a CO meter is available, check CO percentage. It should be 1-3 percent. Adjust if necessary by turning the idle mixture screw (**Figure 55**).
7. If a CO meter is not available, increase engine speed to 740 rpm by turning the idle speed screw. Turn the mixture to get the best, fastest idle. If engine speed increases, reduce it to 740 rpm with the idle speed screw.
8. Once the engine has reached best idle at 740 rpm, turn the *mixture* screw clockwise (for a leaner mixture) until engine speed drops 35-45 rpm.
9. Reconnect the hose to the check valve (U.S. models) or air cleaner (Canadian models).

NOTE: If you own a 1980 or 1981 model, first check the Supplement at the back of the book for any new service information.

# CHAPTER FOUR

# ENGINE

All models use the A14 engine, a four-cylinder pushrod design with cast iron block and aluminum cylinder head. **Figure 1** is an exploded view.

This chapter provides removal, installation, and complete repair procedures for the engine. Specifications (**Table 1**) and tightening torques (**Table 2**) are listed at the end of the chapter.

## ENGINE REMOVAL (F10)

The engine and transmission are removed as a unit and then separated.

1. Remove the hood (Chapter Twelve).
2. Remove the battery.
3. Drain the cooling system. See *Cooling System Flushing*, Chapter Six.
4. Remove the air cleaner. See *Air Cleaner*, Chapter Five.
5. Disconnect the throttle cable, choke wire, and anti-dieseling wire from the carburetor. Disconnect the fuel inlet and return lines.
6. Disconnect the thick wire from the ignition coil. Disconnect the thin wire from the ignition coil negative terminal.
7. Disconnect the engine ground cable.
8. Disconnect the distributor thin wires at the multiple connector.
9. Disconnect the fusible links and engine harness connectors. See **Figure 2**.
10. Disconnect the radiator and heater hoses.
11. Disconnect the brake booster vacuum hose.
12. Remove the air pump air cleaner and carbon canister. See **Figure 3**.
13. Remove the carburetor cooling fan (Chapter Five).
14. Remove the windshield washer tank.
15. Remove the grille (Chapter Twelve).
16. Remove the radiator (Chapter Six).
17. Remove the clutch operating cylinder (Chapter Eight).
18. Remove the left and right buffer rods (**Figure 4**).

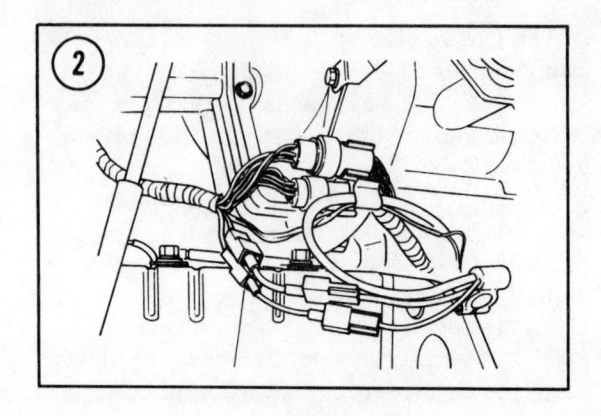

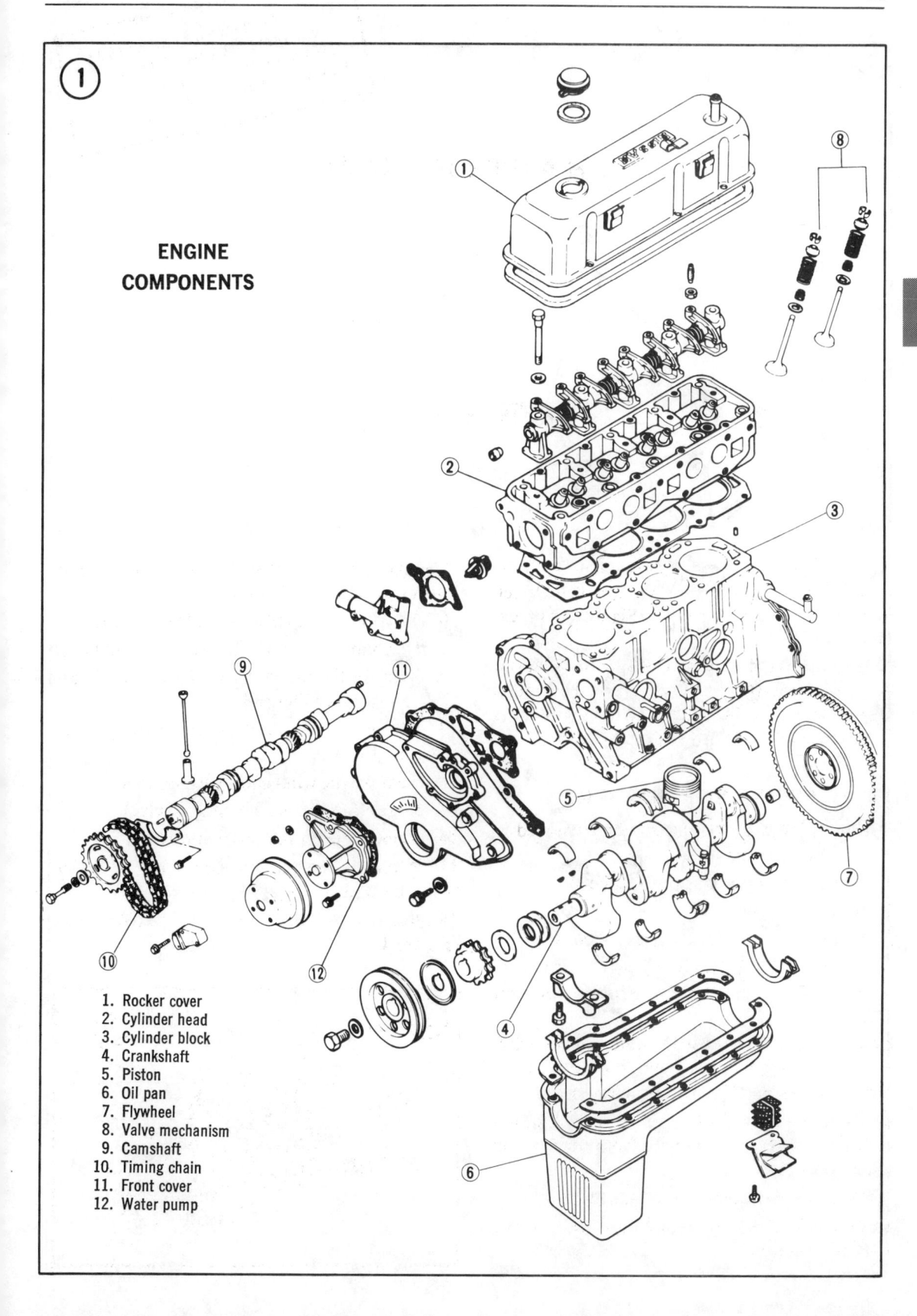

# ENGINE

## ENGINE COMPONENTS

1. Rocker cover
2. Cylinder head
3. Cylinder block
4. Crankshaft
5. Piston
6. Oil pan
7. Flywheel
8. Valve mechanism
9. Camshaft
10. Timing chain
11. Front cover
12. Water pump

# CHAPTER FOUR

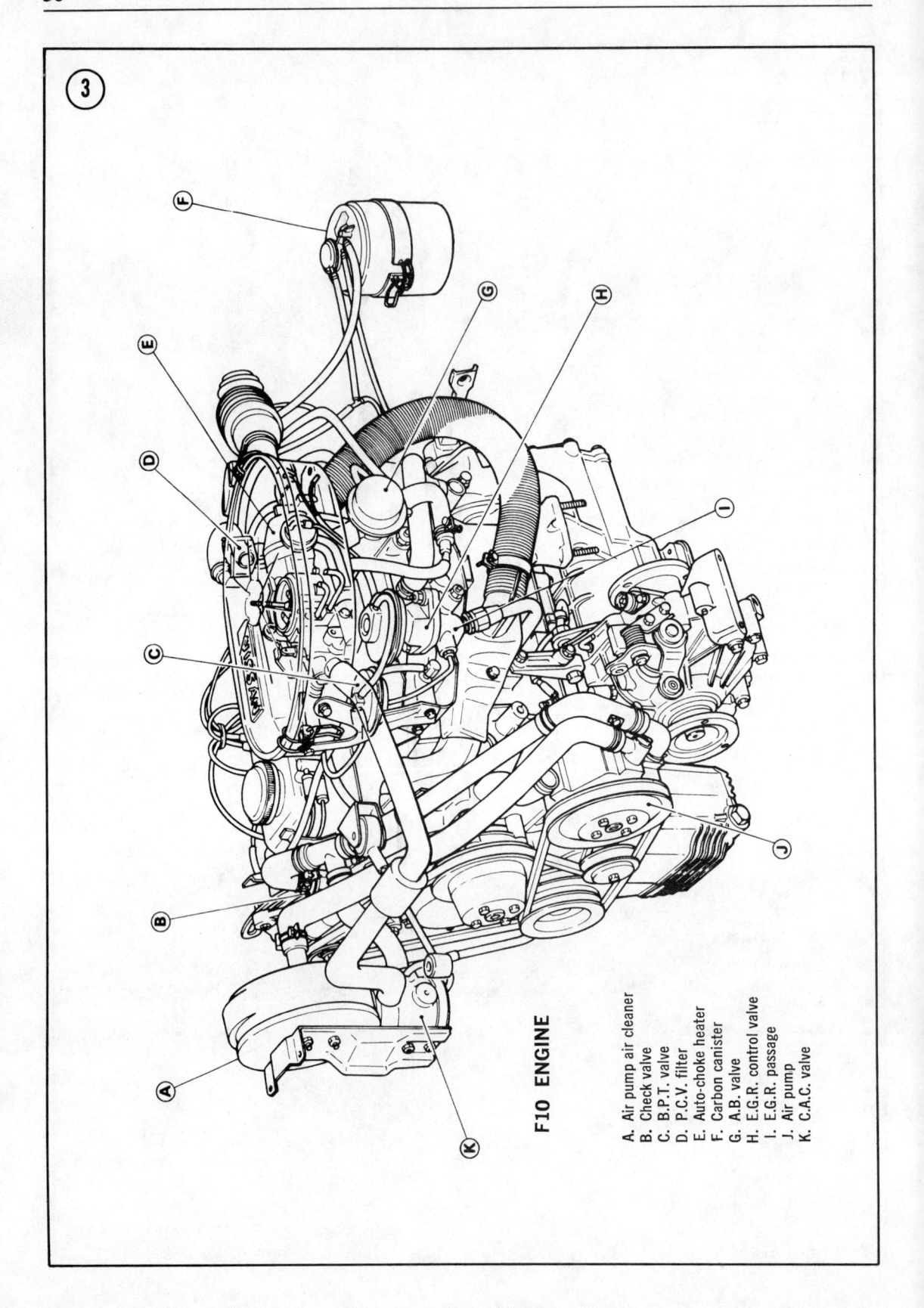

**F10 ENGINE**

A. Air pump air cleaner
B. Check valve
C. B.P.T. valve
D. P.C.V. filter
E. Auto-choke heater
F. Carbon canister
G. A.B. valve
H. E.G.R. control valve
I. E.G.R. passage
J. Air pump
K. C.A.C. valve

# ENGINE

**F10 ENGINE MOUNTING**

1. Buffer rod assembly (R.H.)
2. Engine support bracket (R.H.)
3. Front engine mounting insulator
4. Engine mounting shim
5. Engine support bracket (L.H.)
6. Rear engine mounting insulator
7. Buffer rod assembly (L.H.)

Tightening torque of bolts or nuts kgm (ft.-lb.)

A. 0.8 to 1.2 (5.8 to 8.7)
B. 1.5 to 2.1 (11 to 15)
C. 1.9 to 2.6 (14 to 19)
D. 2.8 to 3.8 (20 to 27)

1. Speedometer cable
2. Shift rod
3. Select rod

1. Adjusting bolt
2. Idler pulley lock nut

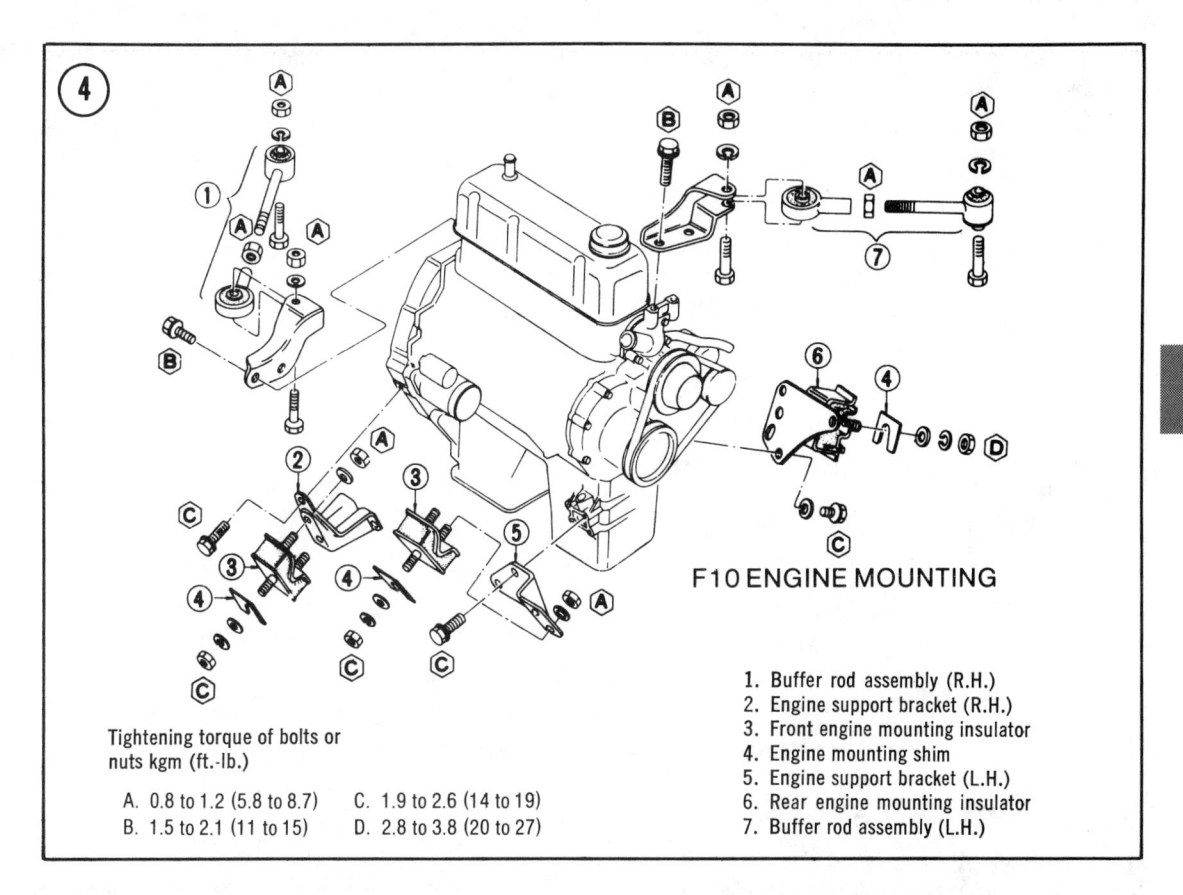

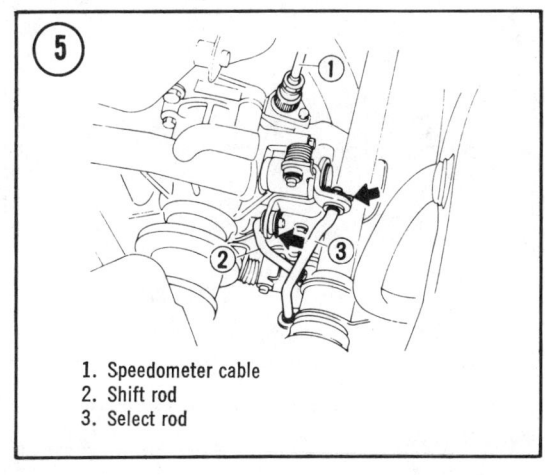

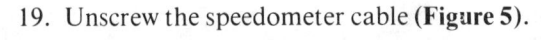

19. Unscrew the speedometer cable (**Figure 5**).

   NOTE: *Steps 20-22 apply to air conditioned cars.*

20. Loosen the idler pulley locknut (**Figure 6**). Back off the adjusting bolt, then remove the compressor belt.

21. Unbolt the compressor from the engine.

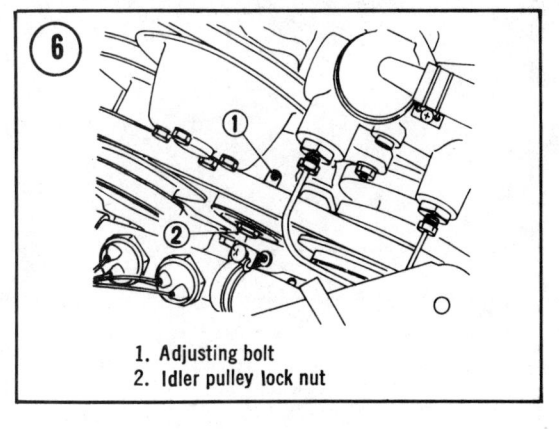

Lay it on the subframe by the steering gear (**Figure 7**) and tie it in place with wire.

*WARNING*
*Do not disconnect the compressor hoses. They contain refrigerant under high pressure which can cause frostbite if it touches skin and blindness if it touches the eyes. If discharged near an open flame the refrigerant can form poisonous gas.*

## CHAPTER FOUR

22. Unbolt the condenser and receiver/drier. Lay them on the right fender well.

23. Disconnect the transmission shift and select rods (**Figure 8**).

24. Attach sling brackets to the engine (**Figure 9**). These are available from Datsun parts departments, or can easily be fabricated.

25. Connect a hoist to the sling brackets. The easiest kind to use is the portable hydraulic crane type, available from many rental dealers. A vehicle with trailer hitch may be necessary to tow the hoist, but this inconvenience is more that made up for by the ease of using the hoist.

nect the pipe from the manifold. See **Figure 10**.

27. Unbolt the axle shafts from the transmission (**Figure 11**).

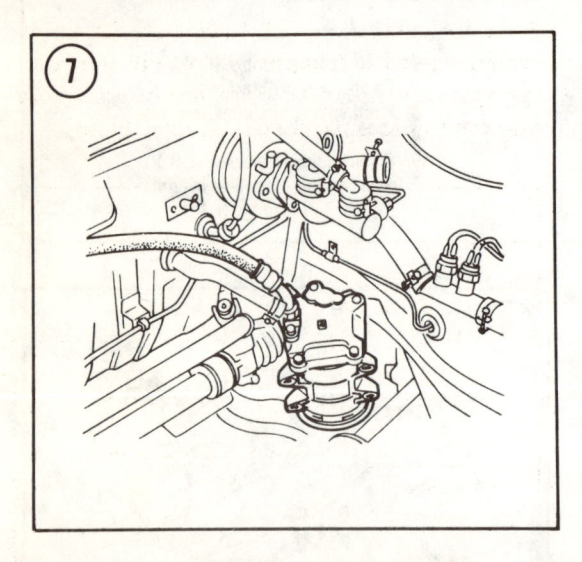

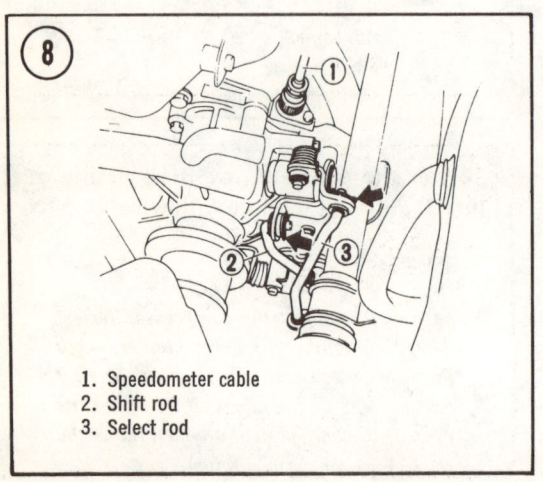

28. Unbolt the link support (**Figure 12**). Lower the shift and select rods.

29. Detach the engine mounting insulators from the engine, then from the subframe. See **Figure 4**.

30. Unbolt the rear mounting insulator from the transmission case.

31. Hoist the engine and transmission out of the car.

### CAUTION
*Do not let the engine strike equipment on the engine compartment sidewalls.*

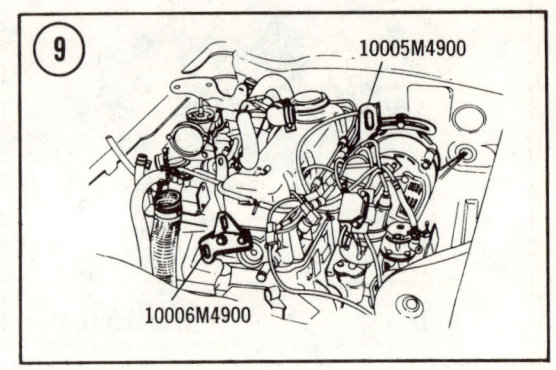

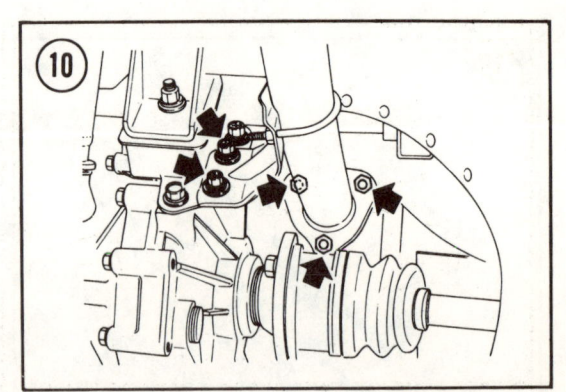

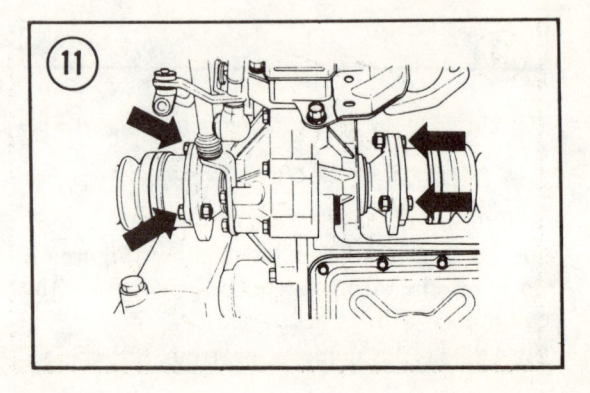

1. Speedometer cable
2. Shift rod
3. Select rod

# ENGINE
59

## ENGINE INSTALLATION (F10)

Installation is the reverse of removal, plus the following.

1. Fasten the engine and transmission securely to their mounts before tightening anything else.

   NOTE: *Be sure the arrow marks on the front insulators point upward (Figure 13). Be sure the locking pawl on the rear insulator (Figure 14) fits correctly into the subframe.*

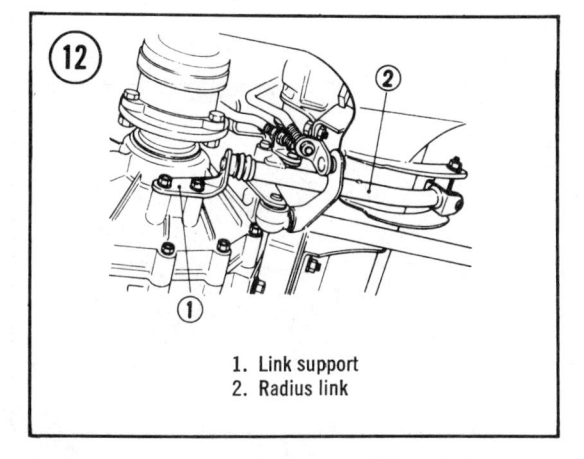

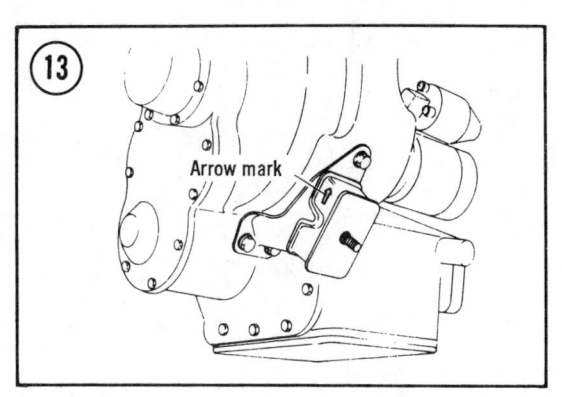

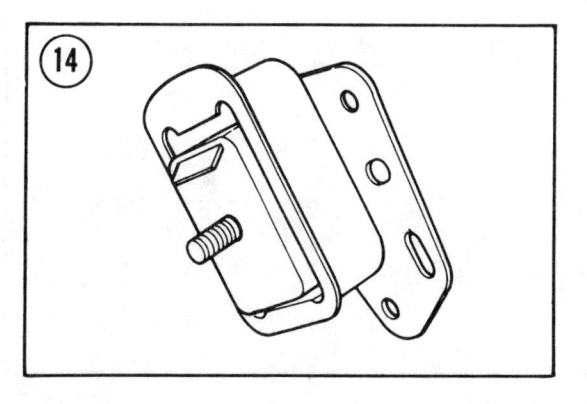

2. Check clearance between the subframe and clutch housing (dimension H, **Figure 15**). It should be 3/8-1/2 in. (10-12mm). If less than 1/4 in. (7mm), add shims between rear and right front insulators and the subframe.

   NOTE: *Do not use more than two shims at each mounting insulator.*

3. Check buffer rod lengths (**Figure 16**) before installing them. The left rod should be 5-3/8 in. (137-139 mm). The right rod should be 8 1/4 to 8 1/3 in. (209-211 mm). If necessary, loosen the locknut and rotate one of the ends to change length. Tighten the locknut.

4. On air conditioned cars, install the compressor right after installing the engine. If the shift linkage and axle shafts are connected first, it will be hard to install the compressor.

5. After installation, fill the engine with an oil recommended in Chapter Three. Fill the cooling system with a 50/50 mixture of antifreeze and water. Bleed the clutch (Chapter Eight).

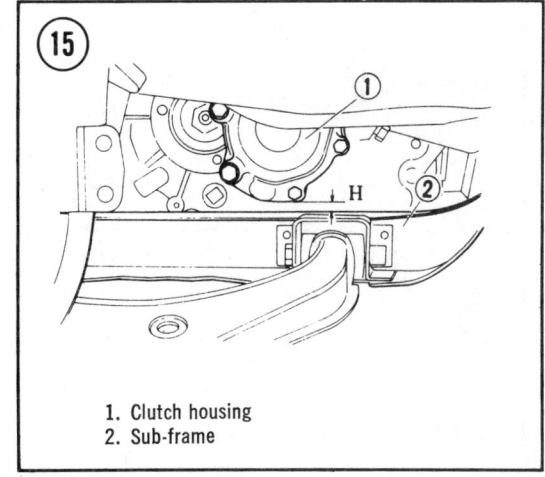

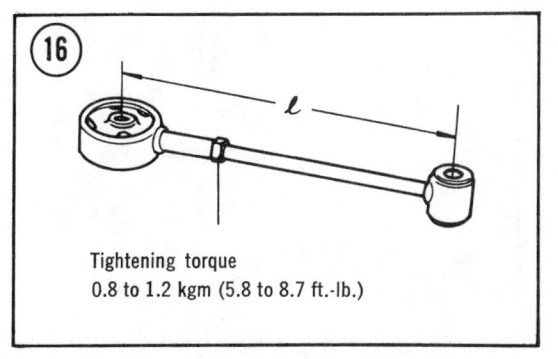

⑫
1. Link support
2. Radius link

⑬
Arrow mark

⑭

⑮
1. Clutch housing
2. Sub-frame

⑯
Tightening torque
0.8 to 1.2 kgm (5.8 to 8.7 ft.-lb.)

CHAPTER FOUR

⑰

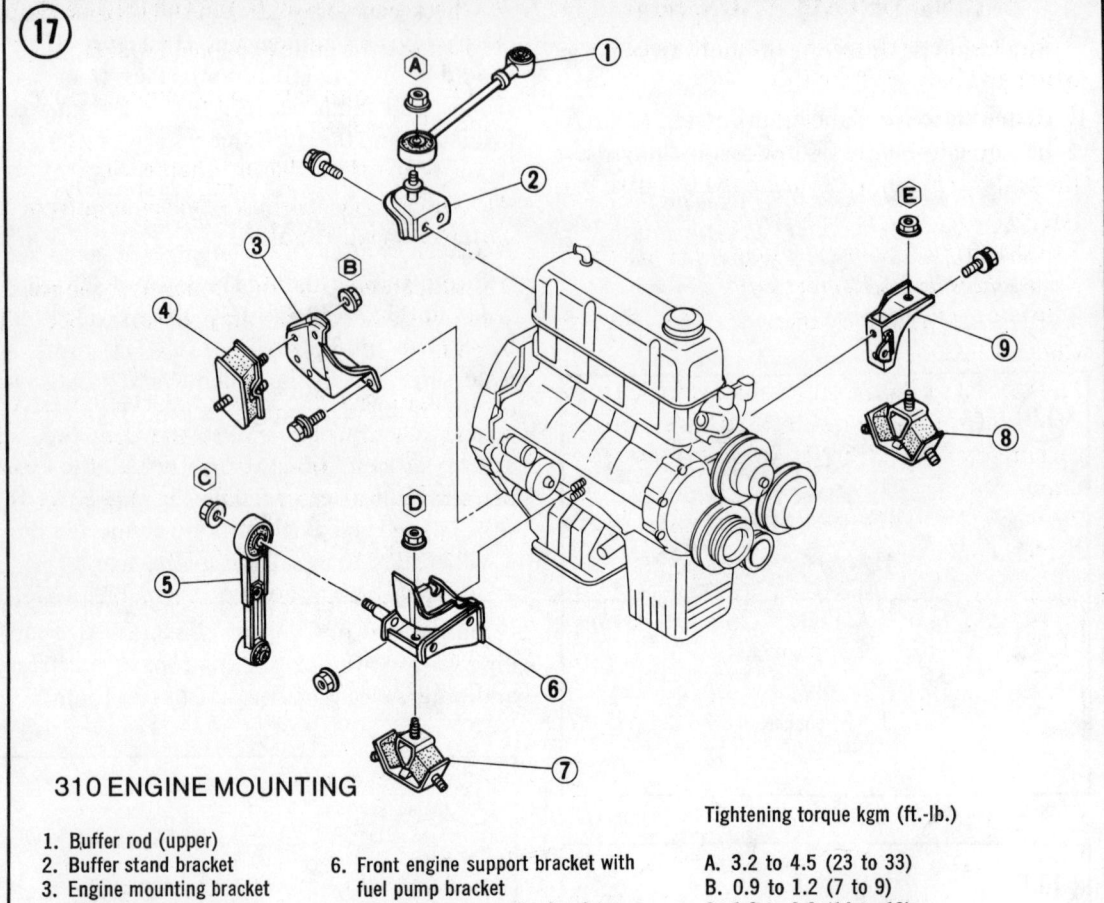

**310 ENGINE MOUNTING**

1. Buffer rod (upper)
2. Buffer stand bracket
3. Engine mounting bracket
4. Front engine mounting insulator (R.H.)
5. Buffer rod (lower)
6. Front engine support bracket with fuel pump bracket
7. Front engine mounting insulator
8. Rear engine mounting insulator
9. Rear engine support bracket

Tightening torque kgm (ft.-lb.)

A. 3.2 to 4.5 (23 to 33)
B. 0.9 to 1.2 (7 to 9)
C. 1.9 to 2.6 (14 to 19)
D. 1.9 to 2.6 (14 to 19)
E. 1.9 to 2.6 (14 to 19)

⑱

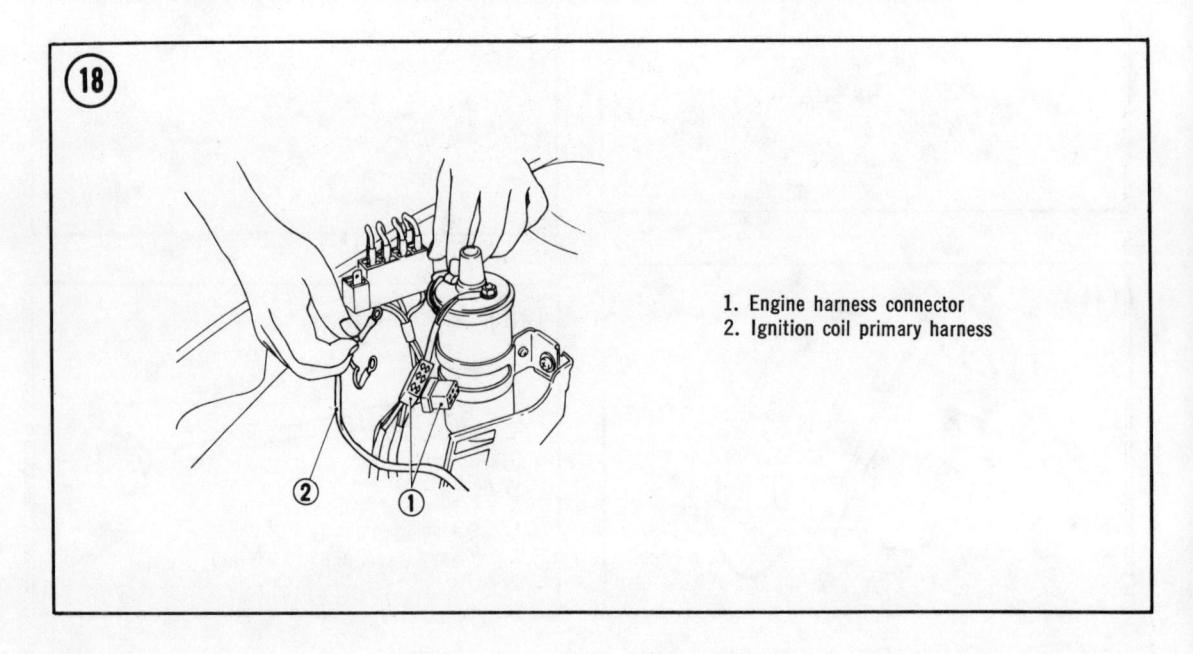

1. Engine harness connector
2. Ignition coil primary harness

# ENGINE

## ENGINE REMOVAL (310)

The engine and transmission are removed as a unit and then separated.

1. Remove the hood (Chapter Twelve).
2. If equipped with air conditioning, remove the battery. If not, disconnect the negative battery cable.
3. Drain the cooling system. See *Cooling System Flushing*, Chapter Six.
4. Remove the air cleaner. See *Air Cleaner*, Chapter Five.
5. Disconnect the throttle cable, electric choke wire, and anti-dieseling solenoid wire from the carburetor. Disconnect the fuel inlet and return lines.
6. Remove the carburetor cooling fan (Chapter Five).
7. Remove the upper buffer rod (**Figure 17**).
8. Remove the grille (Chapter Twelve).
9. On cars with air conditioning, remove the windshield washer tank.
10. Remove the radiator (Chapter Six).
11. Disconnect the thick wire running from ignition coil to distributor.
12. Disconnect the engine harness connectors and ignition coil primary harness. See **Figure 18**.
13. Disconnect the carbon canister hoses. On California models, remove the combined air control valve (CAC). On all models, disconnect the air pump air cleaner hose. See **Figure 19**.
14. Disconnect the heater hoses.

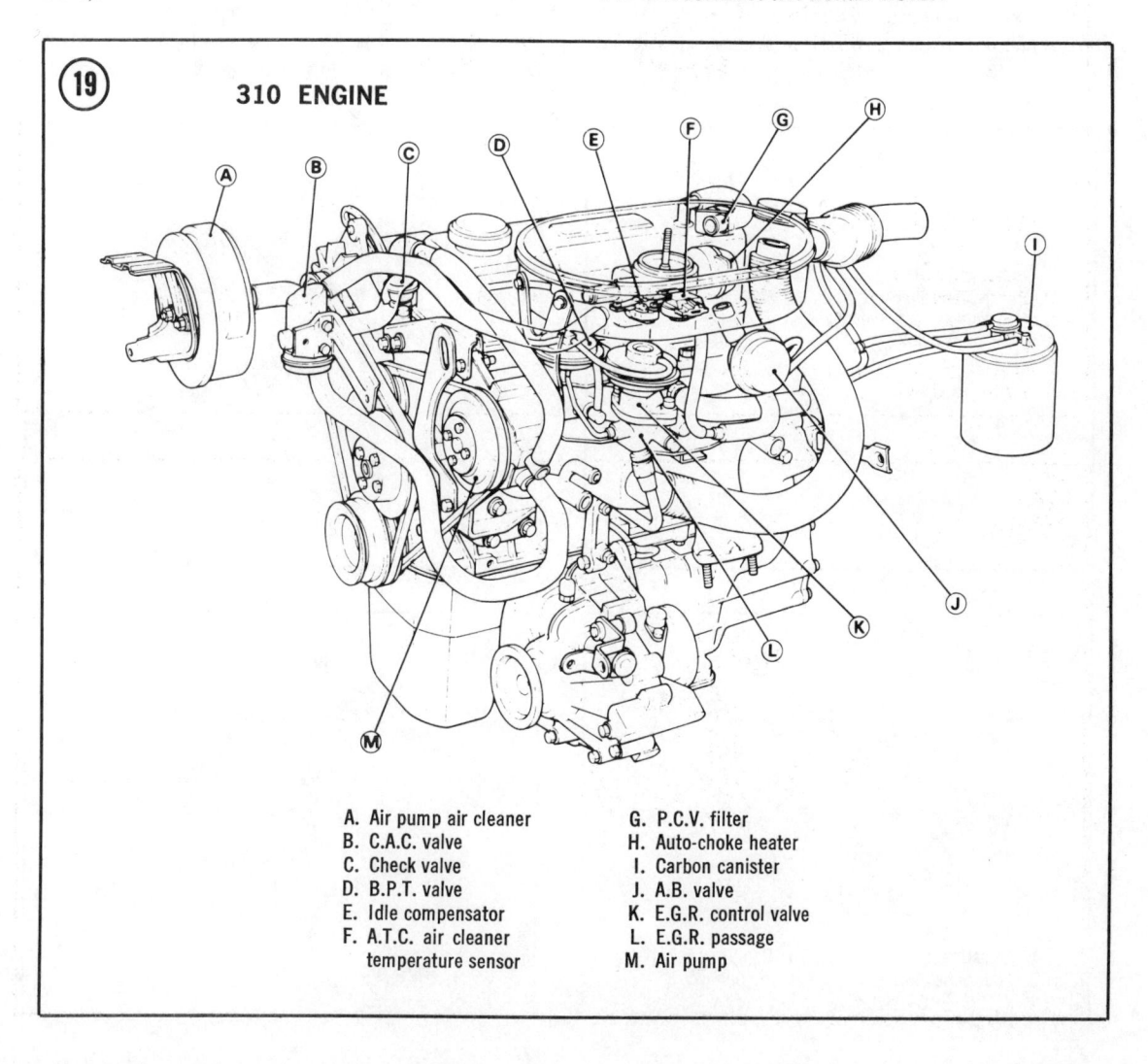

**310 ENGINE**

A. Air pump air cleaner
B. C.A.C. valve
C. Check valve
D. B.P.T. valve
E. Idle compensator
F. A.T.C. air cleaner temperature sensor
G. P.C.V. filter
H. Auto-choke heater
I. Carbon canister
J. A.B. valve
K. E.G.R. control valve
L. E.G.R. passage
M. Air pump

## CHAPTER FOUR

15. Disconnect the engine ground wire (**Figure 20**).

> NOTE: *Steps 16-19 apply to air conditioned cars.*

16. Loosen the air pump and compressor idler pulley nuts (**Figure 21**). Back off the adjusting bolts and remove the belts.

17. Unbolt the air pump upper bracket from the engine. Loosen the lower air pump mounting bolt and pivot the pump toward the engine. This provides access to the compressor.

18. Unbolt the compressor from the engine (**Figure 22**). Move it away from the engine and tie it back with wire or rope.

*WARNING*
*Do not disconnect the compressor hoses. They contain refrigerant under high pressure which can cause frostbite if it touches skin and blindness if it touches the eyes. If discharged near an open flame, the refrigerant can form poisonous gas.*

19. Unbolt the condenser and receiver/drier. Lay them on the battery bracket.

20. Remove the clutch operating cylinder (Chapter Eight).

21. Disconnect the transmission shift and select rods (**Figure 23**).

22. Detach the exhaust pipe from the manifold, and its bracket from the transmission. See **Figure 24**.

23. Detach the axle shafts from the differential side flanges. See **Figure 25**.

24. Attach sling brackets to the engine (**Figure 26**). These are available from Datsun parts departments, or can easily be fabricated.

25. Connect a hoist to the sling brackets. The easiest kind to use is the portable hydraulic crane type, available from many rental dealers.

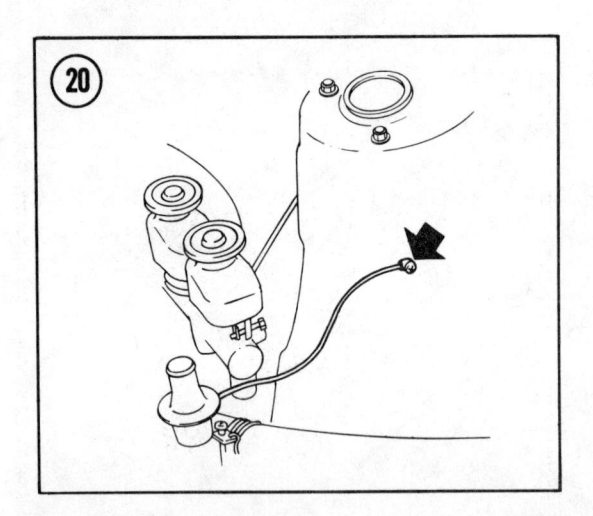

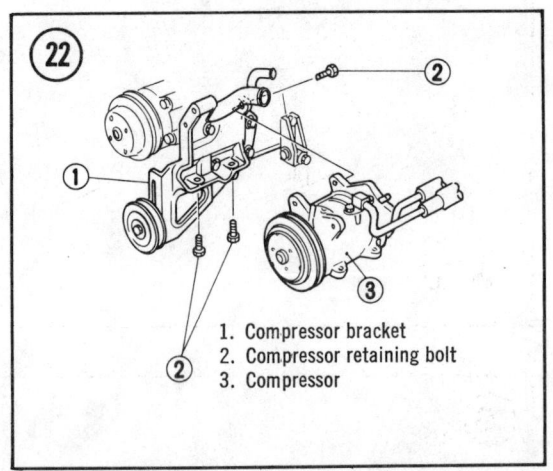

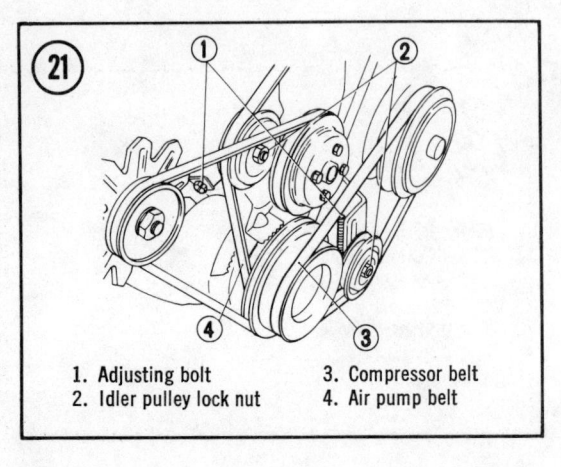

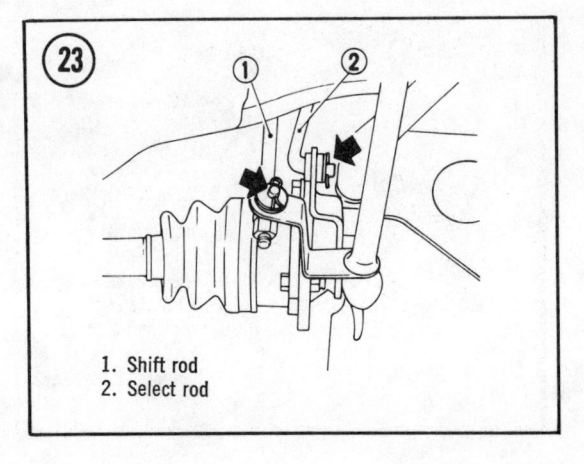

**20**

**22**
1. Compressor bracket
2. Compressor retaining bolt
3. Compressor

**21**
1. Adjusting bolt
2. Idler pulley lock nut
3. Compressor belt
4. Air pump belt

**23**
1. Shift rod
2. Select rod

# ENGINE

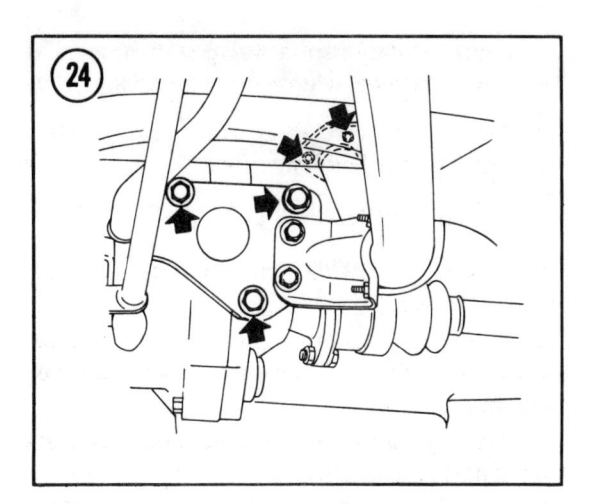

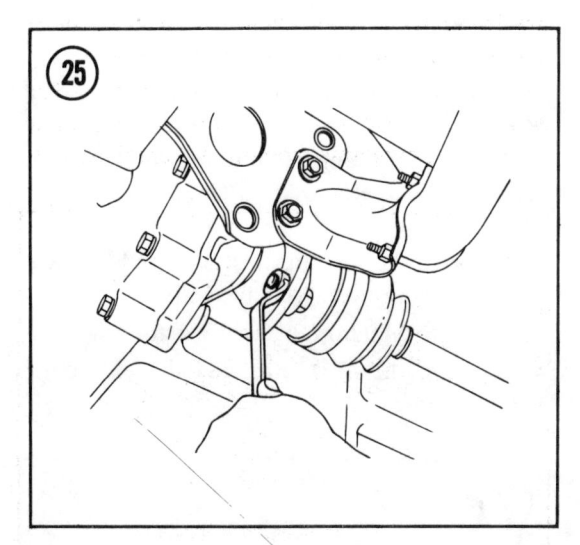

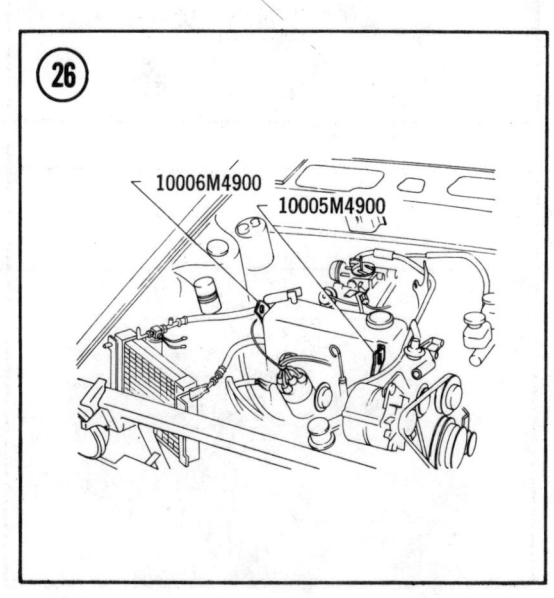

A vehicle with trailer hitch is necessary to tow the hoist, but this inconvenience is more than made up for by the ease of using the hoist.

26. Remove the lower buffer rod (**Figure 27**).

27. Detach the engine mounting insulators from the engine brackets. See **Figure 17**.

28. Hoist the engine and transmission out of the car.

### CAUTION
*Do not let the engine strike equipment on the engine compartment walls.*

## ENGINE INSTALLATION (310)

Installation is the reverse of removal, plus the following.

1. Fasten the engine securely to its mounts before tightening anything else. Tighten the rear mount before the front mounts.

NOTE: *The arrow on the right front mount points upward (**Figure 28**).*

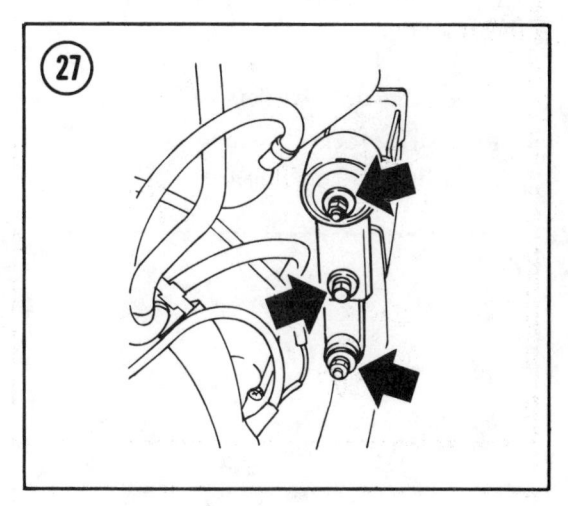

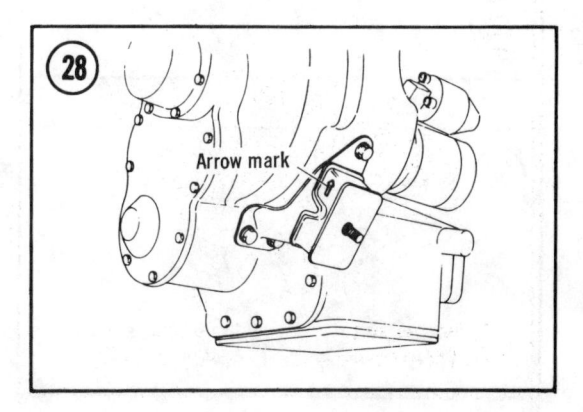

2. On air conditioned cars, install the compressor immediately after installing the engine. If the axle shafts and shift linkage are connected first, it will be hard to install the compressor.

3. When installing the buffer rods, loosen their locknuts (**Figure 29**). Adjust buffer rod length so the rubber end bushings are not deformed when the rods are installed. Tighten the locknuts.

4. Fill the engine with an oil recommended in Chapter Three. Fill the cooling system with a 50/50 mixture of antifreeze and water.

## DISASSEMBLY SEQUENCES

The following sequences are checklists that tell how much of the engine to remove and disassemble to do a specific type of service.

To use the sequences, remove and inspect each part mentioned. Then go through the sequence backwards, installing the parts. Each part is covered in detail in this chapter, unless otherwise noted.

### Decarbonizing or Valve Service

1. Remove the rocker assembly.
2. Remove the cylinder head.
3. Remove and inspect valves. Inspect valve guides and valve seat inserts, removing when necessary.
4. Assemble by reversing Steps 1-3.

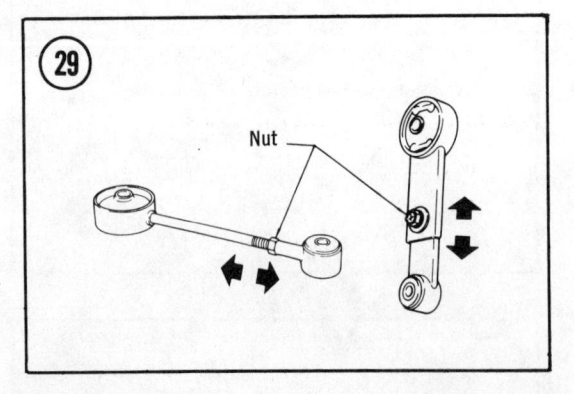

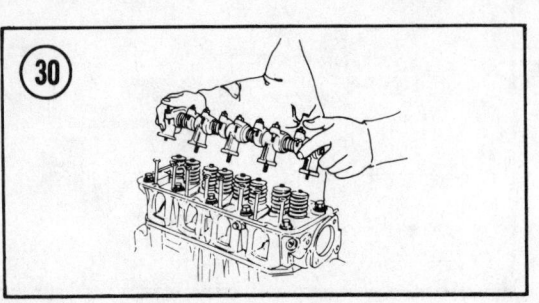

### Valve and Ring Service

1. Remove the engine.
2. Perform Steps 1-3 for valve service.
3. Remove the oil pan.
4. Remove the pistons together with the connecting rods.
5. Remove the piston rings. It is not necessary to separate the pistons from the connecting rods unless a piston, connecting rod, or piston pin needs repair or replacement.
6. Assemble by reversing Steps 1-5.

### General Overhaul

1. Remove the engine and transmission and separate them. Remove clutch (Chapter Eight) from cars equipped with manual transmission.
2. Remove the fan, fan belt and pulley, water pump and thermostat (Chapter Six).
3. Remove oil pump together with oil filter.
4. Remove the alternator and the distributor (Chapter Seven).

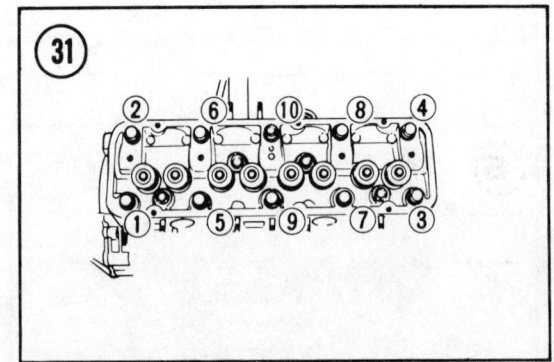

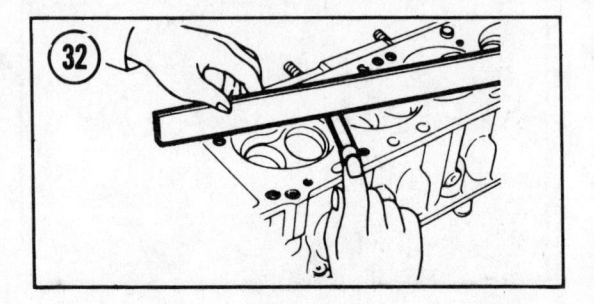

# ENGINE

5. Remove the fuel pump, carburetor, and intake and exhaust manifolds (Chapter Five).
6. Remove rocker assembly and pushrods.
7. Remove the flywheel.
8. Remove the cylinder head.
9. Remove the oil pan and strainer.
10. Remove the crankshaft pulley and engine front cover.
11. Remove the timing chain tensioner. Then remove the crankshaft and camshaft sprockets together with the timing chain.
12. Remove piston-connecting rod assemblies.
13. Remove the main bearing caps, rear main oil seal, and crankshaft.
14. Remove crankcase oil separator.
15. Remove the camshaft and valve lifters.
16. Assembly is the reverse of these steps.

## ROCKER ASSEMBLY

### Removal/Installation

1. Undo the Phillips screws that secure the rocker cover and remove it.
2. Undo the 5 bolts that secure the rocker assembly to the cylinder head. Lift the assembly off (**Figure 30**).
3. Installation is the reverse of removal. Tighten the rocker assembly bolts to 15-18 ft.-lb. (2.0-2.5 mkg). Tighten the bolts gradually, working outward from the center in several stages.

### Inspection

1. Check the rocker shaft and rocker arms for seizure or excessive wear. The clearance between rocker arms and shaft should be 0.0008-0.0021 in. (0.02-0.05mm).
2. Check the valve stem contact surface on the rocker arm for wear. Replace worn rocker arms.

## CYLINDER HEAD

Some of the following procedures must be performed by a dealer or machine shop, since they require special knowledge and expensive machine tools. Others, while possible for the home mechanic, are difficult or very time consuming. A general practice among those who do their own service is to remove the cylinder head, perform all disassembly except valve removal, and take the head to a machine shop for inspection and service. Since the cost is relatively low in proportion to the required effort and equipment, this may be the best approach, even for more experienced owners.

### Removal

1. Drain the cooling system by opening the tap at the bottom of the radiator and removing the plug from the back of the engine.
2. Detach the spark plug wires, then remove the spark plugs.
3. Remove the air cleaner.
4. Remove the valve rocker cover.
5. Disconnect the heater hoses and upper radiator hose from the cylinder head.
6. Remove the intake and exhaust manifolds (Chapter Five).
7. Remove the rocker assembly as described earlier in this chapter.
8. Remove pushrods. Number each pushrod so it can be returned to its original hole.
9. Remove the cylinder head bolts in the order shown in **Figure 31**. Go through the sequence in several stages, loosening the bolts a little at a time to prevent warping the head.

*NOTE*
*The center front bolt (10, **Figure 31**) is thinner than the others. It is identified by a hollow head.*

10. Tap the cylinder head with a soft-faced mallet to free it, then lift it clear of the engine.

*CAUTION*
*Do not pry the head loose, since the aluminum may be gouged.*

### Inspection

1. Check head for water leaks before cleaning.
2. Clean the cylinder head thoroughly in solvent. While cleaning, check for cracks or other visible damage. Look for corrosion or foreign material in oil and water passages. Clean the passages with a stiff spiral wire brush, then blow them out with compressed air.
3. Check the cylinder head bottom (block mating) surface for flatness (see **Figure 32**).

Place an accurate straightedge along the surface. If there is any gap between the straightedge and cylinder head surface, measure it with a feeler gauge. Normal gap is 0.002 in. (0.05mm) or less. Maximum permissible is 0.004 in. (0.1mm). If the gap is beyond the limit, have the head resurfaced by a dealer or machine shop.

4. Check studs in the cylinder head for general condition. Replace damaged studs.

### Decarbonizing

1. Without removing the valves, remove all deposits from the combustion chambers, intake ports, and exhaust ports. Use a wire brush dipped in solvent. Be careful not to damage the aluminum cylinder head.

2. After all carbon is removed from the combustion chambers and ports, clean the entire head in solvent.

3. Clean away all carbon on the piston tops. Do not remove the carbon ridge at the top of the cylinder bore.

### Installation

1. Be sure the cylinder head, block, and cylinder bores are clean. Check all visible oil passages in both head and block for cleanliness.

2. Install a new cylinder head gasket. Never reuse an old head gasket. Do *not* use gasket sealer on the head gasket.

3. Install the cylinder head bolts. Be sure the hollow-headed bolt (1, **Figure 33**) is in the center front hole. Tighten the bolts to 51-54 ft.-lb. (7.0-7.5 mkg). Tighten in several stages to prevent warping the cylinder head.

4. Install the pushrods in the holes from which they were removed.

5. Install the rocker assembly as described earlier in this chapter.

6. Connect the heater hoses and upper radiator hose to the cylinder head.

7. Install the intake and exhaust manifolds (Chapter Five).

8. Install the rocker cover. Use a new gasket.

9. Install the air cleaner.

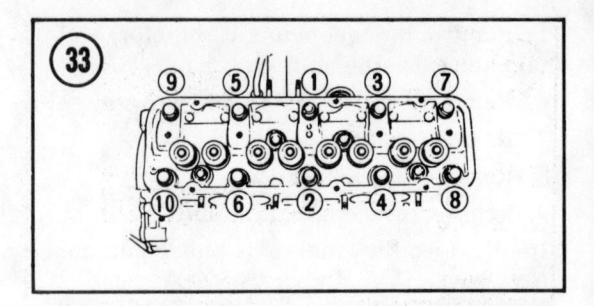

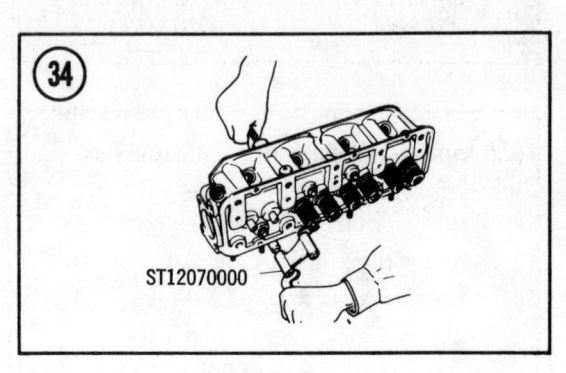

10. Install the spark plugs and reconnect the wires.

11. Fill the cooling system. See *Cooling System Flushing*, Chapter Six.

## VALVES AND VALVE SEATS

### Valve Removal

1. Remove the cylinder head.

2. Compress valve springs with a valve spring compressor (**Figure 34**). Remove the valve keepers and release the compression.

3. Referring to **Figure 35**, remove the valve spring washers, oil seals, springs, and valves. Remove the valves through the combustion chambers. Be sure to keep them in the order of removal.

#### CAUTION
*Remove any burrs from valve stem grooves before removing valves. Otherwise the valve guides will be damaged.*

### Valve Inspection

1. Clean the valves with a wire brush and solvent. Discard cracked, warped, or burned valves.

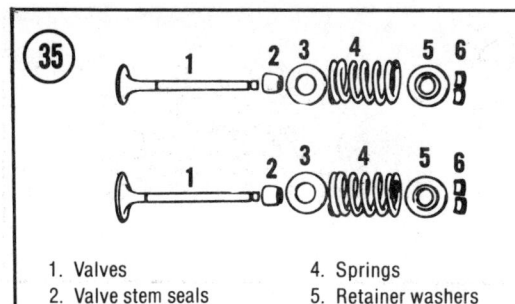

# ENGINE

**35**
1. Valves
2. Valve stem seals
3. Spring seats
4. Springs
5. Retainer washers
6. Valve keepers

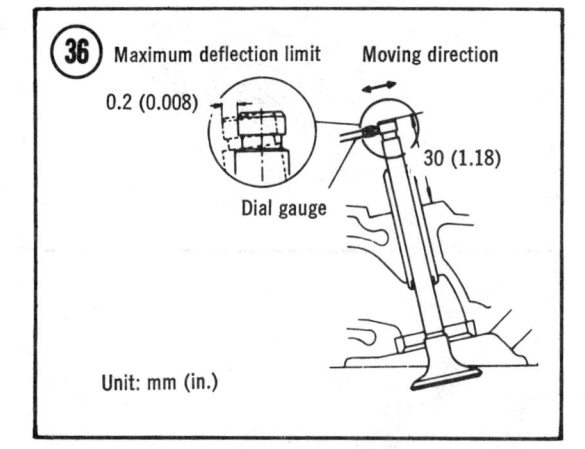

**36** Maximum deflection limit 0.2 (0.008), Moving direction, 30 (1.18), Dial gauge
Unit: mm (in.)

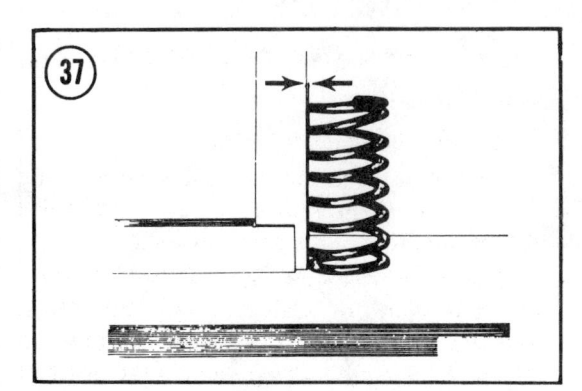

**37**

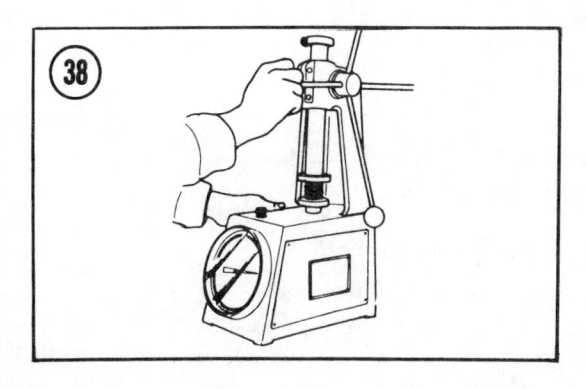

**38**

2. Measure the valve stems at the top, bottom, and center for wear. Compare with specifications at the end of the chapter. Replace valves with worn stems.

3. Examine valve stem ends. Worn ends may be refaced. Do not remove more than 0.020 in. (0.5mm).

4. Remove all carbon and varnish from valve guides with a stiff spiral wire brush.

5. Insert the valves in the valve guides from which they were removed. Hold each valve just slightly off its seat and rock it sideways (**Figure 36**). If it rocks more than 0.008 in. (0.2mm), the valve guide is worn and should be replaced.

6. Measure height of valve springs. Replace springs that are shorter than specifications (**Table 1**).

7. Check springs for bending with a square (**Figure 37**). Replace springs that are bent more than 0.063 in. (1.6mm).

8. Test the springs under load with a spring tester (**Figure 38**). Replace any that fail to meet specifications at the end of the chapter.

9. Inspect valve seats. If worn or burned, they must be reconditioned. This should be performed by the dealer or a machine shop, although the procedure is described later in this section.

**Valve Guide Replacement**

This procedure requires a press capable of 4,400 pounds (2 metric tons). If you do not have such a press, the valve guides can be replaced by a dealer or machine shop. Replacement will be much cheaper if you remove the cylinder head yourself and bring it to the shop.

1. Press out worn guide with a press and suitable drift. This procedure is easier if the cylinder head is heated first. However, it can be done at room temperature.

2. With the cylinder head at room temperature, ream the valve guide hole to the specification at the end of the chapter.

3. Heat head to 302-392°F (150-200°C). Then press the valve guide in with a hammer and a drift such as ST11320000.

4. Ream the valve guide bore to the specification at the end of this chapter.

5. Check that the valves move freely in the new guides.

### Valve Seat Reconditioning

This job is best left to your dealer or local machine shop. They have the special knowledge and equipment required for this precise job. The following procedure is provided in the event you are not near a dealer and your local machine shop is not familiar with the Datsun. Valve seat inserts that are too badly worn to be refaced can be replaced as follows.

1. Remove the old valve seat insert by boring it out until it collapses. Be sure not to cut the cylinder head during boring.
2. Select a valve seat insert.
3. Make sure the recess for the valve seat insert in the cylinder head is concentric with the valve guide bore.
4. Heat the cylinder head to a temperature of 302-329 °F (150-200 °C).
5. Install the valve seat insert, making sure it is bedded securely in the cylinder head.
6. Cut the valve seats to the dimensions shown in **Figure 39**.
7. Coat the corresponding valve face with Prussian blue dye.
8. Insert the valve into the guide.
9. Rotate valve under light pressure about ¼ of a turn.
10. Lift the valve out. If it seats properly, the blue will transfer evenly to the valve seat.

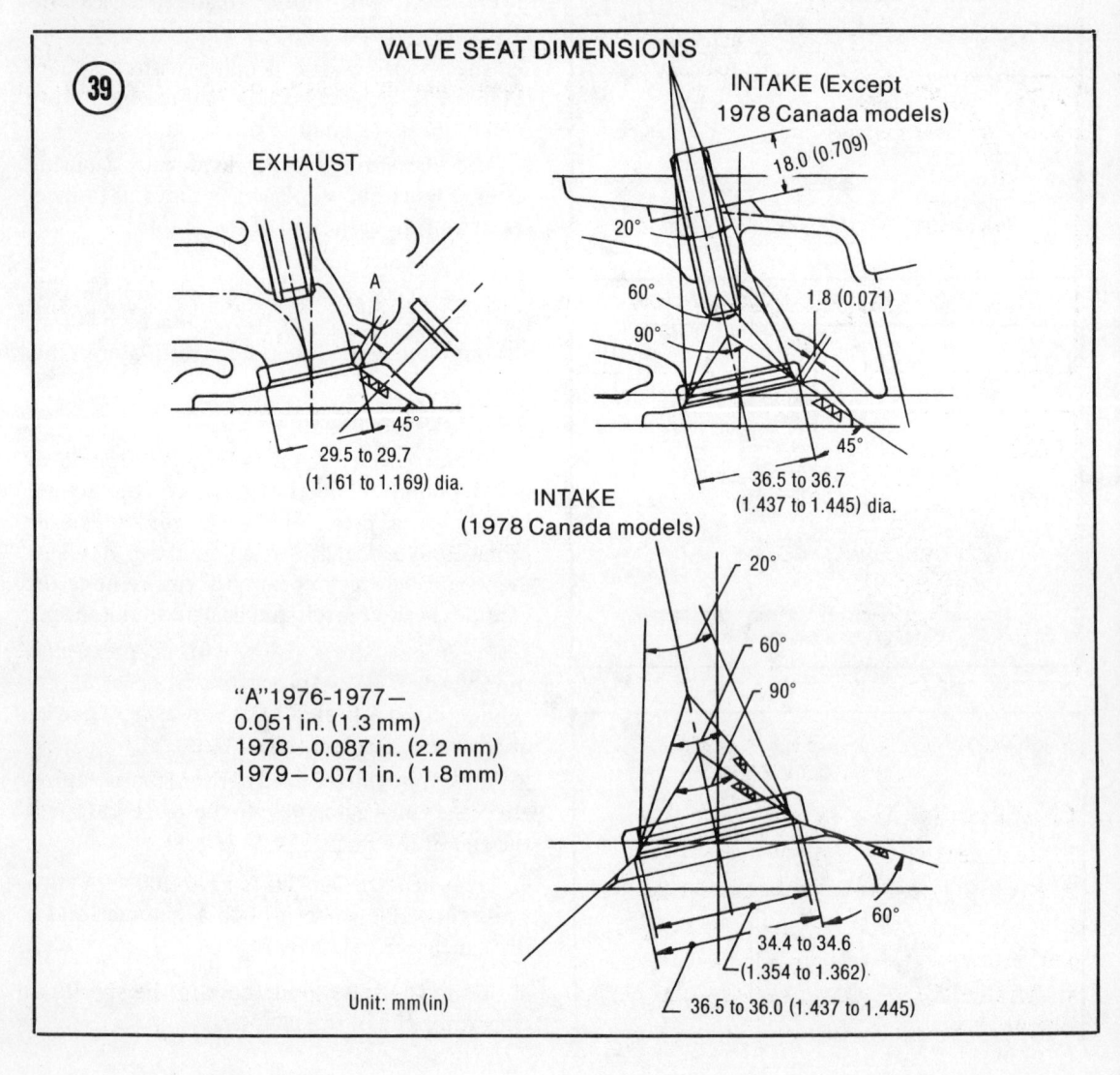

**VALVE SEAT DIMENSIONS**

㊴

EXHAUST

45°
29.5 to 29.7
(1.161 to 1.169) dia.

INTAKE (Except 1978 Canada models)

18.0 (0.709)
20°
60°
90°
1.8 (0.071)
45°
36.5 to 36.7
(1.437 to 1.445) dia.

INTAKE (1978 Canada models)

20°
60°
90°
60°
34.4 to 34.6
(1.354 to 1.362)
36.5 to 36.0 (1.437 to 1.445)

"A" 1976-1977—
0.051 in. (1.3 mm)
1978—0.087 in. (2.2 mm)
1979—0.071 in. (1.8 mm)

Unit: mm(in)

# ENGINE

5. Set the handbrake and place the transmission in gear. Remove the crankshaft pulley bolt with a socket and long breaker bar.

6. Remove the crankshaft pulley. Use a gear puller or pry it off with 2 large screwdrivers.

CAUTION
*If the pulley is pried off, do not gouge the aluminum front cover.*

7. Unbolt the cover from the engine. Tap it with a soft rubber mallet to break the gasket seal, then take it off (**Figure 40**).

8. Check the front cover oil seal for wear or damage. Replace if its condition is in doubt. It is generally a good idea to replace the seal whenever the front cover is removed. Pry the seal out, taking care not to gouge the front cover. Position the new seal on the cover, lay a block of wood on top of it, and tap the seal in with a hammer. The wood keeps the seal from tilting sideways.

NOTE: *The lip of the seal faces into the front cover.*

9. Install by reversing Steps 1-7. Use a new front cover gasket, thinly coated on both sides with gasket sealer. Coat the oil seal lip with clean engine oil before installation.

## Sprocket and Chain Removal

1. Remove the spark plugs to make turning the engine easier.

2. Remove the front cover as described in the preceding section.

3. Before removing the chain and sprockets, check tensioner spindle projection (**Figure 41**). If it exceeds 0.6 in. (15mm), replace the timing chain.

4. Turn the engine so the chain and sprocket match marks, camshaft dowel hole, and crankshaft key groove align as shown in **Figure 42**. To turn the engine, use a socket on the camshaft sprocket (**Figure 43**).

NOTE: *Be sure the transmission is in NEUTRAL before trying to turn the engine.*

5. Place the transmission in gear so the engine won't turn.

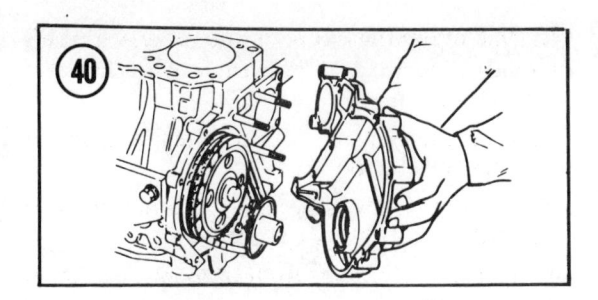

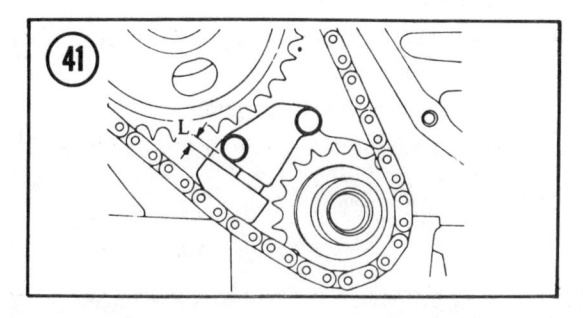

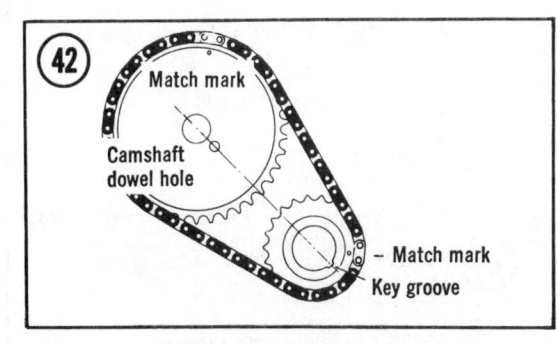

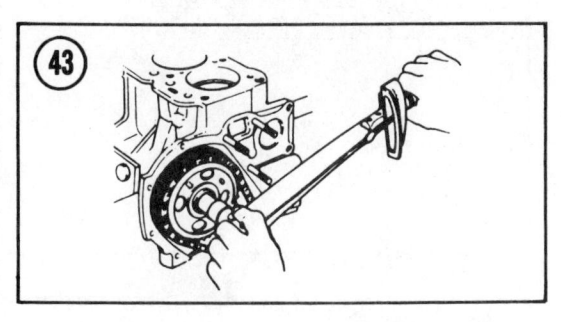

## FRONT COVER, TIMING CHAIN, AND SPROCKETS

### Cover Removal/Installation

1. Remove the alternator (Chapter Seven).
2. Remove the air pump (Chapter Five).
3. If air conditioned, remove the compressor belt. See *Drive Belts*, Chapter Three.
4. Remove the water pump. See *Water Pump*, Chapter Six.

6. Unbolt the chain tensioner and take it off.

7. Slide the oil thrower off the front of the crankshaft.

8. Remove the camshaft sprocket bolt.

9. Remove the chain and both sprockets simultaneously.

**Sprocket and Chain Inspection**

1. Check the sprocket teeth for wear or damge. If these are evident, replace the sprockets.

2. Check the timing chain for wear, damage, or stretching. Replace as needed.

3. Check the distance of both sprockets from the block (**Figure 44**). Measure from the block to the outer edge of the sprocket teeth. Compare the distance of each sprocket. The difference between them should be 0.020 in. (0.5mm) or less. If the difference is greater than this amount, insert adjusting shims behind the crankshaft sprocket as needed.

**Chain and Sprocket Installation**

1. If adjusting shims are needed, install them on the crankshaft sprocket.

2. Referring to **Figure 42**, install the sprockets and chain on the camshaft and crankshaft. Place a straightedge across the centers of both sprockets. Make sure the dowel hole in the camshaft sprocket, and the key groove in the crankshaft sprocket, are in the same line as the sprocket centers. Also be sure the match marks on the timing chain align with the marks on the sprockets.

3. Install retaining bolt on camshaft sprocket. Tighten to 29-35 ft.-lb. (4.0-4.8 mkg).

4. Bolt on the timing chain tensioner. Tighten to 4-6 ft.-lb. (0.6-0.8 mkg). Slide the oil thrower onto the crankshaft. The concave side faces forward.

5. If the distributor has been removed from the engine, turn the camshaft as shown in **Figure 45**. This places No. 1 piston at top dead center on its compression stroke, where it should be for distributor installation.

6. Pour clean engine oil over the chain and sprockets.

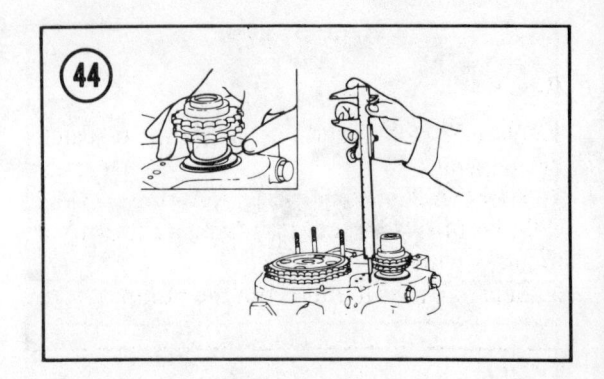

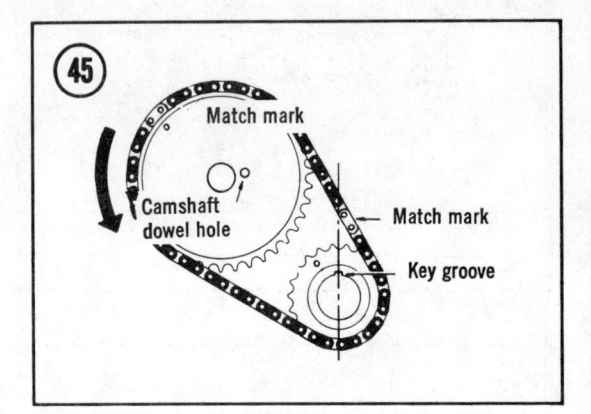

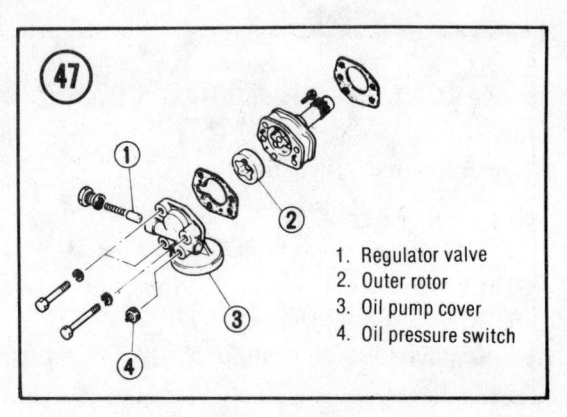

# ENGINE

## OIL PUMP

### Removal

1. Place a pan under the oil pump to catch dripping oil.
2. Remove 3 bolts securing the pump to the cylinder block (**Figure 46**). Remove the pump together with the oil filter.
3. Separate the oil filter from the pump.

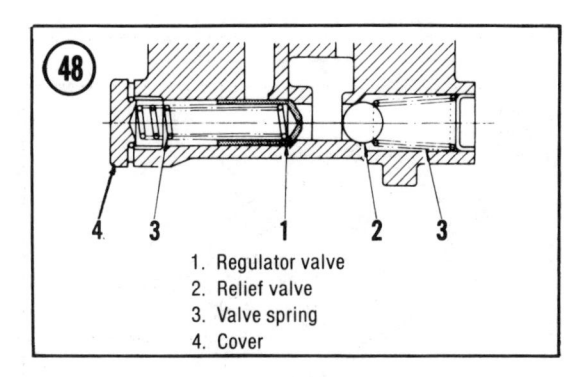

**48**
1. Regulator valve
2. Relief valve
3. Valve spring
4. Cover

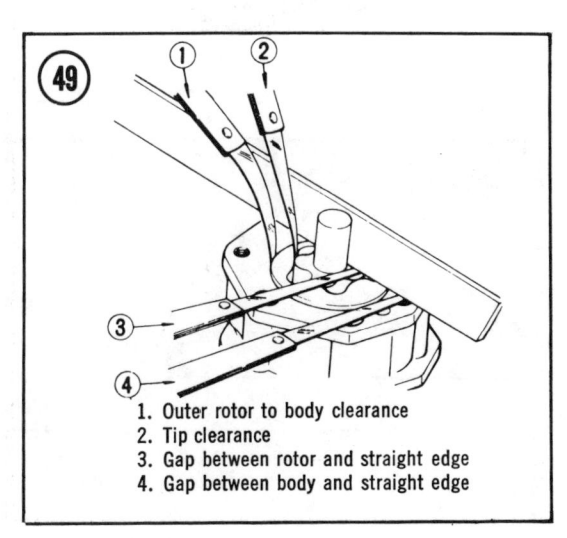

**49**
1. Outer rotor to body clearance
2. Tip clearance
3. Gap between rotor and straight edge
4. Gap between body and straight edge

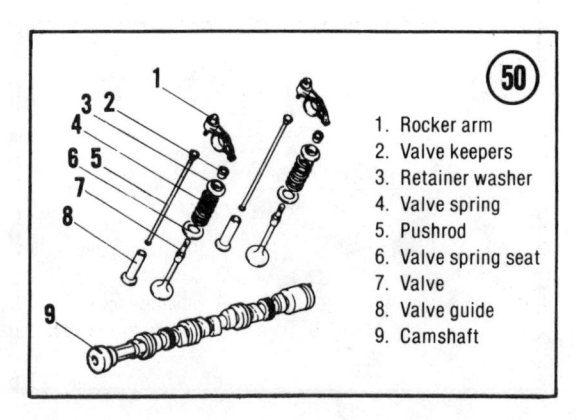

**50**
1. Rocker arm
2. Valve keepers
3. Retainer washer
4. Valve spring
5. Pushrod
6. Valve spring seat
7. Valve
8. Valve guide
9. Camshaft

### Disassembly, Inspection, and Assembly

1. Unbolt the pump cover and take it off. Take the outer rotor out of the pump body. See **Figure 47**.

*NOTE*
*The inner rotor, drive shaft, and pump body are not separable.*

2. Remove the regulator valve cover (**Figure 48**). Take out the valve and spring.
3. Thoroughly clean all parts in solvent. Discard parts that show obvious wear or damage.
4. Install the outer rotor in the pump body.
5. Measure tip clearance (**Figure 49**). It should be 0.005 in. (0.12mm) or less. If more than 0.008 in. (0.2mm), replace the oil pump.
6. Measure outer rotor to body clearance (**Figure 49**). It should be 0.006-0.008 in. (0.15-0.21mm). If more than 0.020 in. (0.5mm), replace the oil pump.
7. Lay a straightedge across the rotors and pump body (**Figure 49**). Measure the gap between rotors and straightedge, or body and straightedge. It should be 0.0012-0.0024 in. (0.03-0.06mm). If not, replace the oil pump.
8. Measure free length of the regulator valve spring. It should be 1.712 in. (43.49mm). If shorter than this, replace the spring.
9. Insert a probe into the regulator valve hole. Push on the relief valve (2, **Figure 48**) to make sure it opens and closes properly.
10. Assemble by reversing Steps 1 and 2. Use a new gasket, coated on both sides with gasket sealer. Tighten the cover bolts to 3-4 ft.-lb. (0.39-0.52 mkg). Tighten the regulator valve plug to 29-36 ft.-lb. (4-5 mkg).

## CAMSHAFT

### Removal

**Figure 50** shows the camshaft and related components. Refer to it as needed for this procedure.

1. Remove the engine.
2. Remove the front cover, timing chain, and sprockets as described earlier.
3. Remove the rocker assembly and pushrods as described earlier. Make a holder to keep the pushrods in order.

4. Remove the fuel pump (Chapter Five), distributor (Chapter Seven) and oil pump.

5. Turn the engine upside down to prevent the valve lifters from obstructing camshaft removal.

6. Reinstall the sprocket on the camshaft (**Figure 51**). Set up a dial indicator as shown and rotate the sprocket. If runout exceeds 0.004 in. (0.1 mm), replace the sprocket.

7. Unbolt camshaft locating plate (**Figure 52**), and remove the camshaft. Rotating the camshaft while pulling it out will make removal easier.

8. To remove the valve lifters, first remove the oil pan, then take the lifters out of the block. Make a holder to keep the lifters in order. They must be replaced exactly as they were.

**Inspection**

1. Check the camshaft bearing clearance. To do this, measure the diameter of each camshaft journal and compare it with the inside diameter of the corresponding camshaft bearing. If the difference between these figures exceeds 0.006 in. (0.15mm), the camshaft bearings must be replaced.

*NOTE*
*Camshaft bearing replacement is a job requiring special tools and experience. The tools alone would cost more than bearing replacement. Take the cylinder block to a dealer or machine shop and have them replace the bearings.*

2. Check the camshaft for bending. Rotate it between accurate centers such as V-blocks or a lathe with a dial gauge contacting the center journal (**Figure 53**). Actual bend is half the reading shown on the gauge. If the camshaft is bent more than 0.002 in. (0.05 mm), replace it.

3. Measure the height of the cam lobes and compare with the specifications at the end of this chapter. If the lobes are worn more than specified, replace the camshaft.

4. Measure the clearance between the camshaft and locating plate as shown in **Figure 54**. Replace the thrust washer if the clearance exceeds 0.004 in. (0.1mm).

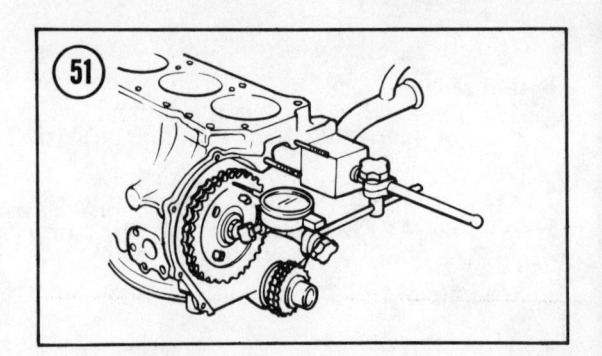

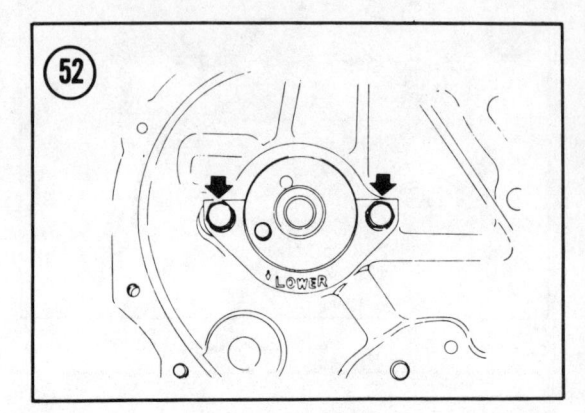

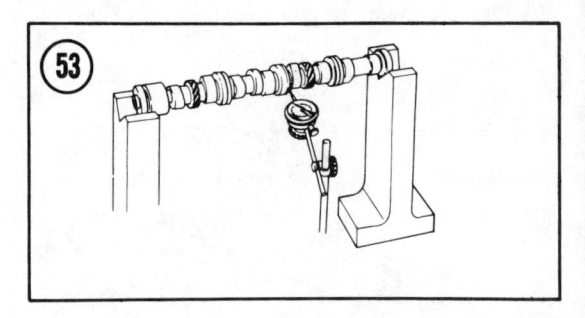

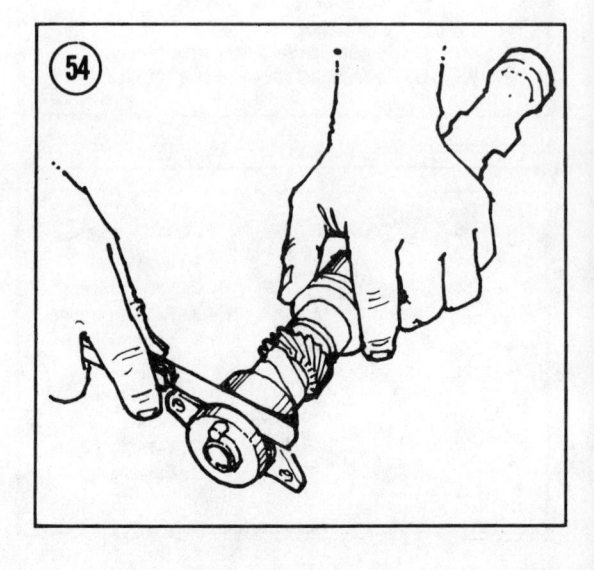

# ENGINE

5. Inspect the lifters for visible wear and scores. Replace as needed. Measure the outside diameter of the valve lifters and compare with the inside diameter of the corresponding holes in the cylinder block. If the difference between these figures exceeds 0.006 in. (0.15mm), replace the valve lifters.

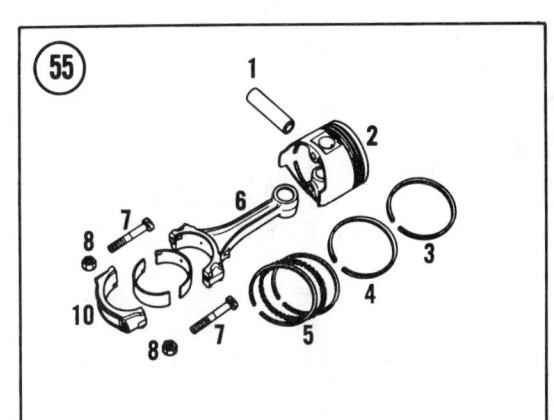

**55**

1. Piston pin
2. Piston
3. Top compression ring
4. Second compression ring
5. Oil ring (rails and spacer)
6. Connecting rod
7. Connecting rod bolts
8. Connecting rod nuts
9. Connecting rod bearings
10. Connecting rod cap

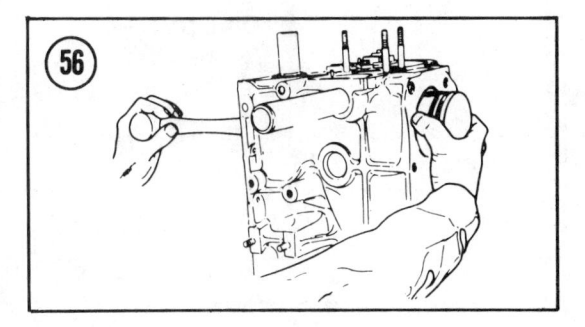

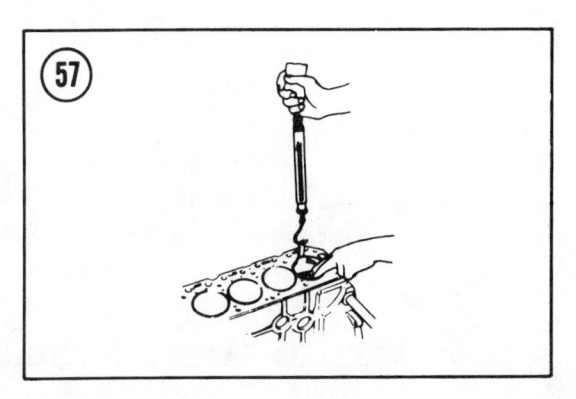

**56**

**57**

## Installation

1. Coat the camshaft and bearings with clean engine oil. Install the camshaft in the block.
2. Install the locating plate so the side marked LOWER is toward the bottom (**Figure 51**).
3. Coat the lifters with clean engine oil. Install them in the block.
4. Install the timing chain and sprockets as described earlier. Be sure the timing marks are lined up as shown in **Figure 42**.
5. Install the oil pump as described earlier in this chapter.
6. Install the fuel pump (Chapter Five) and distributor (Chapter Seven).
7. Install the oil pan, front cover, pushrods, rocker assembly, and engine.

## CONNECTING RODS AND PISTONS

**Figure 55** shows a piston-connecting rod assembly. Refer to it as needed for the following procedures.

### Piston Removal

1. Remove the engine.
2. Remove the cylinder head and oil pan.
3. Remove the carbon ridge at the top of the cylinder bores with a ridge reamer. These are available from tool rental dealers.
4. Rotate the crankshaft so the connecting rod is centered in the bore.
5. Unbolt the connecting rod cap and push the connecting rod and piston out the top of the bore (**Figure 56**). Tap the connecting rod with a hammer handle if needed.

### Piston Clearance Check

This procedure should be performed at normal room temperature (68° F). The pistons and cylinder walls must be clean and dry. Specified clearance is listed at the end of the chapter.

1. Referring to **Figure 57**, attach a spring scale to a feeler gauge. The gauge should be 0.0016 in. (0.04mm) thick.

2. Insert the piston upside down in the cylinder bore.

3. Insert the feeler gauge between the piston and cylinder wall. Pull on the spring scale. The force required to pull the feeler gauge out should be 1.1-3.3 lb. (0.5-1.5 kg). If the force is less than this figure, piston clearance is greater than specified. If the force is greater, piston clearance is less.

4. Repeat the procedure for all 4 pistons and cylinders.

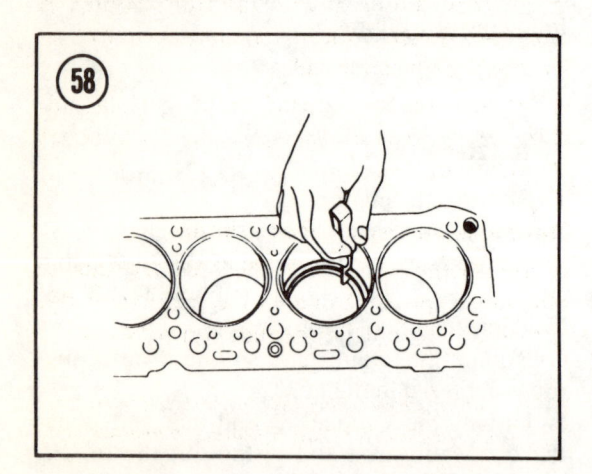

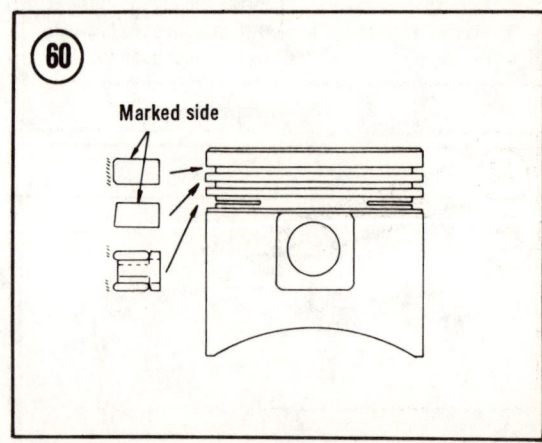

**Piston Ring Fit/Installation**

1. Check the ring gap of each piston ring. To do this, first press the ring about one inch down the bore and square it by tapping gently with an inverted piston.

> NOTE: *If the cylinders have not been rebored, check the gap at the bottom of the ring travel, where the bore is smallest.*

2. Measure the ring gap with a feeler gauge as shown in **Figure 58**. Compare with specifications at the end of the chapter.

3. Check side clearance of the ring as shown in **Figure 59**. Place the feeler gauge alongside the ring all the way into the groove. Specifications (**Table 1**) are given at the end of the chapter.

4. Check the fit of the piston pin in the piston. It should be loose enough to be pressed in by hand at room temperature, but tight enough so it does not wobble. If in doubt, measure outside diameter of pin and inside diameter of pin hole in piston. The difference between the 2 figures should be 0.0003-0.0005 in. (0.008-0.012mm).

5. Using a ring expander tool, carefully install the oil control ring, then the compression rings. The top sides of the rings are marked and must be up. See **Figure 60**.

6. Position the ring gaps as shown in **Figure 61**.

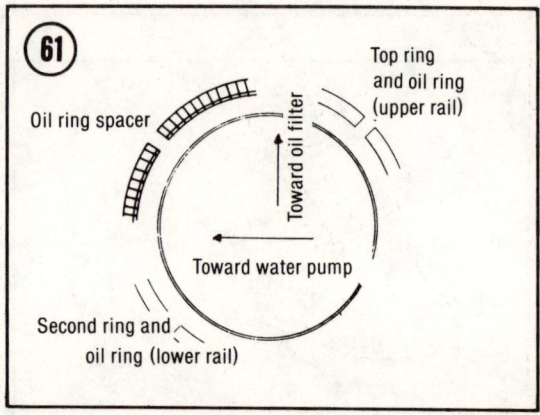

# ENGINE

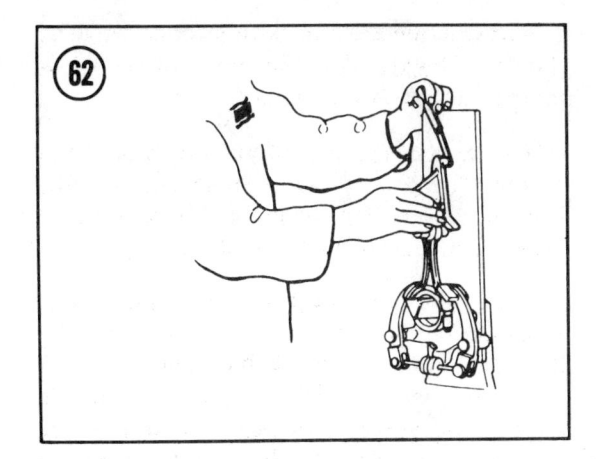

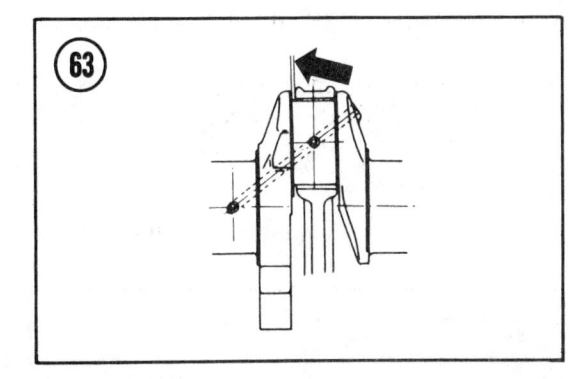

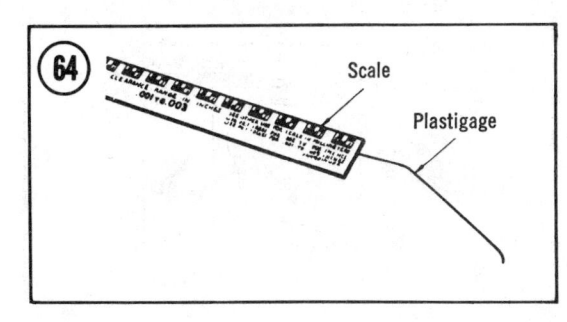

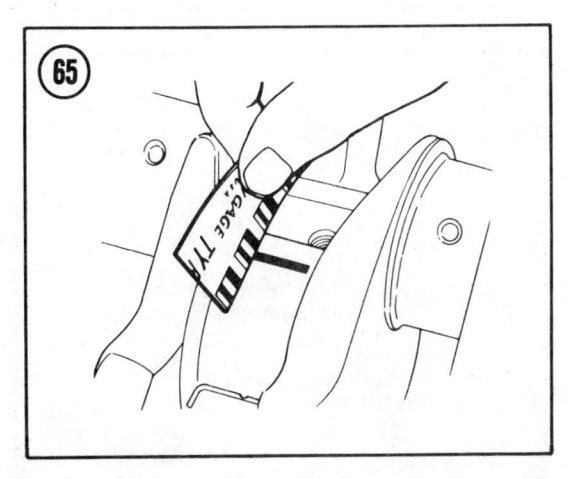

## Piston Pin Removal/Installation

The piston pins are press fitted to the connecting rods and hand fitted to the pistons. Removal requires a pressing force of 2,200-6,600 pounds (1-3 metric tons). This is a job for a dealer or machine shop, which is equipped to fit the pistons and pins, ream the pin bushings to the correct diameter, and align the pins with the connecting rods.

## Connecting Rods

1. Check all connecting rods for bends or twisting with a connecting rod aligner (**Figure 62**). If you do not have such a tool, take the rods to a dealer or machine shop for checking. Maximum tolerance for bend or twisting is listed at the end of the chapter.

2. Install the connecting rods and bearings on the crankshaft. Insert a feeler gauge between the connecting rod big end and crankshaft and measure the clearance (**Figure 63**). Compare with specifications at the end of the chapter. If clearance exceeds the maximum tolerance, replace the connecting rod.

3. Weigh the connecting rods. Weight difference between rods should be 0.18 ounce (5 grams) or less.

## Installing Connecting Rod Bearings

Install the bearings in the rods and caps so the small formed tongs in the bearings fit the grooves machined in the rods and caps.

*NOTE*
*Always replace bearings in complete sets. Never file a rod, cap or bearing to improve the fit.*

## Measuring Bearing Clearance

1. Assemble connecting rods with bearings on the proper crankpins. Do not tighten.

2. Cut Plastigage (**Figure 64**) the width of bearing and insert between crankpin and bearing.

3. Install the bearing cap and tighten to 23-27 ft.-lb. (3.2-3.8 mkg).

4. Remove bearing cap. Bearing clearance is determined by comparing the width of the flattened Plastigage (**Figure 65**) to markings on the envelope. Compare clearance with specifications (**Table 1**) given at the end of the chapter.

## Installing Piston/Connecting Rod Assembly

1. Position the ring gaps as shown in **Figure 61**.

2. Make sure the number (**Figure 66**) or notch (**Figure 67**) faces the water pump end of the engine. Make sure the oil hole in the connecting rod's big end faces the oil filter side of the engine.

3. Immerse the entire piston in engine oil. Slide a ring compressor (**Figure 68**) over the rings and compress the rings into the grooves.

4. Install the piston downward into the bore. Be sure the piston and connecting rod are aligned with the cylinder block as described in Step 1. Tap the piston into the bore with a wooden hammer handle.

> **CAUTION**
> *If the crankshaft is in place, be careful not to nick the connecting rod journals.*

## CRANKSHAFT

### Removal/Inspection

1. Remove the flywheel as described later in this chapter.

2. Unbolt the main bearing caps. Loosen the bolts in 2 or 3 stages, following the sequence shown in **Figure 69**.

3. Remove the crankshaft rear oil seal.

4. Lift the crankshaft and bearings out of the cylinder block.

5. Check the crankshaft for bending with a dial gauge as shown in **Figure 70**. Mount the crank-

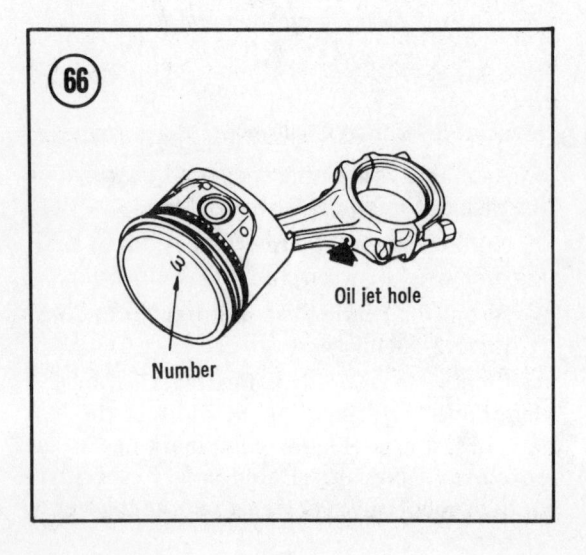

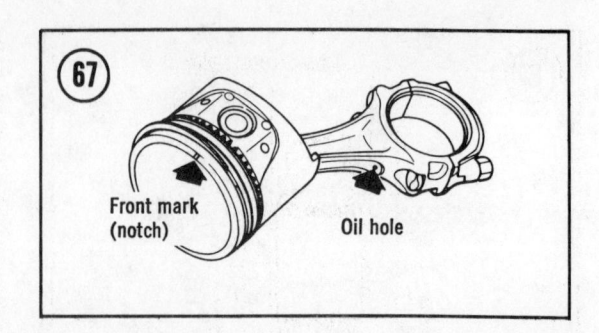

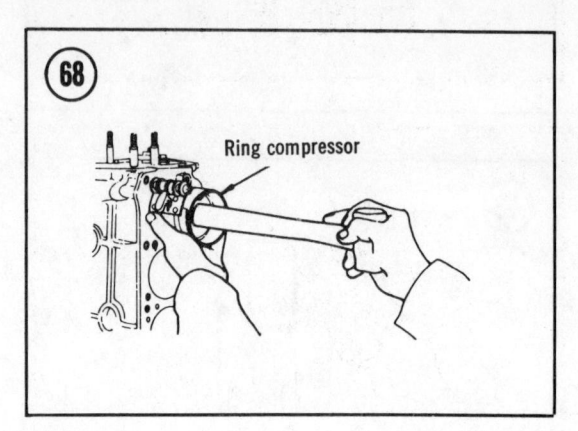

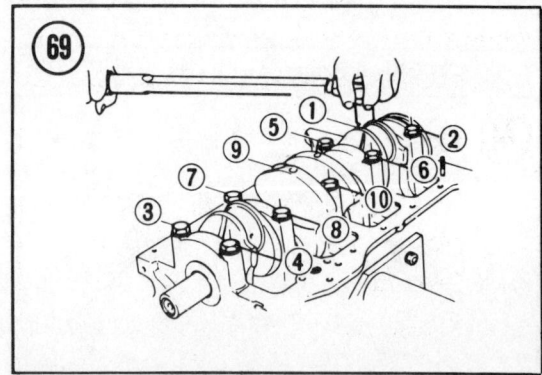

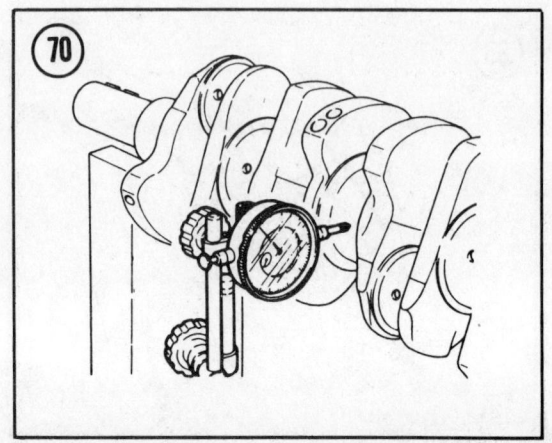

# ENGINE

**71** Crankshaft end-play

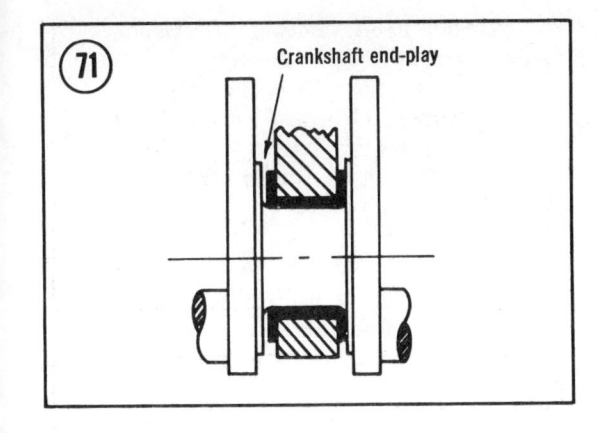

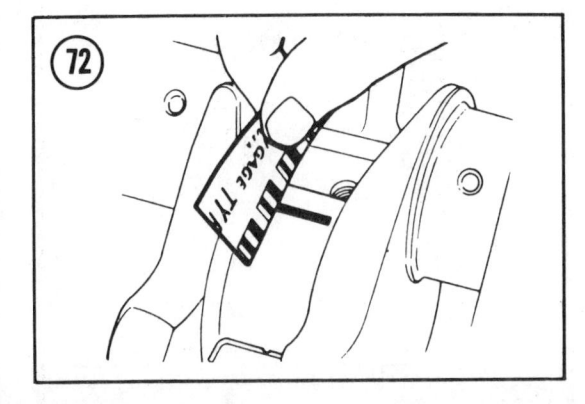

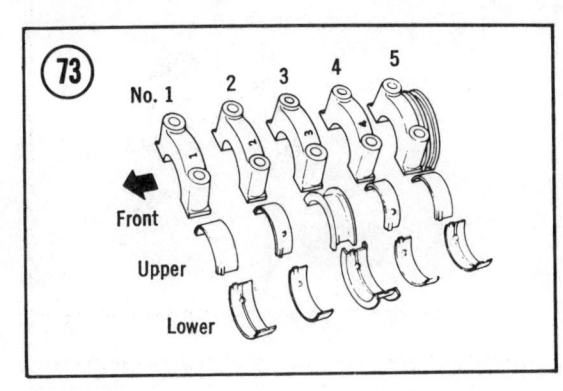

**72**

**73** No. 1  2  3  4  5 / Front / Upper / Lower

**74**

shaft in V-blocks or a lathe and rotate it with the dial gauge contacting the center journal. Actual bend is half the reading shown on the gauge. Compare with specifications (**Table 1**) at end of chapter.

6. Check all journals against the specifications at the end of the chapter for out-of-roundness, taper, and wear. Inspect the journals for scores. If necessary, grind to the next undersize.

7. Measure crankshaft end play. Insert a feeler gauge between the crankshaft center journal and the flange on the center bearing (**Figure 71**). If end play exceeds specifications (**Table 1**) at the end of the chapter, replace the center bearing.

## Measuring Main Bearing Clearance

1. Install bearings and crankshaft in the block.
2. Install the bearing caps with bearings in place, but do not tighten.
3. Cut Plastigage the width of the bearing and insert it between the bearing and crankshaft. Tighten bearing caps to 36-44 ft.-lb. (5-6 mkg).
4. Remove the bearing caps. Bearing clearance is determined by comparing the width of the flattened Plastigage to the markings on the envelope (**Figure 72**). Compare to the specifications (**Table 1**) at the end of the chapter.

## Installation

1. Place the bearing shells in the block so the small formed tongs in the bearings fit the grooves machined for them in the block. The center bearing is flanged. Make sure the oil holes in the bearings line up with oil holes in the block.
2. Coat the bearings and crankshaft journals with clean engine oil.
3. Place the crankshaft in position and install the rear oil seal. Coat the lip of the oil seal with lithium grease.
4. Place the bearing caps (**Figure 73**) in position according to the numbers on them. No. 1 goes at the front of the engine. The arrows on the caps point toward the front of the engine.

NOTE: *Apply gasket sealer to the contact surfaces of the rear main bearing cap and cylinder block (Figure 74).*

5. Tighten the bearing caps. Start at the center bearing and work outward, tightening gradually in 2 or 3 stages. Specified torque is 36-44 ft.-lb. (5-6 mkg). Rotate the crankshaft during tightening to make sure it isn't binding. If the crankshaft becomes difficult to turn, stop and find out why before continuing. Make absolutely certain that bearings are the correct size for the crankshaft, especially if it has been reground. Never use undersize bearings if the crankshaft has not been reground.

## Pilot Bushing

The pilot bushing is located inside the rear end of the crankshaft (**Figure 75**). It should be inspected for damage or excessive wear. If it is worn or damaged, remove it with a puller as shown in **Figure 76** (Datsun No. KV10102800; Kent-Moore No. J25657). Tap the new bushing in until dimension A, **Figure 77**, is 0.11 in. (2.8mm).

> CAUTION
> *Do not tap the bushing in too far. Do not distort the edge.*

## CYLINDER BLOCK INSPECTION

1. Clean the block thoroughly and check all freeze plugs for leaks. Replace any freeze plugs that are suspect. It is a good idea to replace all of them.

> NOTE: *If there is extensive corrosion in water passages, or sludge in oil passages, it is a good idea to have the cylinder block boiled out by a dealer or machine shop. Boiling necessitates replacement of all camshaft bearings and freeze plugs. However, an engine dirty enough to need a block boil-out almost certainly needs new camshaft bearings and freeze plugs anyway.*

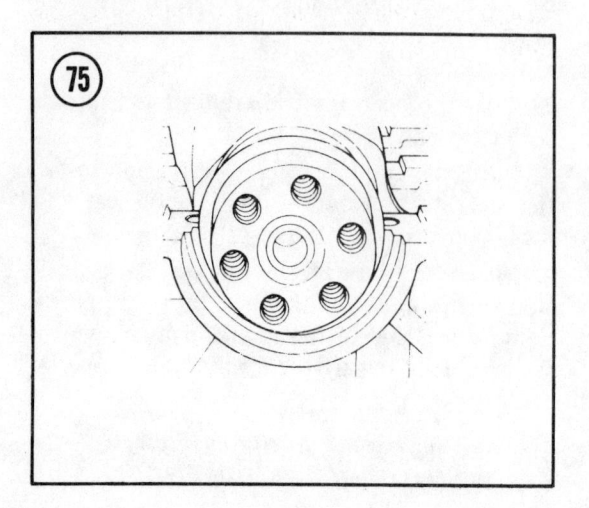

2. Examine the block for cracks.

3. Check the cylinder head mating surface on the block for warping, using a straightedge and feeler gauge as shown in **Figure 78**. Maximum permissible warp is 0.004 in. (0.1mm).

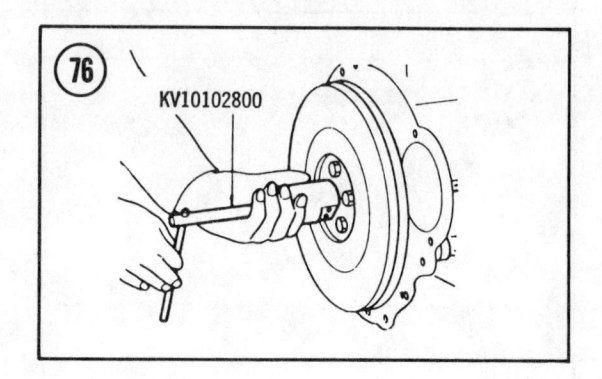

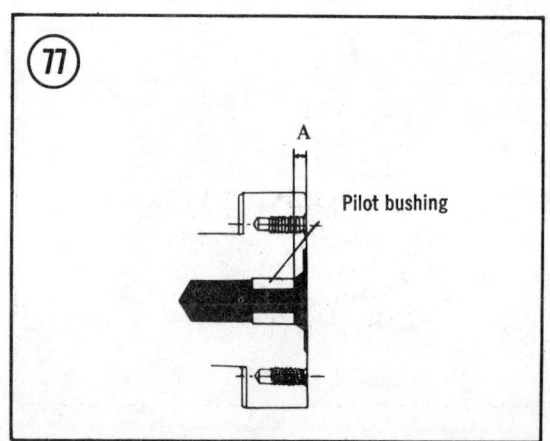

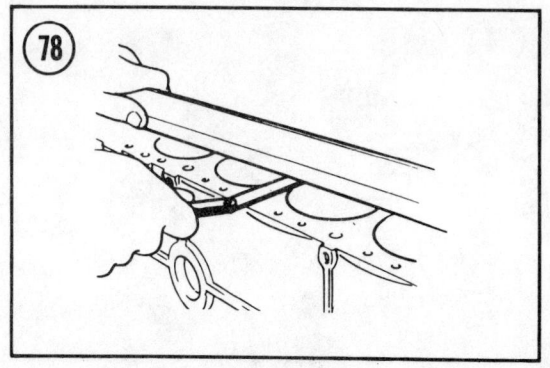

# ENGINE

4. Check the cylinder bores for out-of-roundness and taper as shown in **Figure 79**. Measure both in front-rear and sideways directions. Compare the measurements to specifications at the end of this chapter. If the cylinders exceed the maximum wear tolerance, they must be rebored. Reboring is also necessary if the cylinder walls are badly scuffed or scored.

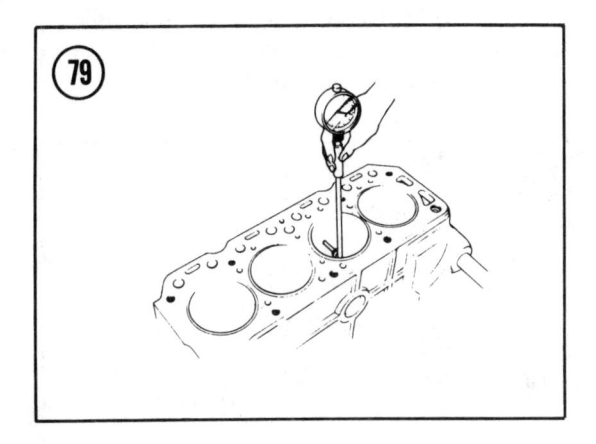

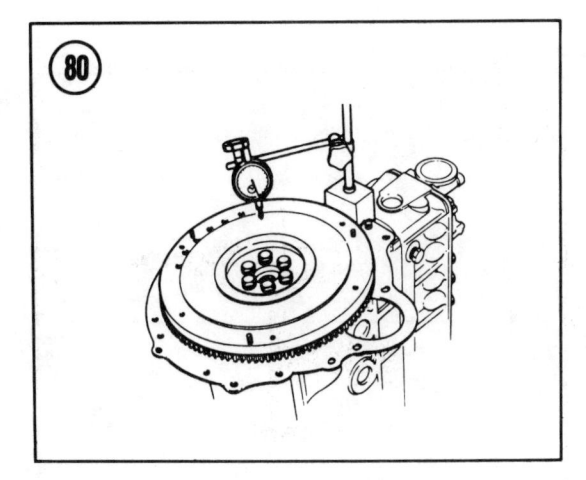

NOTE: *If one cylinder is bored out, all cylinders must be bored to the same diameter. Bore the cylinders in one of the following orders: 2-4-1-3 or 3-1-4-2. When the cylinders are too badly worn to be bored further, undersize cylinder liners can be fitted.*

## FLYWHEEL

### Removal/Installation

1. Remove the engine. Separate the engine and transmission.
2. Remove the clutch (Chapter Eight).
3. Check flywheel runout with a dial gauge **(Figure 80)**. The dial gauge pointer should touch the flywheel as far as possible from the center. Turn the flywheel one full turn and note the dial gauge reading. If more than 0.006 in. (0.15mm), replace the flywheel.
4. Unbolt the flywheel from the crankshaft.
5. Installation is the reverse of removal. Tighten the flywheel bolts to specifications gradually, in a diagonal pattern.
6. After installation, check runout as described in Step 3. If it is excessive, check for foreign material between flywheel and crankshaft flange.

### Inspection

1. Check the flywheel friction surface for scoring, glazing, or burn marks (blue-tinted areas). If these are found, have the flywheel resurfaced by a machine shop.
2. Inspect the flywheel ring gear teeth. If the teeth are chipped, broken, or excessively worn, have a new ring gear shrunk on by a machine shop.

## Table 1   ENGINE SPECIFICATIONS

**Valves**
Intake
   Head diameter — 1.46 in. (37mm)
   Stem diameter — 0.3138-0.3144 in. (7.970-7.985 mm)
   Length — 4.079-4.094 in. (103.6-104.0mm)
   Seat angle (F10) — 45°
   Seat angle (310) — 45°30'
   Spring free length — 1.831 in. (46.5mm)
   Spring loaded length — 1.189 in. @ 129 lb. (30.2mm @ 58.5kg)
   Stem-to-guide clearance — 0.0006-0.0018 in. (0.015-0.045mm)

Exhaust
   Head diameter — 1.18 in. (30mm)
   Stem diameter — 0.3128-0.3134 in. (7.945-7.960mm)
   Length — 4.079-4.094 in. (103.6-104.0mm)
   Seat angle (F10) — 45°
   Seat angle (310) — 45°30'
   Spring free length — 1.831 in. (46.5mm)
   Spring loaded length — 1.189 in. @ 129 lb. (30.2mm @ 58.5 kg)
   Stem-to-guide clearance — 0.0016-0.0028 in. (0.040-0.070mm)

**Valve Guides**
Inside diameter — 0.3150-0.3156 in. (8.000-8.015mm)
Outside diameter (replacement guides) — 0.4816-0.4820 in. (12.233-12.244mm)
Interference fit — 0.0009-0.0017 in. (0.022-0.044mm)
Height from head surface
   1976-1977 — 0.728 in. (18.5mm)
   1978 and later — 0.709 in. (18mm)

**Lifters**
Lifter-to-bore clearance
   Standard — 0.0008-0.0020 in. (0.02-0.05mm)
   Maximum — 0.006 in. (0.15mm)

**Rocker assembly**
Rocker shaft outer diameter — 0.7866-0.7874 in. (19.979-20.0mm)
Rocker arm inner diameter — 0.7882-0.7887 in. (20.020-20.033mm)
Rocker arm to shaft clearance — 0.0008-0.0021 in. (0.020-0.054mm)

**Crankshaft**
Main bearing journal diameter — 1.9666-1.9671 in. (49.951-49.964mm)
Connecting rod journal diameter — 1.7701-1.7706 in. (44.961-44.974mm)
Crankshaft end play
   Standard — 0.002-0.006 in. (0.05-0.15mm)
   Maximum — 0.012 in. (0.3mm)
Crankshaft bend
   Standard — Less than 0.001 in. (0.025mm)
   Maximum — 0.002 in. (0.05mm)
Journal taper and out-of-round, maximum — 0.0012 in. (0.03mm)
Bearing clearance, maximum — 0.004 in. (0.1mm)

**Cylinder block**
Bore diameter — 2.9921-2.9941 in. (76.0-76.050mm)
Maximum wear — 0.004 in. (0.1mm)
Maximum difference between bores — 0.008 in. (0.2mm)
Maximum out-of-round and taper — 0.0006 in. (0.015mm)
Maximum surface warp — 0.004 in. (0.1mm)

# ENGINE

**Table 1  ENGINE SPECIFICATIONS** (continued)

| | |
|---|---|
| **Camshaft** | |
| Bend | |
|   Standard | 0.0006 in. (0.015mm) |
|   Maximum | 0.002 in. (0.05mm) |
| End play | |
|   Standard | 0.0004-0.0020 in. (0.01-0.05mm) |
|   Maximum | 0.004 in. (0.1mm) |
| Lobe lift | |
|   Intake | 0.2224 in. (5.65mm) |
|   Exhaust | 0.2331 in. (5.92mm) |
| Bearing clearance, standard | |
|   First and fifth | 0.0015-0.0024 in. (0.037-0.060mm) |
|   Second and fourth | 0.0011-0.0020 in. (0.027-0.050mm) |
|   Third | 0.0016-0.0025 in. (0.040-0.063mm) |
| Bearing clearance, maximum | 0.006 in. (0.015mm) |
| **Oil pump** | |
| Side clearance (inner to outer rotor) | |
|   Standard | 0.002-0.005 in. (0.05-0.12mm) |
|   Maximum | 0.008 in. (0.2mm) |
| Tip clearance | |
|   Standard | Less than 0.005 in. (0.12mm) |
|   Maximum | 0.008 in. (0.2mm) |
| Outer rotor to body clearance | |
|   Standard | 0.006-0.008 in. (0.15-0.21mm) |
|   Maximum | 0.020 in. (0.5mm) |
| **Pistons and connecting rods** | |
| Piston clearance | 0.001-0.002 in. (0.025-0.045mm) |
| Piston pin clearance | 0.0003-0.0005 in. (0.008-0.012mm) |
| Piston diameter | 2.9908-2.9928 in. (75.967-76.017mm) |
| Available oversizes | 0.020 in. (0.5mm) and 0.039 in. (1mm) |
| Maximum bend (per 3.94 in. or 100mm of connecting rod length) | |
|   1976-1978 | 0.004 in. (0.1mm) |
|   1979 | 0.006 in. (0.15mm) |
| Maximum twist (per 3.94 in. or 100mm of connecting rod length) | |
|   1976-1978 | 0.004 in. (0.1mm) |
|   1979 | 0.012 in. (0.3mm) |
| **Piston rings** | |
| Ring gap, standard | |
|   Top compression | 0.008-0.014 in. (0.20-0.35mm) |
|   Second compression | 0.001-0.002 in. (0.03-0.06mm) |
|   Oil ring rails | 0.012-0.035 in. (0.3-0.9mm) |
| Ring gap, maximum | 0.039 in. (1mm) |
| Ring side clearance, standard | |
|   Top compression | 0.002-0.003 in. (0.04-0.07mm) |
|   Second compression | 0.001-0.002 in. (0.03-0.06mm) |
|   Oil ring | Combined ring — none |
| Ring side clearance, maximum | 0.004 in. (0.1mm) |

Table 2  TIGHTENING TORQUES

|  | Ft.-lb. | Mkg |
|---|---|---|
| Camshaft locate plate bolts | 3-4 | 0.4-0.5 |
| Camshaft sprocket bolt | 29-35 | 4.0-4.8 |
| Carburetor nuts | 4-5 | 0.5-0.7 |
| Connecting rod cap nuts | 23-27 | 3.2-3.8 |
| Crankshaft pulley bolt | 108-145 | 15-20 |
| Cylinder head bolts* | 51-54 | 7.0-7.5 |
| Flywheel bolts |  |  |
|   1976-1977 | 54-61 | 7.5-8.5 |
|   1978-on | 58-65 | 8-9 |
| Fuel pump nuts | 7-10 | 0.9-1.4 |
| Main bearing cap bolts | 36-43 | 5-6 |
| Manifold nuts | 11-14 | 1.5-2.0 |
| Motor mount brackets to engine | 14-18 | 1.9-2.5 |
| Oil pan bolts |  |  |
|   F10 | 11-14 | 1.5-2.0 |
|   310 | 3-4 | 0.4-0.6 |
| Oil pan drain plug | 14-22 | 2-3 |
| Oil pump bolts | 7-10 | 0.9-1.4 |
| Oil strainer bolts | 7-10 | 0.9-1.4 |
| Rocker shaft bracket bolts | 14-18 | 2.0-2.5 |
| Timing chain cover bolts | 4-5 | 0.5-0.7 |
| Water pump bolts | 7-10 | 0.9-1.4 |

*Tighten in 2 or 3 stages. Retighten after warming up engine.

NOTE: If you own a 1980 or 1981 model, first check the Supplement at the back of the book for any new service information.

# CHAPTER FIVE

# FUEL AND EXHAUST SYSTEMS

This chapter provides service procedures for the air cleaner, carburetor, fuel pump, exhaust system, throttle linkage, fuel tank, and fuel system-related emission controls.

## AIR CLEANER

All models use a viscous paper air cleaner element. The element should be replaced at intervals specified in Chapter Three. The air cleaner includes an automatic temperature control system and idle compensator. The 1977-1978 California models use an altitude compensator. See **Figure 1** (1976-1977), **Figure 2** (1978 and later U.S. models), or **Figure 3** (1978 and later Canada models).

### Air Cleaner Removal/Installation

1. Remove the cover nut and clips, lift off the cover, and take the element out.
2. Unbolt the air cleaner from its side bracket.
3. Disconnect the hot air tube from the underside of the air cleaner nozzle.
4. On 1978-1979 models, disconnect the fresh air hose from the side of the air cleaner.
5. Lift the air cleaner up. Carefully note all connections, then disconnect the hoses from the underside. The air cleaner can then be lifted off.
6. Installation is the reverse of removal.

### ATC Inspection

The most likely cause of ATC (automatic temperature control) trouble is that the inlet valve is stuck open. This is not apparent in warm weather, but in cold weather, it results in slow acceleration, engine hesitation, or stalling. The inlet valve may also be stuck shut, which causes extremely high fuel consumption and loss of power.

1. Make sure vacuum hoses are installed correctly. See **Figure 1, 2, or 3**.
2. Check the hoses for cracks, leaks, or plugging.
3. With the engine cold, disconnect the fresh air duct from the air cleaner.
4. Look down the air cleaner inlet with a mirror (**Figure 4**). The valve should be blocking the hot air duct (flap down). When the engine is started, the valve should move to block the fresh air duct (flap up).
5. Warm the engine to normal operating temperature. The valve should move to block the hot air duct. In very cold weather, the valve may take a long time to close.

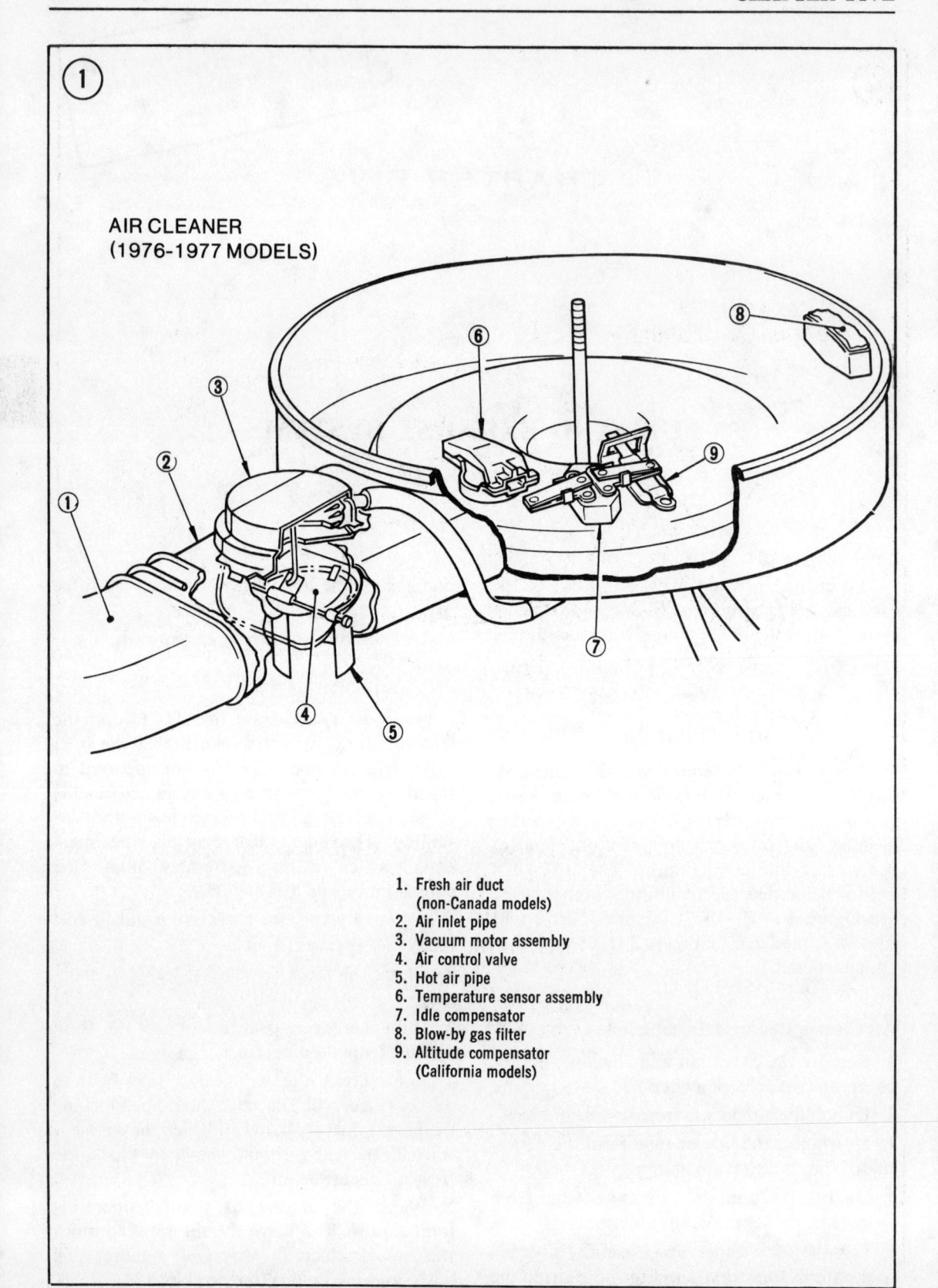

# CHAPTER FIVE

①

**AIR CLEANER  
(1976-1977 MODELS)**

1. Fresh air duct  
   (non-Canada models)
2. Air inlet pipe
3. Vacuum motor assembly
4. Air control valve
5. Hot air pipe
6. Temperature sensor assembly
7. Idle compensator
8. Blow-by gas filter
9. Altitude compensator  
   (California models)

**FUEL AND EXHAUST SYSTEMS**

②

AIR CLEANER
(1978-on U.S. MODELS)

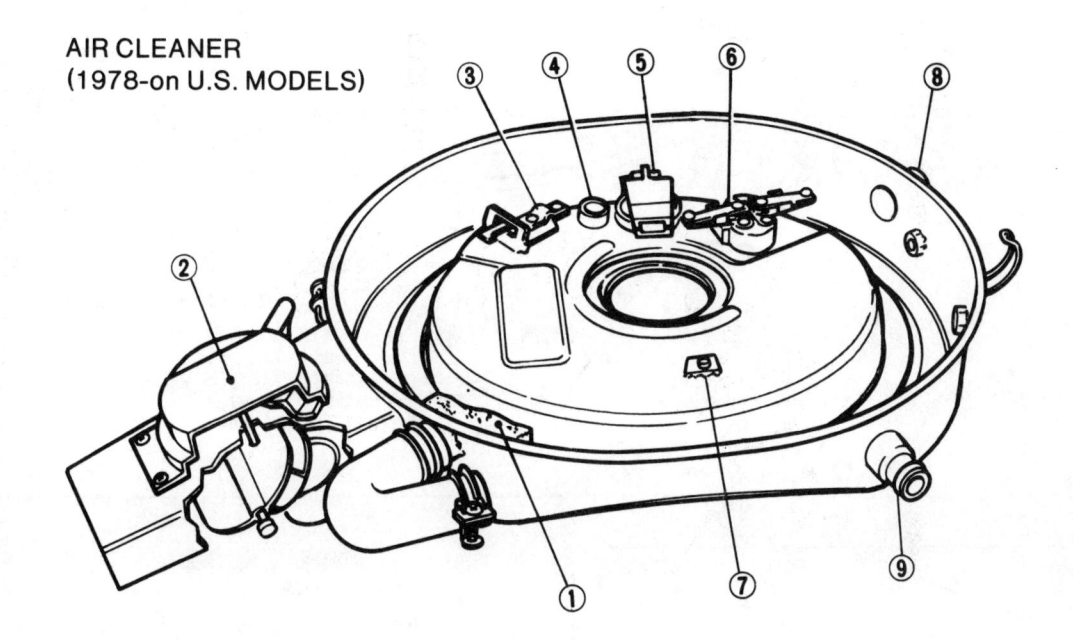

1. PCV filter
2. Heat control valve
3. Altitude compensator (1978 California only)
4. Outlet for A.B. valve
5. Temperature sensor
6. Idle compensator
7. Air hole for throttle opener
8. Outlet for C.A.C. valve
   (California models)
9. Air relief valve
   (non-California models)

**AIR CLEANER
(1978-on CANADA MODELS)**

1. PCV filter
2. Heat control valve
3. Outlet for A.B. valve
4. Temperature sensor
5. Idle compensator
6. Air hole for throttle opener
7. Air induction valve and filter

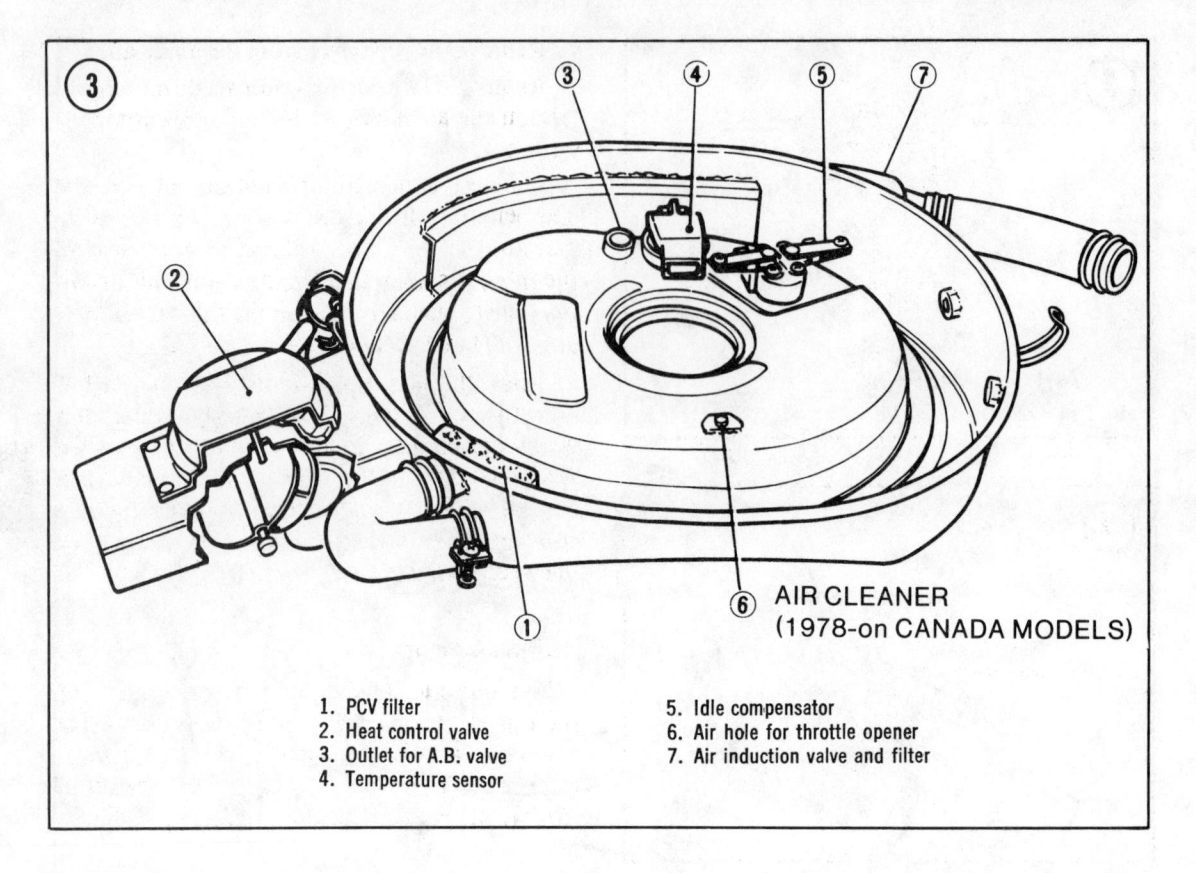

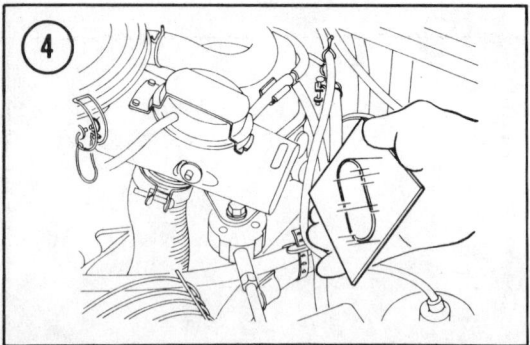

6. If the valve doesn't work properly, disconnect the hose from the vacuum motor. Attach another hose and suck on it (**Figure 5**). The valve should move to block the fresh air duct. When the vacuum hose is pinched off, the valve should stay in place for at least 30 seconds. If it moves to block the hot air duct, replace the vacuum motor.

7. Reconnect the hose to the vacuum motor.

8. If the vacuum motor is good, test the temperature sensor. This requires a cold engine. Start the engine and let it idle. Detach the vacuum hose from the vacuum motor and place a thumb over it. If you can't feel vacuum, replace the temperature sensor.

## Idle Compensator

The idle compensator is a thermostatic valve controlled by underhood temperature. It prevents overrich fuel mixture during hot idle conditions by admitting extra air to the intake manifold. One valve opens at 140-158°F (60-70°C); the other opens at 158-194°F (70-90°C). Test as follows.

# FUEL AND EXHAUST SYSTEMS

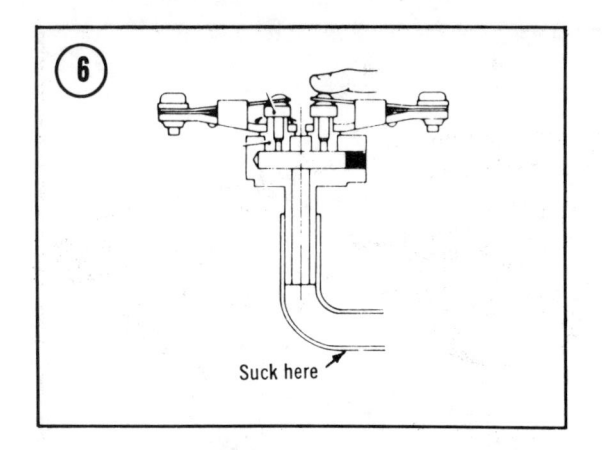

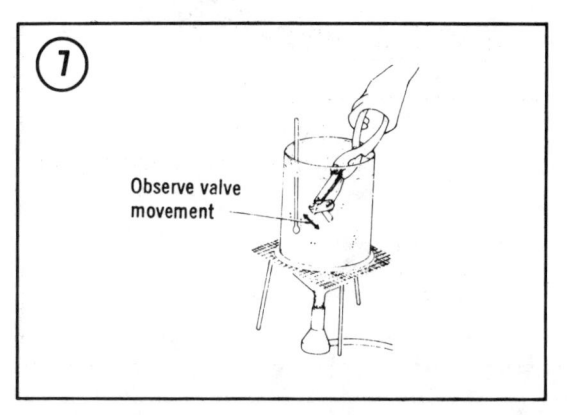

1. Remove the top cover from the air cleaner.
2. Remove 2 idle compensator securing screws. Detach the air hose and take the idle compensator out.
3. Connect a tube to the bottom of the idle compensator. Block one side of the idle compensator with a finger (**Figure 6**) and suck on the tube. It should be extremely difficult or impossible to suck air through the tube at temperatures below 140°F (60°C).
4. Place the idle compensator in water with a thermometer (**Figure 7**). Heat the water and watch the valves. One should open between 140-158°F (60-70°C); the other should open between 158-194°F (70-90°C). Replace the idle compensator if a valve fails to open or opens at the wrong temperature.

## Altitude Compensator

The altitude compensator compensates for the thin air at high altitudes by admitting extra air to the carburetor. Above 4,000 ft. (1,219m), the valve should be set to the "H" position (**Figure 8**). Below that altitude, it should be set to the "L" position.

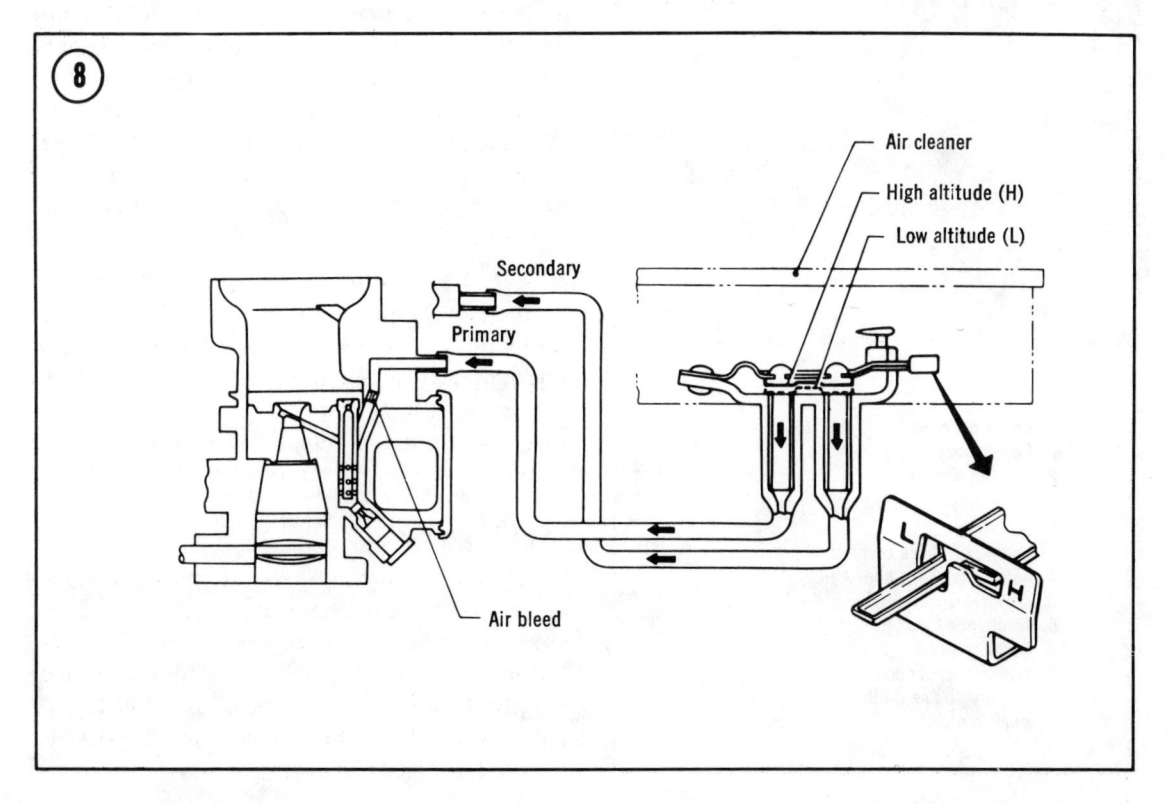

# CHAPTER FIVE

**1976 MODEL**
**CARBURETOR**

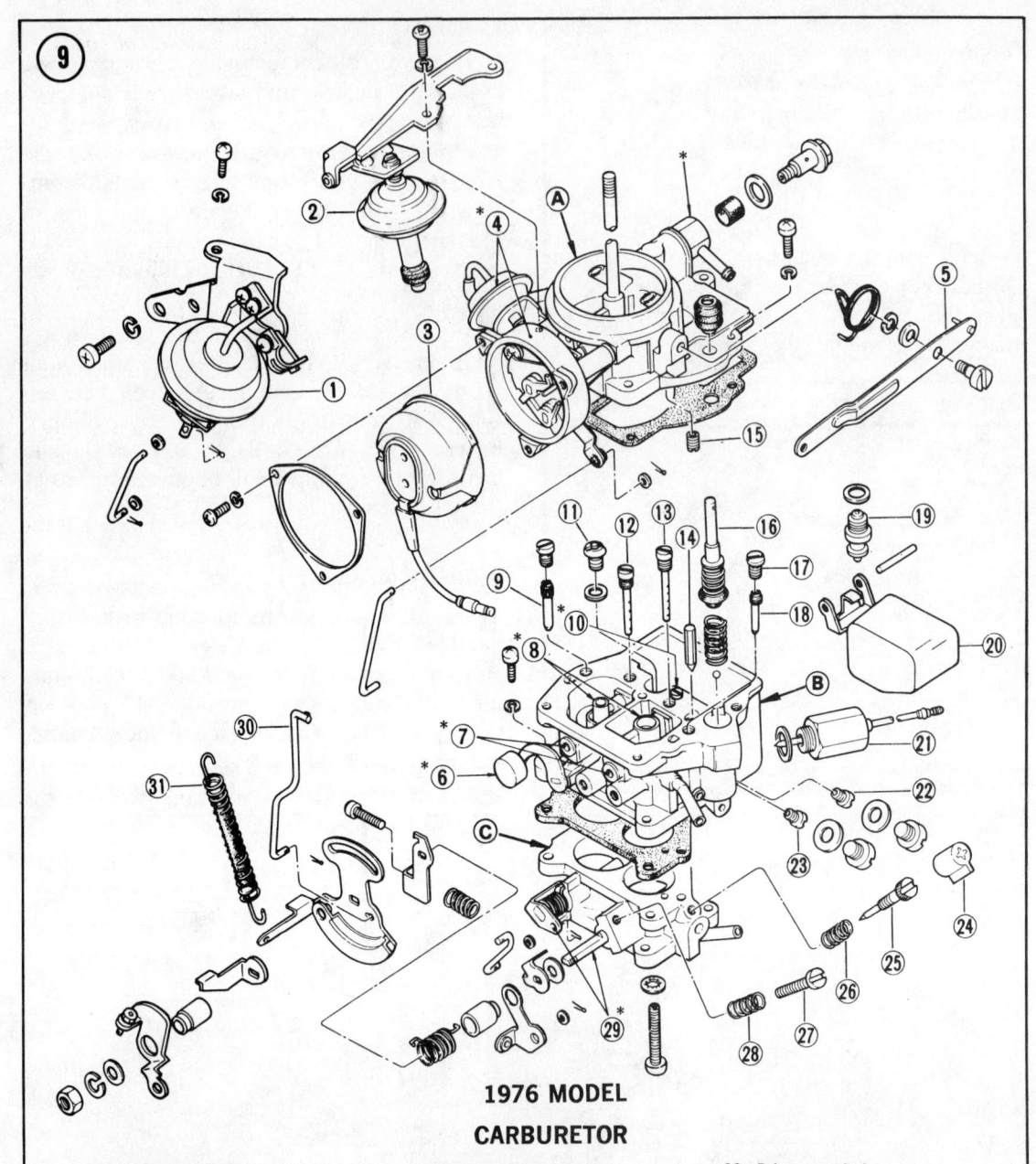

- Ⓐ Choke chamber
- Ⓑ Center body
- Ⓒ Throttle chamber
- 1. Servo diaphragm of throttle opener
- 2. Dash pot
- 3. Automatic choke cover
- *4. Automatic choke body and diaphragm chamber
- 5. Accelerating pump lever
- *6. Auxiliary valve
- *7. Venturi stopper screw
- *8. Primary and secondary small venturi
- 9. Secondary slow jet
- *10. Safe orifice
- 11. Power valve
- 12. Secondary main air bleed
- 13. Primary main air bleed
- 14. Injector weight
- 15. Primary slow air bleed
- 16. Accelerating pump
- 17. Plug
- 18. Primary slow jet
- 19. Needle valve
- 20. Float
- 21. Anti-dieseling solenoid valve
- 22. Primary main jet
- 23. Secondary main jet
- 24. Idle limiter cap
- 25. Idle adjust screw
- 26. Spring
- 27. Throttle adjust screw
- 28. Spring
- *29. Primary and secondary throttle valve
- 30. Accelerating pump rod
- 31. Throttle return spring

Note: Do not remove the parts marked with an asterisk "*"

# FUEL AND EXHAUST SYSTEMS

## CARBURETOR

All models use a 2-barrel, downdraft carburetor. The carburetor includes a throttle opener, electric automatic choke, and anti-dieseling solenoid. The throttle opener admits extra fuel-air mixture to the intake manifold during deceleration, when the mixture is normally too lean to burn. This reduces unburned hydrocarbons in the exhaust. The anti-dieseling solenoid blocks the carburetor slow circuit when the engine is shut off. This prevents the engine from running with the igintion off.

### Removal/Installation

1. Remove the air cleaner as described earlier.
2. Label and disconnect the fuel and vacuum lines. Plug the fuel lines so they won't drip.
3. Disconnect the wires for choke heater and anti-dieseling solenoid.
4. Disconnect the throttle cable.
5. Remove 4 nuts securing the carburetor to the intake manifold. Take the carburetor off.
6. Installation is the reverse of removal. Use a new gasket.

### Disassembly

Refer to **Figure 9** (1976) or **Figure 10** (1977-on). During disassembly, observe the following.

1. When removing jets, note their locations and the number stamped in each jet.
2. Make sure wrenches and screwdrivers fit exactly.
3. Lay all parts in order to ease reassembly.
4. Do not remove parts marked with an asterisk.
5. Do not remove linkage parts from the throttle shafts unless they are bent or otherwise damaged. Be sure replacement parts are available before removing.

### Inspection

1. Thoroughly clean all metal parts (except diaphragms and solenoid) in solvent or dip-in carburetor cleaner. O-rings, gaskets, and diaphragms should be replaced if they are included in the repair kit. If not, clean the old ones with a lint-free cloth.

#### CAUTION
*Do not insert objects such as drill bits or pieces of wire into jets and passages while cleaning. These openings are carefully calibrated, and scratching them may seriously affect carburetor performance.*

2. If jets and passages are difficult to clean, blow them out with compressed air. If a compressor is not available, use a spray-on carburetor cleaner. These usually come with plastic tubes that fit into the can's nozzle, making it possible to spray the cleaner into small openings.
3. Check the needle valve and seat for wear. Replace as needed.
4. Check all castings for cracks.
5. Check the idle mixture screw for wear at the tip. Replace the screw if wear is detected.
6. Check the accelerator pump piston seal and cover for wear, damage, or deterioration. Replace as needed.
7. Inspect the vacuum break diaphragm (4, **Figure 9 or 10**). To do this, disconnect the vacuum break hose from the carburetor body. Hold the choke valve shut and suck on the hose. There should be a strong pull on the choke valve.
8. Test the solenoid. Connect a 12-volt battery between the solenoid wire and solenoid body. The solenoid should click each time current is applied.

### Assembly

Assembly is the reverse of disassembly, plus the following.

1. Use new gaskets and seals.
2. Be sure the main jets and air bleeds are installed in the correct holes. Refer to the numbers written down during disassembly, or see **Table 1**.
3. After assembly, adjust float level, fast idle, choke housing, and dashpot.

### Float Adjustment

1. Raise the float all the way. Measure the gap between float and float chamber (dimension

**CHAPTER FIVE**

**Table 1  FUEL SYSTEM SPECIFICATIONS**

**Jets and air bleeds**

1976
- Primary main jet — #103
- Primary main air bleed — #95
- Primary slow jet — #45
- Secondary main jet — #145
- Secondary main air bleed — #80
- Secondary slow jet — #50
- Power valve — #40

1977-1978
- Primary main jet — #106 (California)
  #105 (49 states, 1977 Canada)
  #103 (1978 Canada)
- Primary main air bleed — #95
- Primary slow jet — #45
- Secondary main jet — #145
- Secondary main air bleed — #80
- Secondary slow jet — #50
- Power valve — #40

1979
- Primary main jet — #107 (California)
  #105 (49 states, Canada)
- Primary main air bleed — #95 (California, Canada)
  #110 (49 states)
- Primary slow jet — #45
- Secondary main jet — #145
- Secondary main air bleed — #80
- Secondary slow jet — #50
- Power valve — #43 (California)

**Fast idle gap**

| | |
|---|---|
| 1976 | 0.031-0.035 in. (0.8-0.88mm) |
| 1977 and later | 0.029-0.034 in. (0.73-0.87mm) |

**Fast idle speed**

| | |
|---|---|
| 1976 | 2,450-2,650 rpm |
| 1977 and later | 1,900-2,700 rpm |

**EGR operating temperature**

| | |
|---|---|
| 1976-1977 | 135-145 °F (57-63 °C) |
| 1978 and later | |
| U.S. | 122-145 °F (50-63 °C) |
| Canada | 104-145 °F (40-63 °C) |

**Dash Pot Adjusting Speed**

| | |
|---|---|
| 1976-1977 | |
| California | 2300-2400 rpm |
| Non-California | 1900-2000 rpm |
| 1978-on | |
| California | 1900-2100 rpm |
| Non-California | 2300-2500 rpm |

# FUEL AND EXHAUST SYSTEMS

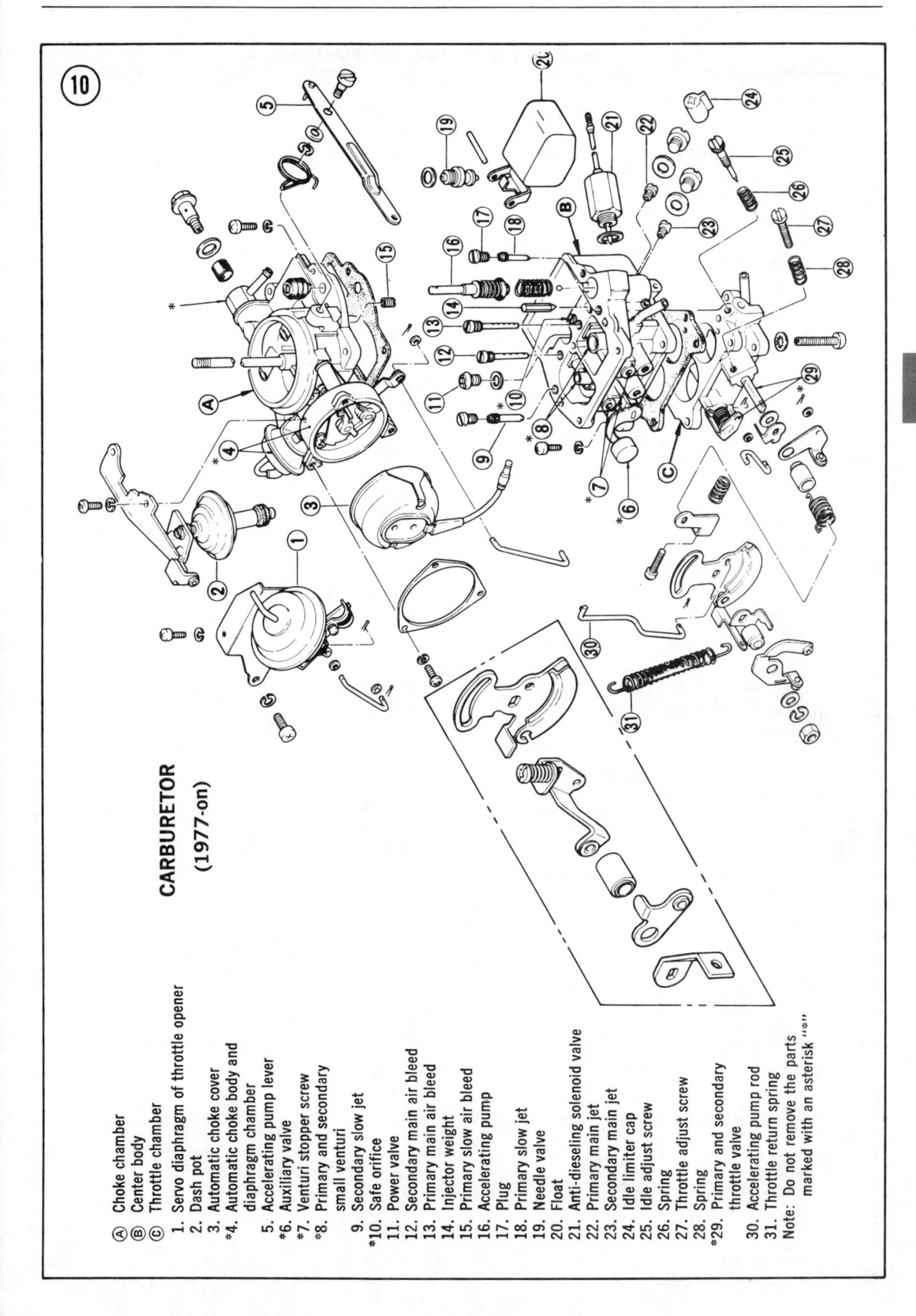

## CARBURETOR (1977-on)

Ⓐ Choke chamber
Ⓑ Center body
Ⓒ Throttle chamber
1. Servo diaphragm of throttle opener
2. Dash pot
3. Automatic choke cover
*4. Automatic choke body and diaphragm chamber
5. Accelerating pump lever
*6. Auxiliary valve
*7. Venturi stopper screw
*8. Primary and secondary small venturi
9. Secondary slow jet
*10. Safe orifice
11. Power valve
12. Secondary main air bleed
13. Primary main air bleed
14. Injector weight
15. Primary slow air bleed
16. Accelerating pump
17. Plug
18. Primary slow jet
19. Needle valve
20. Float
21. Anti-dieseling solenoid valve
22. Primary main jet
23. Secondary main jet
24. Idle limiter cap
25. Idle adjust screw
26. Spring
27. Throttle adjust screw
28. Spring
*29. Primary and secondary throttle valve
30. Accelerating pump rod
31. Throttle return spring
Note: Do not remove the parts marked with an asterisk "*"

"H," **Figure 11**). It should be 0.59 in. (15mm). If necessary, slide the float off its pivot and bend the float seat to change it.

2. Let the float hang down. Raise the needle valve with a knife blade or similar tool. Measure the gap between needle valve and float seat (dimension "h," **Figure 11**) with a gauge rod or drill bit. It should be 0.051-0.067 in. (1.3-1.7mm). Adjust if necessary by bending the float stopper.

3. If the float level is adjusted correctly, fuel level will be even with the sight glass indicator (**Figure 12**) when the engine is idling.

## Choke Cover Adjustment

1. With the engine cold, floor the accelerator. The choke valve should close. It should open without binding when pushed with a finger.

2. Align the groove on the choke cover with the center mark on the choke housing. See **Figure 13**. Then tighten the 3 cover screws.

## Fast Idle Adjustment

Fast idle is the speed at which the engine operates when it is cold and the choke is closed. As the engine warms up, idle speed automatically returns to normal.

1. With the carburetor off the engine, place the fast idle arm on the second step of the fast idle cam. See **Figure 14**. Measure the gap between primary throttle valve and bore as shown. If incorrect (**Table 1**), turn the fast idle screw to change it.

2. With the carburetor on the engine, warm the engine to normal operating temperature. Place the fast idle arm on the second step of the fast idle cam and measure engine speed. If incorrect (**Table 1**), turn the fast idle screw to change it.

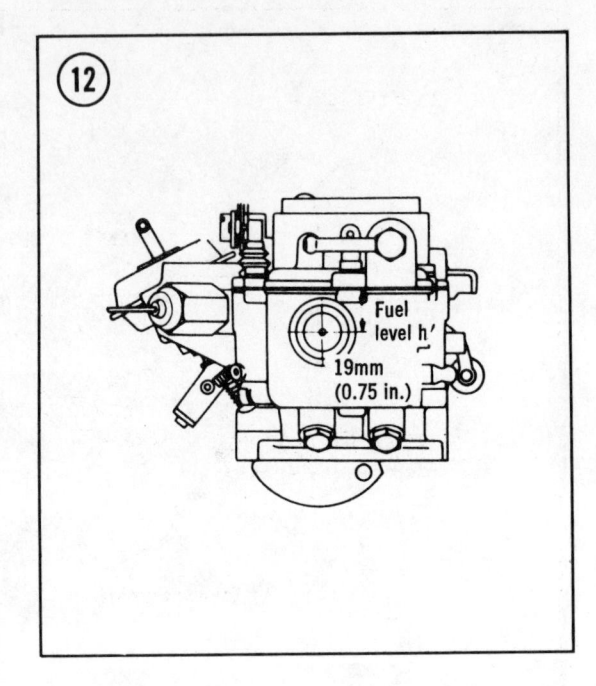

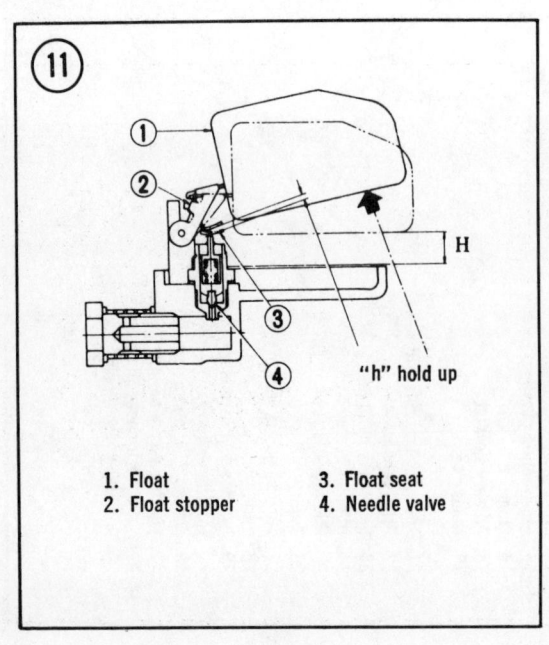

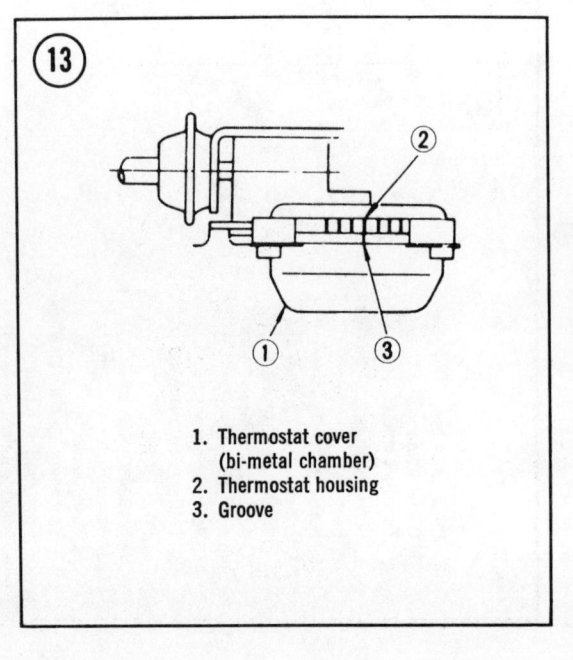

⑫

Fuel level h' 19mm (0.75 in.)

⑪

"h" hold up

1. Float
2. Float stopper
3. Float seat
4. Needle valve

⑬

1. Thermostat cover (bi-metal chamber)
2. Thermostat housing
3. Groove

# FUEL AND EXHAUST SYSTEMS

## Dashpot Adjustment

1. Make sure the engine is tuned. Warm the engine to normal operating temperature. Connect a tune-up tachometer to the engine.

2. Turn the throttle by hand until the throttle lever just touches the dashpot **(Figure 15)**. Note engine speed.

3. If engine speed is incorrect **(Table 1)**, loosen the locknut. Turn the dashpot **(Figure 16)** to adjust engine speed. Then tighten the locknut.

## AUTOMATIC CHOKE CIRCUIT

The choke is heated by an electrical coil. If the choke doesn't open properly, check the linkage for binding. If the linkage is OK, check the choke circuit as follows.

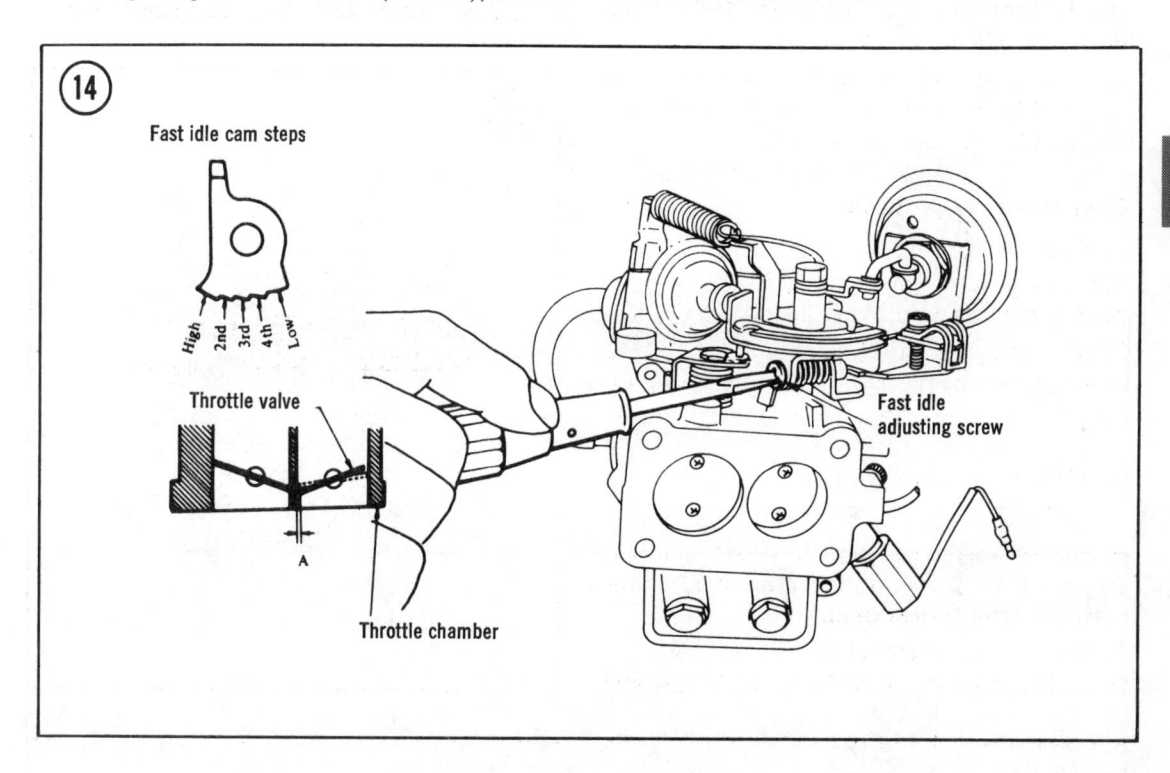

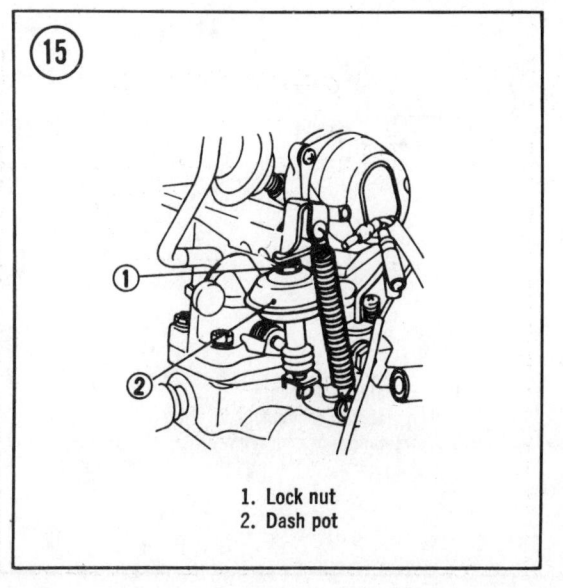

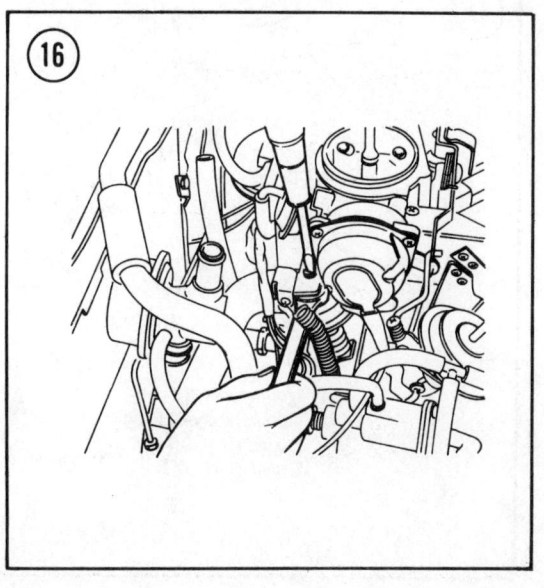

⑭ Fast idle cam steps
High 2nd 3rd 4th Low
Throttle valve
A
Throttle chamber
Fast idle adjusting screw

⑮
1. Lock nut
2. Dash pot

⑯

CHAPTER FIVE

⑰

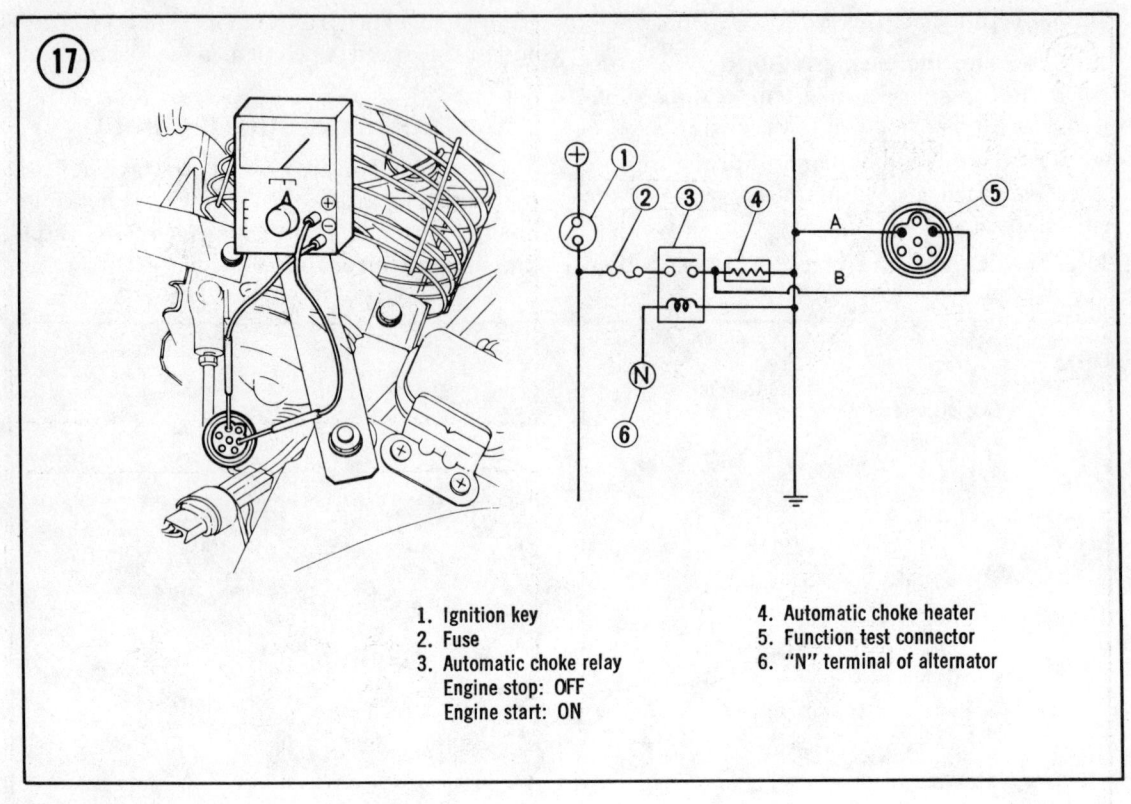

1. Ignition key
2. Fuse
3. Automatic choke relay
   Engine stop: OFF
   Engine start: ON
4. Automatic choke heater
5. Function test connector
6. "N" terminal of alternator

⑱

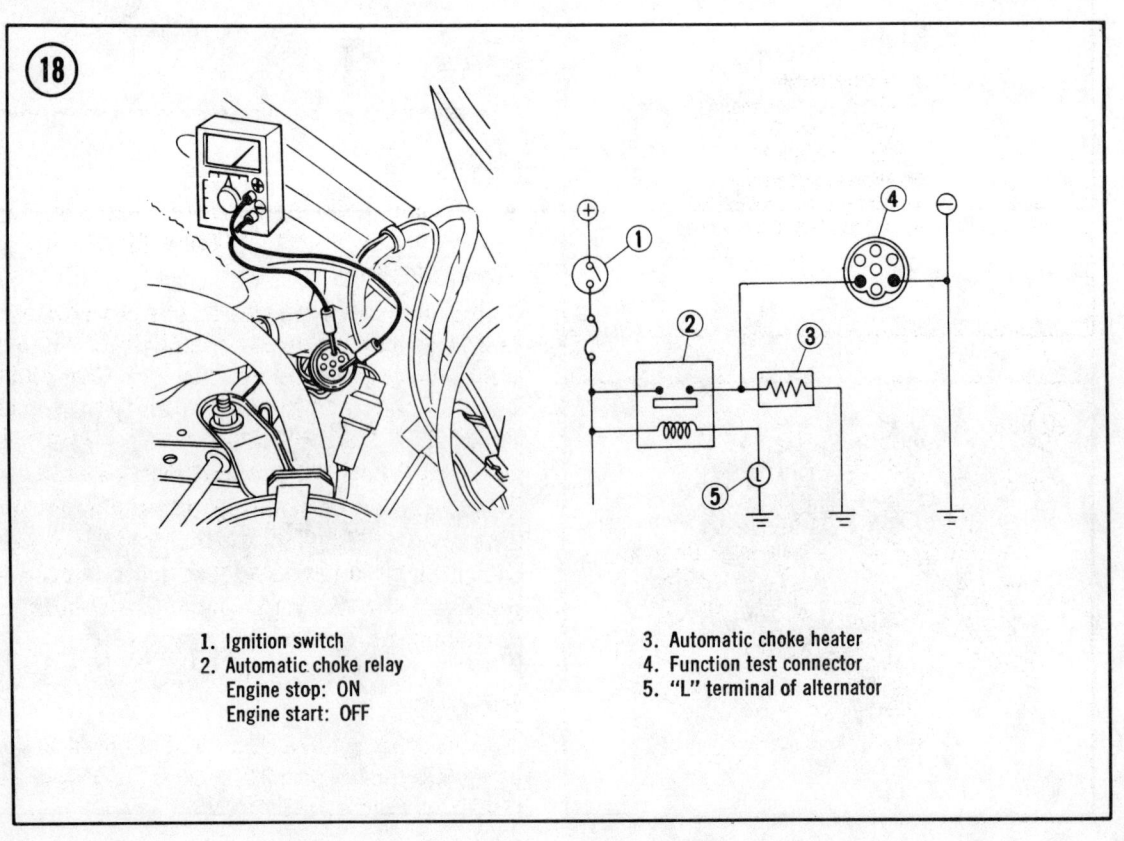

1. Ignition switch
2. Automatic choke relay
   Engine stop: ON
   Engine start: OFF
3. Automatic choke heater
4. Function test connector
5. "L" terminal of alternator

# FUEL AND EXHAUST SYSTEMS

1. Ignition switch
2. Automatic choke relay
   Engine stop: ON
   Engine start: OFF
3. Automatic choke heater
4. Function test connector
5. "L" terminal of alternator

Automatic choke relay

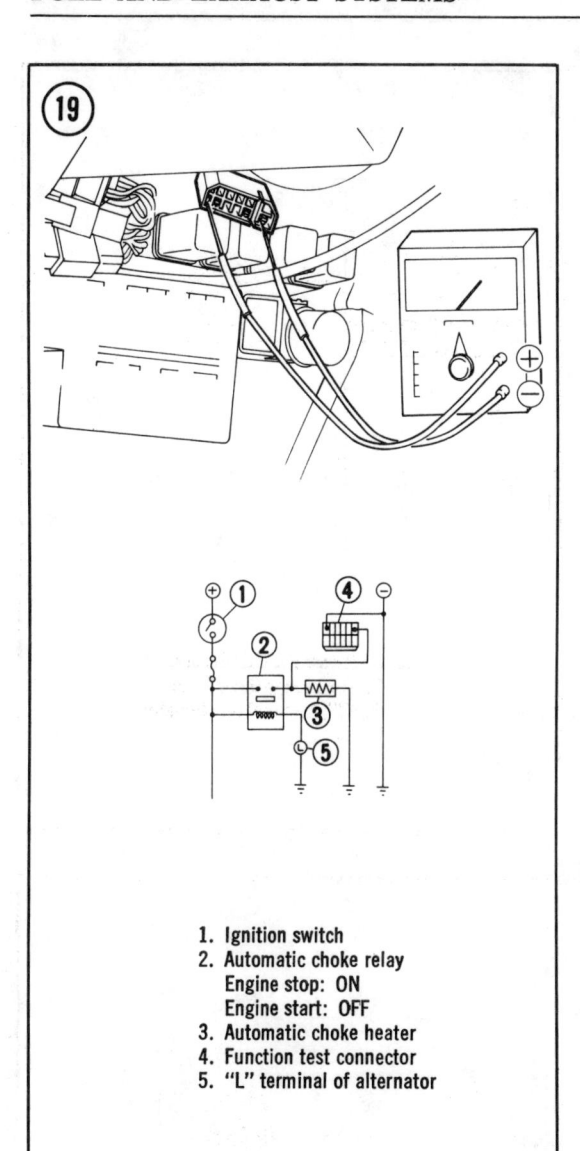

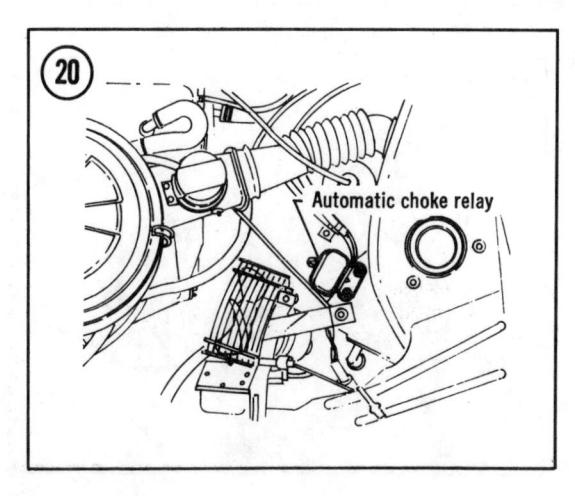

## Circuit Test

1. Disconnect the function test connector. See **Figure 17** (1976-1977), **Figure 18** (1978), or **Figure 19** (1979).

2. Connect an ohmmeter or self-powered test lamp to the test connector terminals shown. Do not connect to any other terminals. The ohmmeter should show continuity (test lamp should light). If it does, the choke circuit is good. If not, look for a broken or disconnected wire.

3. Disconnect the ohmmeter or test lamp. Connect a voltmeter in its place. With the engine idling, there should be 12 volts across the 2 terminals. If not, the cause may be a broken wire, bad connection, or faulty choke relay.

## Relay Test

1. Locate the choke relay and disconnect its connector. See **Figure 20** (1976-1977), **Figure 21** (1978), or **Figure 22** (1979).

2. On 1976-1977 models, connect a 12-volt power source (such as the car's battery) between terminals 3 and 4 (**Figure 23**). An ohmmeter connected between terminals 1 and 2 should show continuity. When the 12-volt source is disconnected, there should be not continuity. If the relay doesn't perform properly, replace it.

3. On 1978 and later models, connect an ohmmeter between terminals 2 and 4 (**Figure 24**). There should be continuity. When a 12-volt battery is connected between terminals 5 and 6, continuity should stop. If not, replace the relay.

### Heater Coil Test

To test the heater coil, connect an ohmmeter between the coil body and coil wire terminal (**Figure 25**). Resistance should be 3.7-8.9 ohms. If not, replace the coil.

## CARBURETOR COOLING FAN

An auxiliary fan blows air on the carburetor to cool it during hot underhood conditions. When coolant temperature reaches 158-165°F (70-74°C), the fan comes on. If the fan doesn't work, check the fuse, then the wiring, then the fan, then the temperature sensor. **Figure 26** shows the system.

### Fan Test

To test the fan disconnect its wiring connector. Hook the fan directly to the battery with jumper wires. If the fan runs, it is good; if not, replace it.

### Temperature Sensor Test

1. Drain the cooling system. See *Cooling System Flushing*, Chapter Six.
2. On F10's, lift the windshield washer reservoir off its bracket.
3. Disconnect the wires from the switch (**Figure 27**). Unscrew it from the hose.

> NOTE: *Be sure to remove the auxiliary fan switch, not the radiator fan switch.*

4. Connect an ohmmeter or self-powered test lamp to the switch terminals. Immerse the

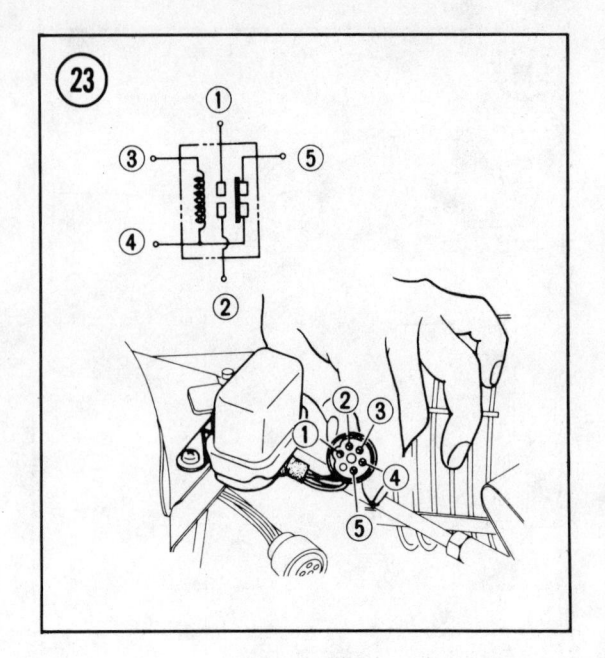

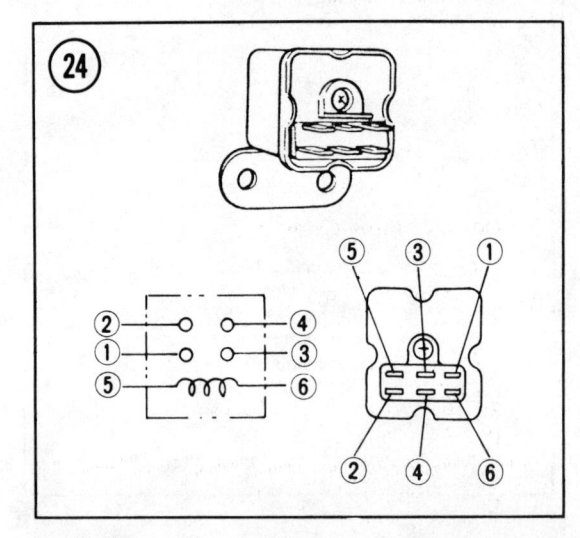

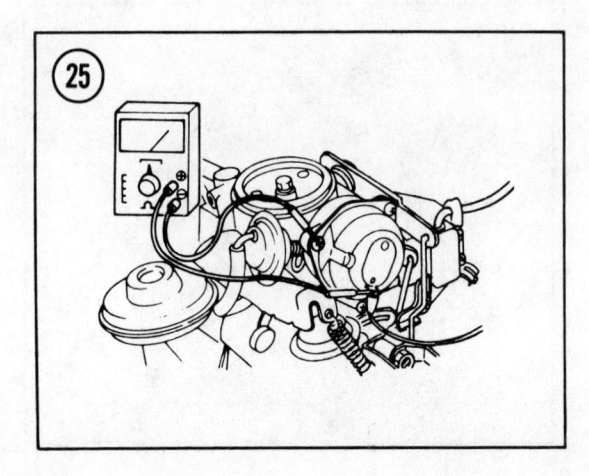

# FUEL AND EXHAUST SYSTEMS

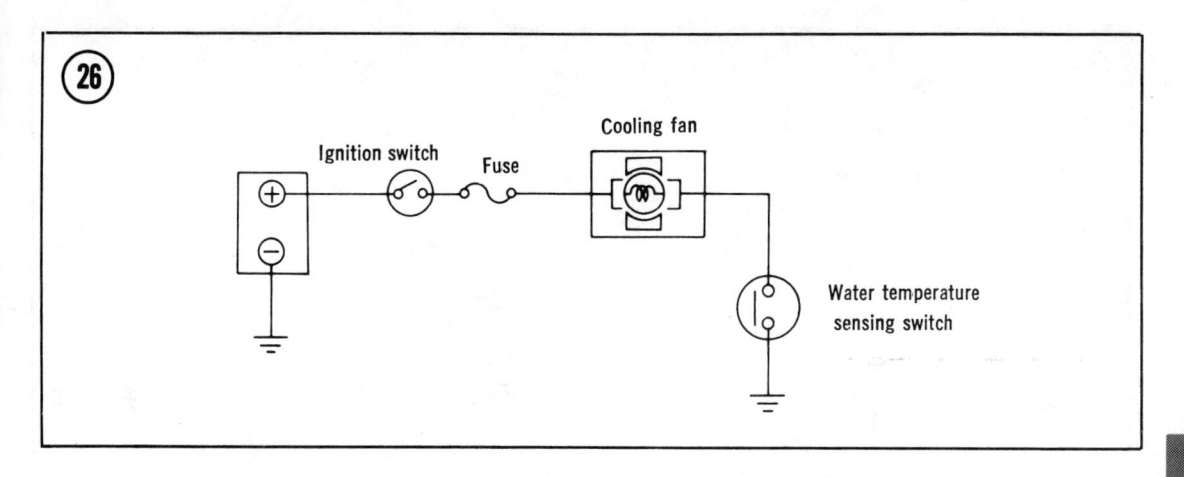

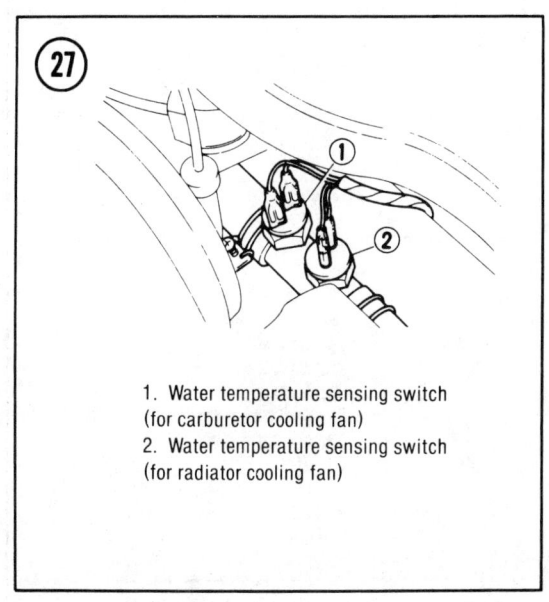

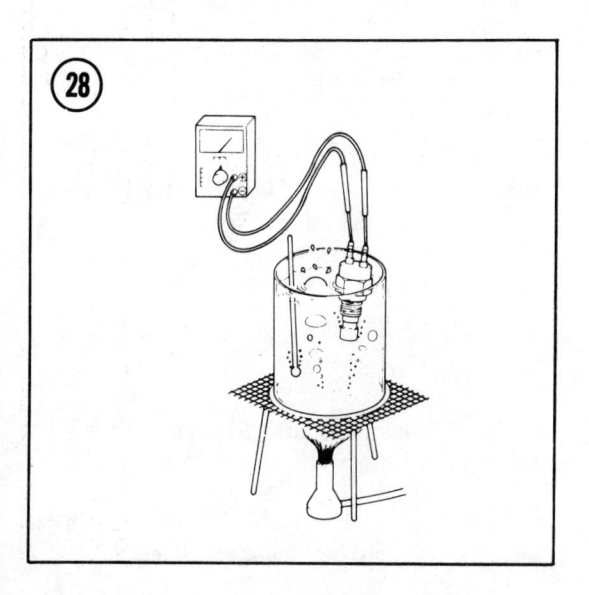

**(26)**

switch in water with a thermometer (**Figure 28**). Heat the water. At some point between 158 and 165°F (70-74°C), the switch should come on (ohmmeter should show continuity or test lamp should light). Below 145-160°F (63-71°C), the switch should go off. If the switch doesn't perform properly, replace it.

## FUEL PUMP

The fuel pump is a mechanical type mounted on the right side of the engine block and driven by the camshaft. Testing is described in Chapter Two, under *Fuel System*.

On 1976-1977 models, capacity is 1 1/4 pt. (600cc) of fuel in one minute at 1,000 rpm. On 1978 and later cars, capacity is one pt. (450cc) of fuel in one minute at 1,000 rpm.

On all models, static pressure is 3.0-3.8 psi (0.21-0.27 kg/cm²).

**(27)**

1. Water temperature sensing switch (for carburetor cooling fan)
2. Water temperature sensing switch (for radiator cooling fan)

### Removal/Installation

1. Place a container beneath the fuel pump to catch dripping fuel.
2. Disconnect the lines from the pump.

### CAUTION
*Plug the inlet line so it doesn't siphon fuel from the tank.*

**(28)**

3. Remove the fuel pump installation nuts. Take the pump out.
4. Installation is the reverse of removal. Use new gaskets, coated on both sides with gasket sealer. Tighten the fuel pump nuts to 7-10 ft.-lb. (0.9-1.4 mkg).

**FUEL PUMP**

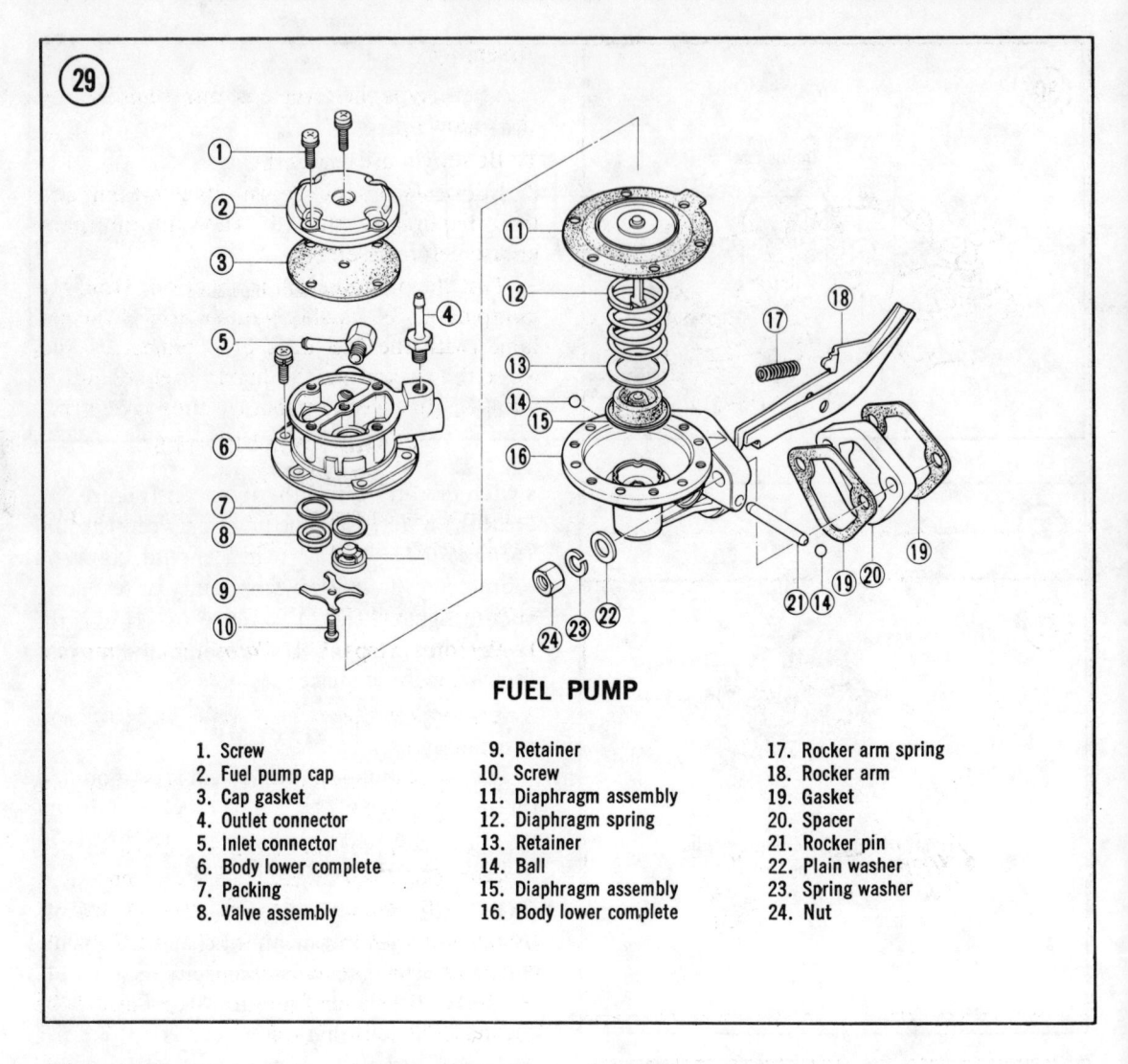

1. Screw
2. Fuel pump cap
3. Cap gasket
4. Outlet connector
5. Inlet connector
6. Body lower complete
7. Packing
8. Valve assembly
9. Retainer
10. Screw
11. Diaphragm assembly
12. Diaphragm spring
13. Retainer
14. Ball
15. Diaphragm assembly
16. Body lower complete
17. Rocker arm spring
18. Rocker arm
19. Gasket
20. Spacer
21. Rocker pin
22. Plain washer
23. Spring washer
24. Nut

## Disassembly

Refer to **Figure 29** for this procedure.

1. Remove the body setscrews and washers. Separate the upper and lower halves of the pump body.

2. Remove the top cover screws. Take the top cover off.

3. Unscrew the fuel line fittings.

4. Remove the valve retainer screw, then remove both valves.

5. Unhook the diaphragm. To do this, push the diaphragm downward and tilt it until the diaphragm rod slips off the rocker arm. The diaphragm can then be lifted out.

6. Tap the pivot pin out of the operating lever with a hammer and pin punch.

## Inspection

1. Clean all parts in solvent before inspection. Discard all gaskets.

2. Check the valves and springs for wear or damage. Blow through the valves. If they are good, air will travel only one way through them. If air can be blown through both sides of a valve, replace it.

3. Check the diaphragm for small holes, cracks, and wear.

4. Check the rocker arm for wear at its pivot point and at both ends.

5. Check the pivot pin in the rocker arm for wear. A worn pin can cause an oil leak.

# FUEL AND EXHAUST SYSTEMS

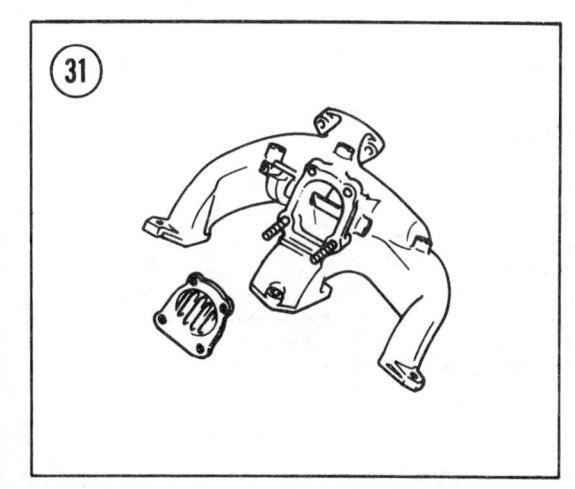

## Assembly

Assembly is the reverse of disassembly, plus the following.

1. Be sure to use new gaskets.
2. Lubricate the rocker arm, its pivot pin, and the diaphragm pullrod with multipurpose grease before assembly.
3. Test the pump before installation. Hold the pump about 3 ft. higher than the top of the gas tank (with the fuel inlet line connected) and work the rocker arm by hand. Fuel should be pumped after several strokes of the rocker arm.

## INTAKE AND EXHAUST MANIFOLDS

**Figure 30** shows the intake manifold; **Figure 31** shows the exhaust manifold.

### Removal/Installation

1. Perform Steps 1-4, *Carburetor Removal/Installation*.
2. Disconnect the exhaust pipe from the exhaust manifold.
3. Label and disconnect the intake manifold vacuum lines.
4. Disconnect the hoses from the EGR valve. See *Exhaust Gas Recirculation System* later in this chapter. If the manifolds need to be separated, detach the exhaust gas tube running from intake manifold to exhaust manifold.
5. Remove the manifold attaching nuts. Take the manifolds off (**Figure 32**).
6. Install in the reverse order. Use a new gasket. Tighten the manifold nuts to 11-14 ft.-lb. (1.5-2.0 mkg).

## THROTTLE OPENER

The throttle opener is designed to open the throttle slightly when the engine is decelerating. This allows air-fuel mixture into the engine, so sufficient combustion can take place to minimize the unburned hydrocarbons in the exhaust.

The system does not normally require adjustment. If there is a problem, it usually causes the engine to idle too quickly. System adjustment is a complicated procedure that should be left to a dealer or mechanic familiar with Datsun emission controls.

## EXHAUST GAS RECIRCULATION SYSTEM

The EGR system recirculates part of the exhaust gas into the combustion chambers for burning. This lowers combustion temperature, reducing the oxides of nitrogen in the exhaust. The 1976 non-California models, and 1977 Canadian models, require periodic inspection and cleaning of the system. This is because they run on leaded gasoline, which can form deposits. Cars which use unleaded gasoline do not require periodic service.

### System Inspection

Refer to the following illustrations:

**Figure 33** — All 1976 models, 1977 Canadian models

**Figure 34** — 1977 U.S. models

**33** EGR SYSTEM (ALL 1976, 1977 CANADA)

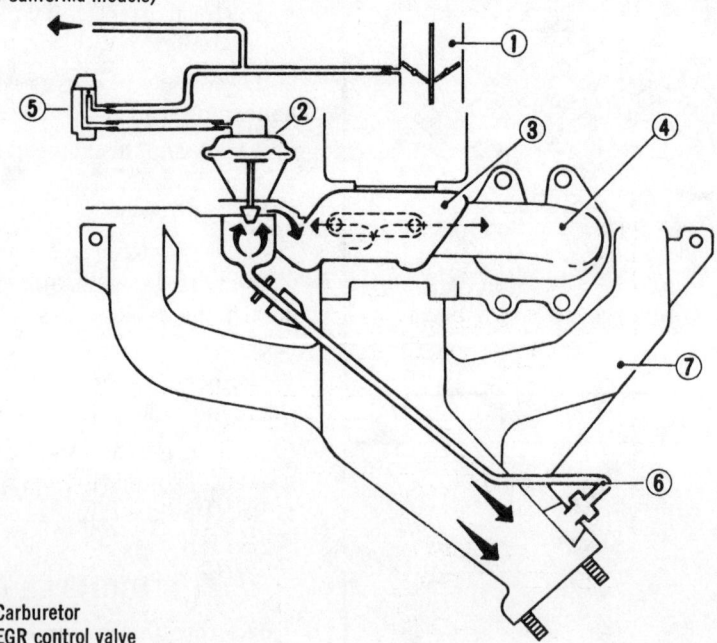

1. Carburetor
2. EGR control valve
3. EGR passage
4. Intake manifold
5. Thermal vacuum valve
6. EGR tube
7. Exhaust manifold

# FUEL AND EXHAUST SYSTEMS

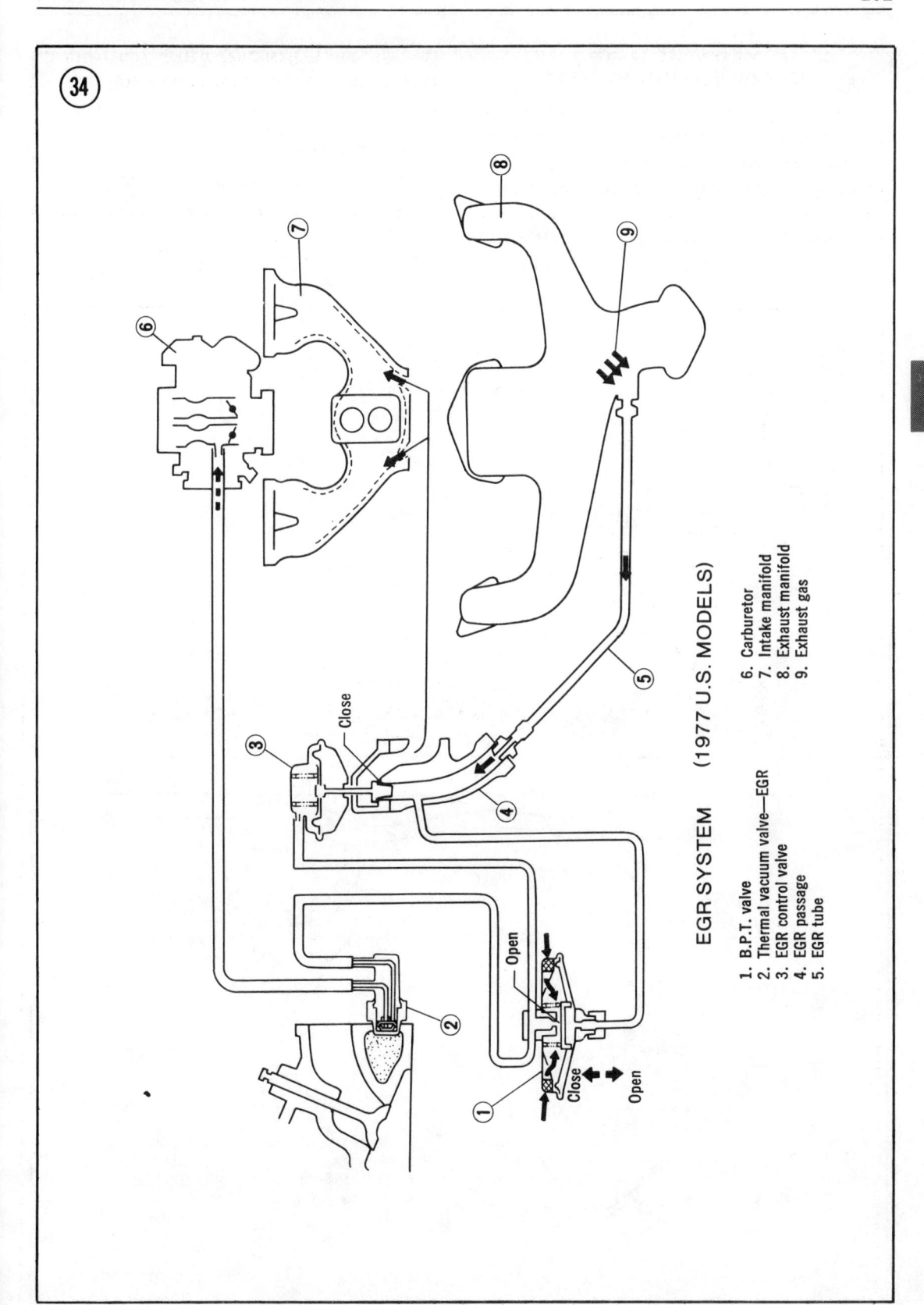

**EGR SYSTEM (1977 U.S. MODELS)**

1. B.P.T. valve
2. Thermal vacuum valve—EGR
3. EGR control valve
4. EGR passage
5. EGR tube
6. Carburetor
7. Intake manifold
8. Exhaust manifold
9. Exhaust gas

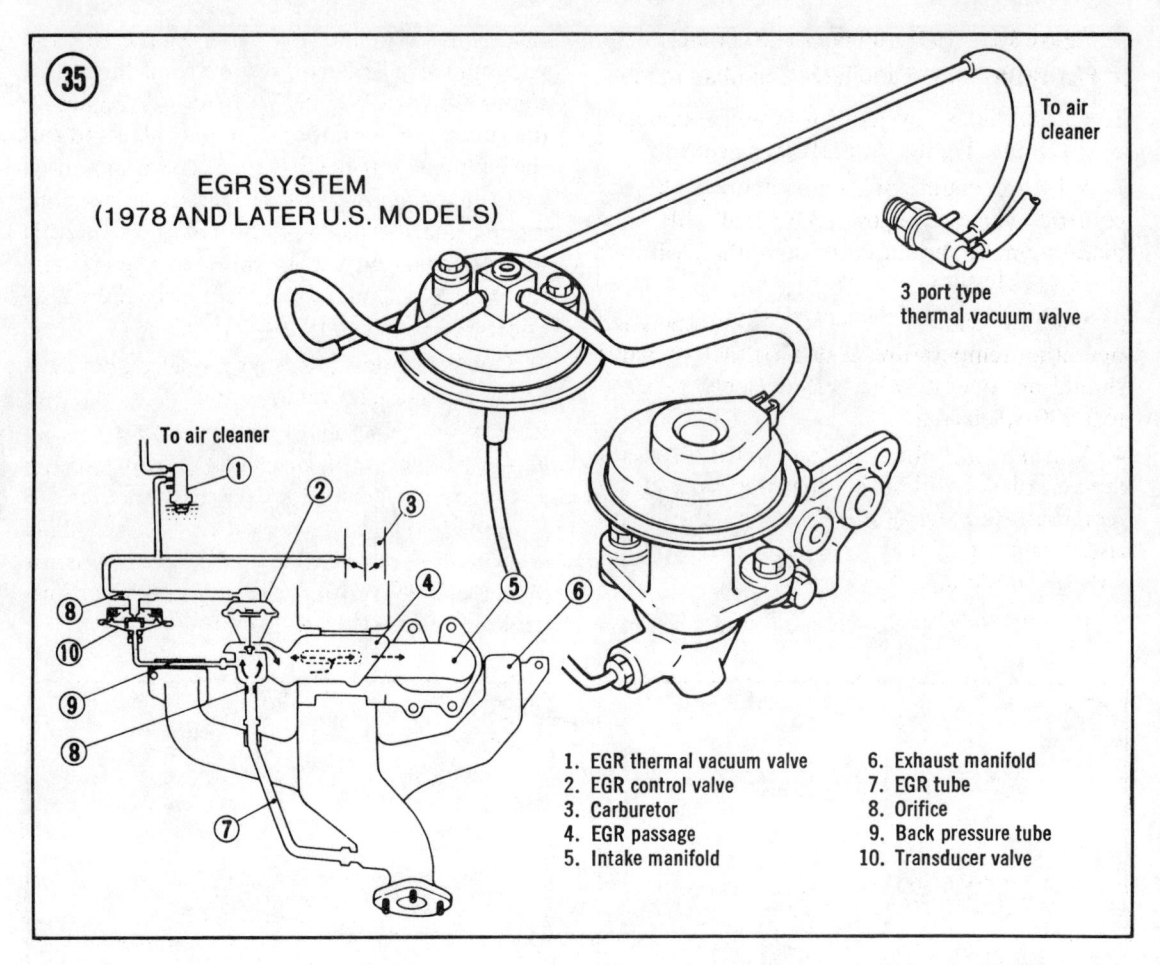

CHAPTER FIVE

③⑤ **EGR SYSTEM (1978 AND LATER U.S. MODELS)**

To air cleaner

3 port type thermal vacuum valve

To air cleaner

1. EGR thermal vacuum valve
2. EGR control valve
3. Carburetor
4. EGR passage
5. Intake manifold
6. Exhaust manifold
7. EGR tube
8. Orifice
9. Back pressure tube
10. Transducer valve

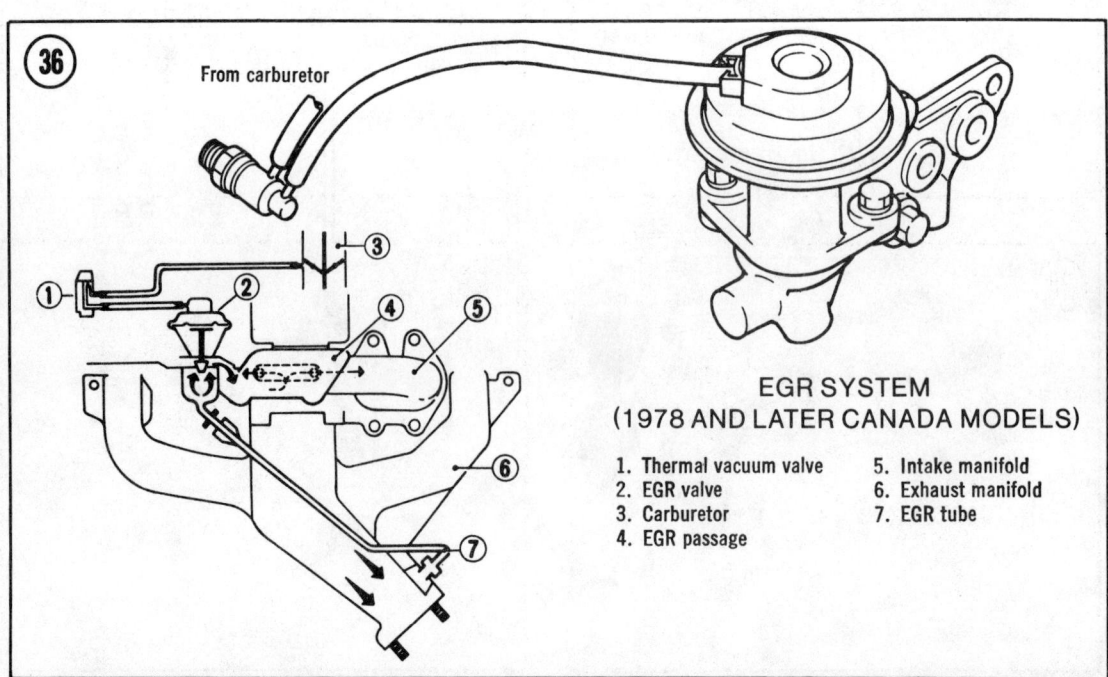

③⑥ From carburetor

**EGR SYSTEM (1978 AND LATER CANADA MODELS)**

1. Thermal vacuum valve
2. EGR valve
3. Carburetor
4. EGR passage
5. Intake manifold
6. Exhaust manifold
7. EGR tube

# FUEL AND EXHAUST SYSTEMS

**Figure 35** — 1978 and later U.S. models
**Figure 36** — 1978 and later Canadian models

1. Check the EGR system for loose or damaged connections. Tighten or replace as needed.

2. With the engine off, reach beneath the EGR control valve (**Figure 37**) and lift the diaphragm. It should move smoothly, without sticking or binding.

3. Start the engine. While it is below specified operating temperature (**Table 1**), the EGR valve should not operate when engine speed is raised to 3,000-3,500 rpm.

4. Warm up the engine until coolant temperature is above specified operating temperature. The EGR valve's diaphragm should rise when engine speed is increased to 3,000-3,500 rpm.

5. If the valve doesn't rise, disconnect the hose that runs from the EGR valve to the thermal vacuum valve or back pressure transducer. See **Figure 38** (all 1976; 1977 Canada) or **Figure 39** (all others). Place a thumb on the valve and run the engine at 3,000-3,500 rpm. There should be suction at the hose end. If there is no suction, replace the thermal vacuum valve. If there is suction, or if you are not sure, remove and test the thermal vacuum valve, back pressure transducer (if so equipped), and EGR valve.

6. On 1976 non-California models, and 1977 Canadian models, remove and clean the EGR valve as described later.

7. On 1976 non-California models, turn off the EGR warning light. To do this, remove the grommet from the detector drive counter. This is located at the right rear of the engine compartment (**Figure 40**). Insert a screwdriver into the hole and push the reset button.

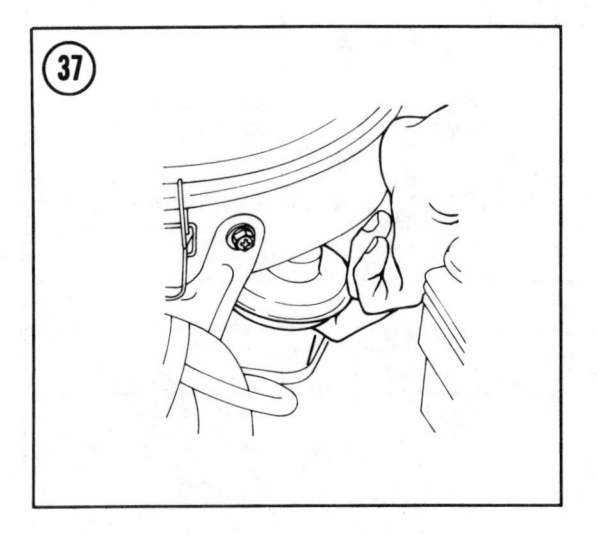

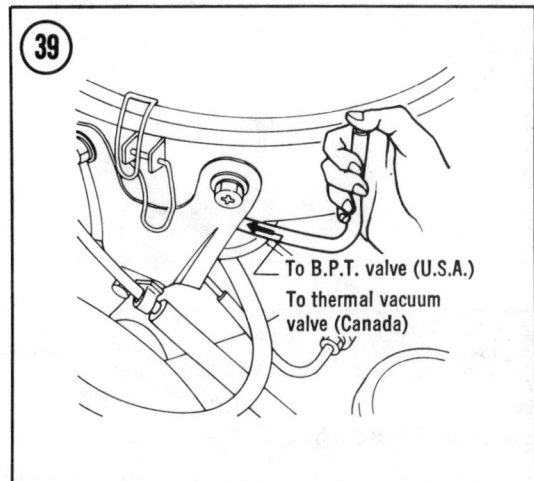

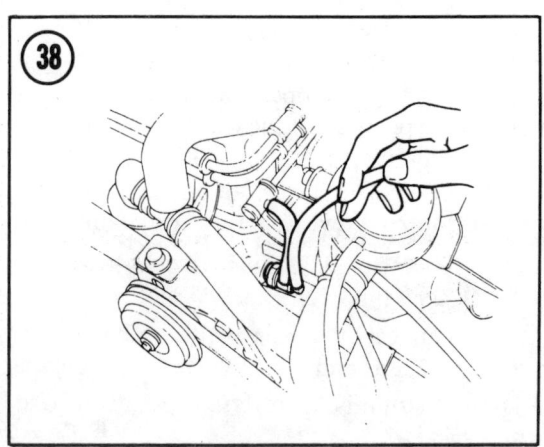

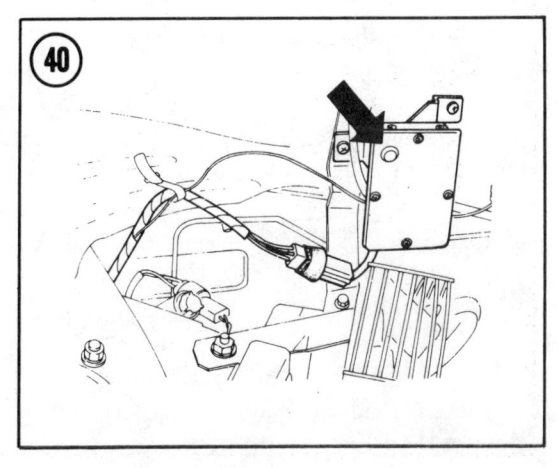

## Thermal Vacuum Valve Removal/Installation

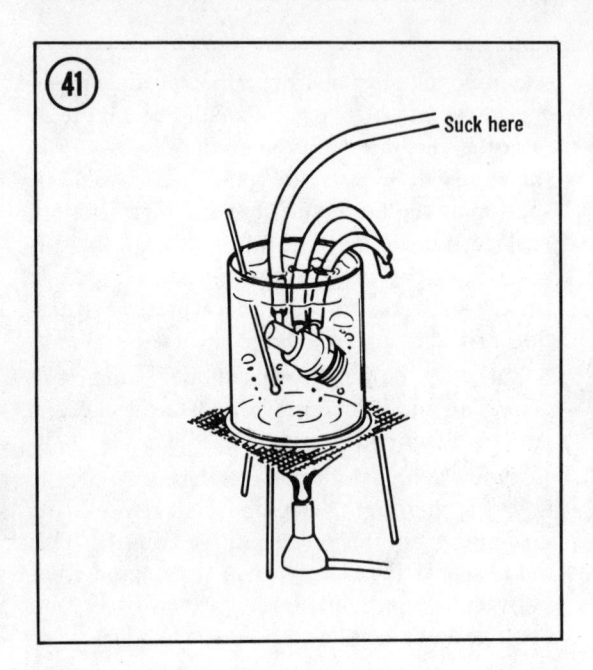

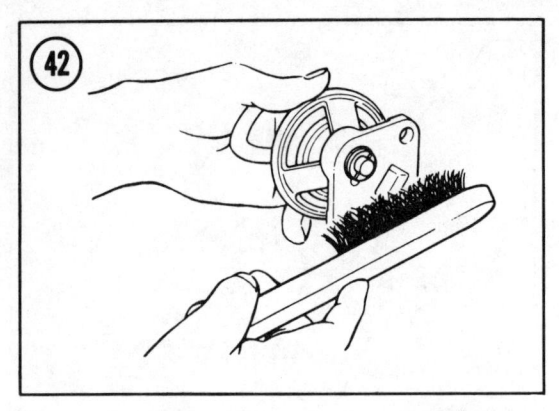

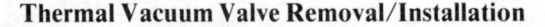

1. Drain the cooling system by opening the tap at the bottom of the radiator. If the coolant is clean, drain it into a clean container for reuse.

2. On 1977 and later models, remove one manifold nut and take off the heat shield.

3. Disconnect the thermal vacuum valve hoses. Unscrew the valve from the engine and lift it out.

4. Installation is the reverse of removal. Use gasket sealer on the valve threads. Tighten to 16 ft.-lb. (2.2 mkg) or less.

## Thermal Vacuum Valve Test

1. Place the valve in water with a thermometer (**Figure 41**).

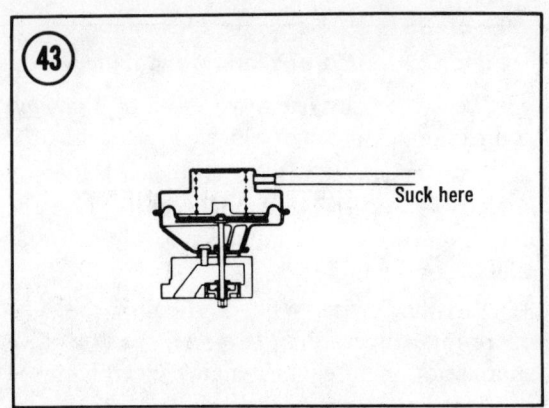

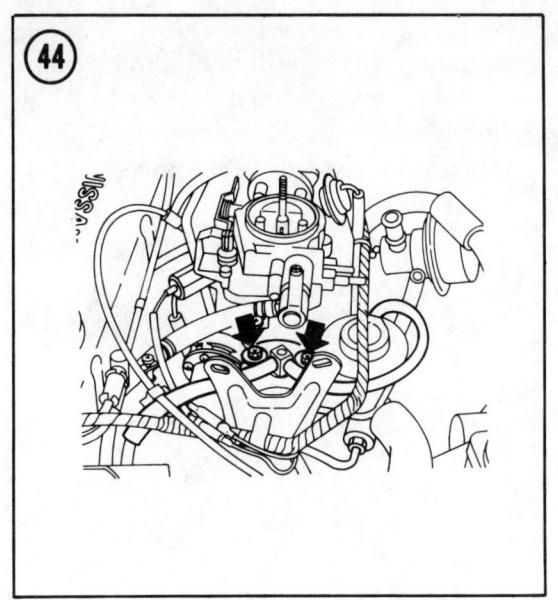

CAUTION
*Keep water out of the valve.*

2. Heat the water and try to suck air through the valve. This should be impossible below specified operating temperature (**Table 1**). Above specified operating temperature, it should be possible. If the valve doesn't perform properly, replace it.

## EGR Control Valve Removal/Installation

1. Remove the air cleaner as described under *Air Cleaner* in this chapter.

2. Disconnect the EGR valve vacuum line. Remove 2 nuts and lift the valve off the engine.

3. Installation is the reverse of removal. Use a new gasket.

# FUEL AND EXHAUST SYSTEMS

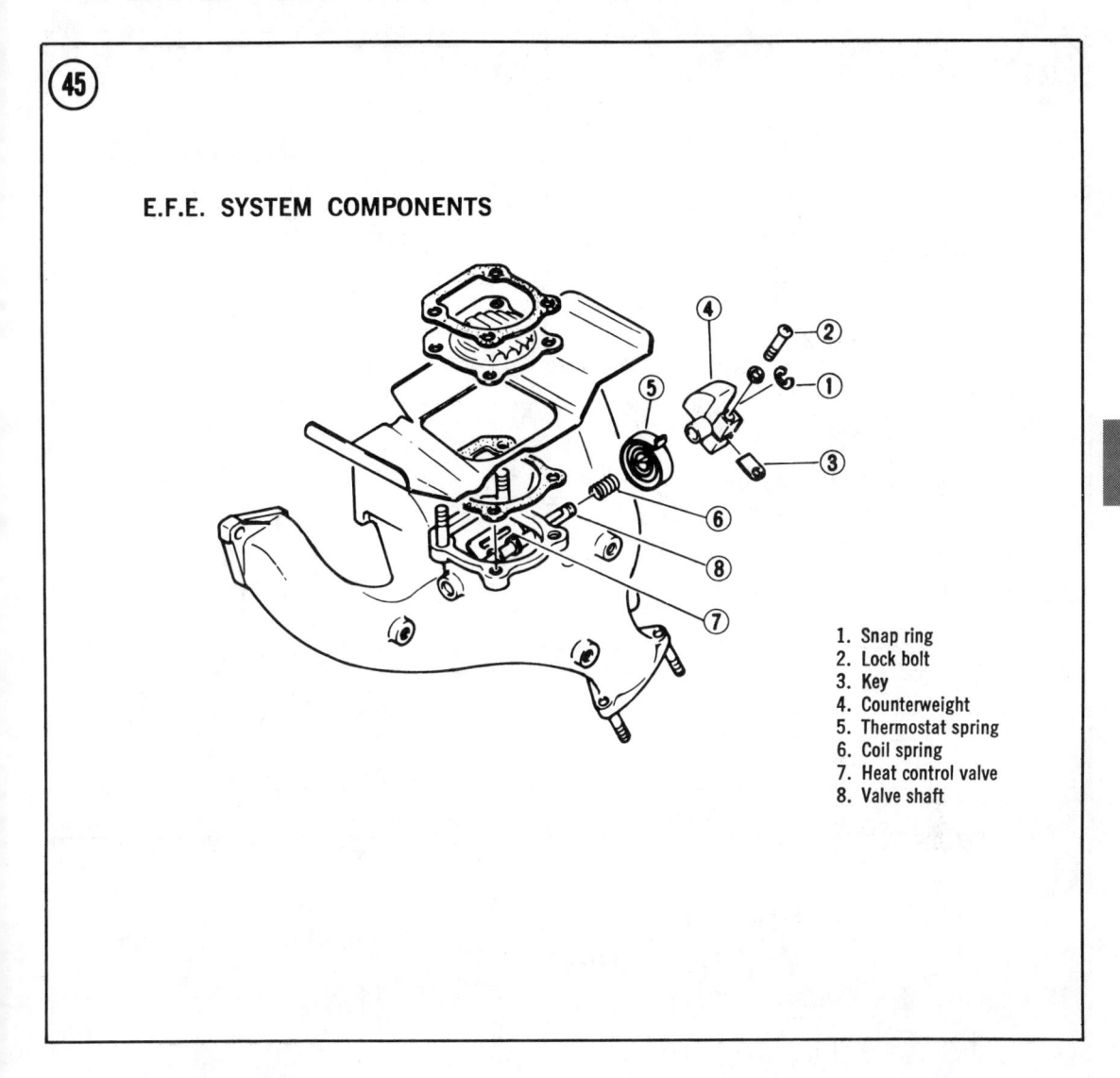

**E.F.E. SYSTEM COMPONENTS**

1. Snap ring
2. Lock bolt
3. Key
4. Counterweight
5. Thermostat spring
6. Coil spring
7. Heat control valve
8. Valve shaft

### EGR Control Valve Test

1. Remove the valve from the engine.

2. Clean the base of the valve with a wire brush and solvent. See **Figure 42**. Replace the valve if visibly damaged.

3. Connect a vacuum line to the valve and suck on it. See **Figure 43**. The valve diaphragm should rise, then stay up by itself for at least 30 seconds after vacuum is released. If it does not, replace the valve.

### Back Pressure Transducer Removal/Installation

1. Remove the air cleaner as described under *Air Cleaner* in this chapter.

2. Disconnect the vacuum lines from the transducer. Remove 2 screws (**Figure 44**) and lift it out.

3. Installation is the reverse of removal.

### EARLY FUEL EVAPORATIVE SYSTEM

This system is a thermostatically controlled flap valve in the exhaust manifold (**Figure 45**). When the engine is cold, it routes exhaust gas against the base of the intake manifold. This heats the incoming fuel mixture for more complete combustion. When the engine is warm, the valve routes the exhaust gas straight down the exhaust manifold. The system requires no regular service.

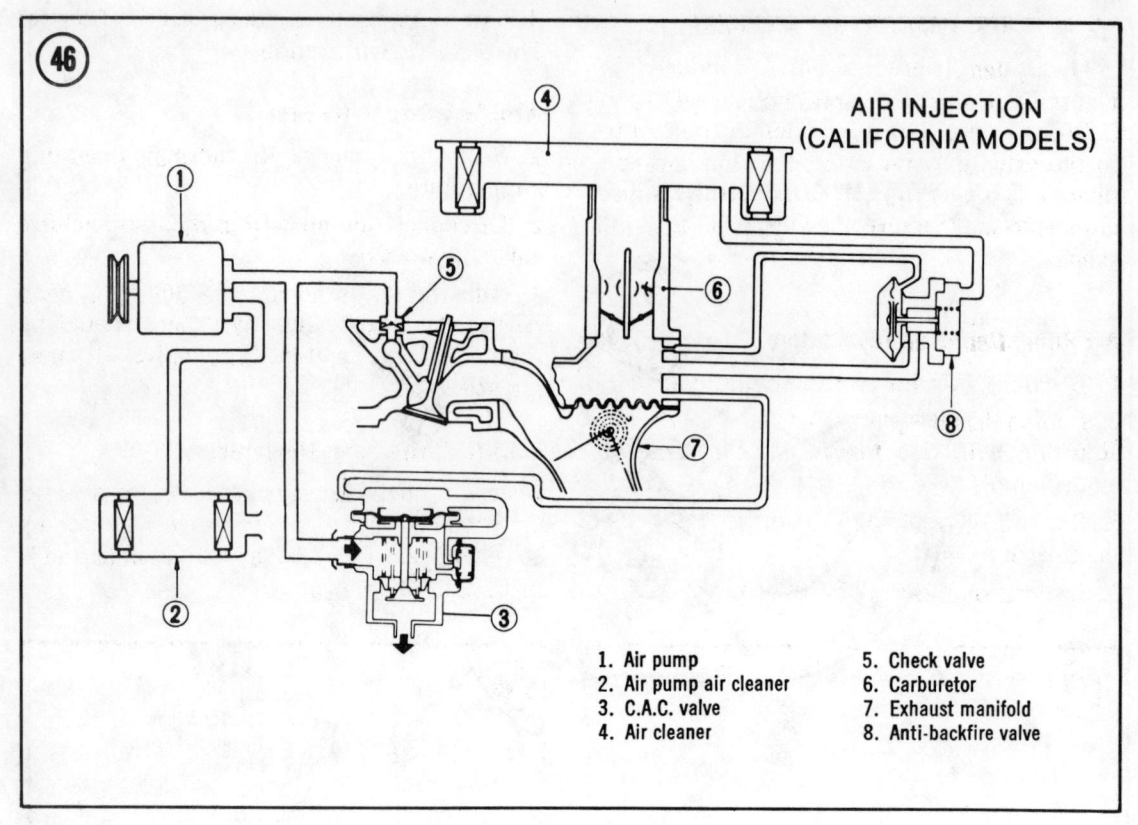

**AIR INJECTION (CALIFORNIA MODELS)**

1. Air pump
2. Air pump air cleaner
3. C.A.C. valve
4. Air cleaner
5. Check valve
6. Carburetor
7. Exhaust manifold
8. Anti-backfire valve

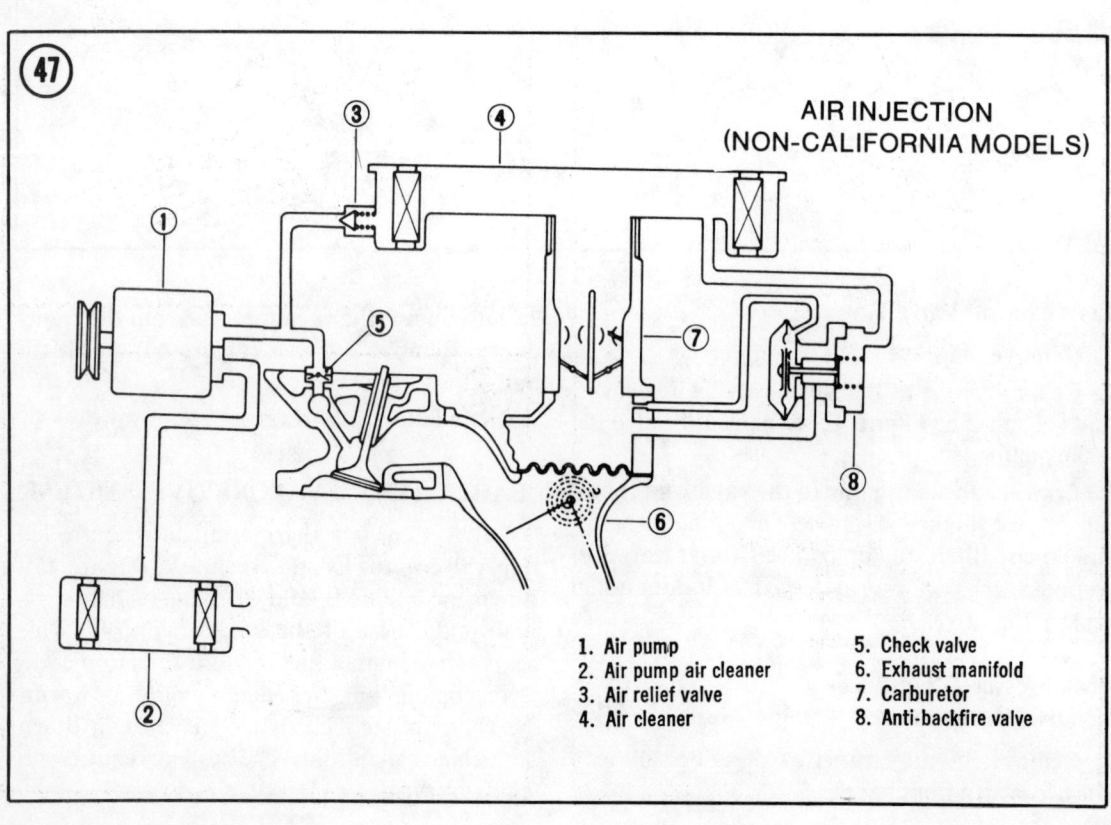

**AIR INJECTION (NON-CALIFORNIA MODELS)**

1. Air pump
2. Air pump air cleaner
3. Air relief valve
4. Air cleaner
5. Check valve
6. Exhaust manifold
7. Carburetor
8. Anti-backfire valve

# FUEL AND EXHAUST SYSTEMS

## AIR INJECTION SYSTEM

This system is used on all U.S. models. See **Figure 46** (California cars) or **Figure 47** (non-California cars). The system pumps fresh air into the exhaust ports so combustion can continue for a longer time. This reduces the carbon monoxide and unburned hydrocarbons in the exhaust.

### Air Pump Removal/Installation

1. Disconnect the hoses from the air pump.
2. Loosen the idler pulley locknut, then the belt adjusting bolt. See **Figure 48**. Take the air pump belt off.
3. Remove the air pump mounting bolts. Lift the air pump out.
4. Installation is the reverse of removal. Adjust the air pump belts as described in Chapter Three, *Drive Belts* section.

### Anti-Backfire Valve Test

1. Warm the engine to normal operating temperature.
2. Disconnect the hose from the anti-backfire valve (**Figure 49**).
3. Run the engine at about 3,000 rpm, then release the throttle quickly. There should be suction at the anti-backfire valve. If not, replace it.

### Anti-Backfire Valve Replacement

The anti-backfire valve is located at the rear of the air cleaner. To remove, disconnect its hoses and take it out (**Figure 50**). Installation is the reverse of removal.

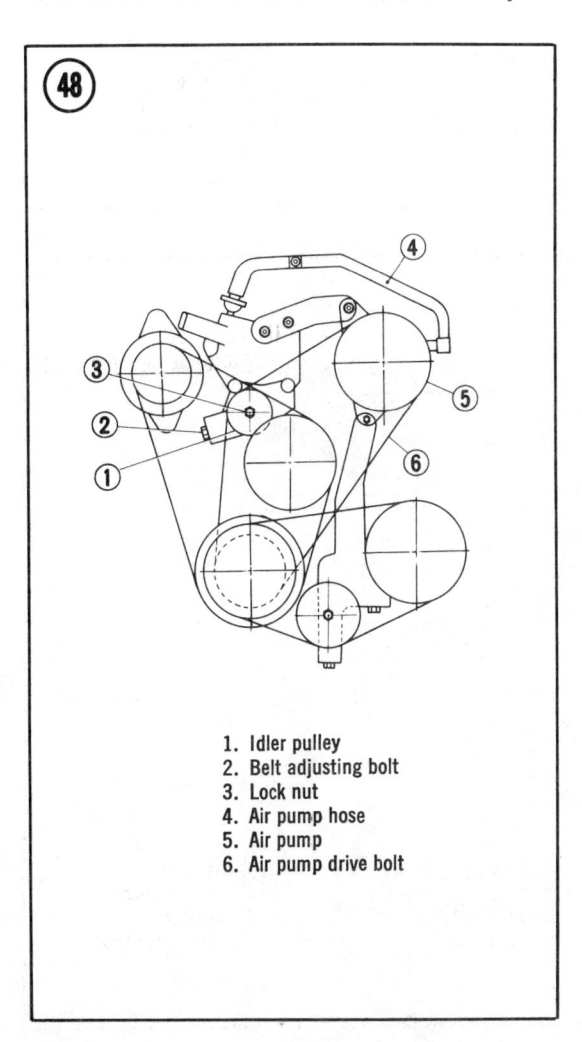

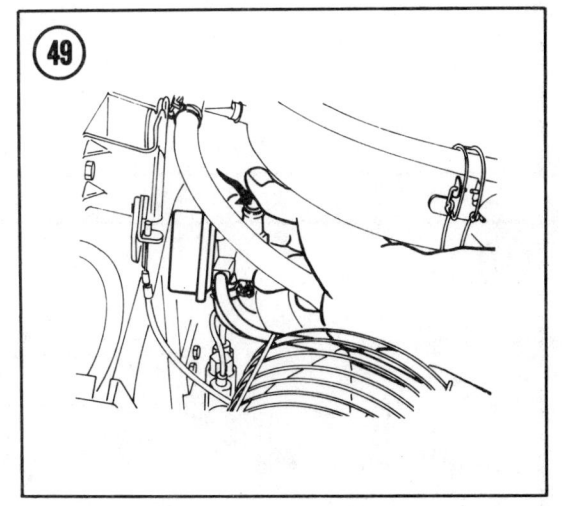

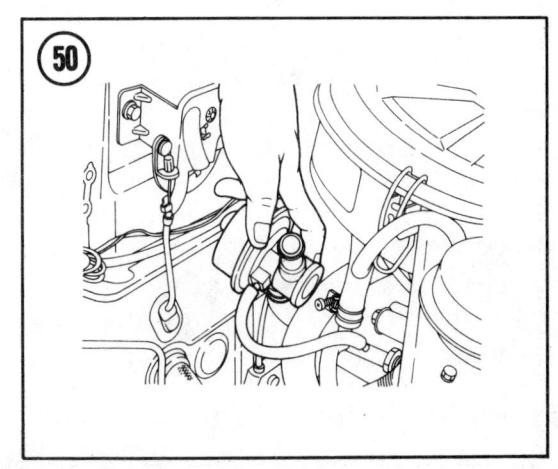

---

**48**

1. Idler pulley
2. Belt adjusting bolt
3. Lock nut
4. Air pump hose
5. Air pump
6. Air pump drive bolt

**49**

**50**

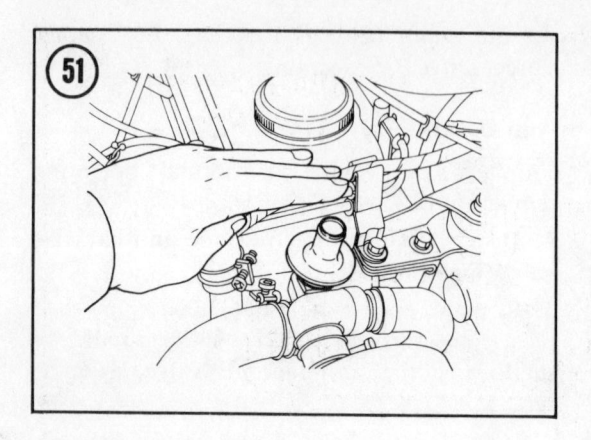

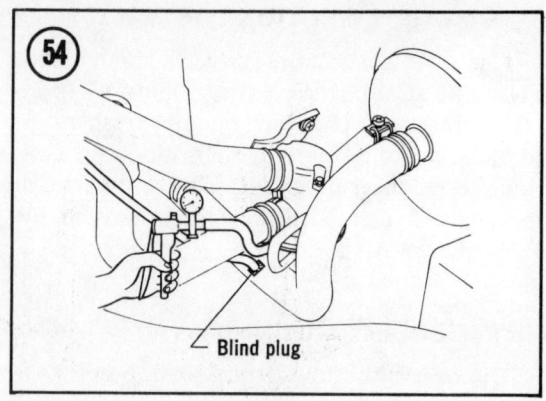

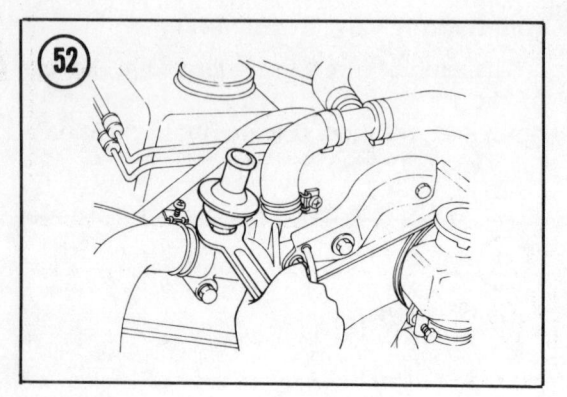

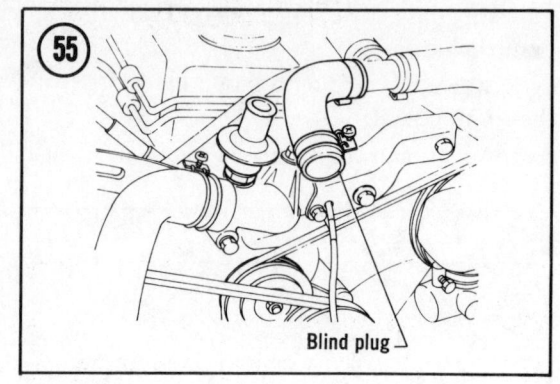

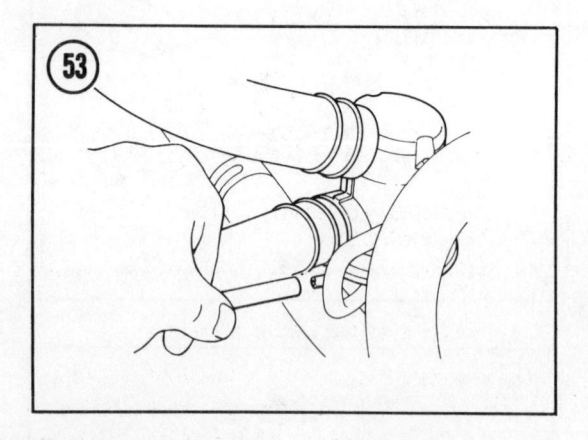

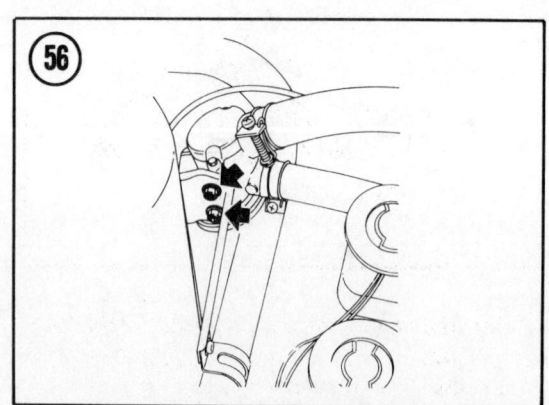

## Check Valve Test

1. Warm the engine to normal operating temperature. Disconnect the hose from the check valve (**Figure 51**).

2. Check for exhaust gas leaks at idle, then at 2,000 rpm. If leaks are detected, replace the valve.

## Check Valve Replacement

To remove, unscrew the valve from the cylinder head (**Figure 52**). Screw new valve in.

NOTE: *If the valve is difficult to remove, soak its threads with penetrating oil.*

## Combined Air Control Valve Test (California Models)

1. Warm the engine to normal operating temperature. Check the CAC valve hoses for cracks or other damage.

# FUEL AND EXHAUST SYSTEMS

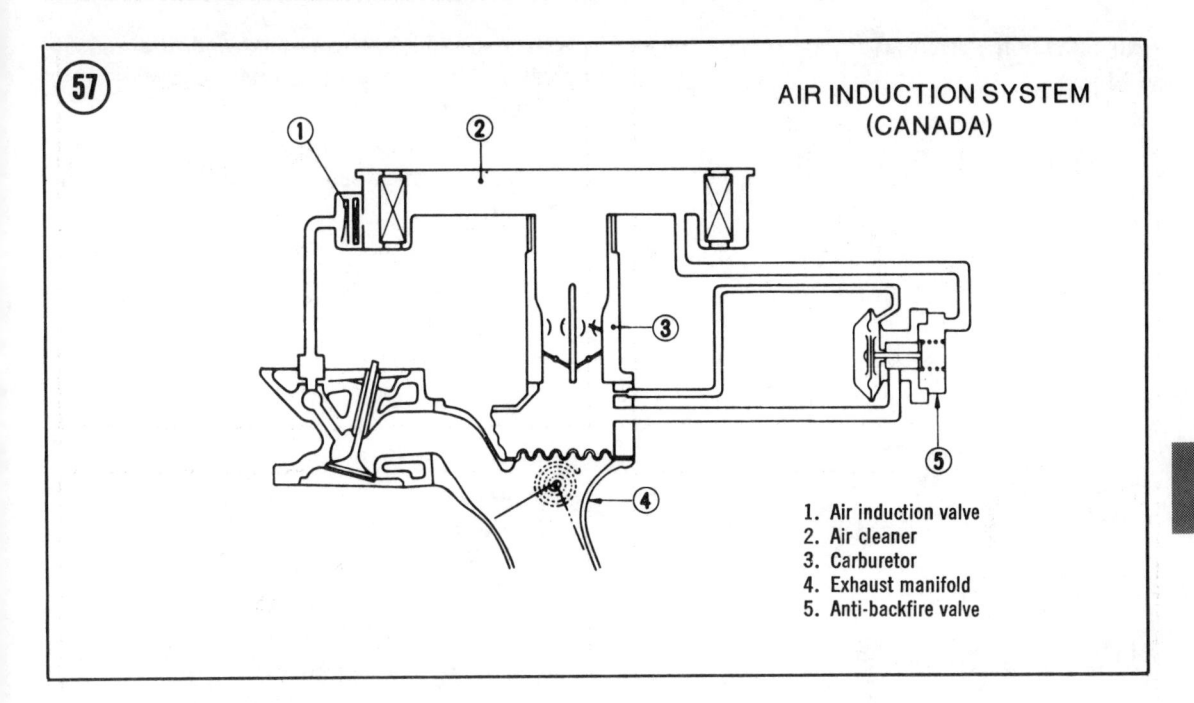

**AIR INDUCTION SYSTEM (CANADA)**

1. Air induction valve
2. Air cleaner
3. Carburetor
4. Exhaust manifold
5. Anti-backfire valve

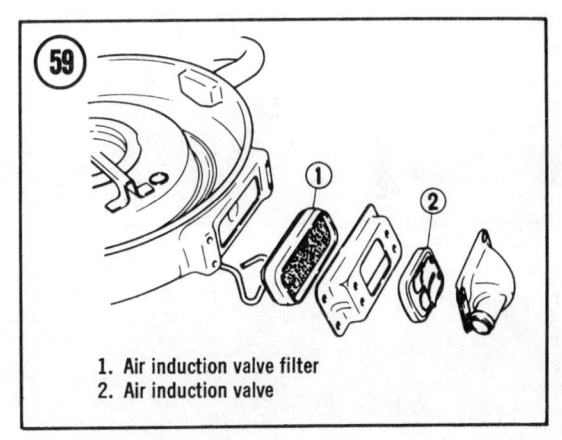

1. Air induction valve filter
2. Air induction valve

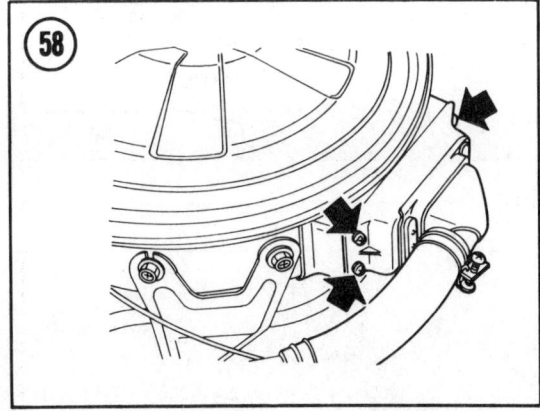

2. With the engine idling, disconnect and plug the CAC valve vacuum hose (**Figure 53**). Air should flow from the CAC valve.

3. Connect a vacuum pump to the CAC valve (**Figure 54**). Pump vacuum to 8-9 in. (200-250mm). Run the engine at 3,000 rpm and make sure no air leaks from the CAC valve.

4. Disconnect the hose from the check valve and plug it (**Figure 55**). Make sure air flows from the CAC valve.

### Combined Air Control Valve Replacement

To replace the valve, disconnect its hoses and remove its mounting screws (**Figure 56**).

### AIR INDUCTION SYSTEM

This system, used on Canadian models, works much like the air injection system. However, it uses exhaust system vacuum to suck air into the exhaust ports, instead of blowing it in with an air pump. See **Figure 57**.

### Air Induction Filter

The filter should be replaced at intervals specified in Chapter Three.

1. Remove the cover securing screws (**Figure 58**).
2. Remove the filter (**Figure 59**).
3. Installation is the reverse of removal.

## CHAPTER FIVE

### Air Induction Valve Test

Disconnect the valve's hose (**Figure 60**). It should be possible to suck air through the hose, but not to blow air into it. If air goes both ways or neither way, replace the valve.

### Anti-Backfire Valve Test

Anti-backfire valve testing and replacement are the same as for U.S. models. See *Air Injection System* earlier in this chapter.

## EVAPORATIVE EMISSION CONTROL SYSTEM

This system is designed to prevent gasoline vapor from escaping into the atmosphere. The system should be inspected, and the carbon canister filter replaced, at intervals specified in Chapter Three.

1. Check the vapor lines (**Figure 61**) for leaks and loose connections. Tighten or replace as needed.

2. Remove the filter from the bottom of the carbon canister (**Figure 62**). Install a new one.

## EXHAUST SYSTEM

Refer to the following illustrations:

**Figure 63** — 1976-1977 California sedan and hatchback

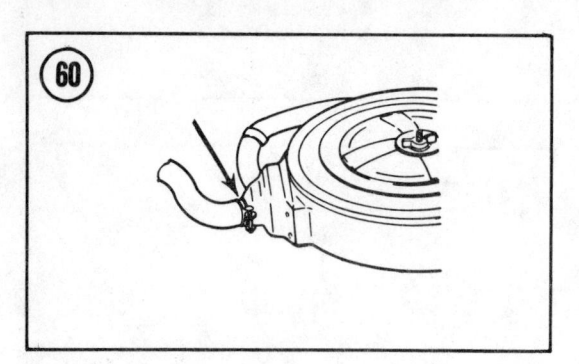

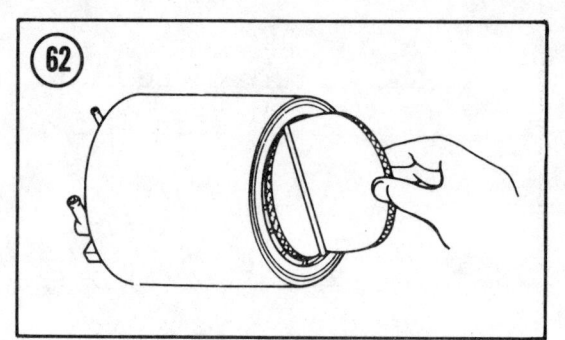

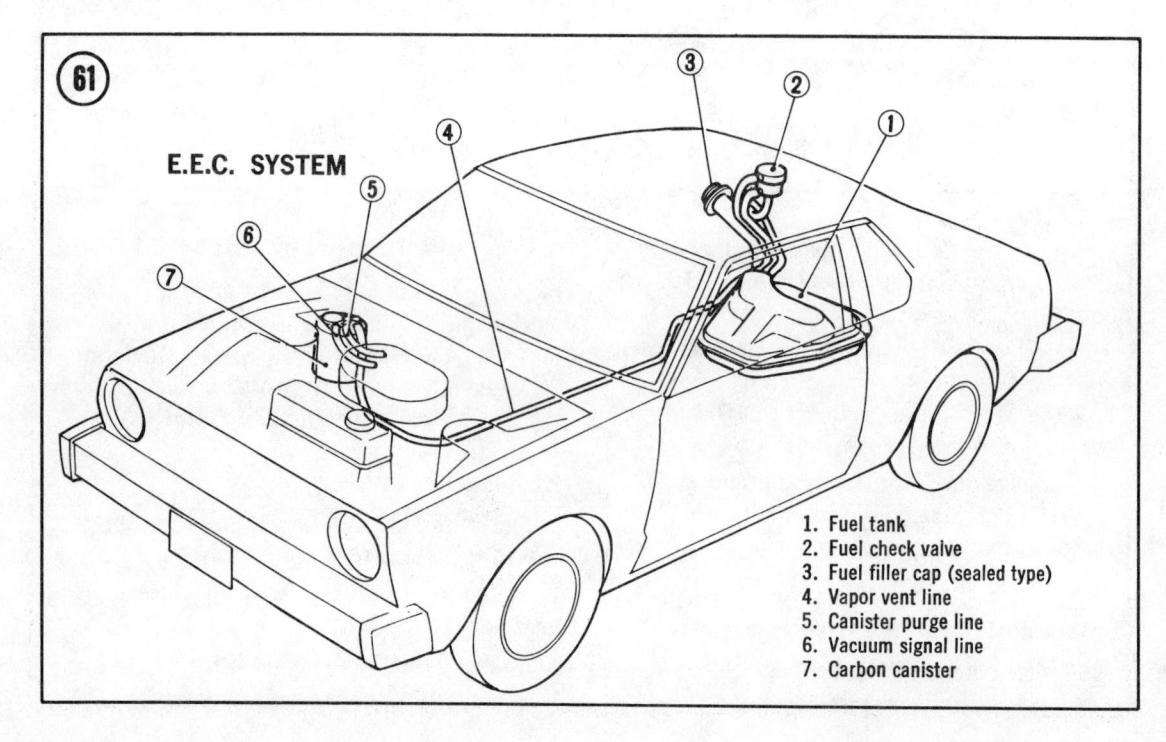

**Figure 64** — 1976-1977 California wagon

**Figure 65** — 1978 California sedan and hatchback

**Figure 61** — E.E.C. SYSTEM
1. Fuel tank
2. Fuel check valve
3. Fuel filler cap (sealed type)
4. Vapor vent line
5. Canister purge line
6. Vacuum signal line
7. Carbon canister

# FUEL AND EXHAUST SYSTEMS

111

EXHAUST SYSTEM
(1976-1977 CALIFORNIA SEDAN AND HATCHBACK)

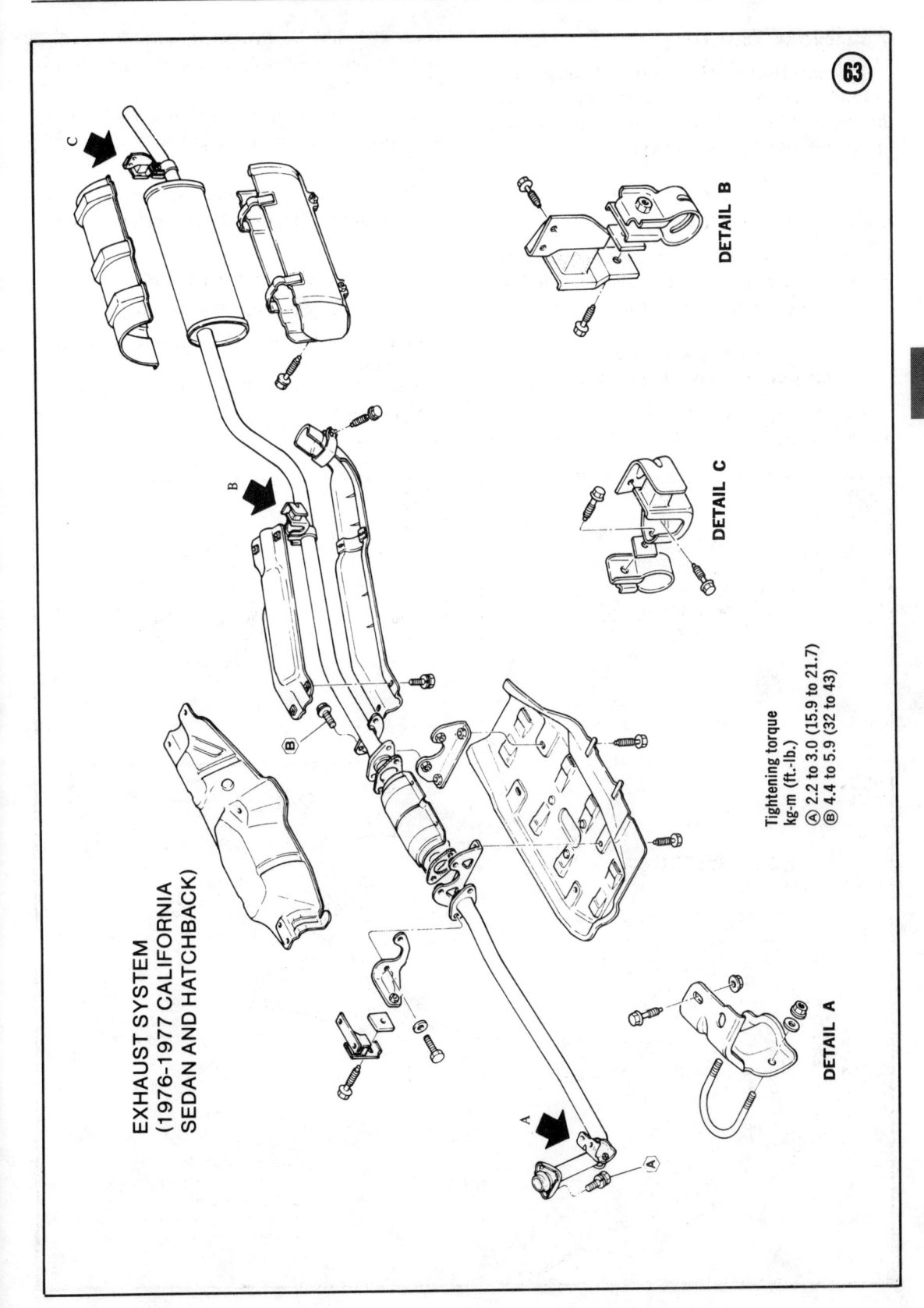

DETAIL B

DETAIL C

DETAIL A

Tightening torque
kg-m (ft.-lb.)
Ⓐ 2.2 to 3.0 (15.9 to 21.7)
Ⓑ 4.4 to 5.9 (32 to 43)

112

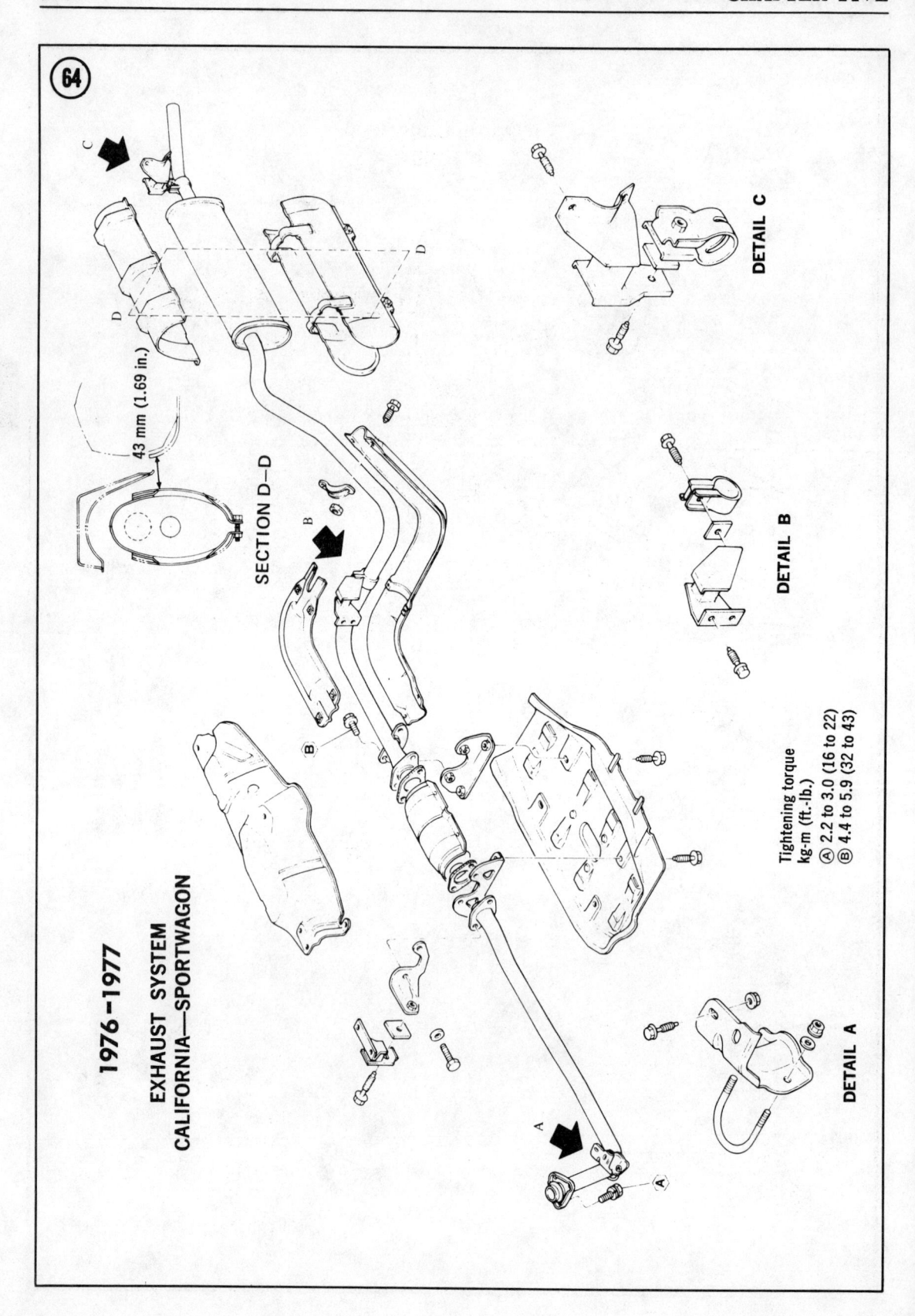

CHAPTER FIVE

# FUEL AND EXHAUST SYSTEMS

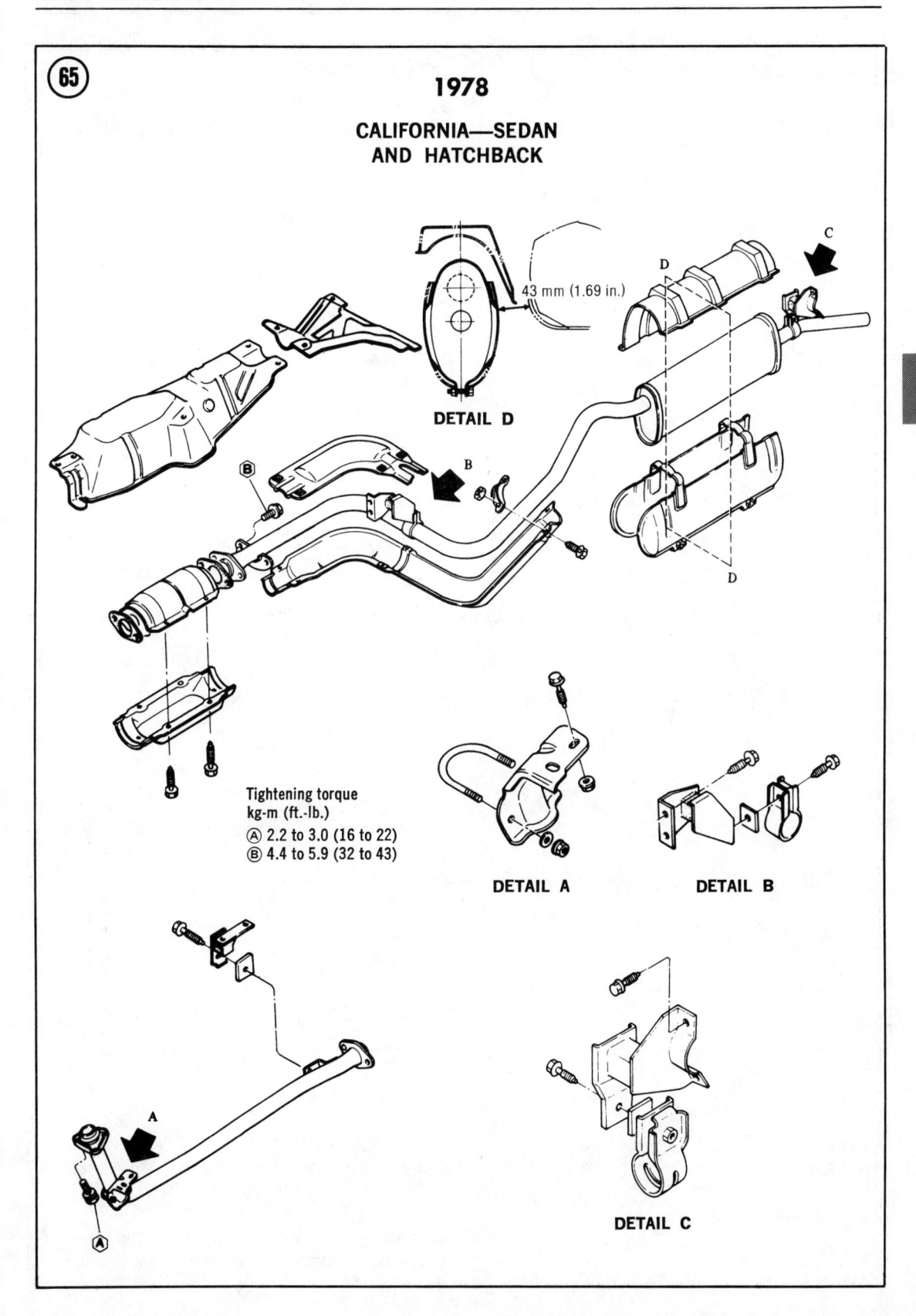

**(65)**

**1978**

**CALIFORNIA—SEDAN AND HATCHBACK**

43 mm (1.69 in.)

DETAIL D

Tightening torque kg-m (ft.-lb.)
Ⓐ 2.2 to 3.0 (16 to 22)
Ⓑ 4.4 to 5.9 (32 to 43)

DETAIL A

DETAIL B

DETAIL C

114                                                                    CHAPTER FIVE

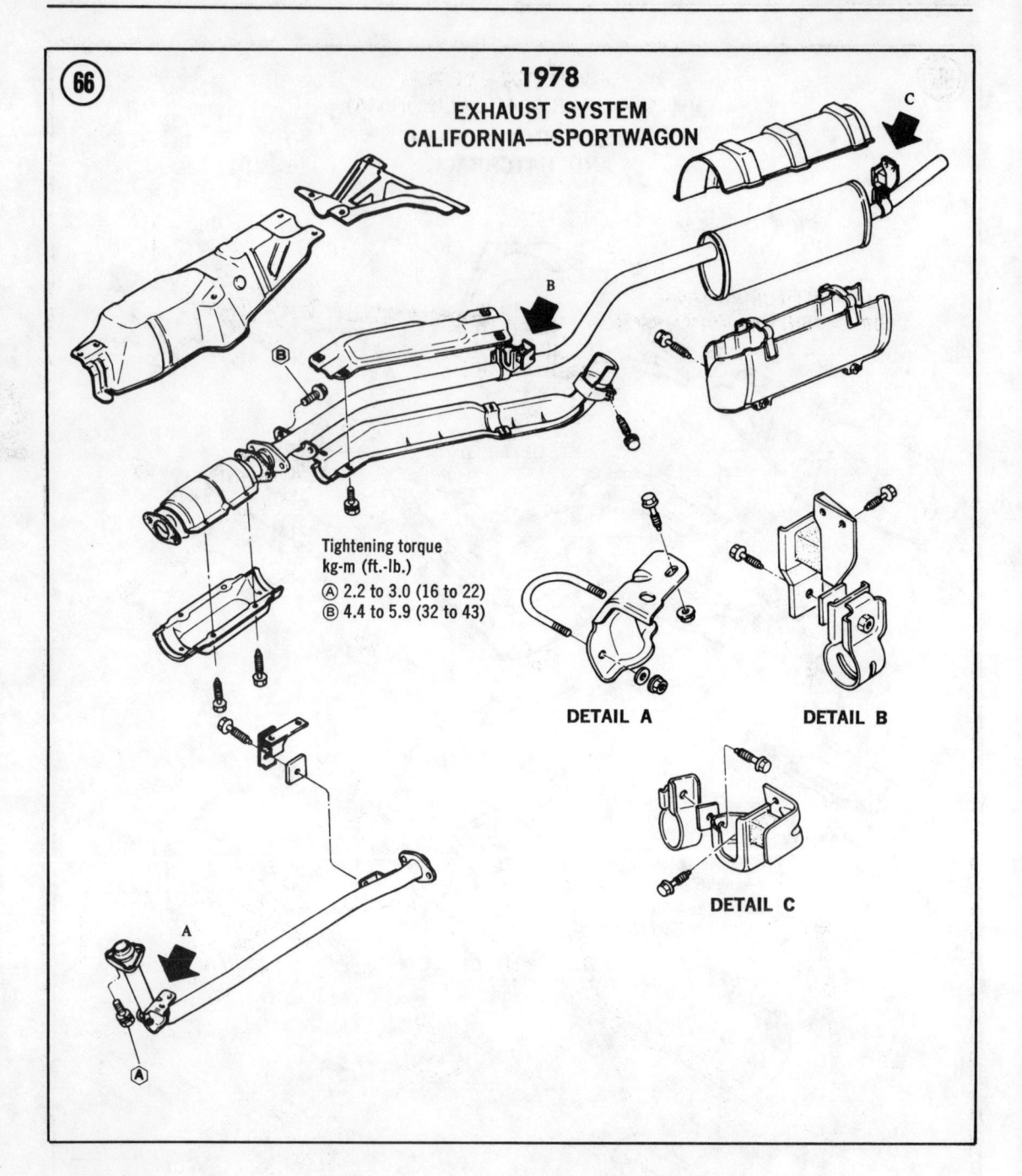

**1978 EXHAUST SYSTEM CALIFORNIA—SPORTWAGON**

Tightening torque kg-m (ft.-lb.)
Ⓐ 2.2 to 3.0 (16 to 22)
Ⓑ 4.4 to 5.9 (32 to 43)

DETAIL A

DETAIL B

DETAIL C

**Figure 66** — 1978 California wagon

**Figure 67** — 1976-1978 non-California models

**Figure 68** — 1979 California models

**Figure 69** — 1979 non-California models

### Removal/Installation

1. Prior to removal, soak all bolts, nuts, and slip-on pipe joints with penetrating oil such as WD-40.

2. Undo the necessary clamps and hanger brackets, referring to the appropriate illustration.

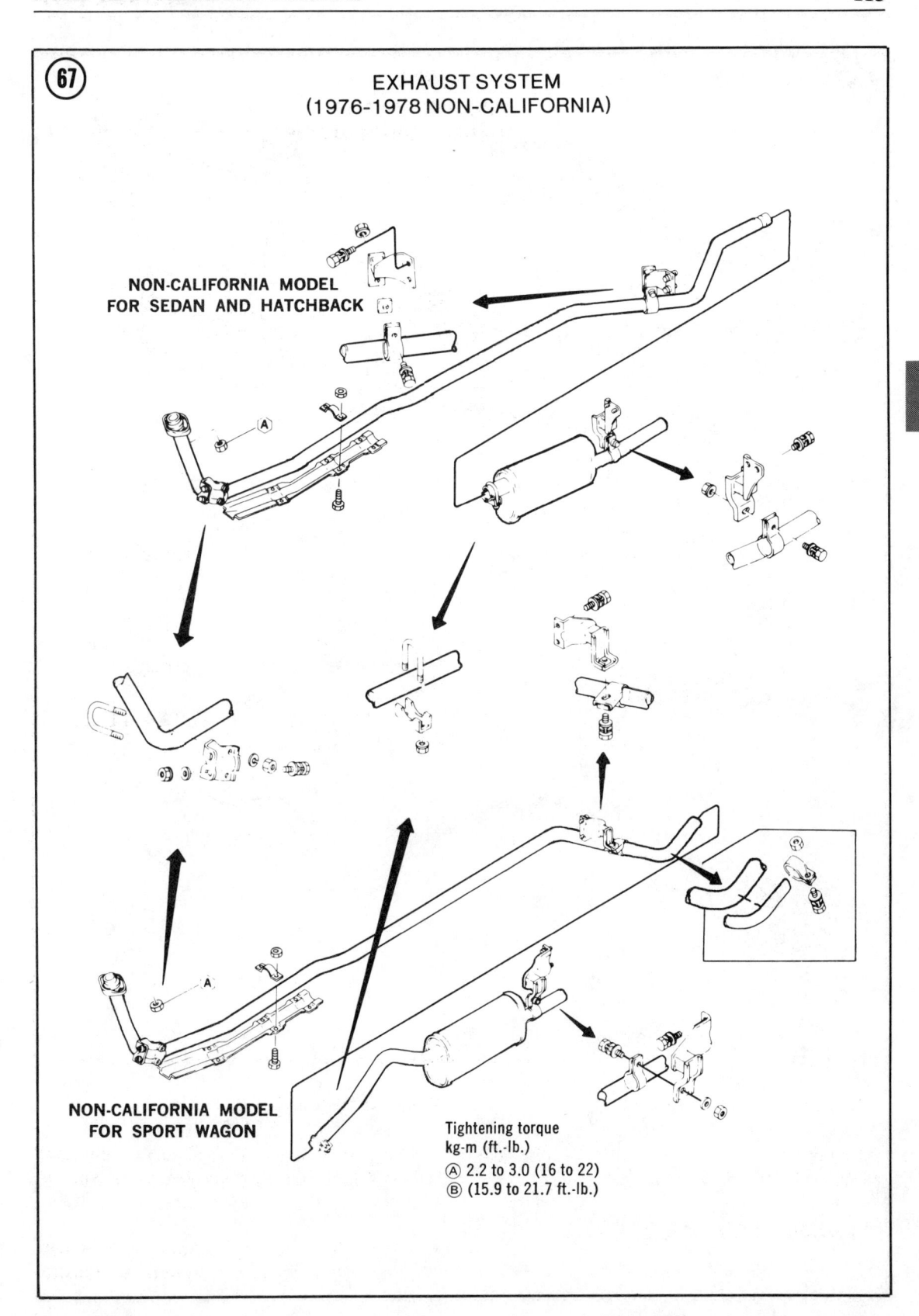

# FUEL AND EXHAUST SYSTEMS

**EXHAUST SYSTEM
(1976-1978 NON-CALIFORNIA)**

NON-CALIFORNIA MODEL FOR SEDAN AND HATCHBACK

NON-CALIFORNIA MODEL FOR SPORT WAGON

Tightening torque
kg-m (ft.-lb.)
Ⓐ 2.2 to 3.0 (16 to 22)
Ⓑ (15.9 to 21.7 ft.-lb.)

## CHAPTER FIVE

⑥⑧ 1979 CALIFORNIA MODELS EXHAUST SYSTEM

1. Front exhaust tube
2. Catalytic converter
3. Converter lower shelter
4. Rear exhaust tube
5. Rear exhaust tube lower shelter
6. Rear exhaust tube upper shelter
7. Exhaust muffler
8. U-bolt clamp
9. No. 1 exhaust mounting bracket
10. No. 2 exhaust mounting insulator
11. No. 2 exhaust mounting bracket
12. No. 3 exhaust mounting insulator
13. No. 3 exhaust mounting bracket
14. No. 4 exhaust mounting insulator
15. No. 4 exhaust mounting bracket

Tightening torque: kg-m (ft.-lb.)
A: 2.0 to 2.5 (14 to 18)
B: 3.2 to 4.3 (23 to 31)
C: 0.8 to 1.1 (5.8 to 8.0)
D: 0.3 to 0.4 (2.2 to 2.9)
E: 0.5 to 0.7 (3.6 to 5.1)

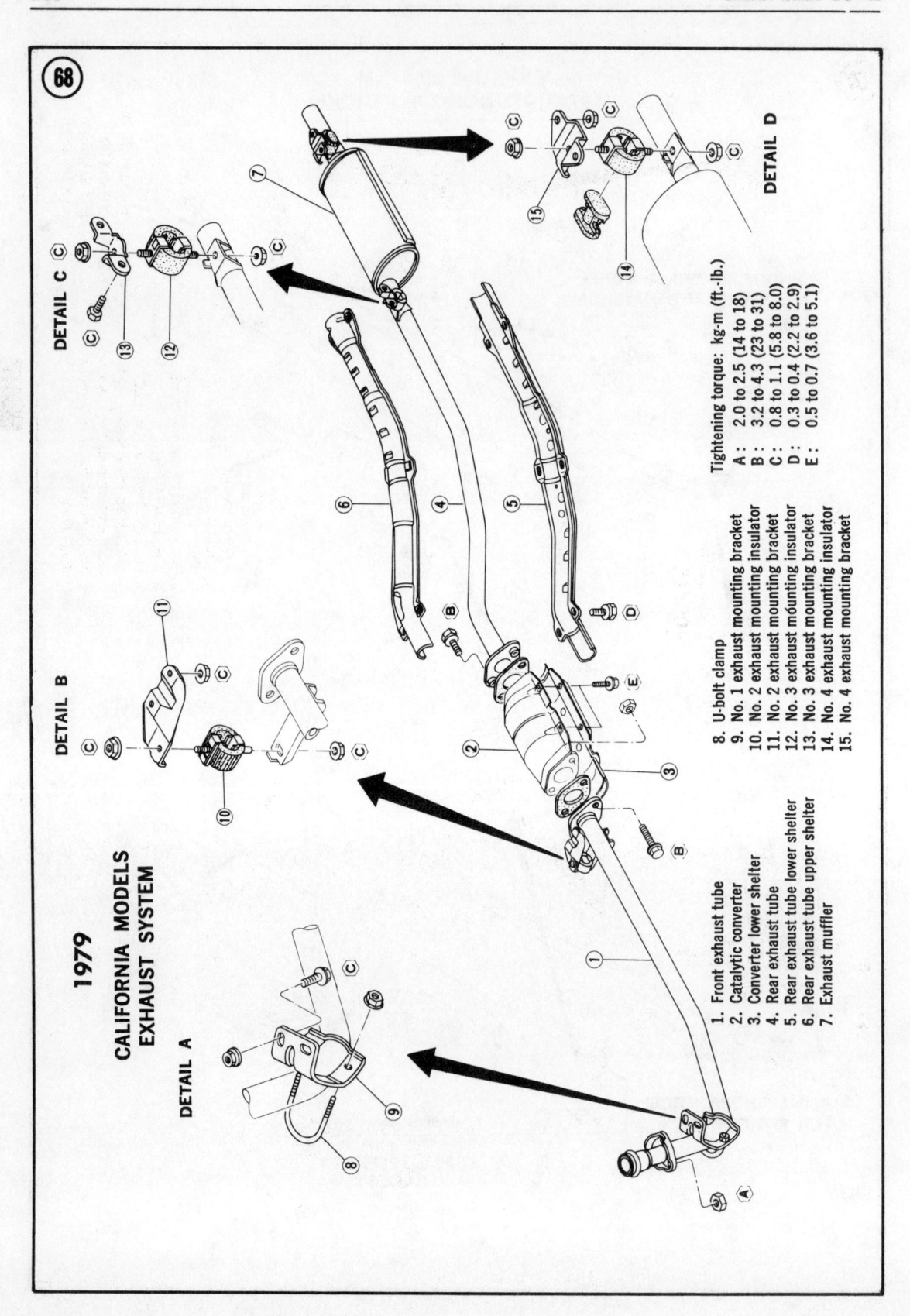

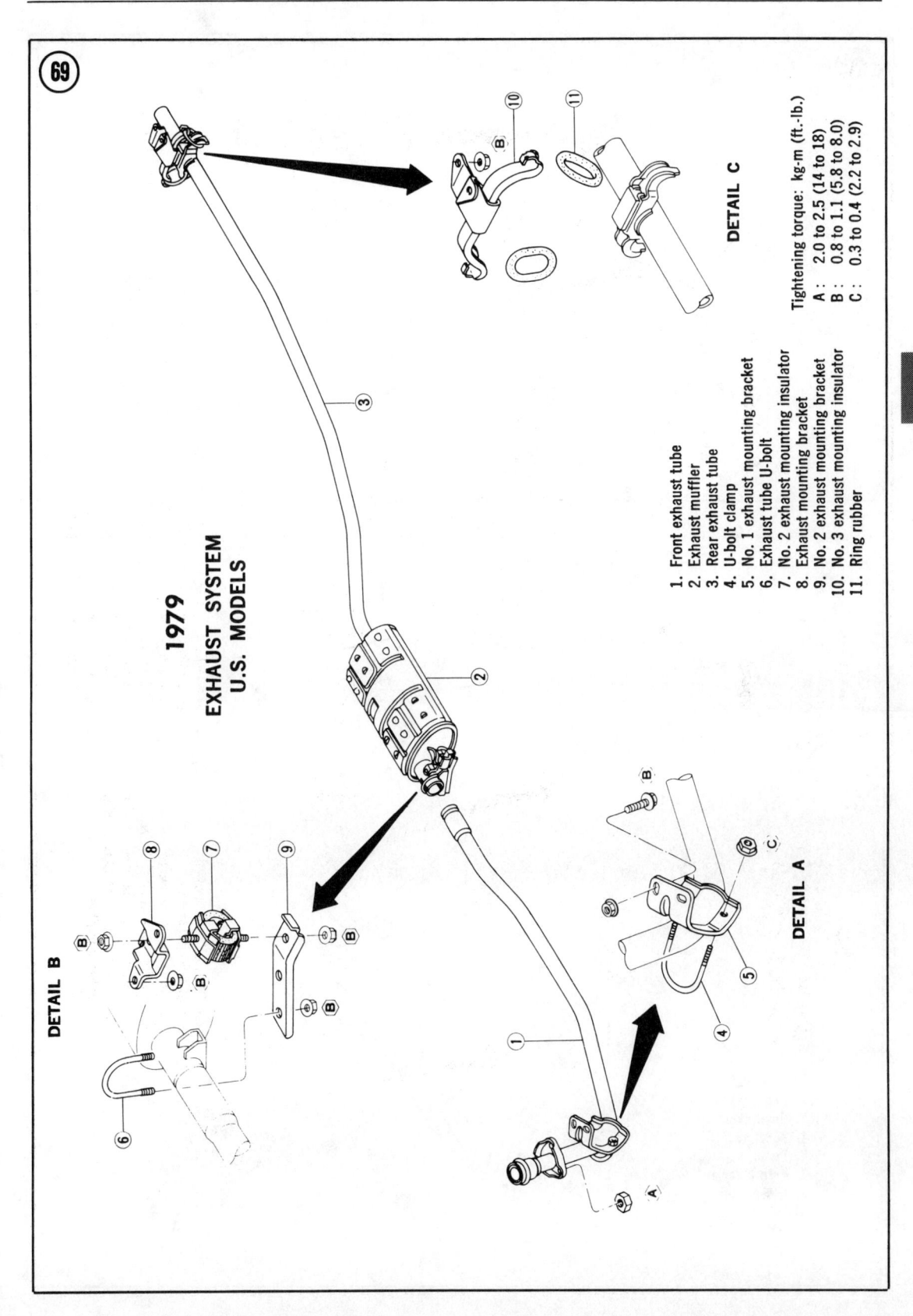

## FUEL AND EXHAUST SYSTEMS

**69**

**1979 EXHAUST SYSTEM U.S. MODELS**

1. Front exhaust tube
2. Exhaust muffler
3. Rear exhaust tube
4. U-bolt clamp
5. No. 1 exhaust mounting bracket
6. Exhaust tube U-bolt
7. No. 2 exhaust mounting insulator
8. Exhaust mounting bracket
9. No. 2 exhaust mounting bracket
10. No. 3 exhaust mounting insulator
11. Ring rubber

Tightening torque: kg-m (ft.-lb.)
A : 2.0 to 2.5 (14 to 18)
B : 0.8 to 1.1 (5.8 to 8.0)
C : 0.3 to 0.4 (2.2 to 2.9)

DETAIL A

DETAIL B

DETAIL C

118                                                                                         CHAPTER FIVE

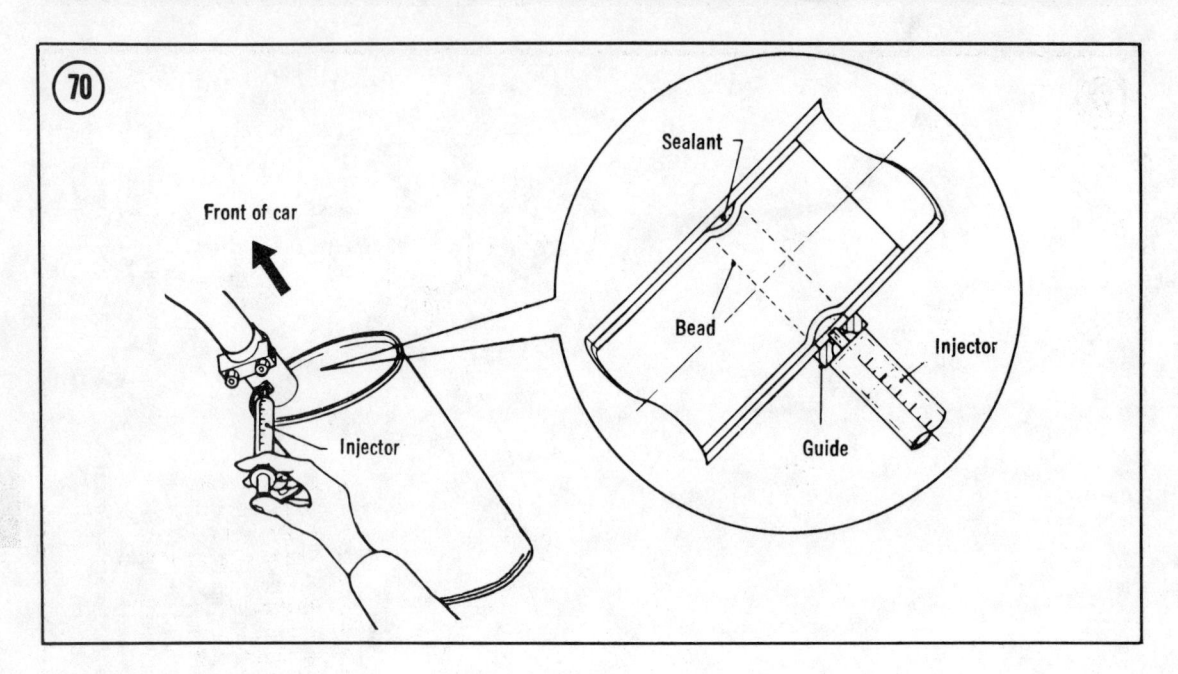

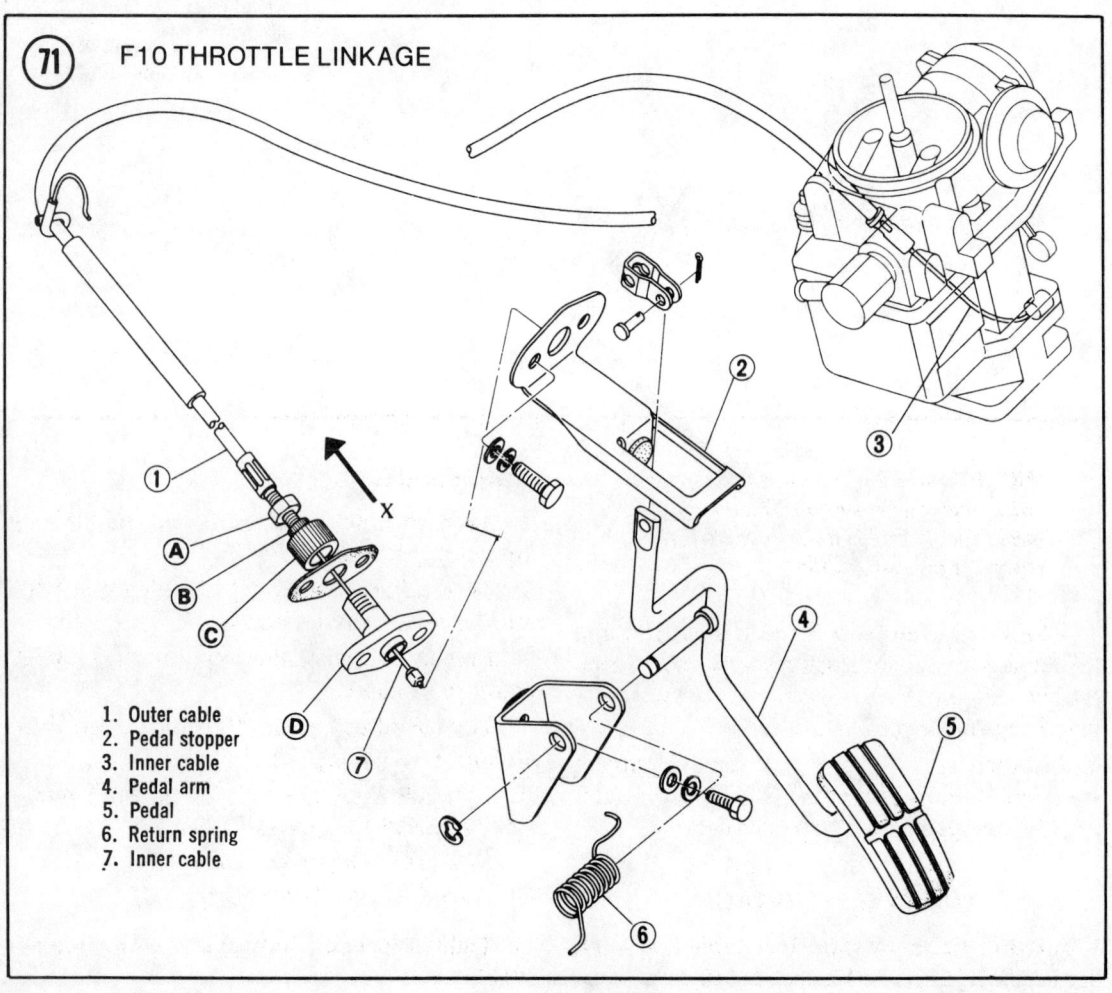

⑦⓪

Front of car

Sealant

Bead

Injector

Guide

Injector

⑦①  F10 THROTTLE LINKAGE

1. Outer cable
2. Pedal stopper
3. Inner cable
4. Pedal arm
5. Pedal
6. Return spring
7. Inner cable

# FUEL AND EXHAUST SYSTEMS

⑫

1. Accelerator pedal
2. Accelerator arm
3. Return spring
4. Accelerator pedal bracket
5. Snap ring
6. Nylon collar
7. Wire end
8. Wire casing end
9. Grommet
10. Accelerator wire

**310 THROTTLE LINKAGE**

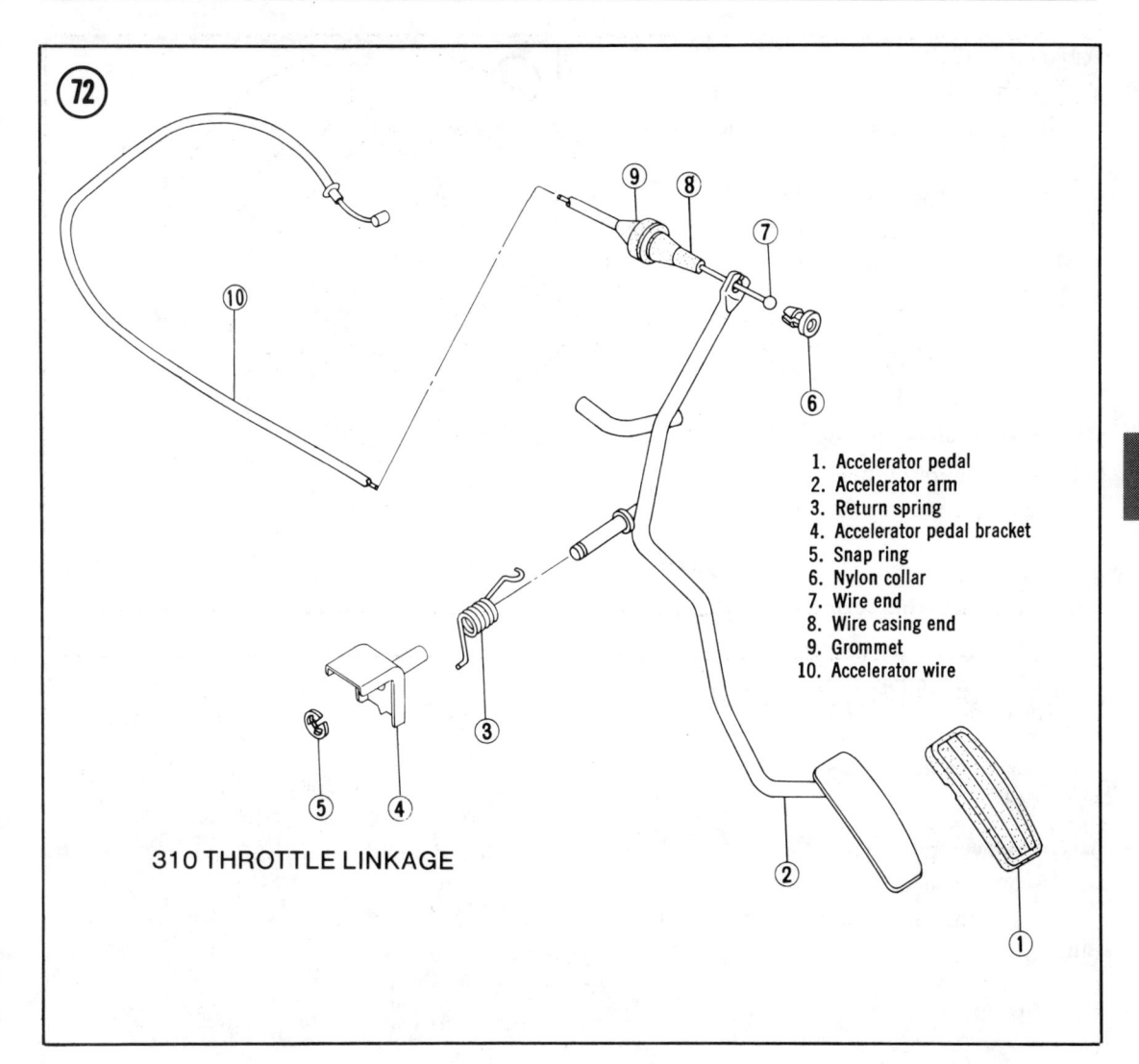

NOTE: *Slip-on pipe joints are injected with sealer (Figure 70). To break the seal, tap lightly with a hammer, then twist the pipes or mufler.*

3. Check removed parts for excessive rust, and for damage caused by bottoming the car. Check rubber mounts for melting, cracks, or deterioration. Replace as needed.

4. Installation is the reverse of removal. Inject sealer into slip-on joints with a Datsun sealer kit. Use new gaskets on flanged joints.

## THROTTLE LINKAGE

The throttle on all models is cable operated. See **Figure 71** (F10) or **Figure 72** (310).

**Adjustment (F10)**

1. Turn locknut "A" as far as possible in direction "X".

2. Turn adjusting nut "C" in direction "X" until it comes off the threads.

3. Pull on the outer cable until there is no slack in the inner cable.

4. Tighten adjusting nut "B" until it is 0-$\frac{1}{8}$ in. (0.3mm) from flange "D".

5. Turn adjusting nut "C" as far onto flange "D" as it will go.

NOTE: *While turning nut "C", keep nut "B" from turning.*

6. Tighten locknut "A" against adjusting nut "B".

## Adjustment (310)

1. Hold the choke valve open and move the throttle by hand. This releases the automatic choke linkage.
2. Loosen the cable clamp (**Figure 73**).
3. Pull the outer cable in direction "P" until the throttle shaft just starts to move.
4. Push the outer cable in direction "Q" $3/64$-$1/8$ in. (1-3mm). Then tighten the cable clamp.

## FUEL TANK AND LINES

### Removal/Installation (F10)

Refer to **Figure 74** for this procedure.

1. Disconnect the negative cable from the battery.
2. Drain the fuel into a sealable container. This will keep fumes from forming.

> **WARNING**
> *Do not drain fuel near any flame, such as a water heater pilot light. Do not drain into an open pan. This will allow fumes to escape and create a fire hazard.*

3. Disconnect the hoses and vent line.
4. Disconnect the gauge wiring connector.
5. Remove the tank mounting bolts and take it out.

> NOTE: *When removing bolts, work from front to rear of the car.*

6. Installation is the reverse of removal.

### Removal/Installation (310)

Refer to **Figure 75** for this procedure.

1. Disconnect the negative cable from the battery.
2. Securely block both front wheels so the car will not roll in either direction.
3. Jack up the right rear corner of the car so the fuel will drain more completely.
4. Drain the fuel into a sealable container. This will prevent fumes from forming.

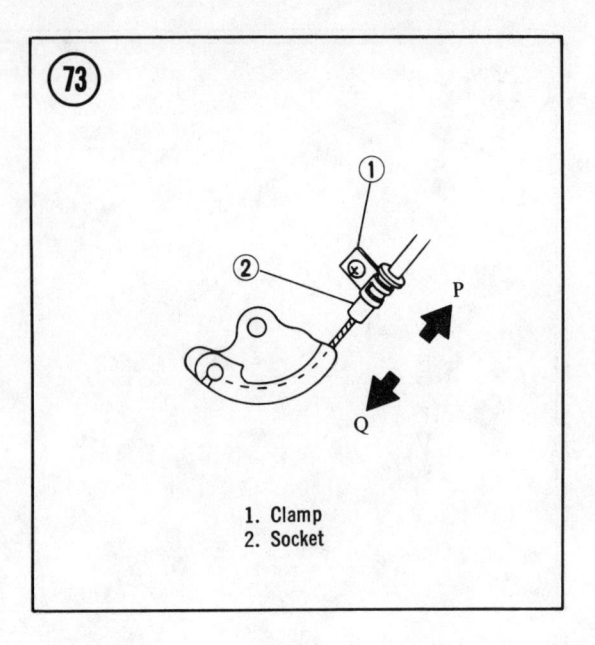

> **WARNING**
> *Do not drain fuel near any flame, such as a water heater pilot light. Do not drain into an open pan. This will allow fumes to escape and create a fire hazard.*

5. Disconnect the hoses from the tank.
6. Release the handbrake. Detach the handbrake from the bracket on the fuel tank.
7. Undo the tank attaching bolts. Remove the tank downward and to the front.
8. Installation is the reverse of removal.

### Repairing Leaks

Fuel tank leaks can be repaired by soldering.

> **WARNING**
> *A fuel tank, even an empty one, is capable of exploding and killing anyone nearby. Always observe the following precautions when repairing a tank.*

1. Have the tank steam cleaned *inside and outside*.
2. Fill the tank with inert gas such as carbon dioxide or nitrogen, or fill the tank *completely* with water. If air space is left in the tank, residual fuel could sweat from the walls and form explosive vapor.
3. Set a fire extinguisher nearby.

After the repair is made, pour the water out, put about a quart of gasoline in the tank, and slosh it around. Pour the gasoline out, blow the tank dry, and install it in the car.

**73**

1. Clamp
2. Socket

# FUEL AND EXHAUST SYSTEMS

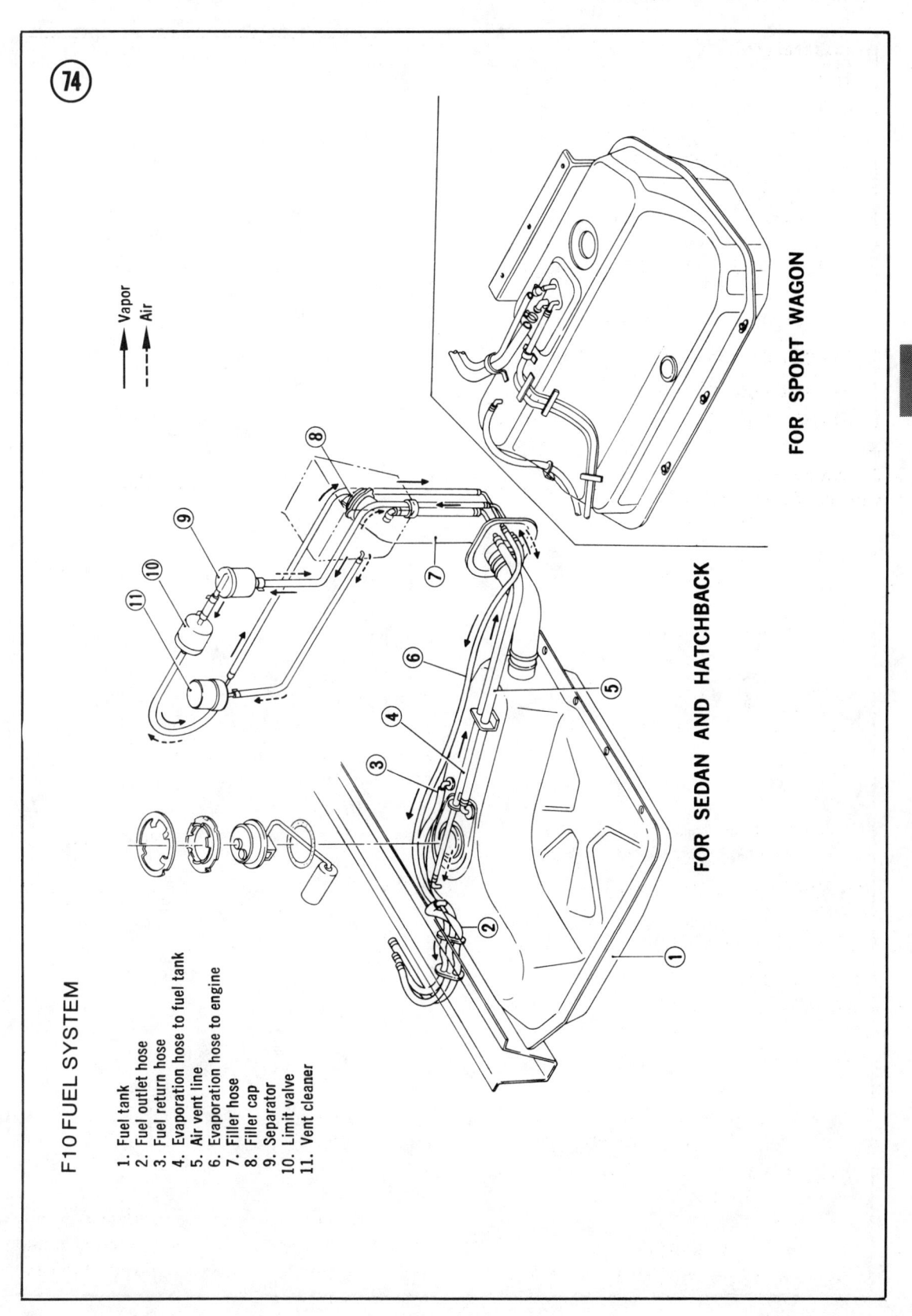

(74)

→ Vapor
--→ Air

FOR SPORT WAGON

FOR SEDAN AND HATCHBACK

F10 FUEL SYSTEM

1. Fuel tank
2. Fuel outlet hose
3. Fuel return hose
4. Evaporation hose to fuel tank
5. Air vent line
6. Evaporation hose to engine
7. Filler hose
8. Filler cap
9. Separator
10. Limit valve
11. Vent cleaner

CHAPTER FIVE

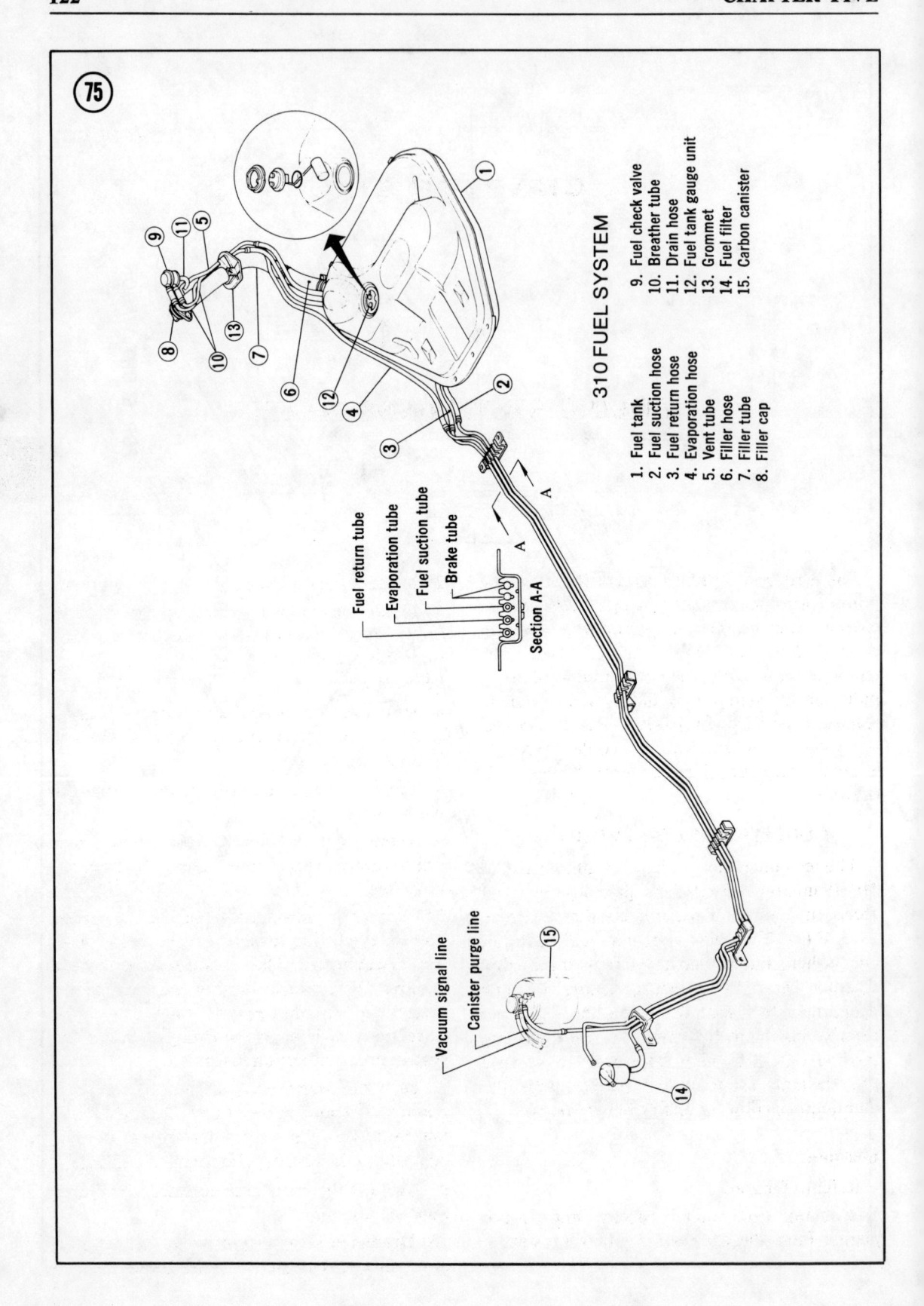

⑦⑤

**310 FUEL SYSTEM**

1. Fuel tank
2. Fuel suction hose
3. Fuel return hose
4. Evaporation hose
5. Vent tube
6. Filler hose
7. Filler tube
8. Filler cap
9. Fuel check valve
10. Breather tube
11. Drain hose
12. Fuel tank gauge unit
13. Grommet
14. Fuel filter
15. Carbon canister

Fuel return tube
Evaporation tube
Fuel suction tube
Brake tube
Section A-A

Vacuum signal line
Canister purge line

# CHAPTER SIX

# COOLING SYSTEM AND HEATER

The F10 and 310 use a centrifugal water pump to propel coolant through the radiator, engine, and heater. A thermostat controls coolant flow. An electric fan, operated by a temperature switch in the bottom radiator hose, pulls air through the radiator when coolant reaches a specified temperature. The heater is a hot water type which circulates coolant through a small radiator (heater core) behind the firewall.

## COOLING SYSTEM FLUSHING

The recommended coolant for all models is a 50/50 mixture of ethylene glycol-based antifreeze and water. This protects the system to $-31°F$ ($-35°C$). The system should be drained, flushed, and refilled at intervals specified in Chapter Three. If desired, a chemical flushing agent may be used, following the manufacturer's instructions, prior to the flushing method described here. However, make sure the flushing agent is compatible with the aluminum parts in the engine before using.

### Flushing

Refer to **Figure 1**.

1. Coolant can stain concrete and injure plants. Park the car over a gutter or similar area.

2. Remove the radiator cap.

3. Drain the cooling system by opening the drain tap at the bottom of the radiator and removing the drain plug from the engine block below the carburetor.

4. Disconnect the heater hose from the connection at the rear of the engine. The hose will be a drain during flushing.

5. Turn the heater control on the instrument panel to maximum.

6. Remove the thermostat as described under *Thermostat* in this chapter. Close the drain tap and plug.

7. Connect a water supply such as a garden hose to the fitting from which the heater hose was disconnected. This does not have to be a positive fit, as long as most of the water enters the engine. If necessary, temporarily connect an extra piece of hose to the fitting to make the garden hose connection easier.

8. Turn the water on and flush for 3 to 5 minutes. Do not run the engine. During the last minute of flushing, repeatedly squeeze the upper radiator hose to expel all trapped coolant.

9. Turn off the water and reconnect the heater hose to the engine.

10. Drain the system by opening the radiator drain tap and removing the block drain plug.

**COOLING SYSTEM**

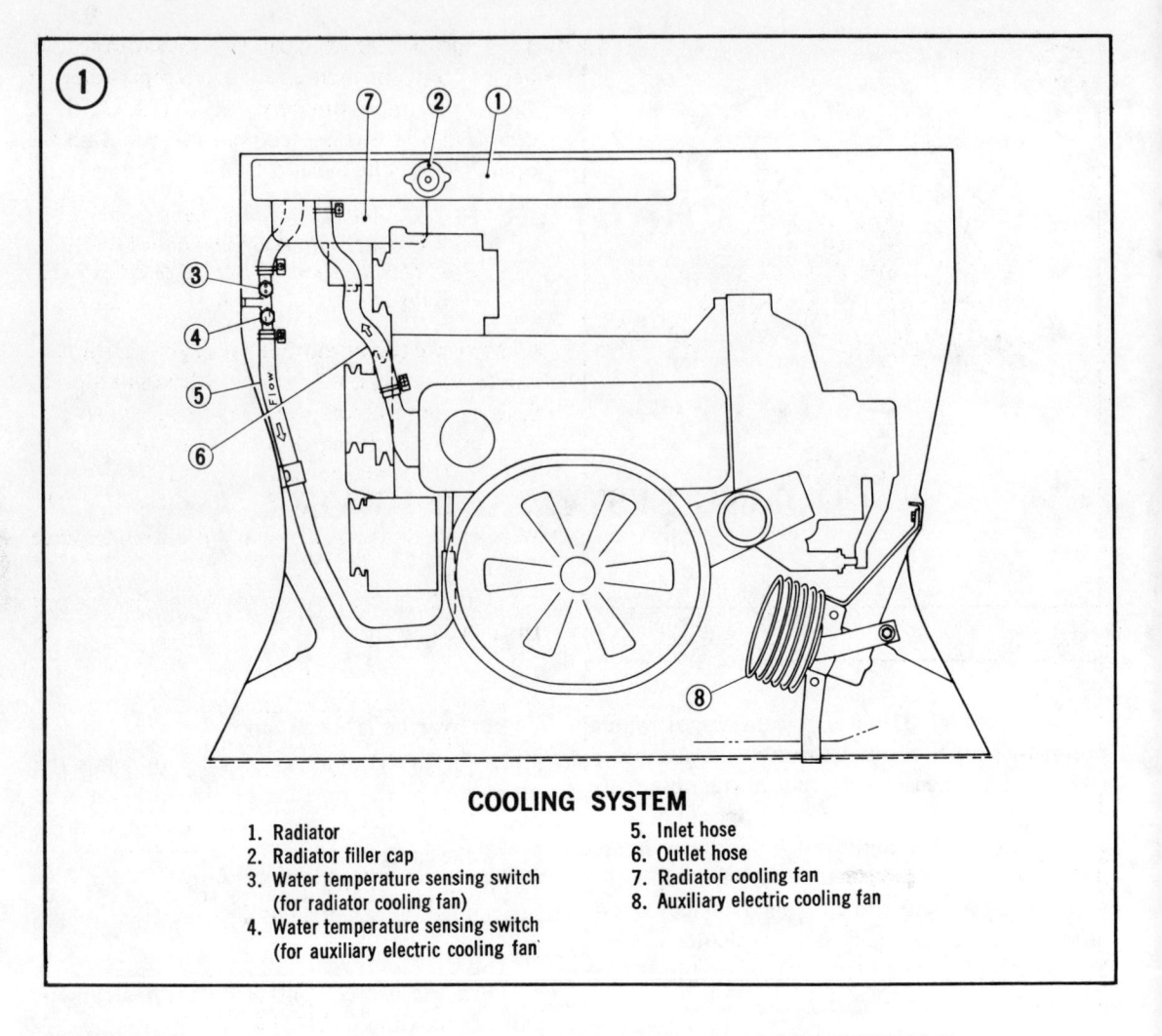

1. Radiator
2. Radiator filler cap
3. Water temperature sensing switch (for radiator cooling fan)
4. Water temperature sensing switch (for auxiliary electric cooling fan)
5. Inlet hose
6. Outlet hose
7. Radiator cooling fan
8. Auxiliary electric cooling fan

## Refilling

1. Be sure all hoses are connected and the drains closed.

2. Fill the cooling system with a 50/50 mixture of ethylene glycol-based antifreeze and water, even if you live in a climate which doesn't require this degree of freeze protection. The antifreeze is also a good corrosion inhibitor. F10 coolant capacity is 7 qt. (6.6 liters); 310 capacity is 6-¼ qt. (5.9 liters).

3. When the system is full, install the radiator cap.

4. Run the engine for several minutes to check for leaks. Then shut off the engine and recheck radiator coolant level. It should be 3/4 to 1 1/4 in. (20-30mm) below the top of the filler neck. If the level is too low, add water, then run the engine for several more minutes.

## THERMOSTAT

The thermostat blocks coolant flow to the radiator when the engine is cold. As the engine warms, the thermostat opens, admitting coolant to the radiator.

### Removal and Testing

1. Drain the cooling system by opening the tap at the bottom of the radiator. If the coolant is clean, drain it into a clean container for reuse.

2. Disconnect the air check valve and water outlet hoses from the outlet elbow (**Figure 2**). Unbolt the outlet elbow and take the thermostat out.

3. Immerse the thermostat in water with a thermometer (**Figure 3**).

# COOLING SYSTEM AND HEATER

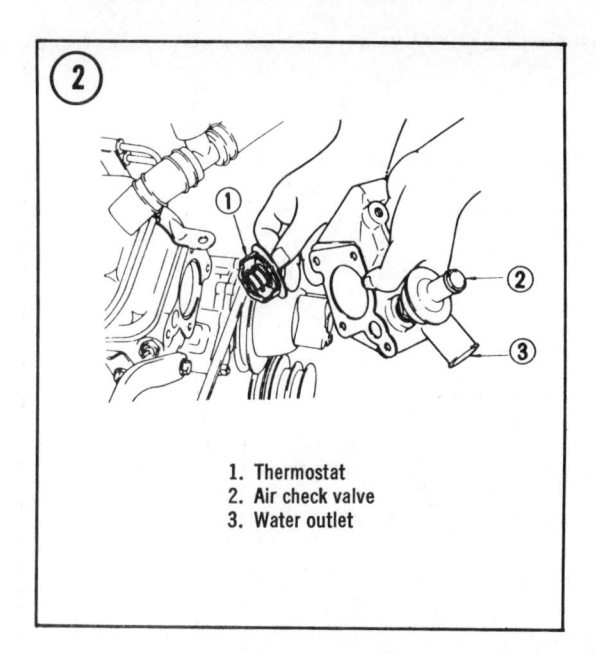

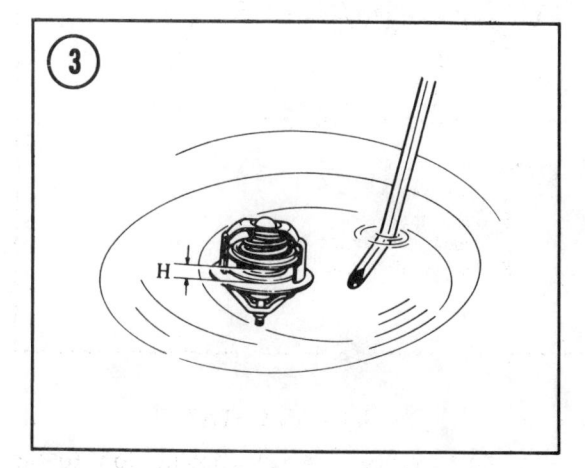

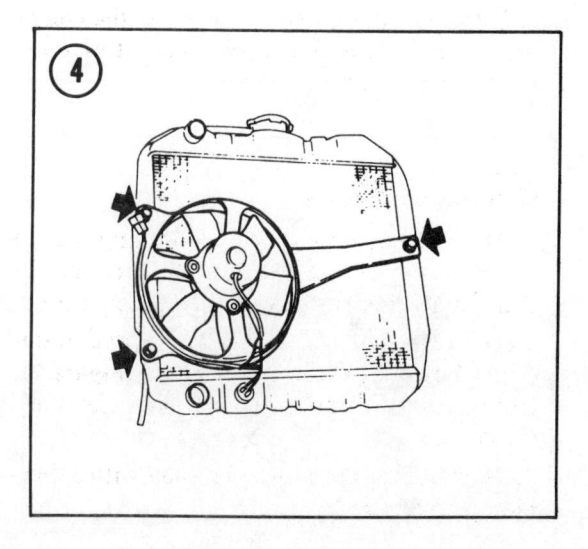

4. Heat the water until the thermostat just begins to open, then check the temperature. It should be approximately 180°F (82°C). If it opens at the wrong temperature, or fails to open, replace the thermostat.

> NOTE: *Actual opening temperature may vary a few degrees from that specified. This does not indicate a defective thermostat.*

5. Measure the maximum lift of the thermostat valve. To do this, mark a screwdriver at a point 0.31 in. (8mm) from the tip. The screwdriver is used as a measuring device. Heat the water to 203°F (95°C) and measure the valve's lift with the marked screwdriver (dimension "H", **Figure 3**). If lift is less than specified, replace the thermostat.

### Installation

1. If a new thermostat is being installed, test it as described in the preceding section.
2. Install the thermostat in the engine.
3. Install the water outlet elbow, using a new gasket coated on both sides with gasket sealer.
4. Tighten the outlet elbow securing bolts. Reconnect the hoses to the elbow.
5. Fill the cooling system. See *Cooling System Flushing*.

### RADIATOR

**Figure 4** shows the radiator.

### Removal/Installation

1. Open the tap at the bottom of the radiator to drain it. If the coolant is clean, drain it into a clean container for reuse.
2. Disconnect the upper and lower hoses from the radiator.
3. Disconnect the fan motor wires. Remove the fan securing screws and lift the fan assembly out.
4. Remove the grille (Chapter Twelve).
5. Remove the radiator mounting bolts. Lift the radiator out.
6. Installation is the reverse of removal. Fill the cooling system as described under *Cooling System Flushing*.

1. Thermostat
2. Air check valve
3. Water outlet

**CHAPTER SIX**

**5**

Washer tank

**6**

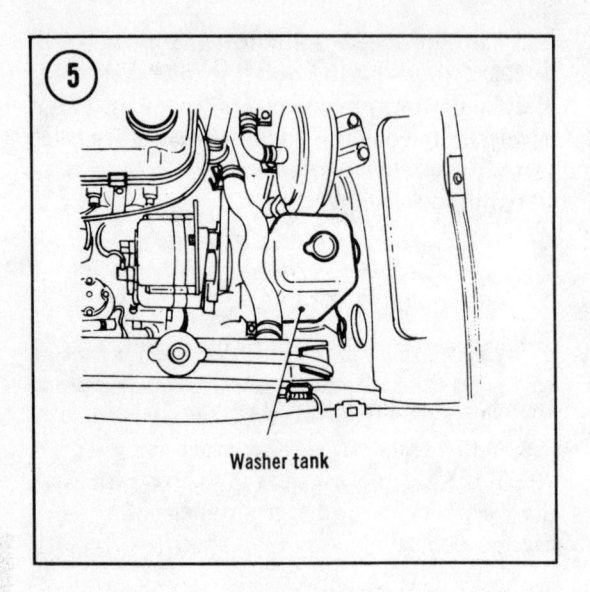

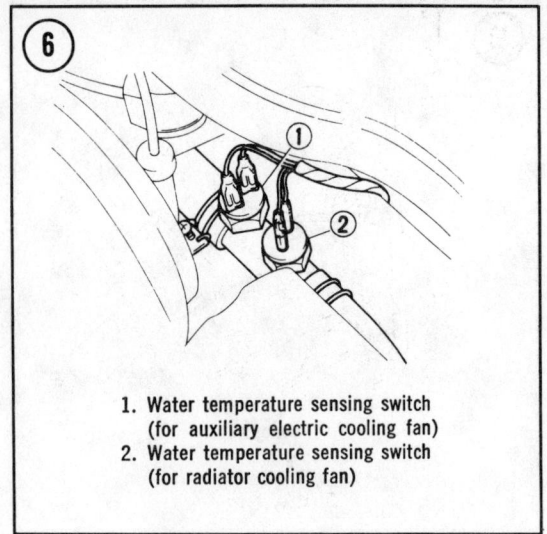

1. Water temperature sensing switch (for auxiliary electric cooling fan)
2. Water temperature sensing switch (for radiator cooling fan)

**7**

**8**

### FAN

The fan should go on when the coolant temperature reaches 181-189°F (83-87°C). If the fan fails to go on, check the fuse, then test the fan and temperature switch.

**Fan Test**

To test the fan, disconnect its wires. Connect the fan wires directly to the battery with jumper wires. If the fan motor runs, it is good. If not, it is defective; replace it.

**Temperature Switch Test**

1. Drain the cooling system by opening the tap at the bottom of the radiator. If the coolant is clean, drain it into a clean container for reuse.
2. On F10's, lift the windshield washer tank off its bracket. See **Figure 5**.
3. Unscrew the temperature switch (2, **Figure 6**) from the lower radiator hose.

> NOTE: *Be sure to remove the right switch. The other switch is for the carburetor fan.*

4. Place the switch in water with a thermometer (**Figure 7**). Connect an ohmmeter or self-powered test lamp to its terminals.
5. Heat the water. At some point between 181-189°F (83-87°C), the ohmmeter should show continuity (test lamp should light). If not, replace the switch.

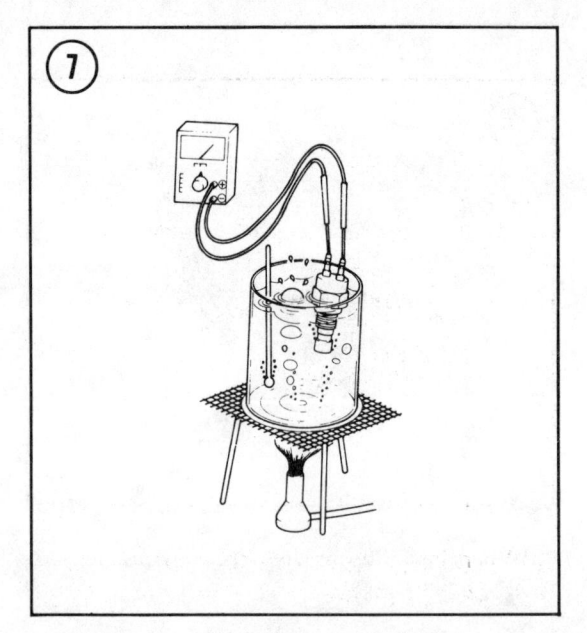

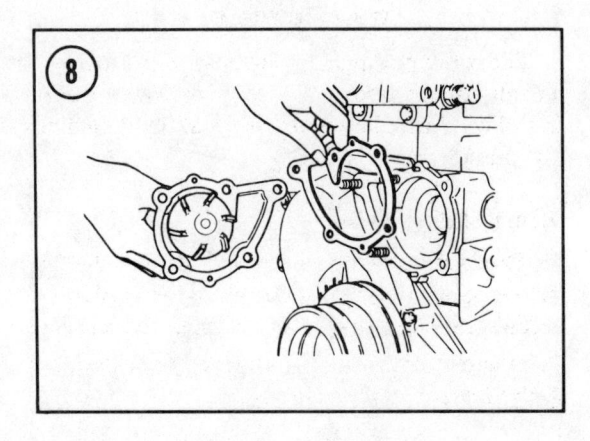

# COOLING SYSTEM AND HEATER

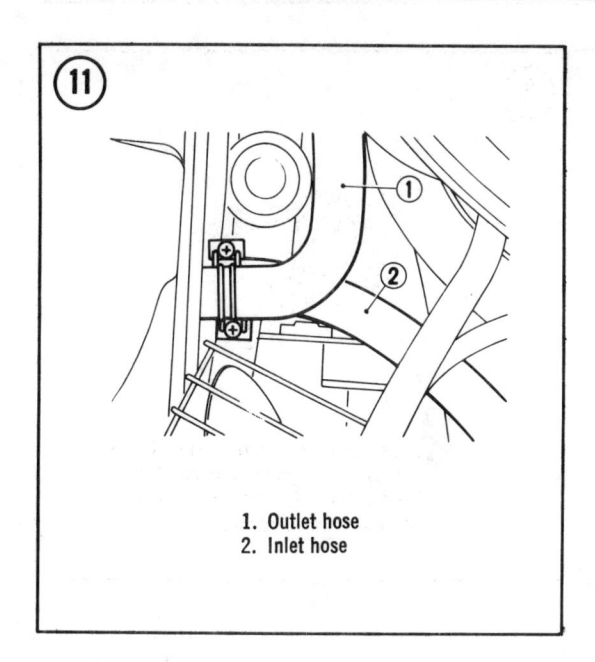

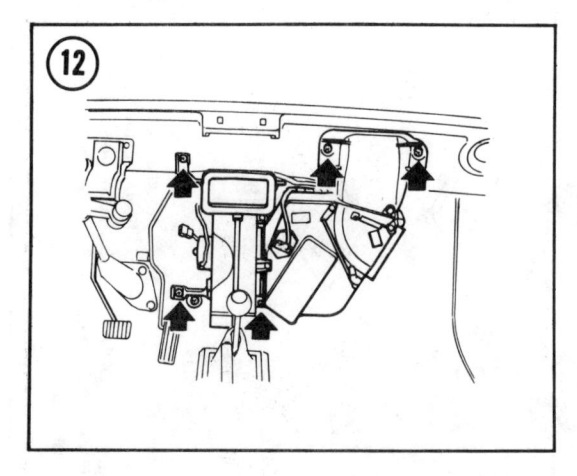

**1. Outlet hose**
**2. Inlet hose**

6. When installing the switch, tighten to 14-18 ft.-lb. (2.0-2.5 mkg).

## WATER PUMP

The water pump is a non-repairable type with an aluminum body. A defective water pump may be indicated by noise or a leak from behind the pump pulley.

### Removal/Installation

1. Drain the cooling system by opening the tap at the bottom of the radiator. If the coolant is clean, drain it into a clean container for reuse.
2. Remove the water pump drive belt. See *Drive Belts*, Chapter Three.
3. Unbolt the water pump from the engine. Remove the pump and gasket (**Figure 8**).
4. Installation is the reverse of removal. Use a new gasket, coated on both sides with gasket sealer. Tighten the water pump bolts to 6½-10 ft.-lb. (0.9-1.4 mkg).

## HEATER

The heater is located under the dash.

### Removal/Installation (F10)

Refer to **Figure 9** for this procedure.
1. Disconnect the negative cable from the battery.
2. Drain the cooling system. See *Cooling System Flushing* earlier in this chapter.
3. Disconnect the water hoses from the heater.
4. Detach the defroster hose from each side of the heater.
5. Remove the cable clamps for heater cock, floor door, and intake door.
6. Disconnect the fan motor wires.
7. Remove 4 heater mounting screws. Remove the heater into the passenger compartment.
8. Installation is the reverse of removal. Fill the cooling system as described under *Cooling System Flushing* earlier in this chapter.

### Removal/Installation (310)

Refer to **Figure 10** for this procedure.
1. Disconnect the negative cable from the battery. Set the temperature lever to HOT.
2. Drain the cooling system. See *Cooling System Flushing* earlier in this chapter.
3. Remove the instrument panel (Chapter Twelve).
4. Disconnect the control cables and rod from the heater.
5. Working in the engine compartment, disconnect the heater hoses (**Figure 11**).
6. Remove the heater mounting screws (**Figure 12**). Remove the heater assembly into the engine compartment.
7. Installation is the reverse of removal. Fill the cooling system as described under *Cooling System Flushing* in this chapter.

**CHAPTER SIX**

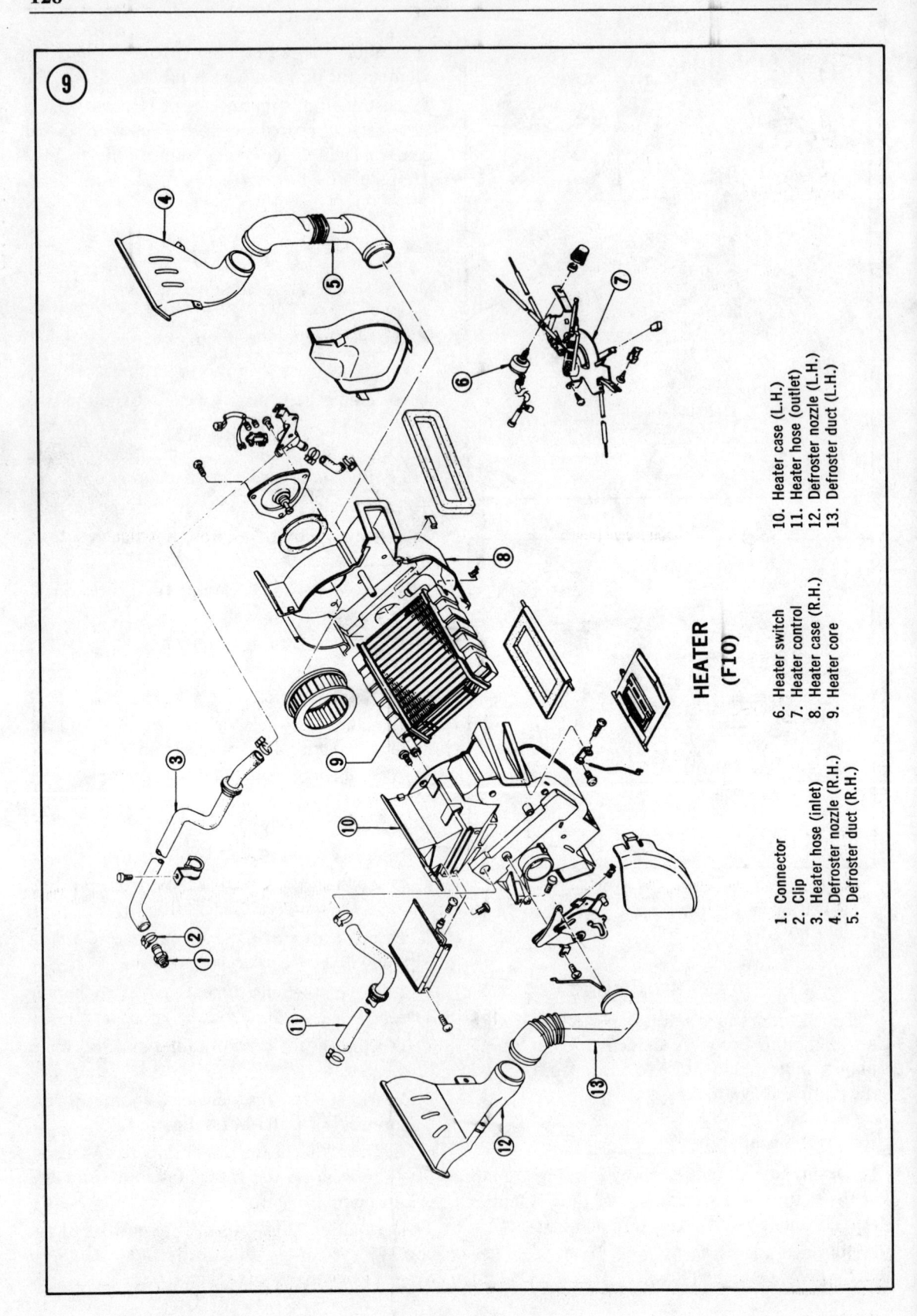

**HEATER (F10)**

1. Connector
2. Clip
3. Heater hose (inlet)
4. Defroster nozzle (R.H.)
5. Defroster duct (R.H.)
6. Heater switch
7. Heater control
8. Heater case (R.H.)
9. Heater core
10. Heater case (L.H.)
11. Heater hose (outlet)
12. Defroster nozzle (L.H.)
13. Defroster duct (L.H.)

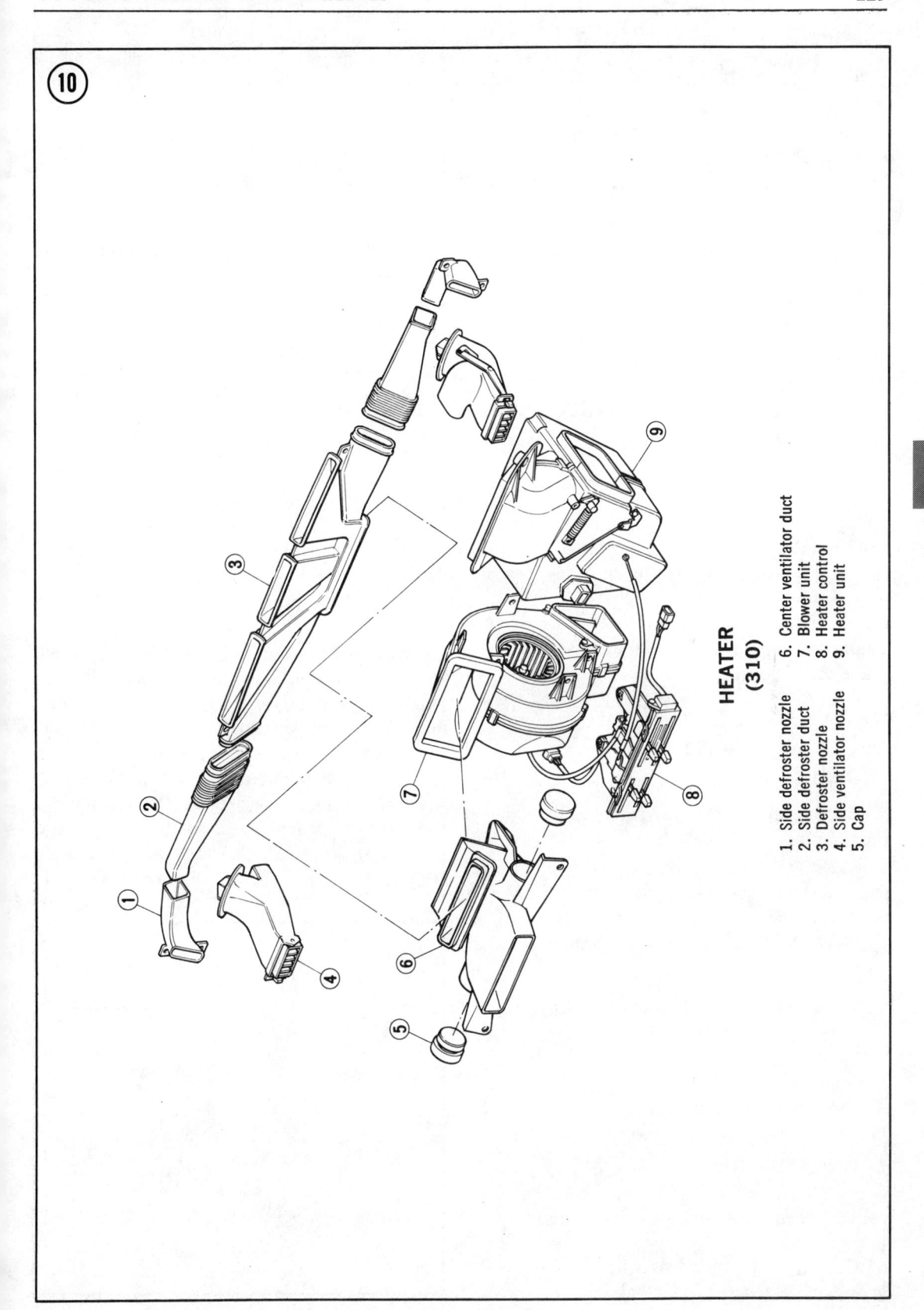

# COOLING SYSTEM AND HEATER

**HEATER (310)**

1. Side defroster nozzle
2. Side defroster duct
3. Defroster nozzle
4. Side ventilator nozzle
5. Cap
6. Center ventilator duct
7. Blower unit
8. Heater control
9. Heater unit

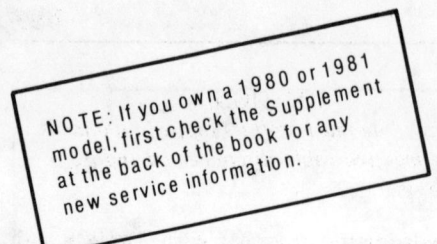

> NOTE: If you own a 1980 or 1981 model, first check the Supplement at the back of the book for any new service information.

# CHAPTER SEVEN

# ELECTRICAL SYSTEM

All models use a 12-volt negative ground electrical system. This chapter includes service procedures for the battery, charging system, starter, lighting and ignition systems, fuses, instruments, and windshield wipers. Wiring diagrams are included at the end of the book.

## BATTERY

### Care and Inspection

1. Disconnect both battery cables and remove the battery.
2. Clean the top of the battery with a baking soda and water solution. Scrub with a stiff bristle brush. Wipe battery clean with a cloth moistened in ammonia or baking soda solution.

> CAUTION
> Keep cleaning solution out of battery cells or the electrolyte will be seriously weakened.

3. Clean battery terminals with a stiff wire brush or terminal cleaning tool.
4. Examine battery case for cracks.
5. Install the battery and reconnect the battery cables. Be sure the cables are connected to the proper terminals.
6. Coat the battery cables with light mineral grease or Vaseline after tightening.
7. Check electrolyte level and top up with distilled water if necessary.

### Testing

Hydrometer testing is the best way to check battery condition. Use a hydrometer with numbered graduations from 1.100-1.300 rather than one with just color-coded bands. To use the hydrometer, squeeze the rubber ball, insert the tip in the cell, and release the ball (**Figure 1**).

Draw enough electrolyte to float the weighted float inside the battery. Note the number in line with the surface of the electrolyte. This is the specific gravity for the cell.

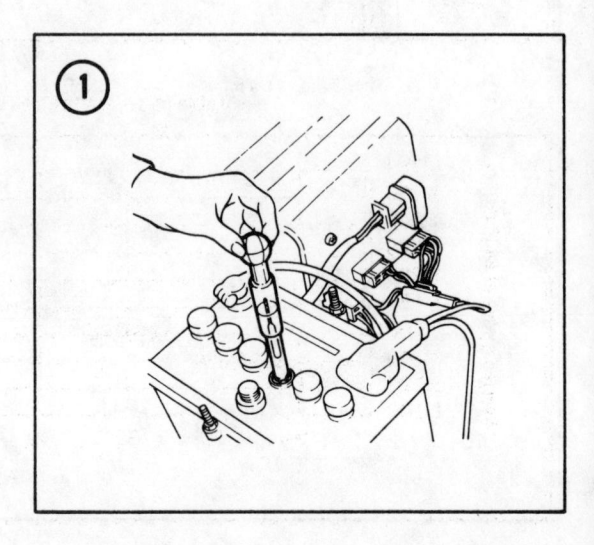

# ELECTRICAL SYSTEM

Return the electrolyte to the cell from which it came.

The specific gravity of the electrolyte in each battery cell indicates that cell's condition. A fully charged cell will read from 1.260-1.280 at 68°F (20°C). If the cells test below 1.200, the battery must be recharged. Charging is also necessary if the specific gravity varies more than 0.025 from cell to cell. **Table 1** converts specific gravity readings into battery charge percentages.

*NOTE*
*For every 10° above 80° F (25° C) electrolyte temperature, add 0.004 to specific gravity reading. For every 10° below 80° F (25° C), subtract 0.004.*

**CAUTION**
*Electrolyte must be fully topped up and the negative cable disconnected before charging.*

The hydrometer, together with **Table 1**, can be used to check the progress of the charging operation.

## ALTERNATOR

The 1976-1977 models use an external regulator, mounted on the relay bracket in the engine compartment. The 1978 and later models use an integrated circuit regulator, mounted inside the alternator.

Service procedures are similar for all models. Differences are noted when they occur. **Figure 2**

**ALTERNATOR (1976-1977)**

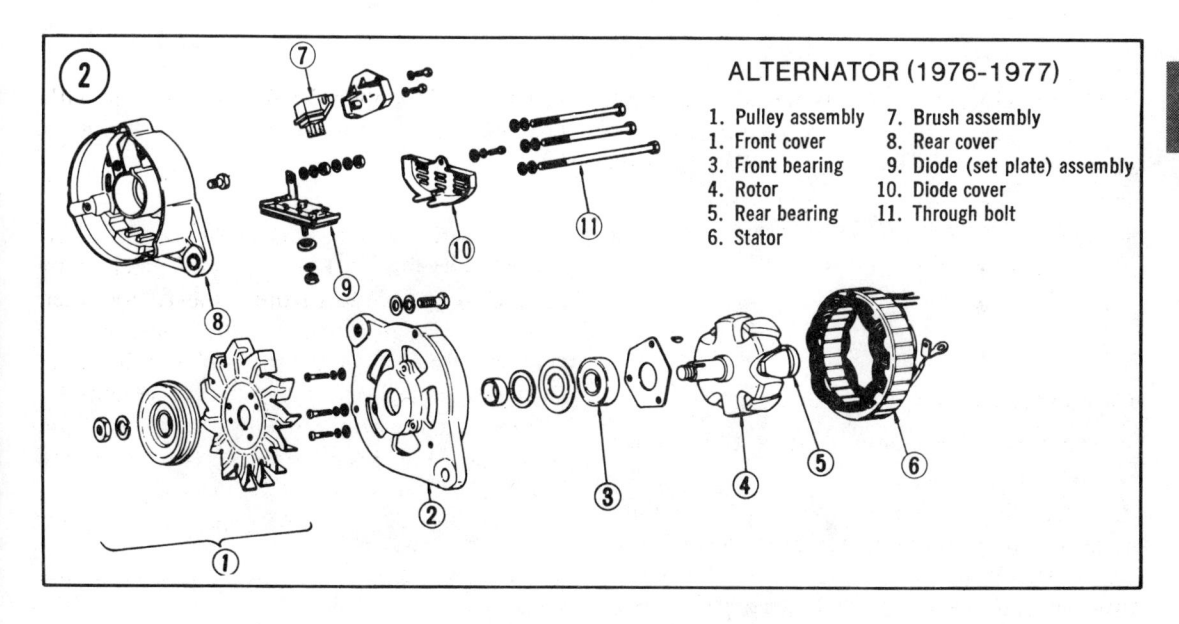

1. Pulley assembly
2. Front cover
3. Front bearing
4. Rotor
5. Rear bearing
6. Stator
7. Brush assembly
8. Rear cover
9. Diode (set plate) assembly
10. Diode cover
11. Through bolt

Table 1   BATTERY CHARGE PERCENTAGE

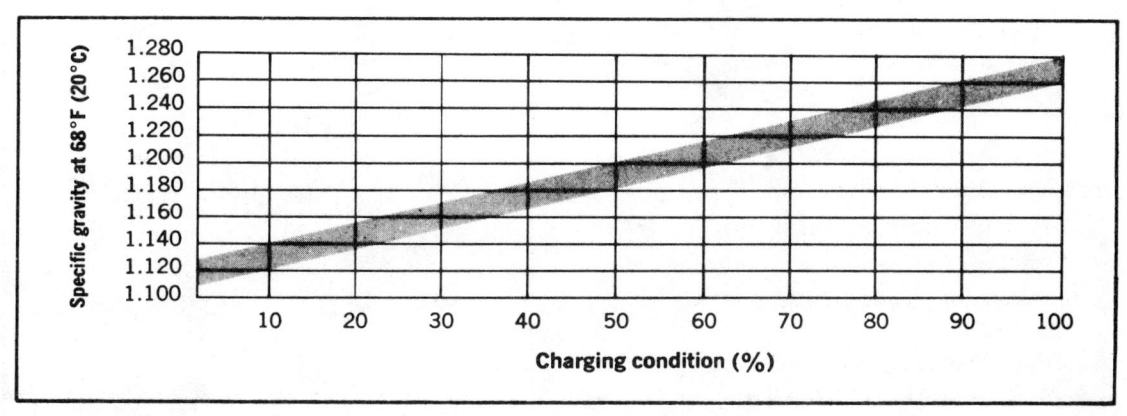

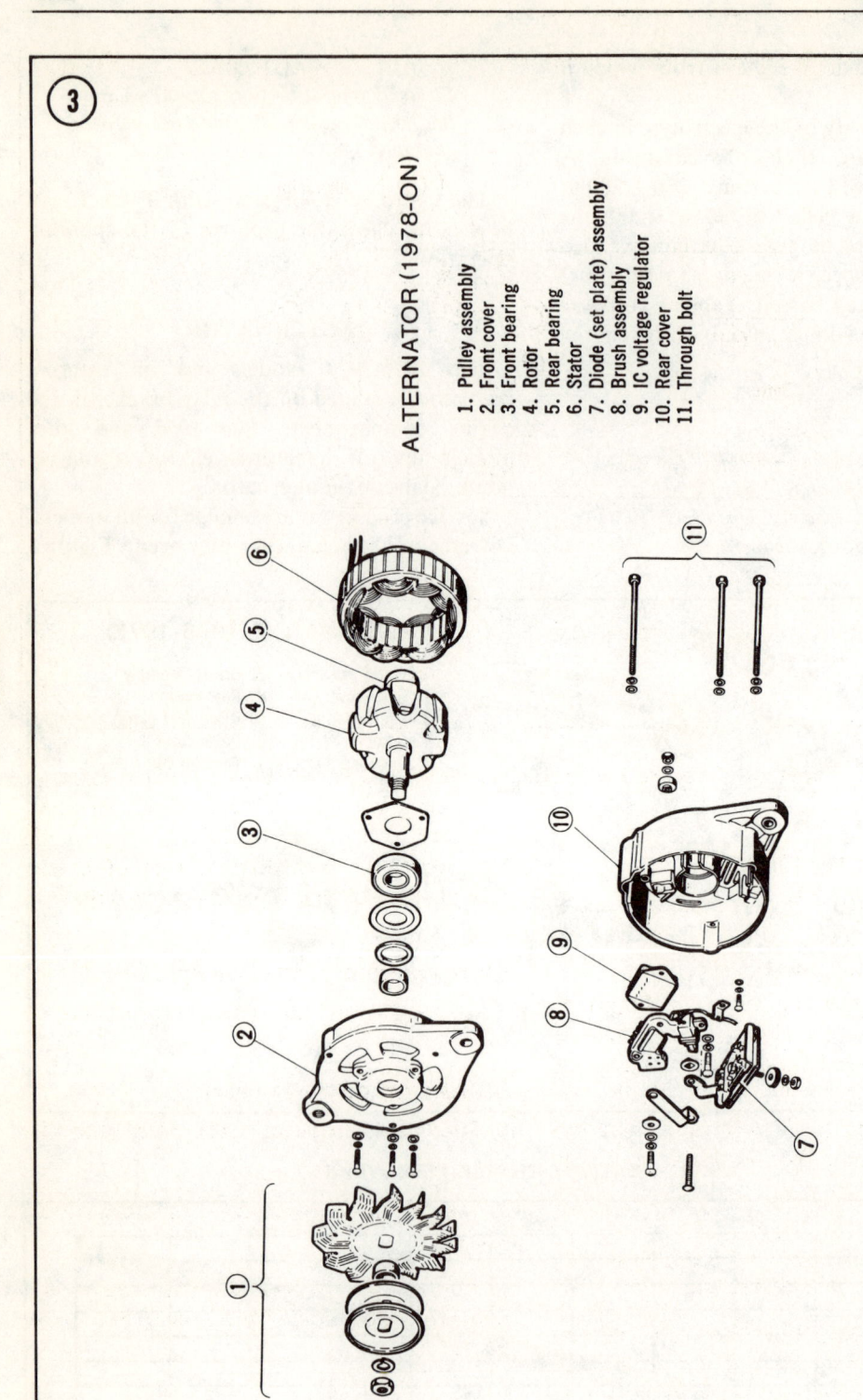

## ALTERNATOR (1978-ON)

1. Pulley assembly
2. Front cover
3. Front bearing
4. Rotor
5. Rear bearing
6. Stator
7. Diode (set plate) assembly
8. Brush assembly
9. IC voltage regulator
10. Rear cover
11. Through bolt

# ELECTRICAL SYSTEM

**④**

*Alternator*
*Battery*
*Voltmeter*

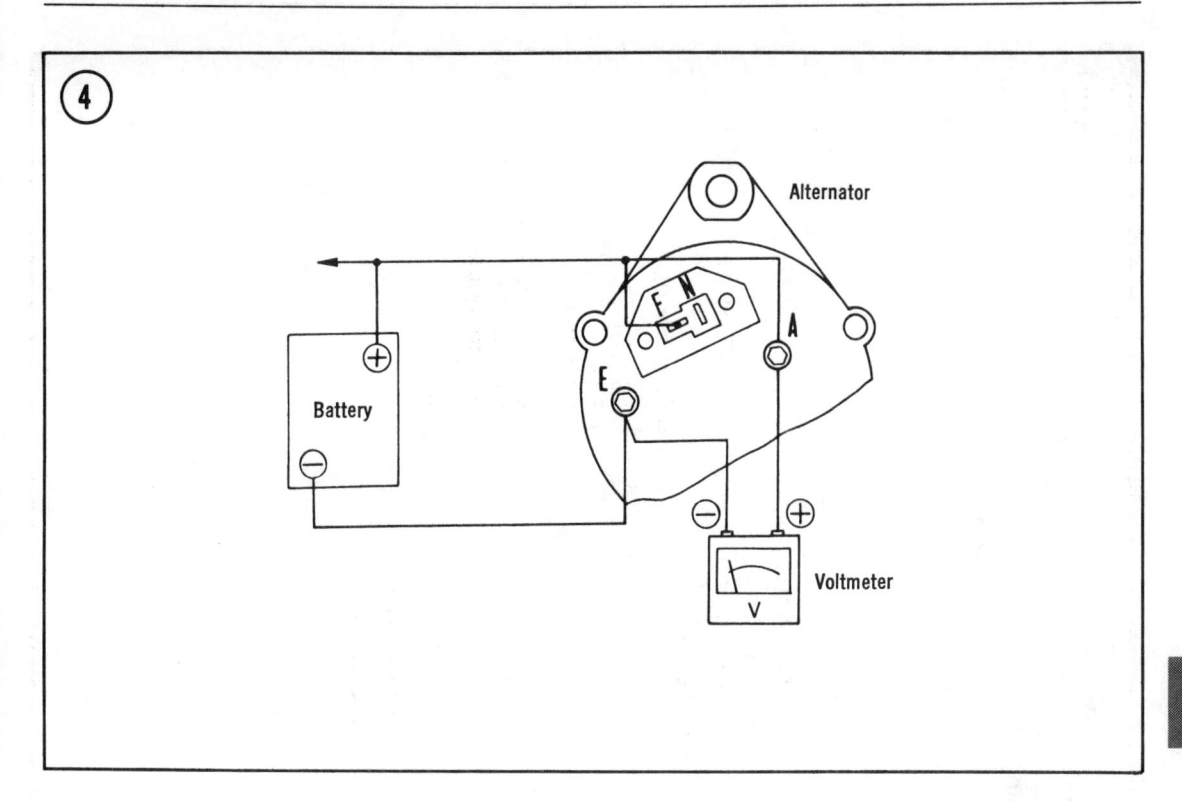

shows the 1976-1977 alternator. **Figure 3** shows the 1978 and later version.

**Alternator Output Test (Through 1977)**

This test requires a 30-volt voltmeter and a fully charged battery.

1. Disconnect the alternator wires.

2. Set up the test circuit shown in **Figure 4**. Measure the voltage. It should be the same as when the voltmeter is connected between the battery terminals.

3. Start the engine. Gradually increase engine speed to 1,100 rpm, then note the voltmeter reading.

> **CAUTION**
> *Do not run the engine at speeds above 1,000 rpm. Do not race the engine.*

The voltmeter should indicate 12.5 volts or more. If the reading is low, the alternator should be disassembled and tested as described later in this chapter.

**Regulator Testing (Through 1977)**

This is a complicated procedure that requires equipment not usually possessed by the home mechanic. Take the job to a Datsun dealer or other competent garage. Some regulator problems can be corrected by adjustment, but the cost should be compared to that of a new regulator before choosing a course of action.

**Alternator Output Test (1978 and Later)**

Refer to **Figures 5 and 6** for this procedure.

**Alternator Removal/Installation**

1. Disconnect the negative cable from the battery.

2. Disconnect the wires from the back of the alternator.

3. Loosen the alternator mounting bolts. Remove the alternator drive belt. See *Drive Belts*, Chapter Three.

4. Remove the alternator mounting bolts. Take the alternator out.

5. Installation is the reverse of removal. Adjust drive belt tension as described in Chapter Three.

## CHAPTER SEVEN

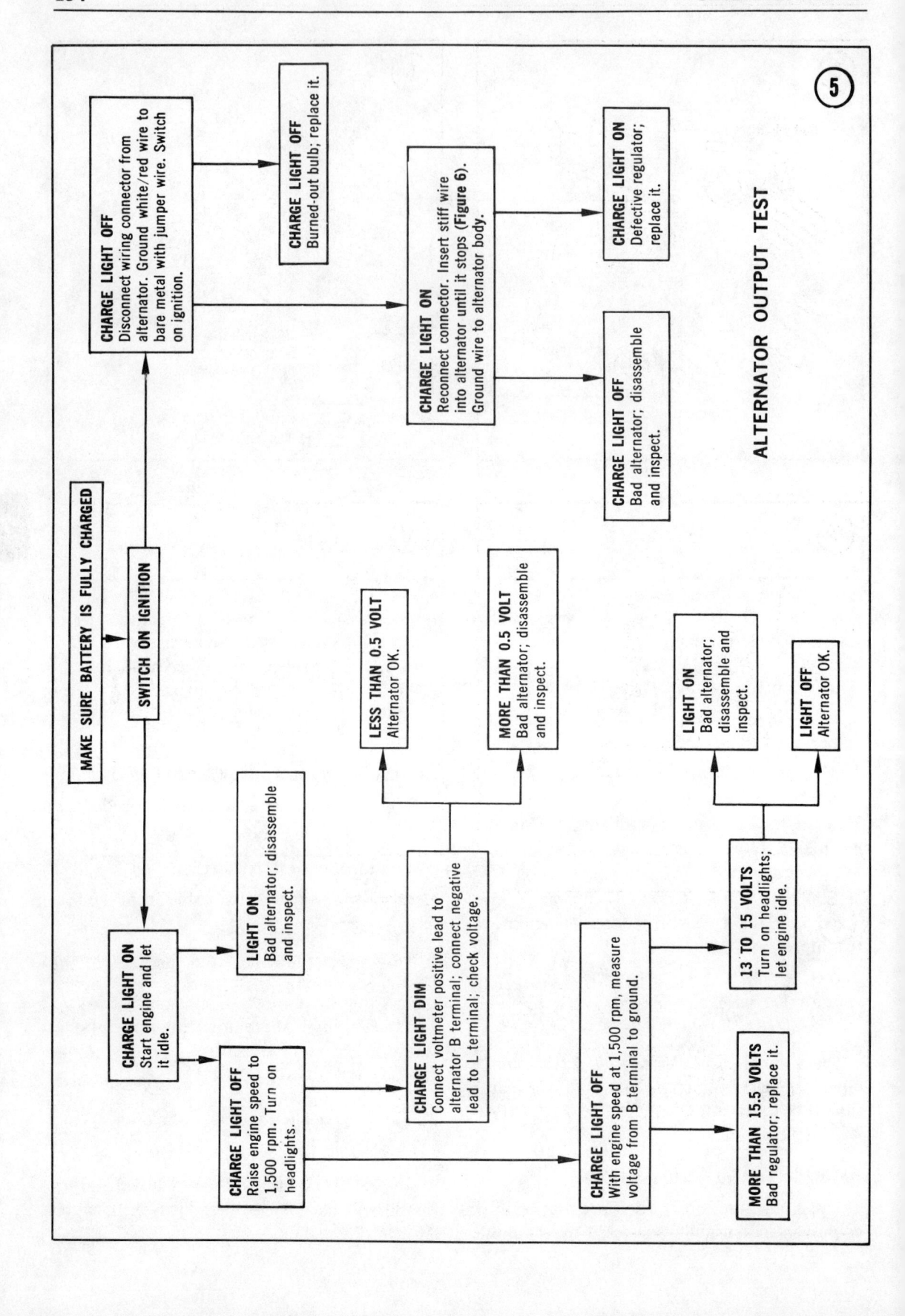

### ALTERNATOR OUTPUT TEST

**MAKE SURE BATTERY IS FULLY CHARGED**

**SWITCH ON IGNITION**

**CHARGE LIGHT OFF** — Disconnect wiring connector from alternator. Ground white/red wire to bare metal with jumper wire. Switch on ignition.

- **CHARGE LIGHT OFF** — Burned-out bulb; replace it.
- **CHARGE LIGHT ON** — Reconnect connector. Insert stiff wire into alternator until it stops (**Figure 6**). Ground wire to alternator body.
  - **CHARGE LIGHT ON** — Defective regulator; replace it.
  - **CHARGE LIGHT OFF** — Bad alternator; disassemble and inspect.

**CHARGE LIGHT ON** — Start engine and let it idle.

- **LIGHT ON** — Bad alternator; disassemble and inspect.
- **CHARGE LIGHT OFF** — Raise engine speed to 1,500 rpm. Turn on headlights.
  - **CHARGE LIGHT DIM** — Connect voltmeter positive lead to alternator B terminal; connect negative lead to L terminal; check voltage.
    - **LESS THAN 0.5 VOLT** — Alternator OK.
    - **MORE THAN 0.5 VOLT** — Bad alternator; disassemble and inspect.
  - **CHARGE LIGHT OFF** — With engine speed at 1,500 rpm, measure voltage from B terminal to ground.
    - **13 TO 15 VOLTS** — Turn on headlights; let engine idle.
      - **LIGHT ON** — Bad alternator; disassemble and inspect.
      - **LIGHT OFF** — Alternator OK.
    - **MORE THAN 15.5 VOLTS** — Bad regulator; replace it.

⑤

# ELECTRICAL SYSTEM

⑥

Suitable wire

⑨

1. "N" terminal
2. Brush holder
3. Brush holder cover

⑦

⑩

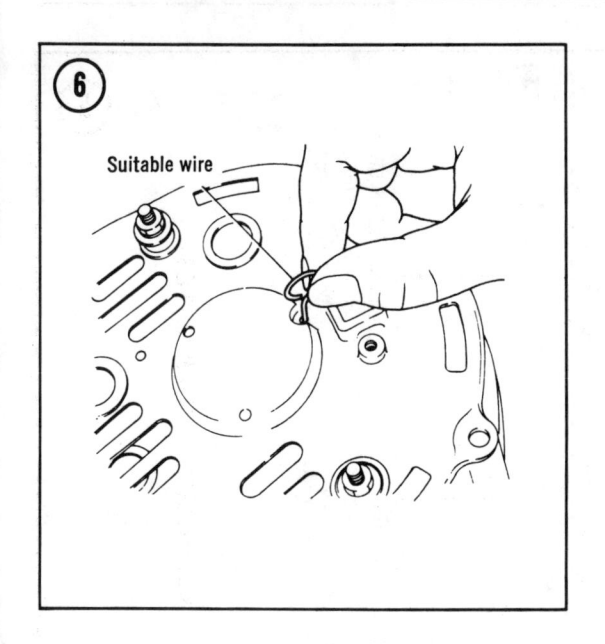

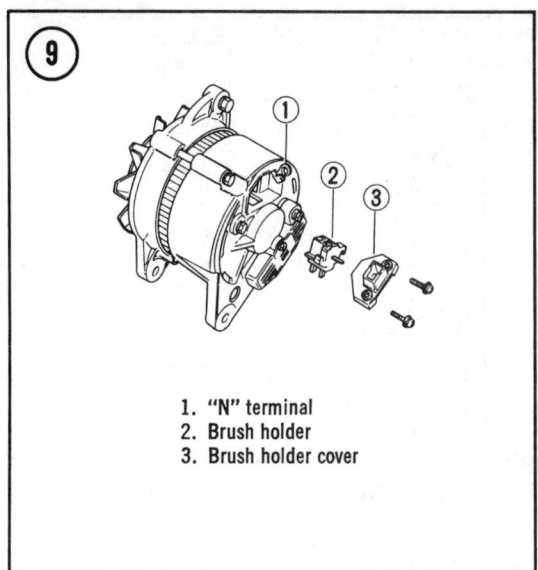

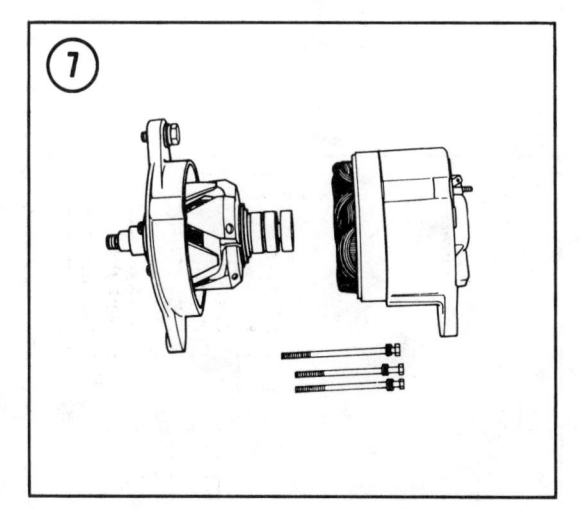

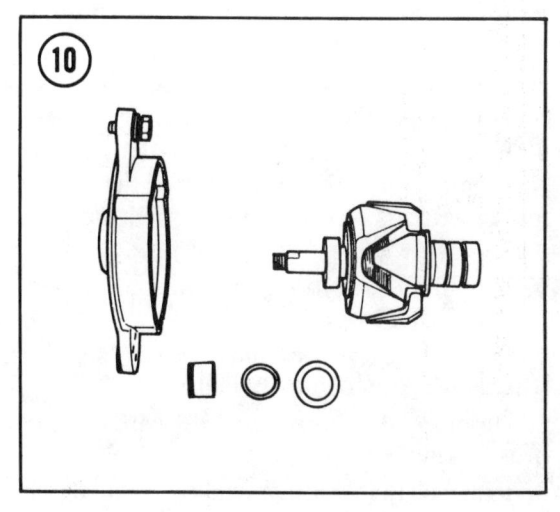

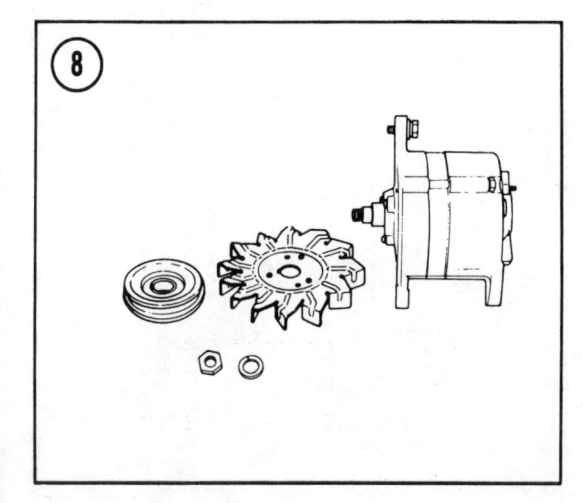

⑧

## Alternator Disassembly (1976-1977)

1. Unscrew 3 through-bolts.

2. Separate the diode end housing from the drive end housing by tapping the drive end housing lightly with a soft-faced mallet. **Figure 7** shows the through-bolts and separated housings.

3. Place the drive end in a soft-jawed vise. Remove the pulley nut, pulley, and fan (**Figure 8**).

4. Remove the brush holder (**Figure 9**).

5. Remove the bearing retainer setscrews, then separate the rotor from the drive end housing. See **Figure 10**.

6. Examine the front and rear bearings. Spin them and check for excessive noise, roughness, or play. Leave the bearings in place if serviceable. If not, remove the front bearing with a press, and the rear bearing with a press or gear puller (**Figure 11**).

7. Remove the diode cover screw and cover (**Figure 12**). Unsolder 3 stator coil leads, remove the A terminal and diode attaching nut, then remove the diode assembly.

8. Separate the diode end housing from the stator.

**Alternator Disassembly (1978 and Later)**

1. Remove 4 through-bolts. Tap the drive end loose from the stator with a soft-faced mallet. See **Figure 13**.

2. Place the drive end in a soft-jawed vise. Remove the pulley nuts, then take off the fan and pulley (**Figure 14**).

3. Remove 3 bearing retainer setscrews. Take the rotor out of the drive end (**Figure 15**).

4. Inspect the bearings. If they are loose, or if rotation is rough or noisy, replace them. Remove the rear bearings with a press or gear puller (**Figure 11**).

5. Unsolder the stator coil wires from the terminals (**Figure 16**).

6. Remove the brush assembly securing screws. Separate the stator from the end cover.

7. Unsolder the brush-to-diode wire at the brush assembly terminal.

8. Remove the brush assembly and diode holder. See **Figure 17**.

**Alternator Inspection and Repair**

For the following tests, use an ohmmeter or make a small continuity tester like the one shown in **Figure 18**.

1. Test for rotor continuity by touching test leads to the rotor slip rings (**Figure 19**). The ohmmeter should indicate continuity (test lamp should light). If the lamp stays off, or if the ohmmeter shows infinite resistance, there is an open field coil. Replace the rotor.

2. Test for a grounded rotor by touching one tester lead to the rotor core and the other to each of the slip rings (**Figure 20**). If the ohm-

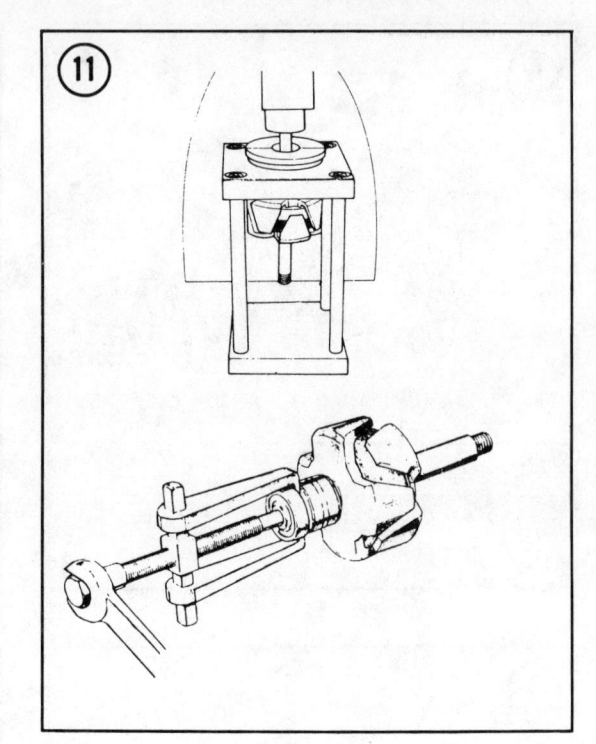

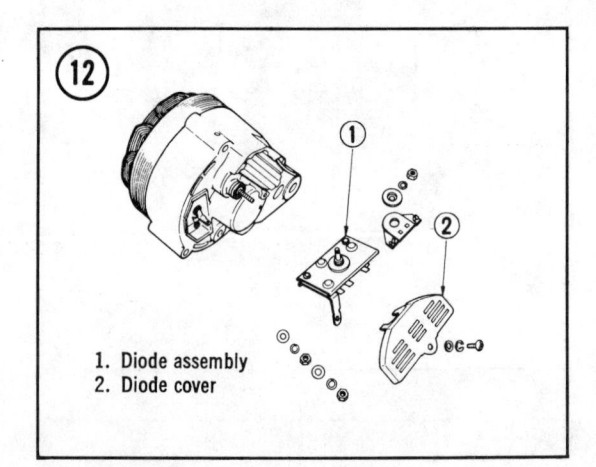

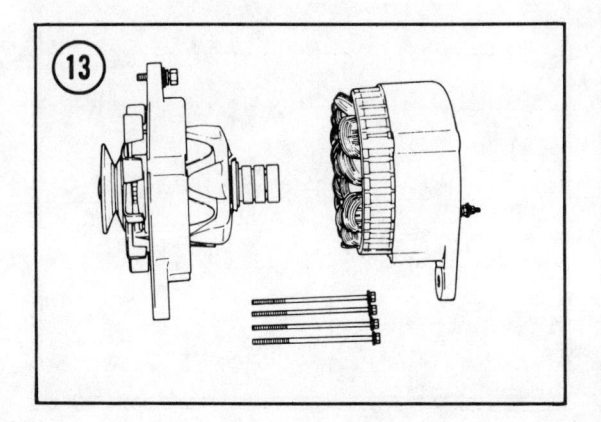

# ELECTRICAL SYSTEM

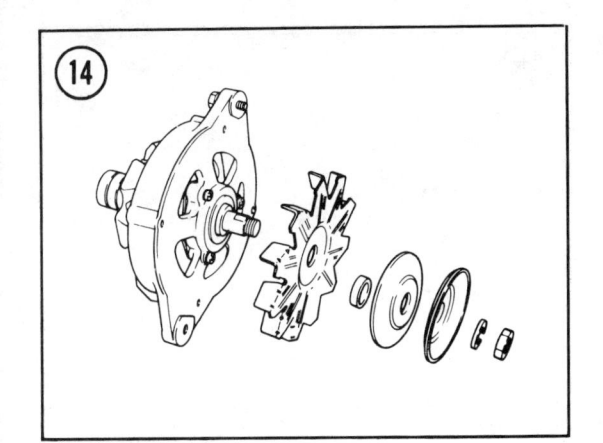

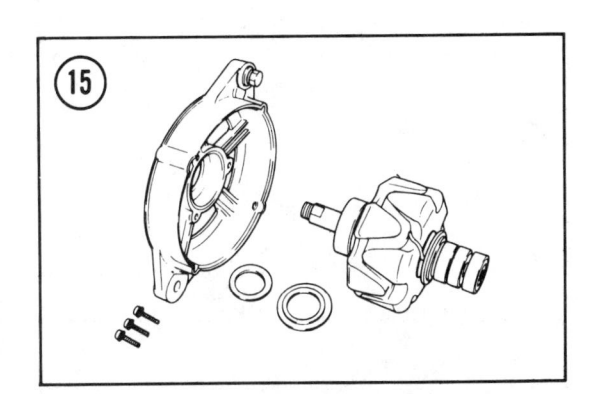

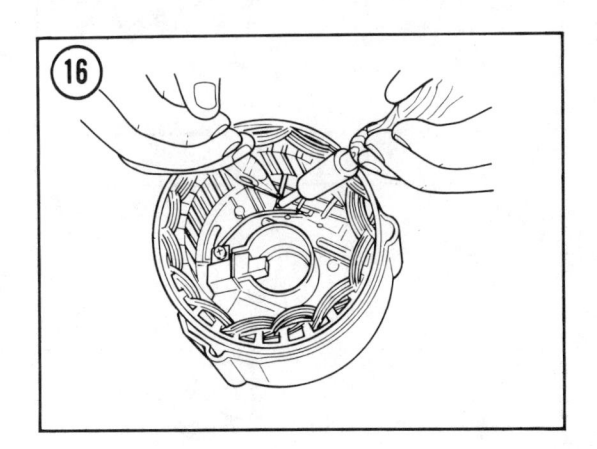

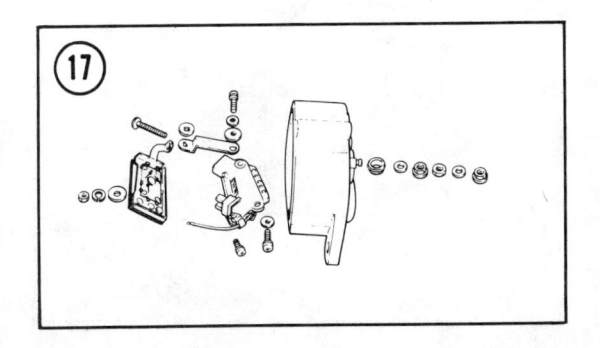

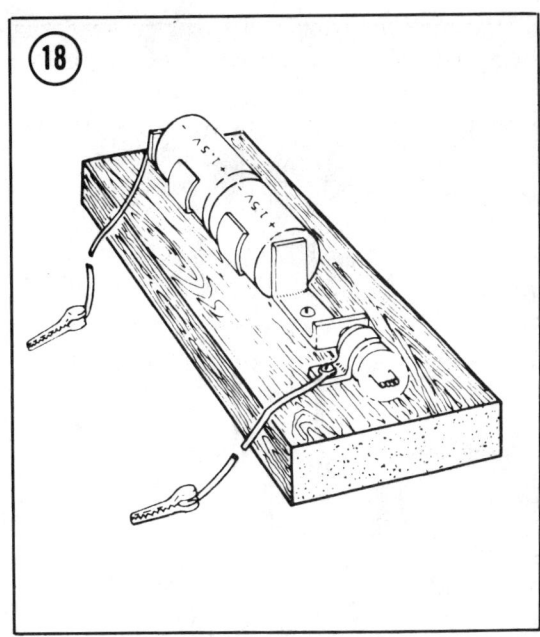

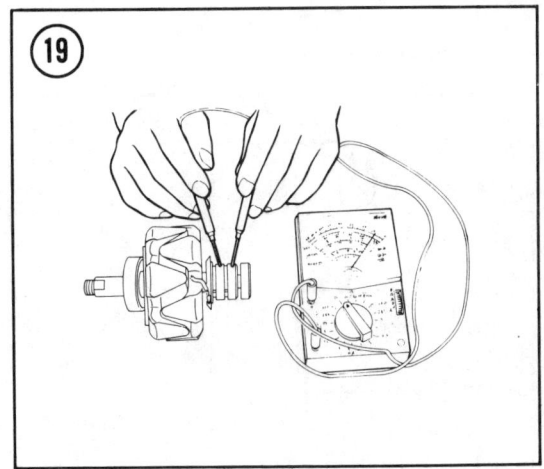

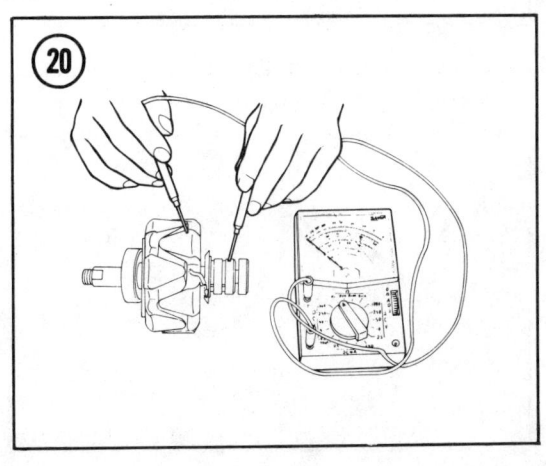

137

**CHAPTER SEVEN**

21

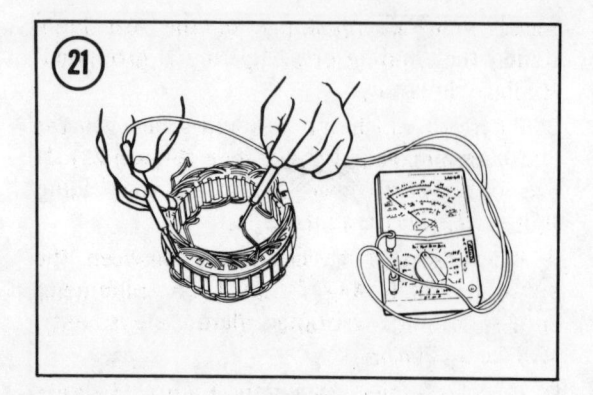

22

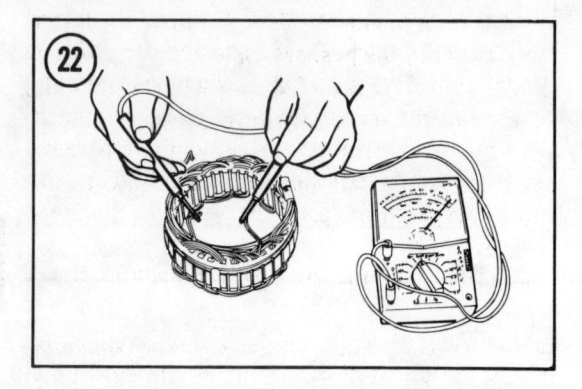

23
Conductive direction

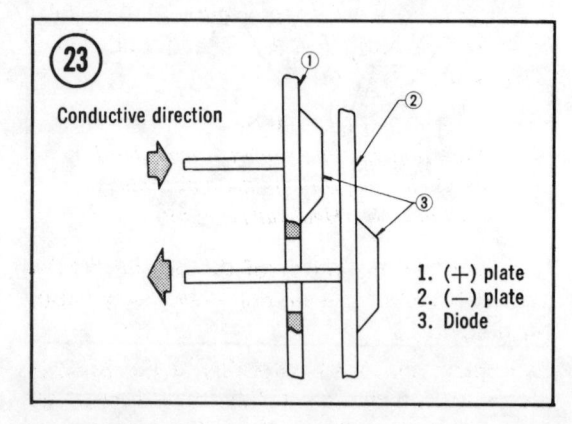

1. (+) plate
2. (−) plate
3. Diode

24

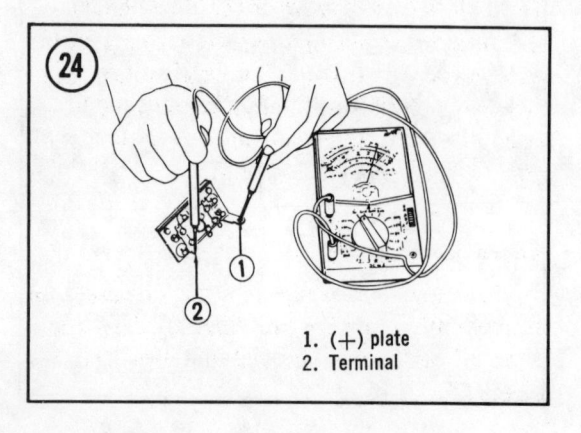

1. (+) plate
2. Terminal

25

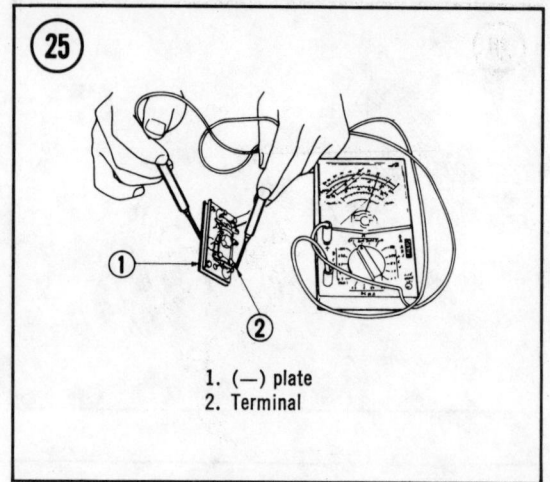

1. (−) plate
2. Terminal

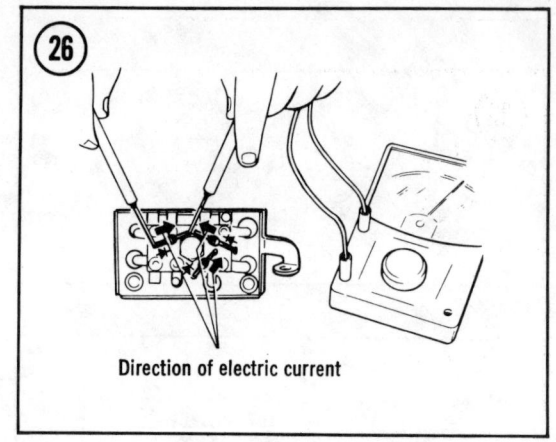

26
Direction of electric current

27
Brush wear limiting line

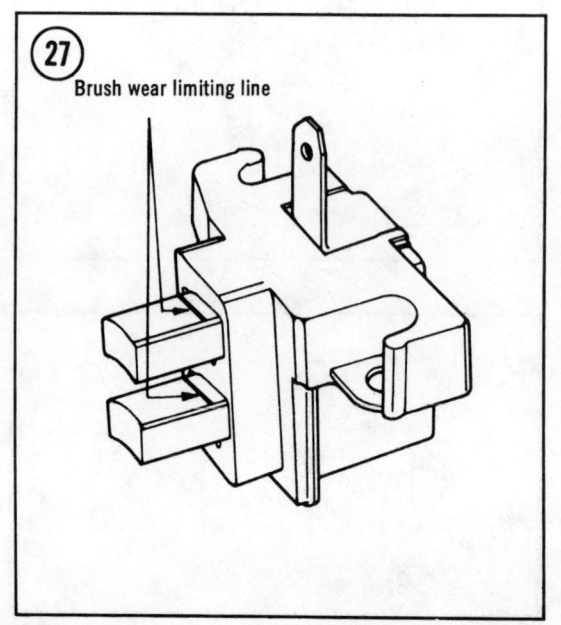

# ELECTRICAL SYSTEM

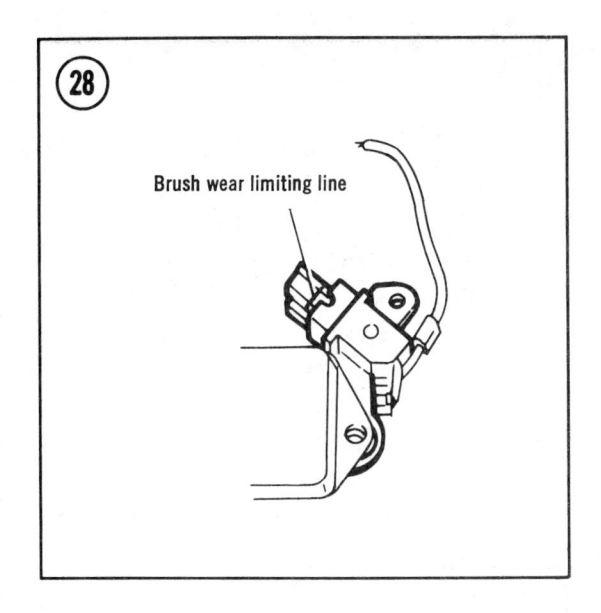

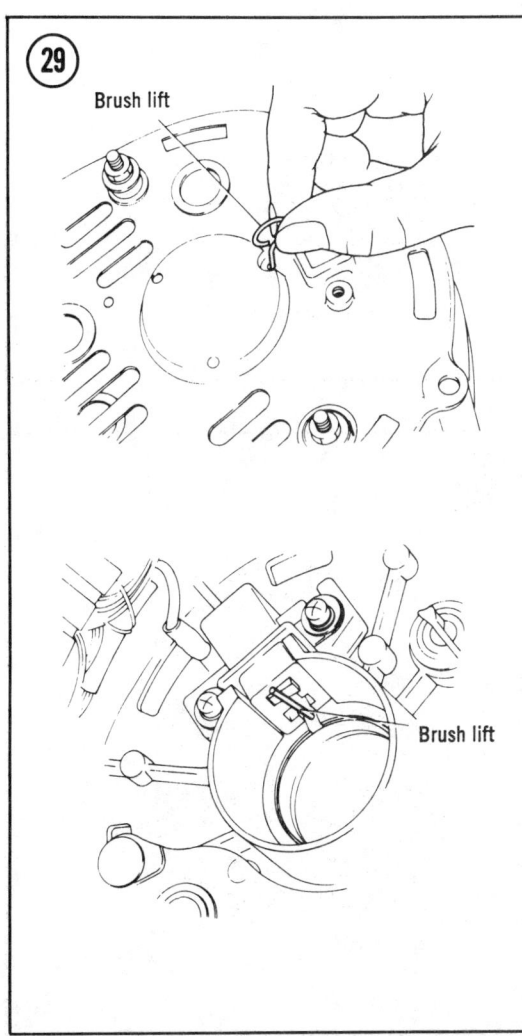

**(28)** Brush wear limiting line

**(29)** Brush lift

Brush lift

meter indicates resistance or the test lamp lights, the winding or a slip ring is grounded. Replace the rotor.

3. There should be no continuity between the stator terminals and stator core (**Figure 21**). If the ohmmeter shows continuity (test lamp lights), replace the stator.

4. There should be continuity between the stator terminals (**Figure 22**). If the ohmmeter shows infinite resistance (lamp stays out), replace the stator.

5. Check the diode assembly. **Figure 23** shows the direction of current flow. Connect the tester leads between the positive plate and the positive diode terminal (**Figure 24**). Current should only flow from the terminals to the plate. Connect the tester leads from the negative plate to each negative diode terminal (**Figure 25**). Current should only flow from the plate to the terminals.

If any diode fails this test, replace the diode plate.

6. On 1978 and later models, check the sub-diodes' current flow. Current should flow only in the directions shown in **Figure 26**. If a sub-diode is defective, unsolder it and solder in a new one.

### CAUTION
*When applying the soldering iron, hold the diode lead with needle-nosed pliers to absorb heat. Heat can ruin diodes.*

7. Check the movement of the brushes in the holder. If the brushes do not move freely, clean the holder.

8. Replace the brush assembly if the brushes are worn to the limit line. See **Figure 27** (through 1977) or **Figure 28** (1978 and later).

9. Check brush spring pressure. Using a spring scale, press the brushes into the holder until they extend 0.079 in. (2mm). Replace the brush assembly if the required pressure is less than 9 oz. (255 gr).

## Alternator Assembly

Alternator assembly is the reverse of disassembly. On 1978 and later models, use a piece of stiff wire to hold the brushes. See **Figure 29**.

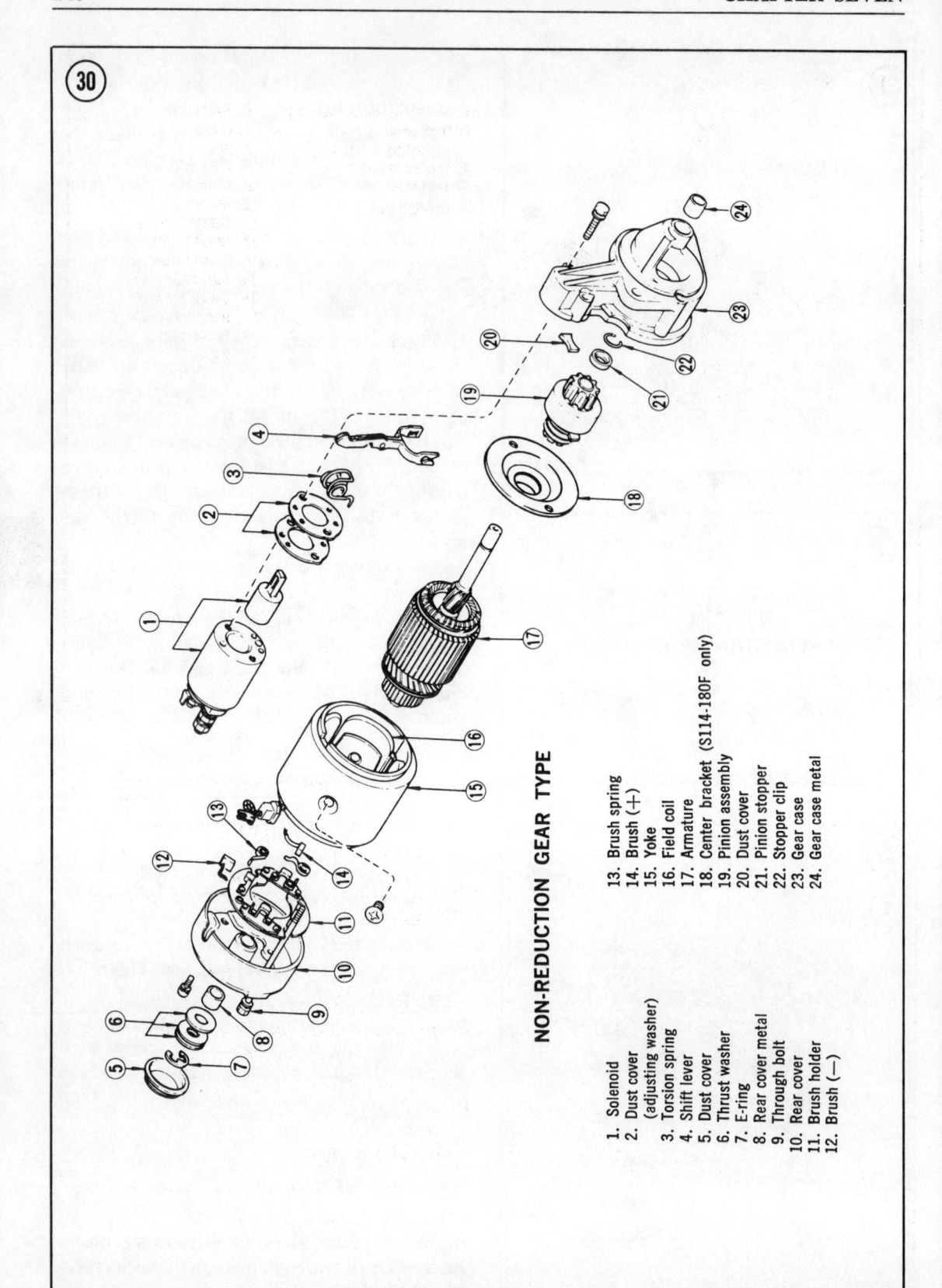

## NON-REDUCTION GEAR TYPE

1. Solenoid
2. Dust cover (adjusting washer)
3. Torsion spring
4. Shift lever
5. Dust cover
6. Thrust washer
7. E-ring
8. Rear cover metal
9. Through bolt
10. Rear cover
11. Brush holder
12. Brush (—)
13. Brush spring
14. Brush (+)
15. Yoke
16. Field coil
17. Armature
18. Center bracket (S114-180F only)
19. Pinion assembly
20. Dust cover
21. Pinion stopper
22. Stopper clip
23. Gear case
24. Gear case metal

# ELECTRICAL SYSTEM

**31**

1. Magnetic switch assembly
2. Dust cover (adjusting washer)
3. Torsion spring
4. Shift lever
5. Through bolt
6. Rear cover
7. O-ring
8. Yoke
9. Field coil
10. Brush
11. Armature
12. Center bearing
13. Brush spring
14. Brush holder
15. Dust cover
16. Center housing
17. Reduction gear
18. Pinion gear
19. Packing
20. Gear case

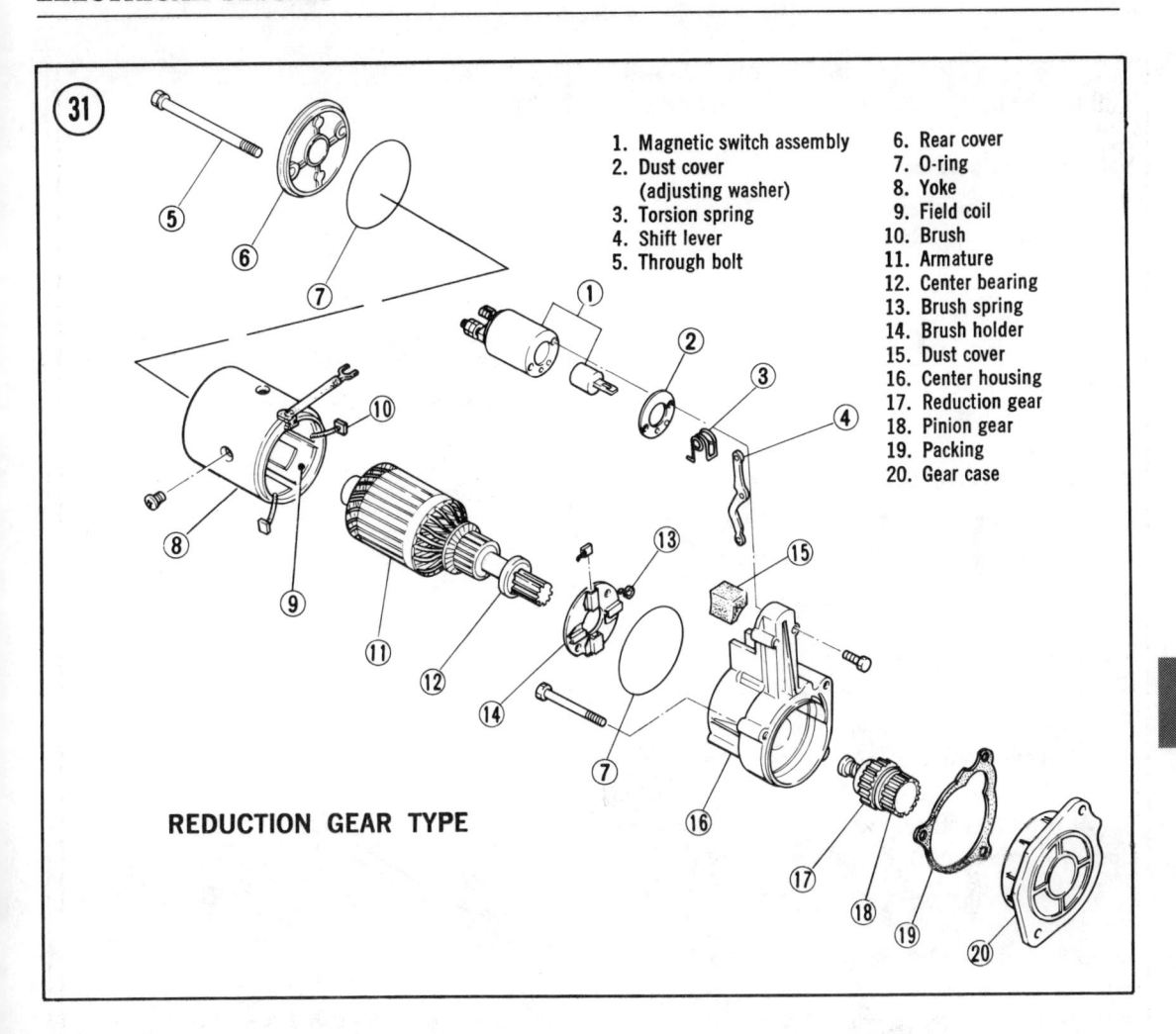

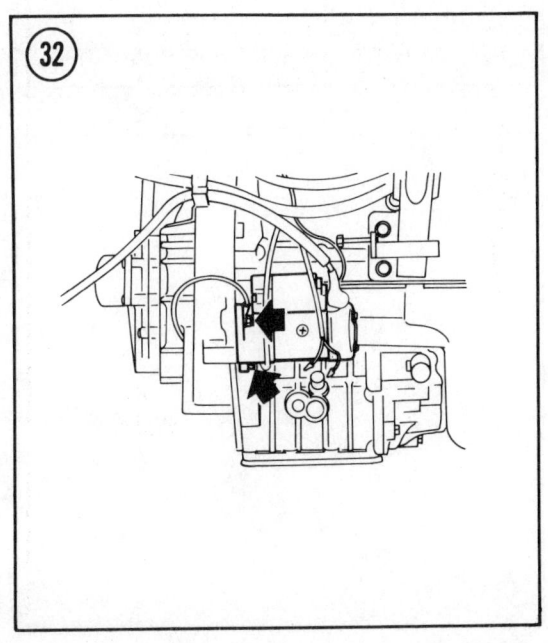

**REDUCTION GEAR TYPE**

**32**

### CAUTION
*When pulling the wire out, lever the outer end toward the center of the alternator. Otherwise, it will drag on the slip rings and scratch them.*

## STARTER

The 1976 and later models use a conventional starter as standard equipment. A reduction gear starter is optional on 1978 and later models. **Figure 30** shows the conventional starter; **Figure 31** shows the reduction gear type.

### Removal/Installation

1. Disconnect the negative cable from the battery.

2. Disconnect the wires from the starter motor (**Figure 32**). Remove the mounting bolts, then take the motor out.

3. Installation is the reverse of removal. Tighten the bolts to 20-27 ft.-lb. (2.7-3.7 mkg).

**Solenoid Replacement**

1. Remove the starter as described earlier.
2. Disconnect the wires running from solenoid to starter.
3. Remove 2 solenoid attaching screws.
4. Unhook the solenoid plunger from the shaft lever inside the starter. Lift the solenoid off.
5. Installation is the reverse of removal.

**Brush Replacement (Conventional Starter)**

1. Remove the starter as described earlier.
2. Remove the dust cover, snap ring, and thrust washer(s) from the front end of the starter. See **Figure 33**.
3. Remove 2 through-bolts and 2 setscrews (**Figure 34**). Take the end cover off the starter.
4. Make a wire hook and pull back the brush springs (**Figure 35**). Slide the brushes out of their slots.
5. Measure brush length. It should be more than 0.47 in. (12mm). Replace worn brushes.
6. Check brush movement in the slots. Clean the slots and brushes if movement is not smooth. Examine brush springs. It should take a pull of 3.1-4.0 lb. (1.4-1.8 kg) to bend the springs. Replace weak springs.
7. Check the brush holder for shorts to ground. Use an ohmmeter or continuity tester such as the one shown in **Figure 18**. Touch one tester lead to the brush holder and the other to the positive brush slots. These are the 2 slots with insulating material under them. If the ohmmeter shows continuity (test lamp lights), a short circuit exists. Replace the brush holder.
8. Install brushes by reversing Steps 1-4.

> CAUTION
> *Use resin core solder when soldering brush leads. Do not use acid core solder.*

**Brush Replacement (Reduction Gear Starter)**

1. Remove the through-bolts and front cover from the starter.
2. Separate the yoke from the center housing. See **Figure 36**.

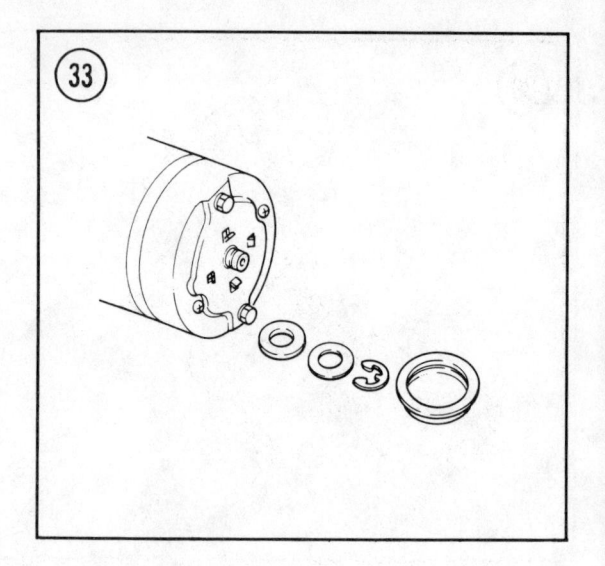

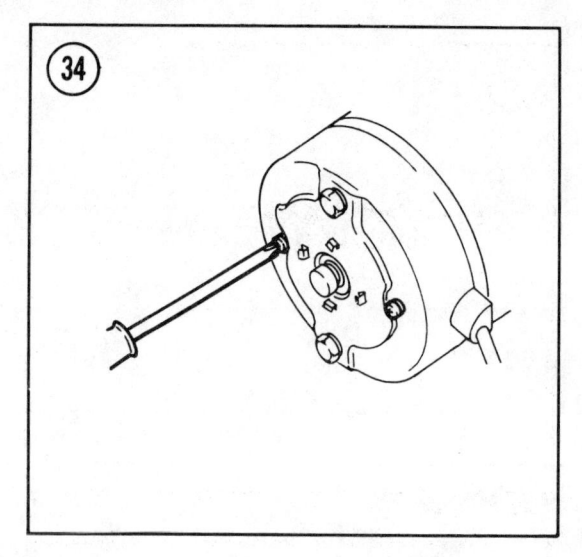

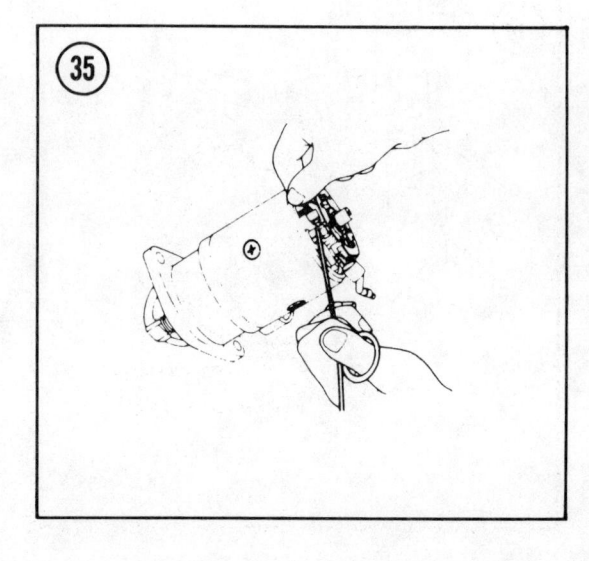

# ELECTRICAL SYSTEM

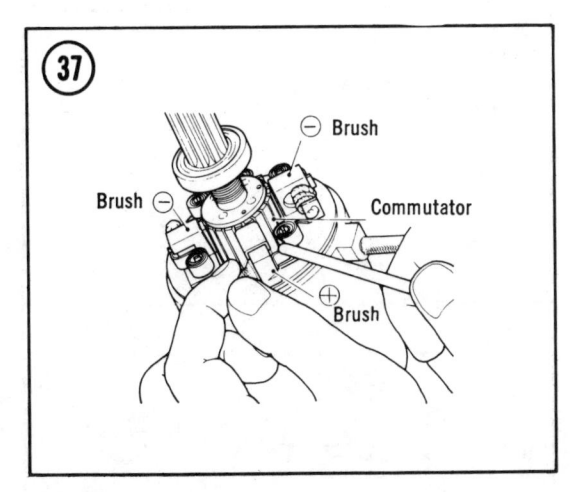

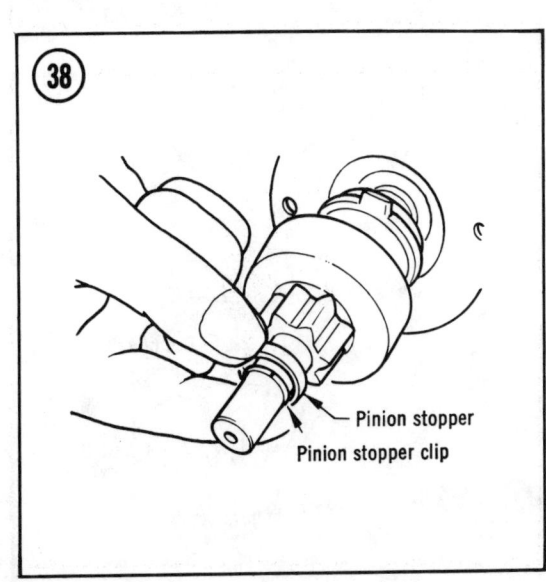

3. Make a wire hook and pull back the positive brush springs. See **Figure 37**. Pull the positive brushes out of the slots.

> NOTE: *The positive brushes are attached to the field coils, not the brush holder.*

4. Slide the brush holder off the commutator. Remove the negative brushes from their slots.

5. Measure brush length. Minimum length is 0.43 in. (11mm). Replace the brushes if they are shorter than the minimum.

6. Check brush movement in the slots. Clean the slots and brushes if movement is not smooth. Examine brush springs. Replace if weak or damaged.

7. Check the brush holder for shorts to ground. Use an ohmmeter or a continuity tester like the one shown in **Figure 18**. Touch one probe to the brush holder, and the other to the positive (insulated) brush slots. If the ohmmeter shows continuity (or test lamp lights), replace the brush holder.

## Disassembly (Conventional Starter)

1. Remove the solenoid and brushes as described earlier.
2. Slide the yoke off the armature.
3. Push the pinion stopper toward the pinion (**Figure 38**). Remove the stopper clip, then the pinion stopper and pinion assembly (overrunning clutch).

## Disassembly (Reduction Gear Starter)

1. Remove solenoid and brushes as described earlier.
2. Take the armature out of the yoke.
3. Unbolt the center housing from the gear case.
4. Remove the reduction gear and pinion gear from the gear case.

## Inspection (All Models)

1. Clean all parts with a lint-free cloth.

> CAUTION
> *Do not clean starter parts with solvent. Solvent will ruin the field coil insulation and melt the grease in the overrunning clutch.*

2. Check electrical terminals for visible wear or damage. Clean or replace as needed.

3. Test field coils for continuity. Use an ohmmeter or a self-powered test lamp like the one shown in **Figure 18**. Connect one probe to the positive terminal and the other to the positive brush leads (**Figure 39**). No reading on the ohmmeter (test lamp stays out) indicates an open field coil, which must be replaced.

4. Test for grounded field coils. Touch one test lead to the yoke and the other to the field coil positive terminal (**Figure 40**). Resistance (lamp lights) indicates a grounded field coil, which must be replaced.

5. Check the armature for visible damage such as burned windings or a worn shaft. Replace if these can be seen.

6. Inspect the commutator surface. If rough, sand it lightly with 500 grit emery paper.

7. Check commutator diameter. On conventional starters, the minimum is 1.26 in. (32mm). On reduction gear starters, the minimum is 1.14 in. (29mm). If less than this, replace the armature.

8. Check the depth of the mica between commutator segments. It should be 0.020-0.032 in. (0.5-0.8mm). If less than 0.008 in. (0.2mm), undercut the mica with a piece of hacksaw blade. **Figure 41** shows right and wrong ways to undercut the mica.

9. Inspect the soldered connections between the armature leads and commutator. Resolder loose connections with resin core solder.

10. Test for a grounded armature. Touch one test lead to the armature shaft and the other to each commutator segment in turn (**Figure 42**). If continuity occurs (test lamp lights), the armature is grounded and must be replaced.

11. Check the armature winding for shorts. To check, use an armature tester (growler). Take this job to an automotive electrical shop if you don't have the tester.

12. Check the armature for opens. Usually these are indicated by burn marks on the commutator caused by brushes bridging the open circuit. Use an ohmmeter or tester like the one shown in **Figure 18**. Place the tester leads on successive pairs of armature segments. If the ohmmeter shows no continuity (test lamp stays off), an open circuit exists.

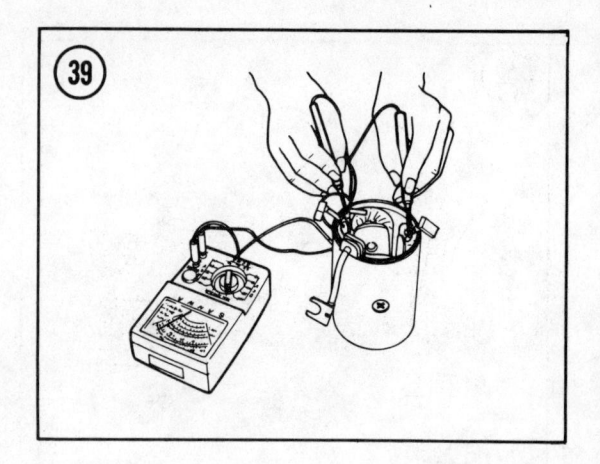

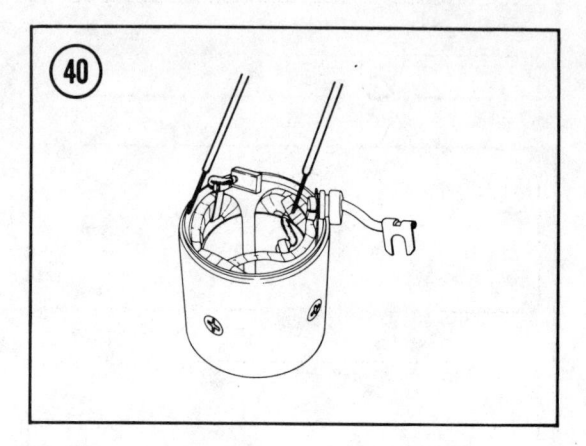

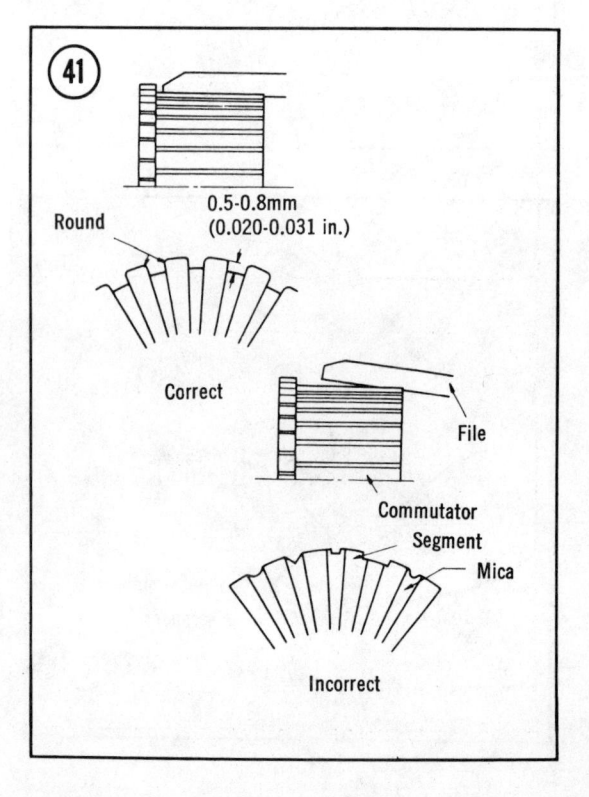

# ELECTRICAL SYSTEM

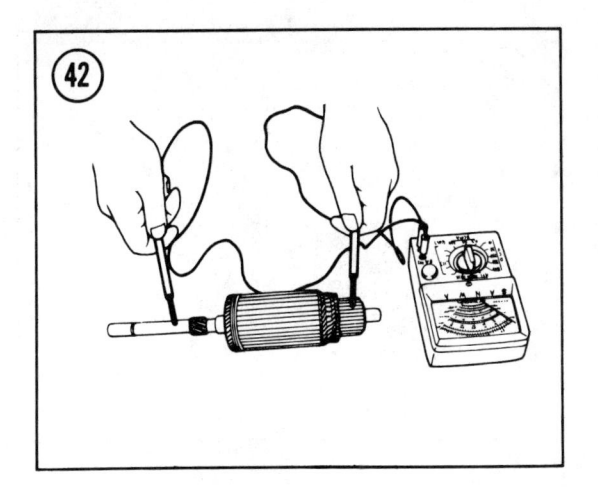

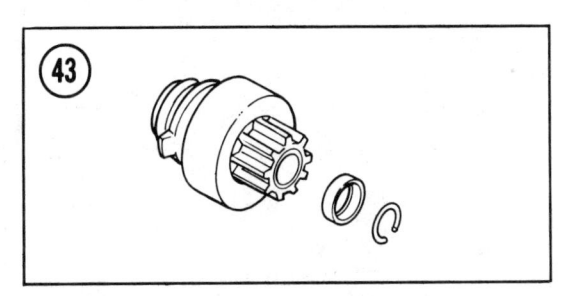

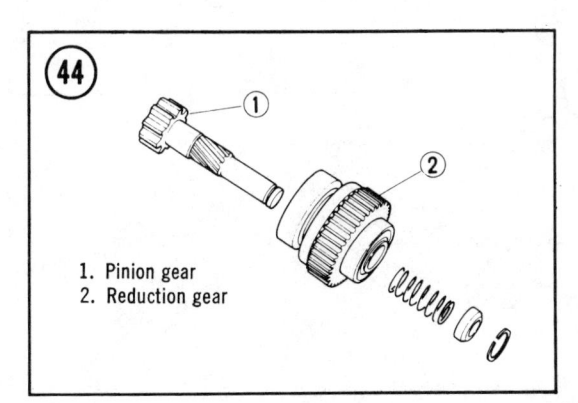

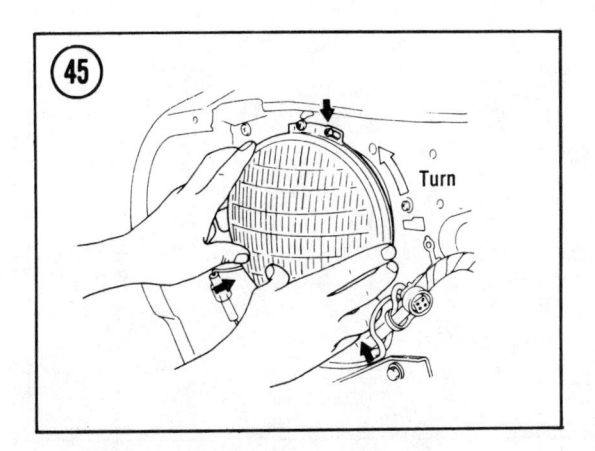

13. On conventional starters, inspect the overrunning clutch (**Figure 43**). The pinion gear should turn easily in one direction, and not at all in the other direction. If the overrunning clutch slips or drags, replace it. The pinion shaft should slide easily along the armature shaft splines.

14. On reduction gear starters, inspect the pinion and reduction gear assembly (**Figure 44**). The pinion gear should slide freely along the armature shaft splines. The pinion gear shaft must slide easily through the reduction gear. If not, replace the pinion and reduction gear assembly.

15. On conventional starters, inspect the bushing at each end of the starter. If worn, tap them out with a suitable drift. Tap in new bushings with the same tool.

16. On reduction gear starters, inspect the ball bearing at each end of the starter. Replace if looseness or rough movement can be detected.

## LIGHTING SYSTEM

### Headlight Replacement

1. Remove the grille (Chapter Twelve).

2. Remove 3 retaining screws (**Figure 45**). Turn the retaining ring counterclockwise, then remove the retaining ring and bulb.

> NOTE: *Don't turn the 2 aiming screws.*

3. Installation is the reverse of removal. Make sure the word TOP, molded in the lens, is up. Have headlight aim checked by a Datsun dealer or certified lamp adjusting station.

### Front Parking/Turn Signal Lights and Side Marker Lights

To replace a bulb, remove the lens securing screws and take off the lens. Push the bulb into the socket, turn counterclockwise and take the bulb out. Installation is the reverse of removal.

### Rear Combination Lights

All rear combination lights (stop, tail, backup, and turn indicators) are accessible from inside the luggage compartment. On F10 hatchbacks and wagons, and all 310's, remove the ac-

146

CHAPTER SEVEN

⑥

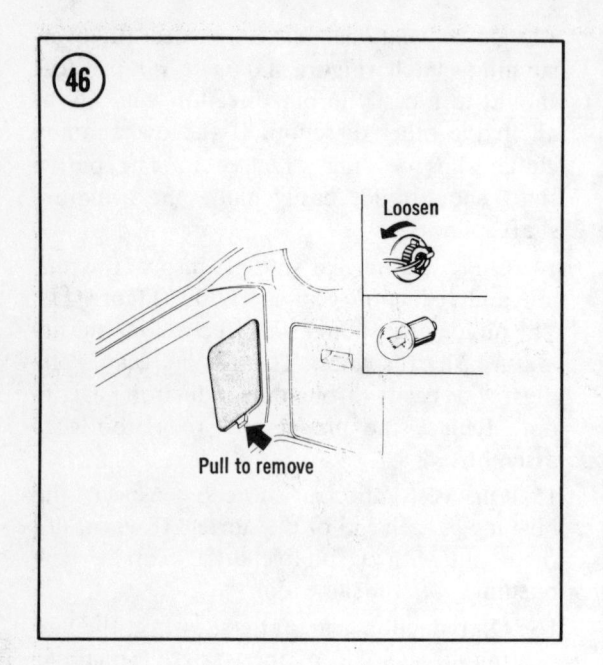

Loosen

Pull to remove

⑦

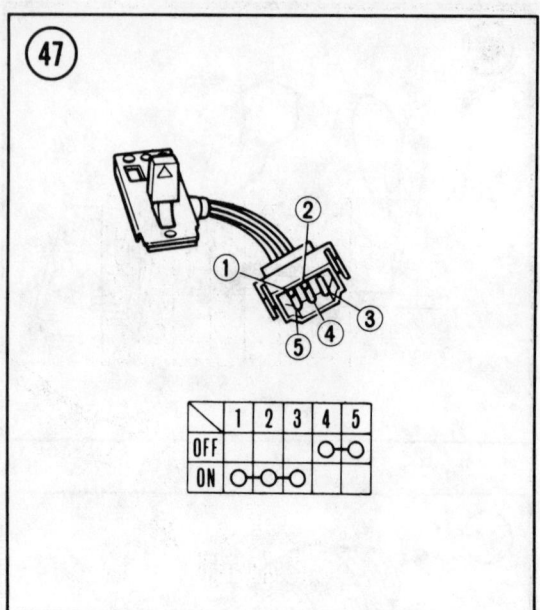

|   | 1 | 2 | 3 | 4 | 5 |
|---|---|---|---|---|---|
| OFF |   |   |   | O—O |
| ON | O—O—O |   |   |   |

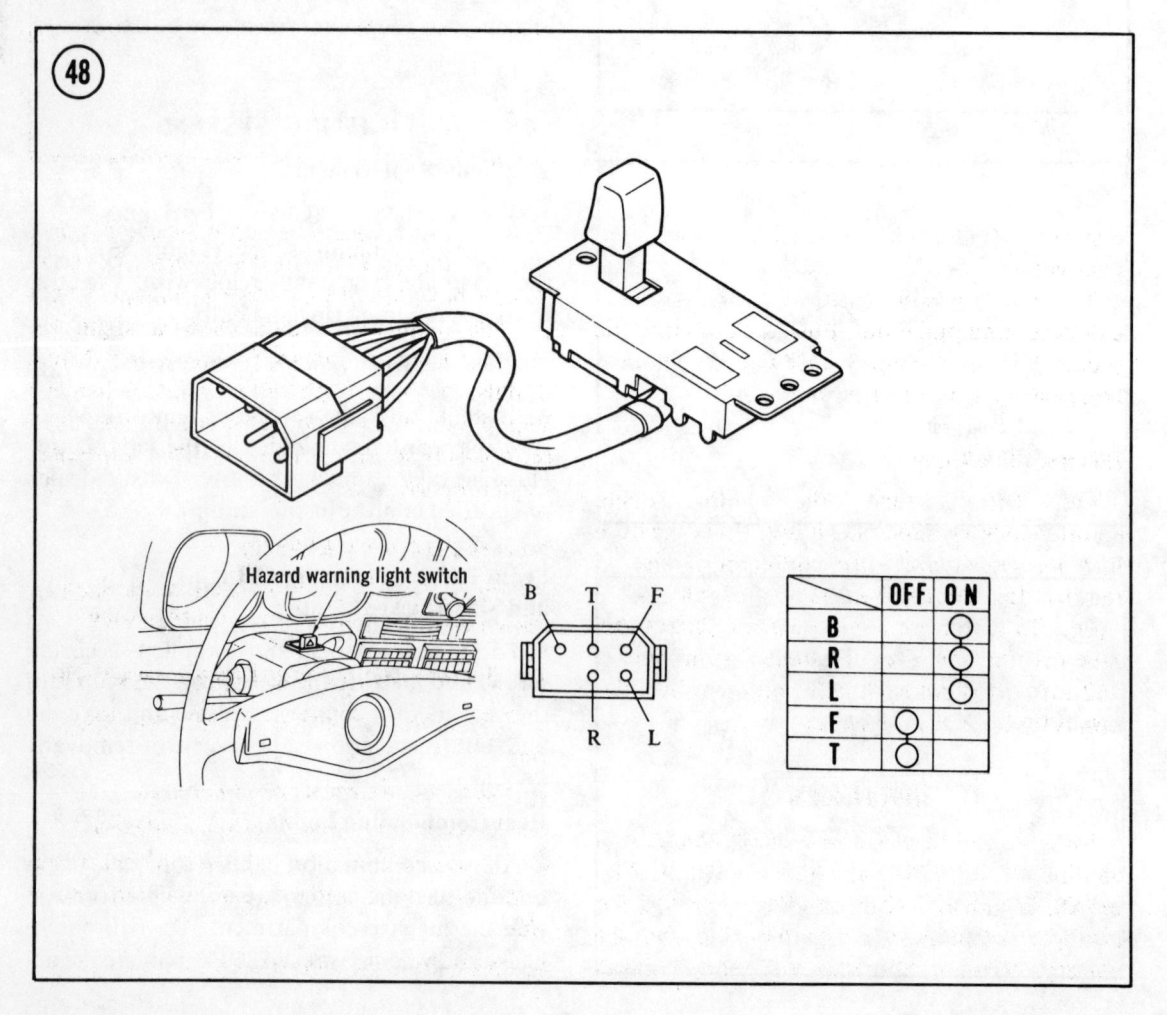

⑧

Hazard warning light switch

B  T  F

R  L

|   | OFF | ON |
|---|---|---|
| B |   | O |
| R |   | O |
| L |   | O |
| F | O |   |
| T | O |   |

# ELECTRICAL SYSTEM

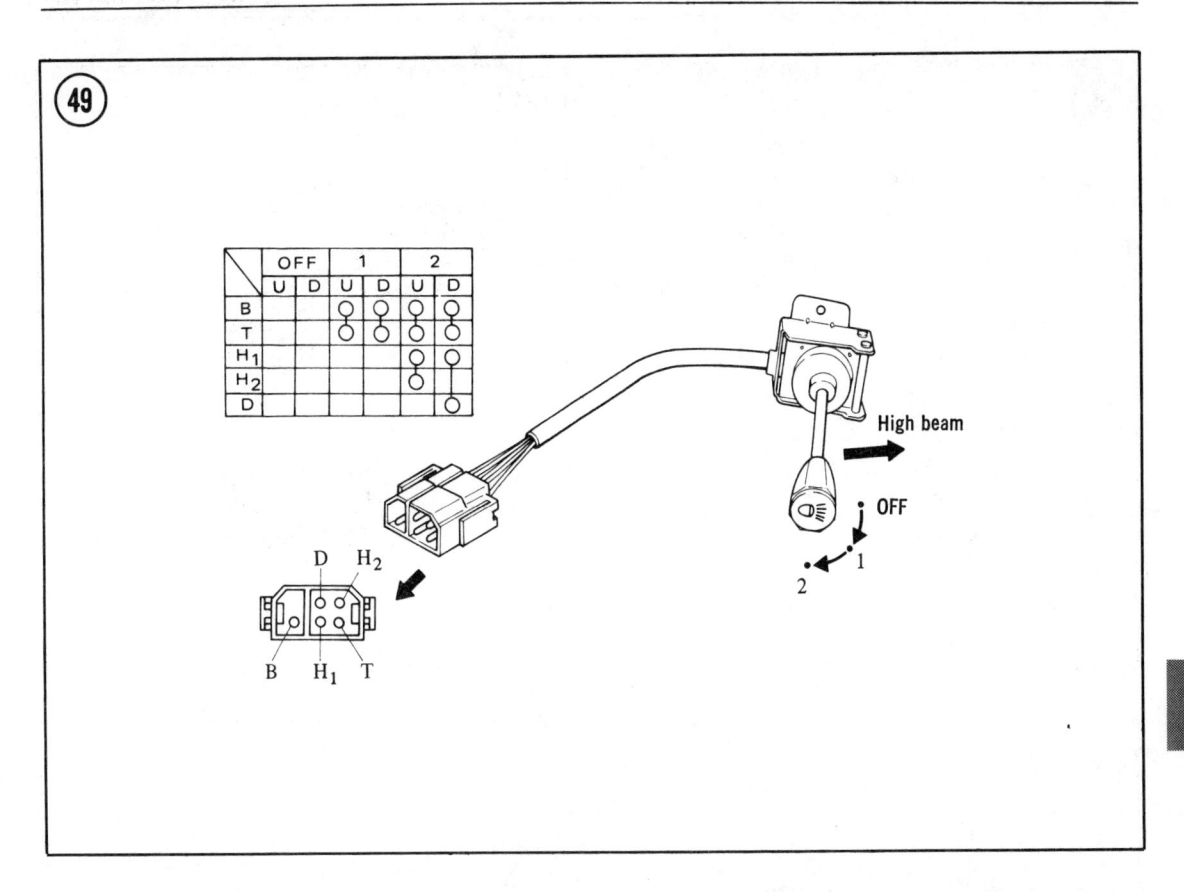

cess cover (**Figure 46**). This isn't necessary on F10 sedans.

To remove a bulb, turn the socket counterclockwise and pull it out. Push the bulb into the socket and turn counterclockwise to remove. Installation is the reverse of removal.

### License Plate Light

On F10's, turn the bulb socket counterclockwise and take it out. Press the bulb into its socket and turn counterclockwise to remove. Installation is the reverse of removal.

On 310's, remove 2 lens securing screws and take off the lens. Press the bulb into its socket and turn counterclockwise to remove. Installation is the reverse or removal.

## SWITCHES

Switches can be tested with an ohmmeter or a continuity tester like the one shown in **Figure 18**. To test an F10 hazard flasher switch, for example, disconnect the negative cable from the battery. Remove the upper steering column cover and disconnect the switch wiring connector. See **Figure 47**. With the switch off, there should be continuity between terminals T and F. The ohmmeter should show a slight resistance when connected between the 2 terminals, or a test lamp should light. When the switch is on, there should be continuity between terminals B, R, and L. If the switch doesn't perform properly, replace it.

### Hazard Flasher Switch (310)

This is tested in the same manner as the F10 switch described above. Refer to **Figure 48**.

### Headlight Switch Testing/Replacement (F10)

1. Disconnect the negative cable from the battery.
2. Remove the steering column cover.
3. Disconnect the switch wiring connector.
4. Test the switch, referring to **Figure 49**. If it is bad, remove its mounting screws and take it off.
5. Installation is the reverse of removal.

## Turn Signal Switch Testing/Replacement (F10)

1. Disconnect negative cable from battery.
2. Remove the steering wheel (Chapter Eleven). Remove the steering column cover.
3. Disconnect the switch wiring connector. Test, referring to **Figure 50**.
4. If the switch is bad, remove its attaching screws and take it off.
5. Installation is the reverse of removal.

## Wiper-Washer Switch Testing/Replacement (F10)

1. Disconnect negative cable from battery.
2. Reach behind the instrument panel. Disconnect the switch connector.
3. Disconnect the fiber optic illuminator from the illumination light. See **Figure 51**.
4. Press the switch knob in, twist, and pull off.
5. Remove the switch ring nut. If you don't have the necessary tool, insert a small screwdriver in one of the nut's notches. Tap gently counterclockwise. Do this to each notch in turn until the nut comes loose.

### CAUTION
*Don't let the screwdriver slip and scratch the instrument panel.*

6. Pull the switch out and test it, referring to **Figure 51**.
7. Installation is the reverse of removal.

## Combination Switch Testing/Replacement (310)

The combination switch controls the lights, turn signals, wipers, and washers.

1. Disconnect the negative cable from the battery.
2. Remove the steering wheel (Chapter Eleven).
3. Remove the steering column cover.
4. Disconnect the switch wiring connectors. Test it, referring to **Figure 52**.

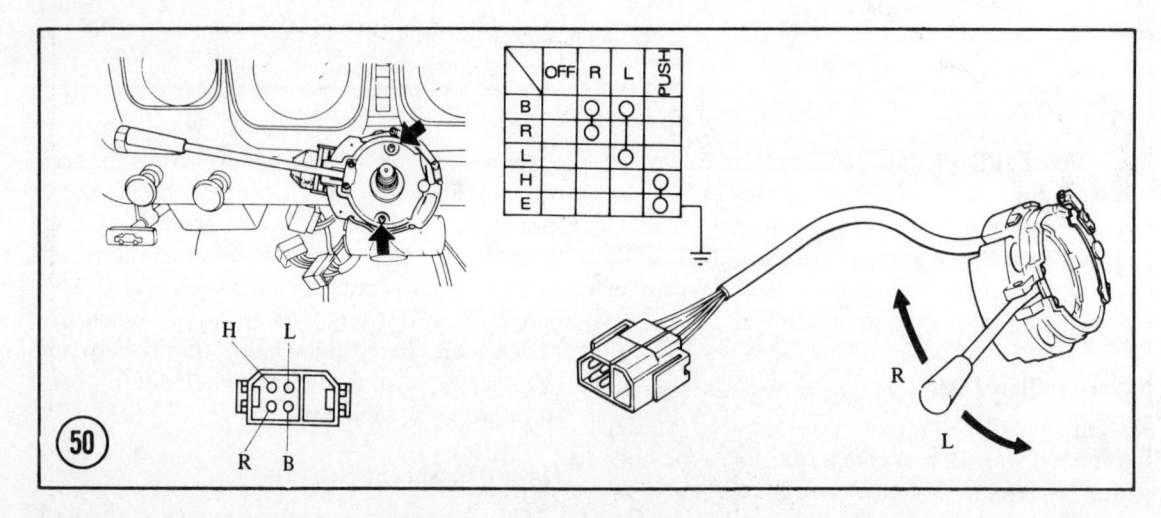

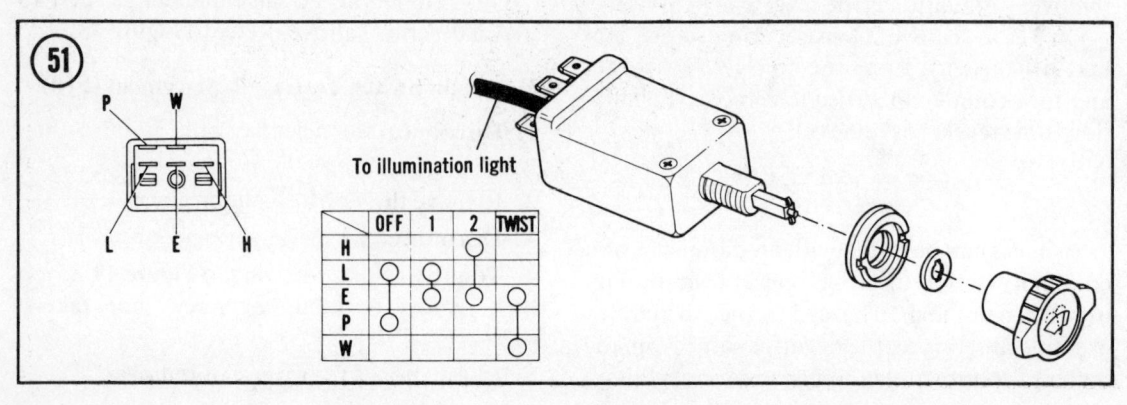

# ELECTRICAL SYSTEM

## (52) COMBINATION SWITCH

### Lighting switch

|   | OFF |   |   | 1ST |   |   | 2ND |   |   |
|---|---|---|---|---|---|---|---|---|---|
|   | H | L | P | H | L | P | H | L | P |
| 1 |   |   |   |   | O | O |   | O | O |
| 2 |   |   |   | O | O | O | O | O | O |
| 3 |   | O |   |   | O | O |   | O | O |
| 4 |   | O |   |   | O |   |   | O |   |
| 5 |   |   |   |   |   |   |   | O |   |

### Turn signal switch

|    | OFF | R | L | H |
|----|-----|---|---|---|
| 6  |     | O | O |   |
| 7  |     | O |   |   |
| 8  |     |   | O |   |
| 9  |     |   | O | O |
| 10 |     |   |   | O ⏚ |

### Wiper and washer switch

|    | OFF | INT | LOW | HI | WASH |
|----|-----|-----|-----|----|------|
| 11 |     | O   |     |    |      |
| 12 |     |     |     |    | O    |
| 13 |     | O   | O   | O  | O    |
| 14 | O   | O   |     |    |      |
| 15 |     |     |     | O  |      |
| 16 | O   | O   | O   |    |      |
| 17 |     |  0  |     |    |      |
| 18 |     |  ⇝  |     |    |      |
| 19 |     | 1000Ω |   |    |      |

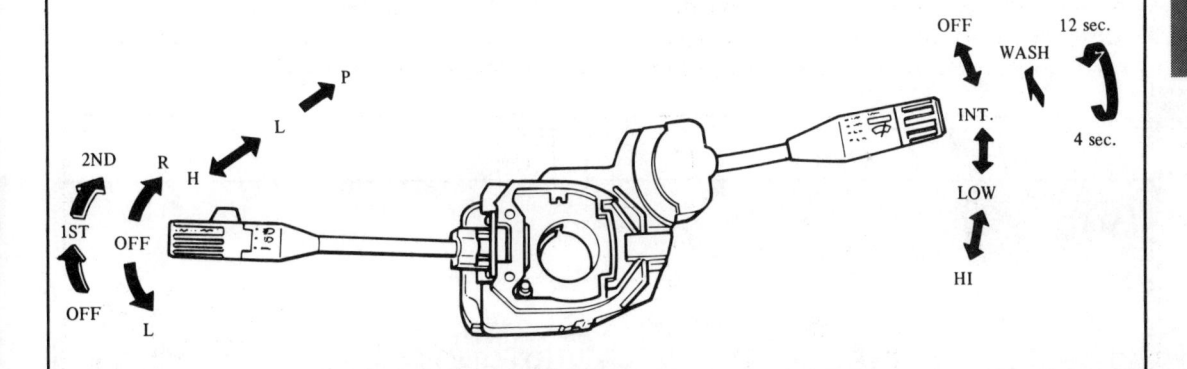

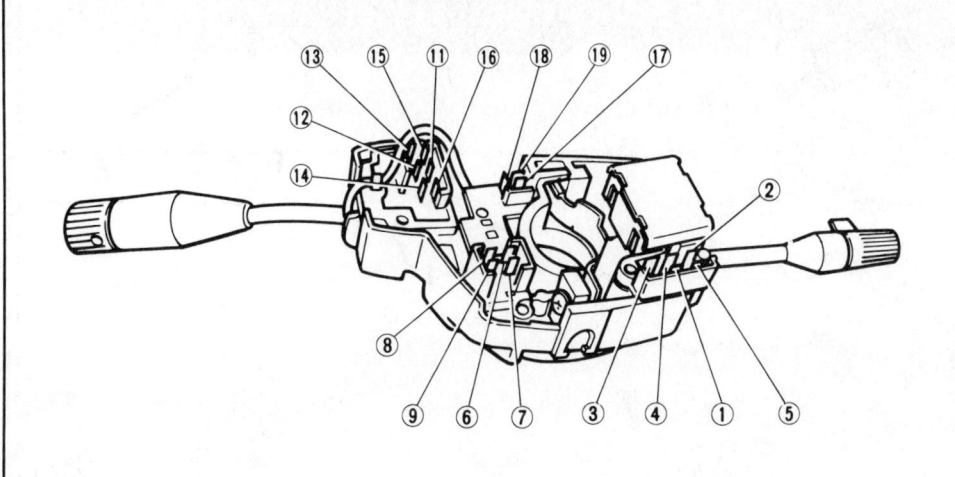

150

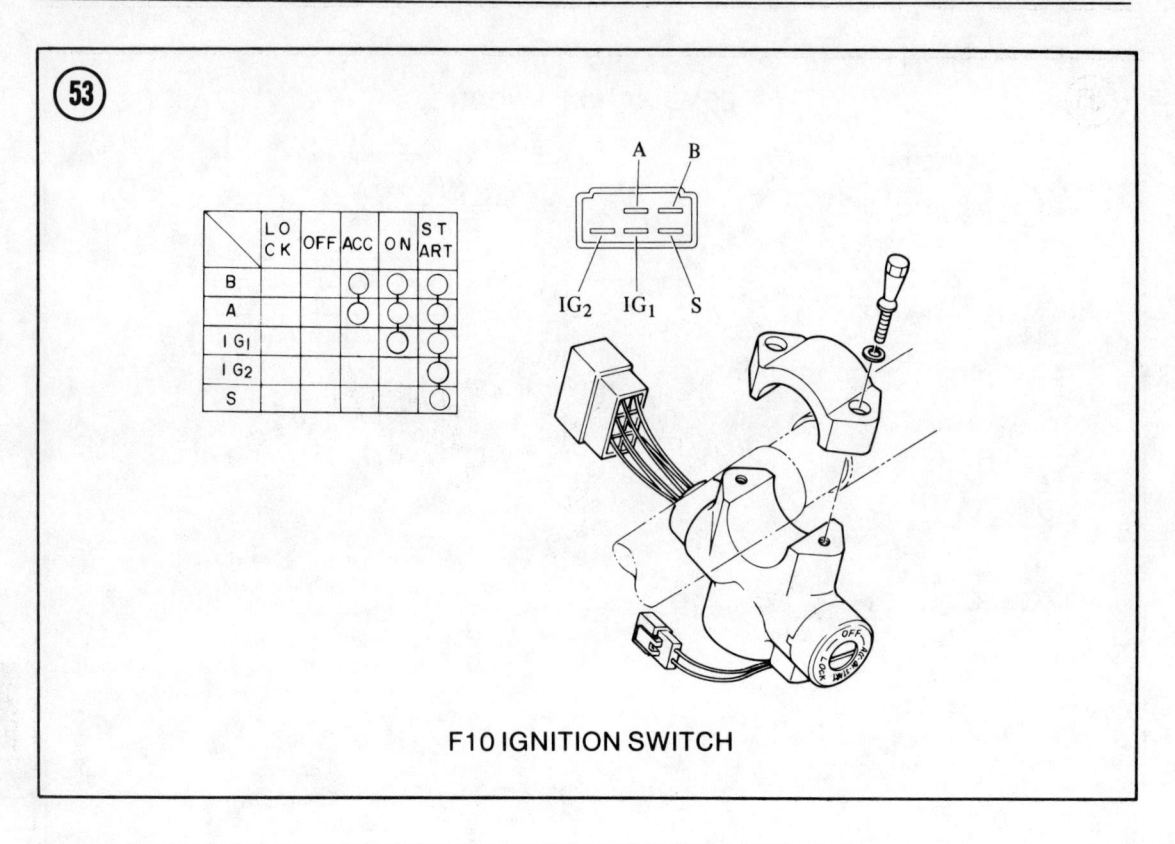

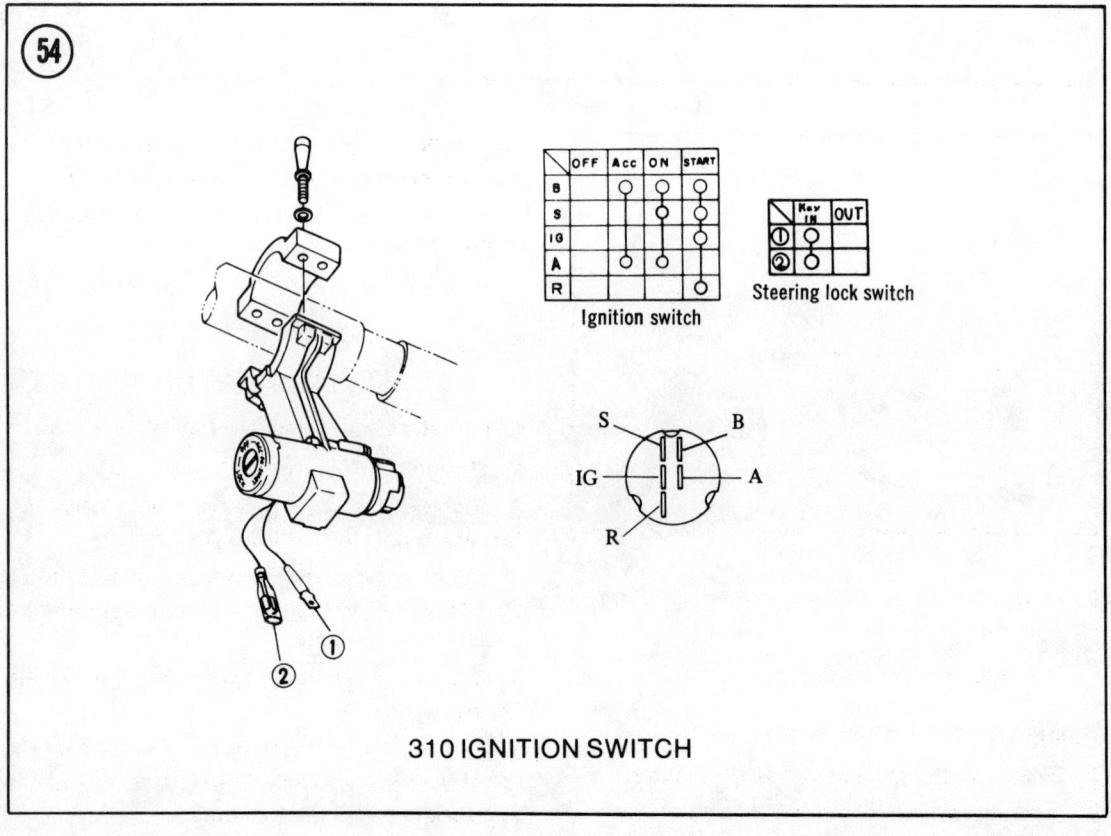

CHAPTER SEVEN

**F10 IGNITION SWITCH**

**310 IGNITION SWITCH**

# ELECTRICAL SYSTEM

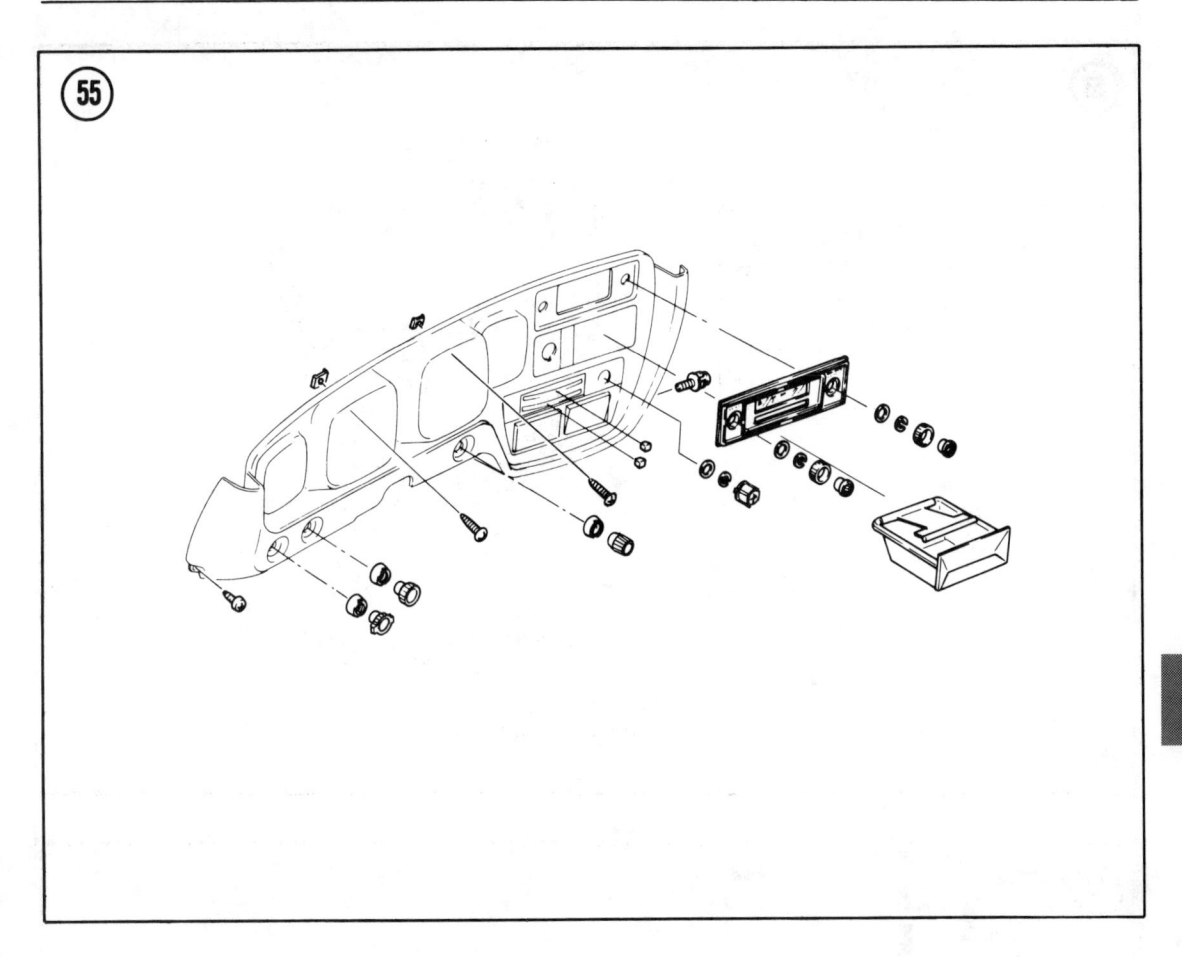

**(55)**

**(56)**

5. If the switch is bad, remove its mounting screws and take it off.

6. Installation is the reverse of removal.

## Ignition Switch Testing/Replacement

1. Disconnect negative cable from battery.
2. Remove the steering column cover.

3. Disconnect the switch wiring connector. To test, see **Figure 53** (F10) or **Figure 54** (310).

4. If the switch is bad, remove its mounting screw and take it off the steering lock.

5. Installation is the reverse of removal.

## INSTRUMENTS

### Cluster Removal/Installation (F10)

1. Disconnect negative cable from battery.

2. Remove the control knobs and ring nuts (**Figure 55**). Remove the ashtray.

3. Remove the cover retaining screws. Pull the cover out, disconnect the cigarette lighter wire, and take the cover off.

4. Remove 4 screws securing the cluster. See **Figure 56**.

5. Pull the cluster partway out. Disconnect the speedometer cable and wiring connector, then pull the cluster all the way out.

152

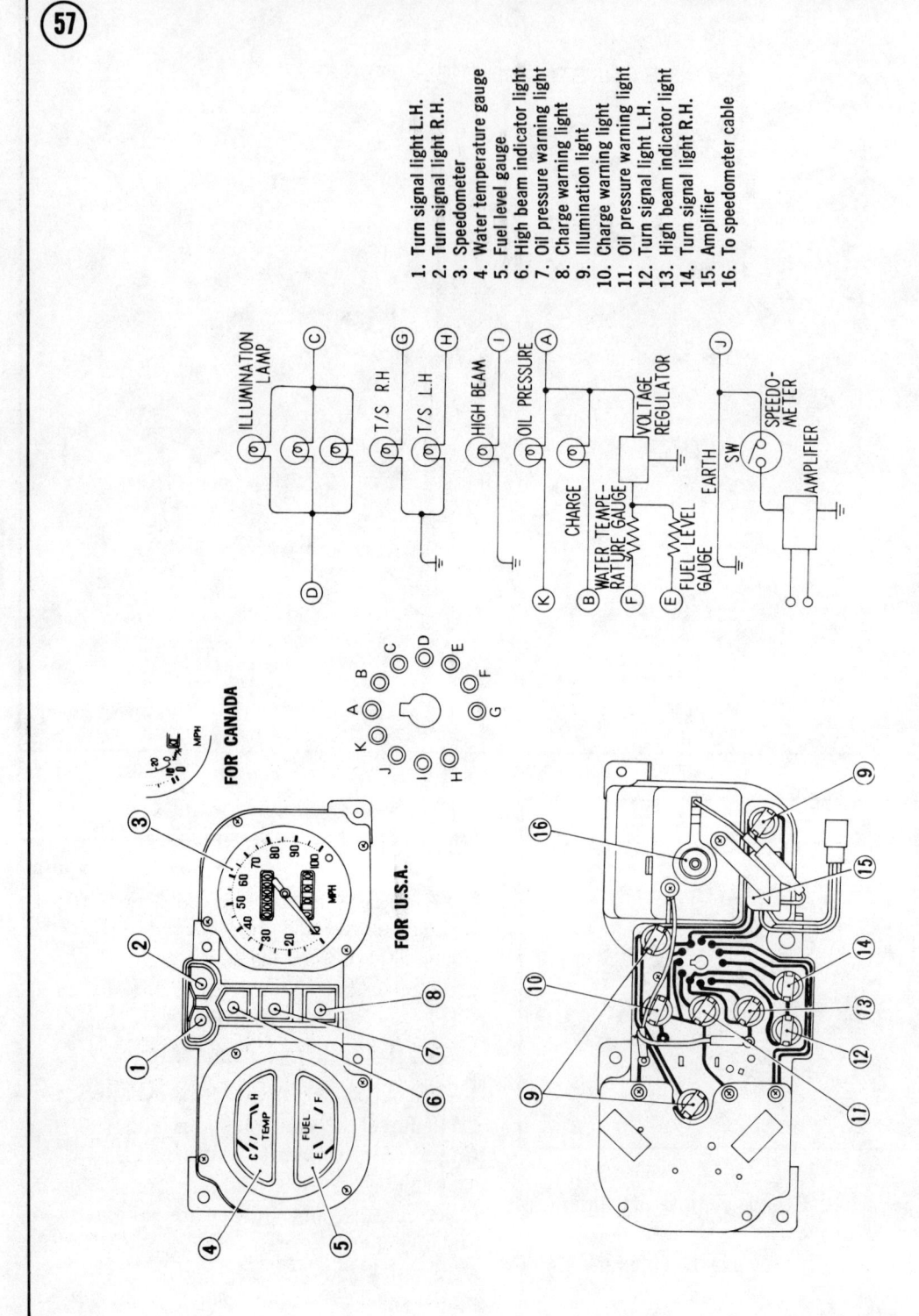

CHAPTER SEVEN

⑤⑦

1. Turn signal light L.H.
2. Turn signal light R.H.
3. Speedometer
4. Water temperature gauge
5. Fuel level gauge
6. High beam indicator light
7. Oil pressure warning light
8. Charge warning light
9. Illumination light
10. Charge warning light
11. Oil pressure warning light
12. Turn signal light L.H.
13. High beam indicator light
14. Turn signal light R.H.
15. Amplifier
16. To speedometer cable

F10 INSTRUMENT PANEL BULBS

# ELECTRICAL SYSTEM

**F10 INSTRUMENT CLUSTER**

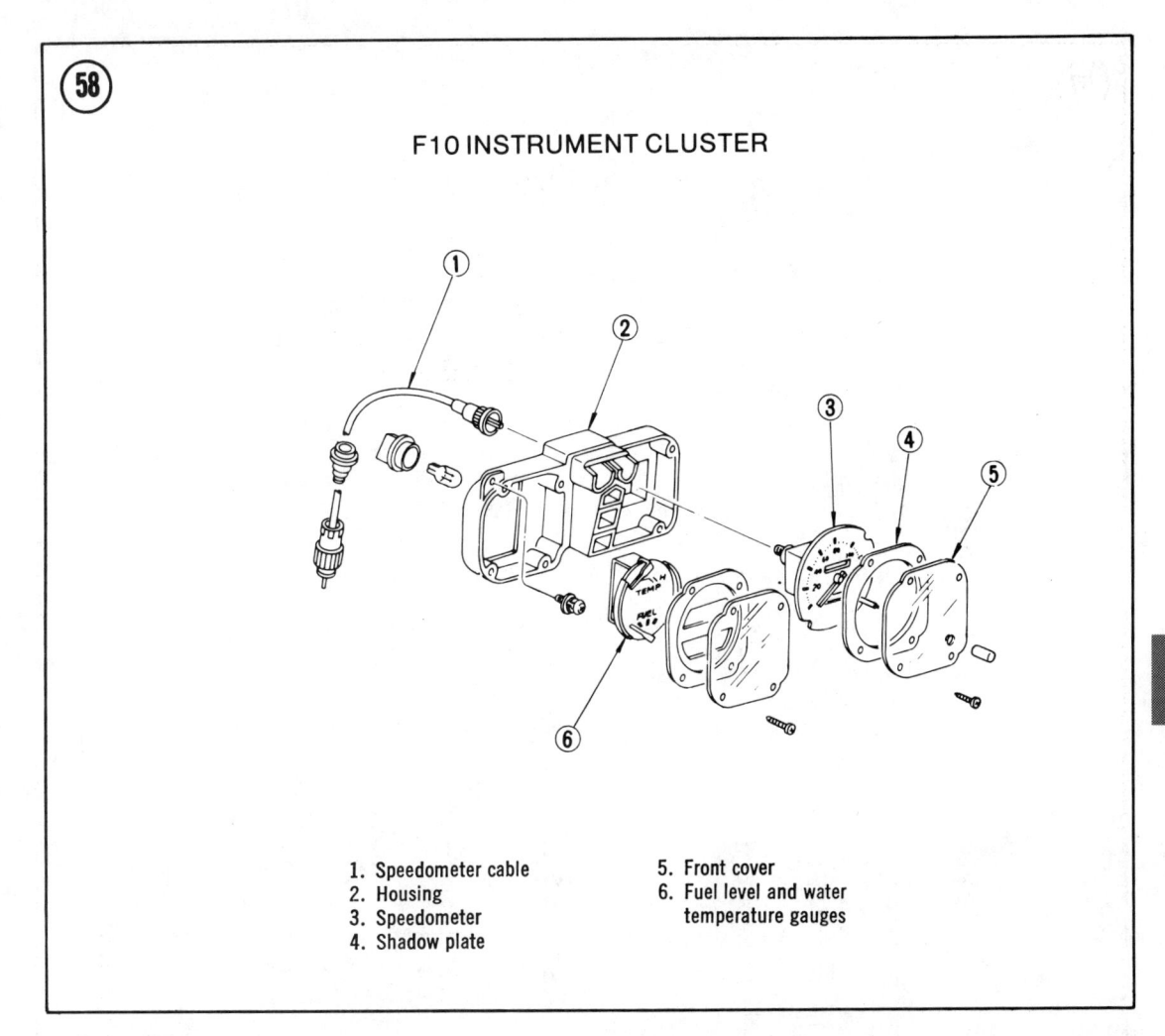

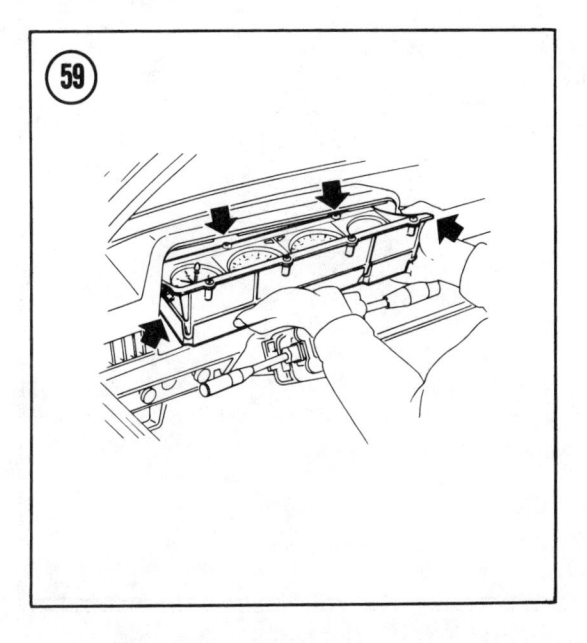

1. Speedometer cable
2. Housing
3. Speedometer
4. Shadow plate
5. Front cover
6. Fuel level and water temperature gauges

6. To replace bulbs, refer to **Figure 57**. To replace individual gauges, disassemble the cluster as shown in **Figure 58**.

7. Installation is the reverse of removal.

**Cluster Removal/Installation (310)**

1. Disconnect negative cable from battery.

2. Remove the steering wheel (Chapter Eleven). Remove the steering column cover.

3. Remove the lower cover from the driver's side of the instrument panel.

4. Remove the cluster lid. See *Instrument Panel*, Chapter Twelve.

5. Disconnect the speedometer cable and wiring connectors from the back of the cluster.

6. Remove the cluster securing screws (**Figure 59**).

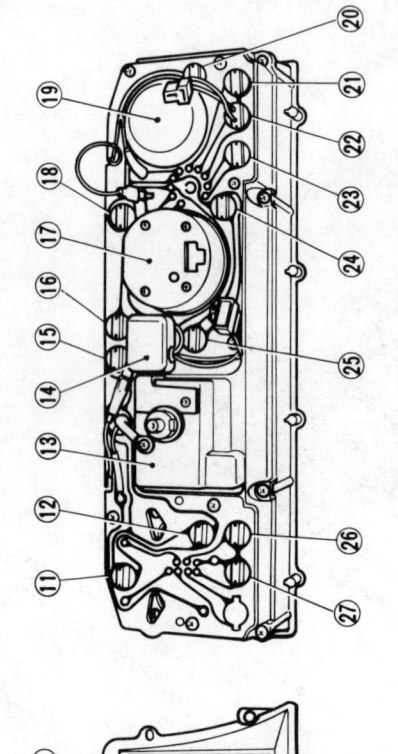

## 310 INSTRUMENT PANEL BULBS

1. Clock
2. Tachometer
3. Speedometer
4. Fuel level gauge
5. Water temperature gauge
6. Charge warning lamp
7. Oil pressure warning lamp
8. High beam pilot lamp
9. Fasten belts warning lamp
10. Brake warning lamp
11. Illumination lamp
12. Illumination lamp
13. Speedometer
14. Speed switch amplifier (except Canadian model)
15. Turn signal indicator lamp (R.H.)
16. Turn signal indicator lamp (L.H.)
17. Tachometer
18. Illumination lamp
19. Clock
20. Illumination lamp
21. Brake warning lamp
22. Fasten belts warning lamp
23. High beam pilot lamp
24. Illumination lamp
25. Illumination lamp
26. Oil pressure warning lamp
27. Charge warning lamp

# ELECTRICAL SYSTEM

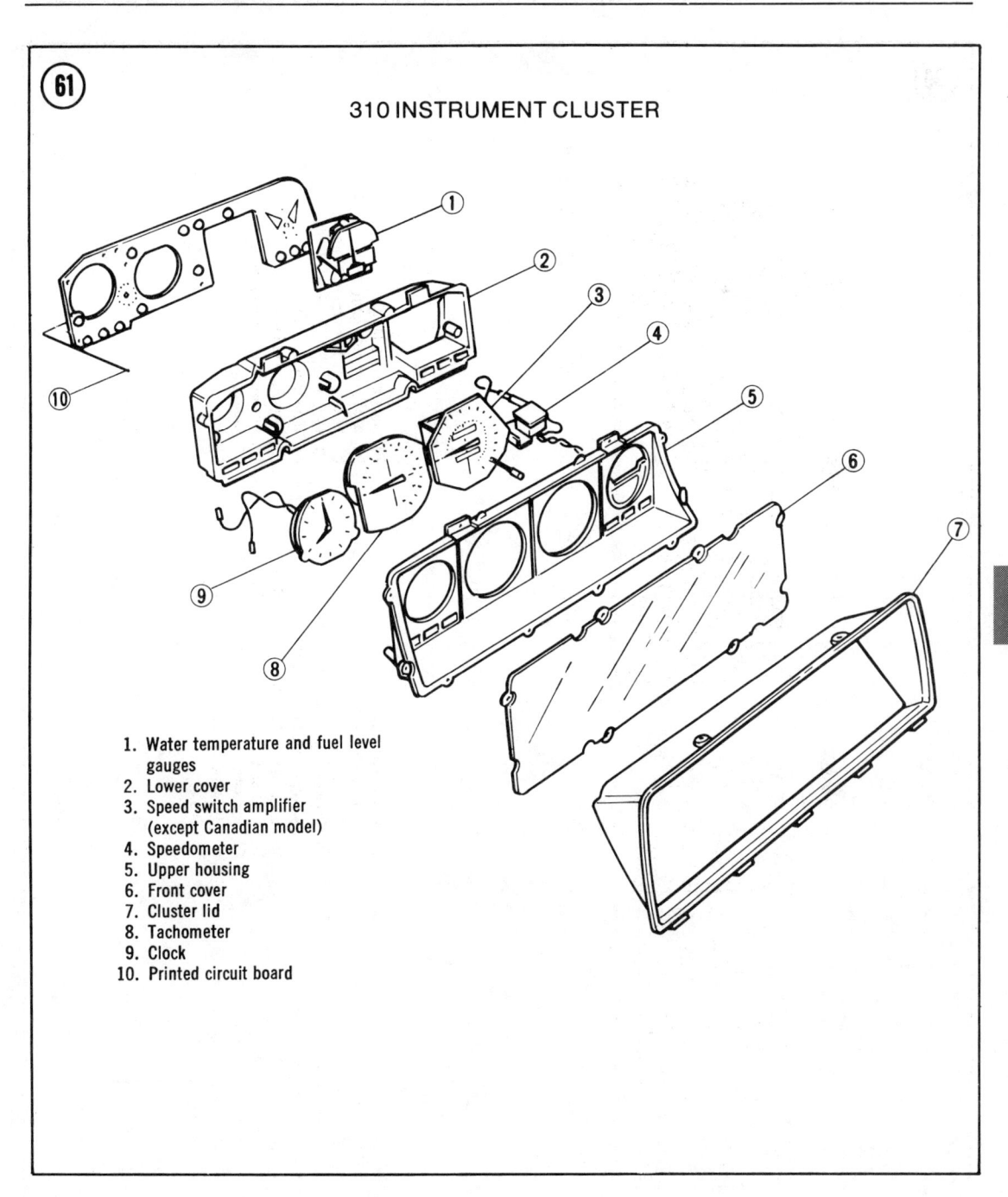

**61**

**310 INSTRUMENT CLUSTER**

1. Water temperature and fuel level gauges
2. Lower cover
3. Speed switch amplifier (except Canadian model)
4. Speedometer
5. Upper housing
6. Front cover
7. Cluster lid
8. Tachometer
9. Clock
10. Printed circuit board

7. To change bulbs, refer to **Figure 60**.
8. To replace individual gauges, disassemble the cluster as shown in **Figure 61**.

## HORN

There two horns, mounted at the front of the engine compartment. If the horns work, but are not loud enough, make sure the wires are making good contact and the horns are properly grounded to the body.

If only one horn works, check the wiring to the non-working horn. If the horn is receiving current and is grounded properly, it is probably defective.

**62**

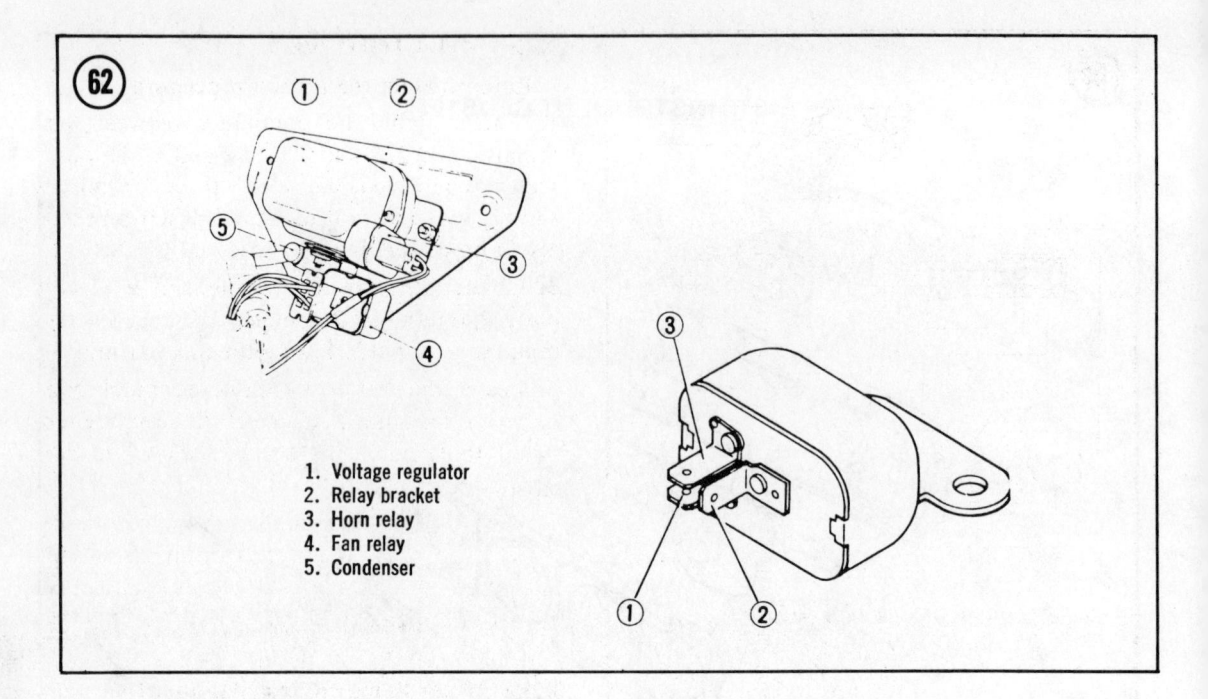

1. Voltage regulator
2. Relay bracket
3. Horn relay
4. Fan relay
5. Condenser

**63**

If neither horn works, check the horn fuse, then the battery. If these are good, remove the horn relay. See **Figure 62** (F10) or **Figure 63** (310). The relay is mounted at the right front corner of the engine compartment.

When a 12-volt battery is connected between terminals 1 and 2, there should be continuity between terminals 1 and 3. An ohmmeter connected between these terminals should show resistance, or a self-powered test lamp should light.

If the horn relay is good, remove the horn pad from the steering wheel. See *Steering Wheel*, Chapter Eleven. Check for broken wires or burned contacts.

## WINDSHIELD WIPERS AND WASHERS

### Wiper Motor Removal/Installation (F10)

1. Disconnect negative cable from battery.

2. Remove the glove compartment. See *Instrument Panel*, Chapter Twelve.

3. Remove the wiper motor mounting bolts. See **Figure 64**.

4. Detach the wiper motor from the linkage ball-joint. Disconnect the motor wires, then take the motor out.

5. Installation is the reverse of removal.

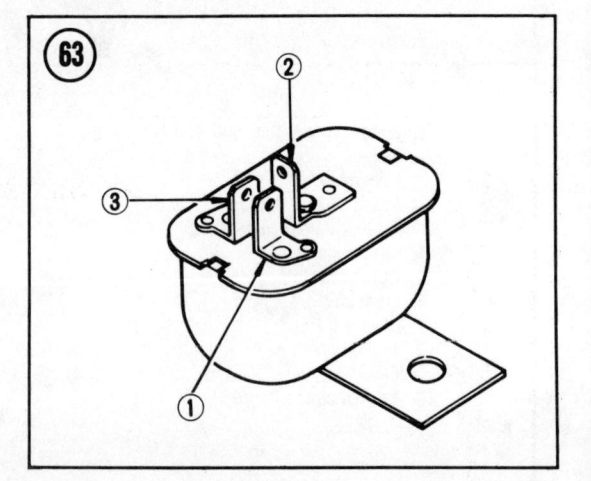

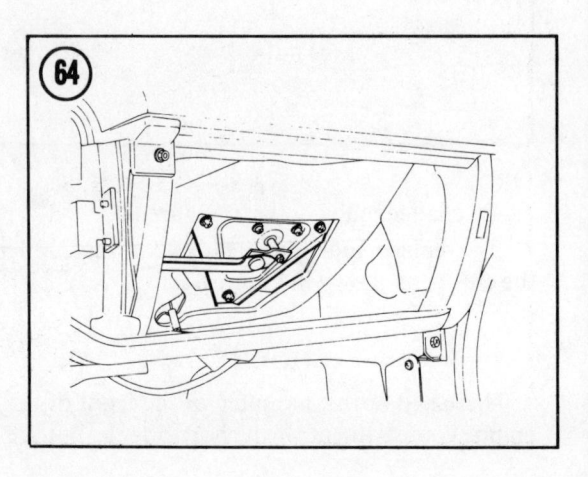

**64**

# ELECTRICAL SYSTEM

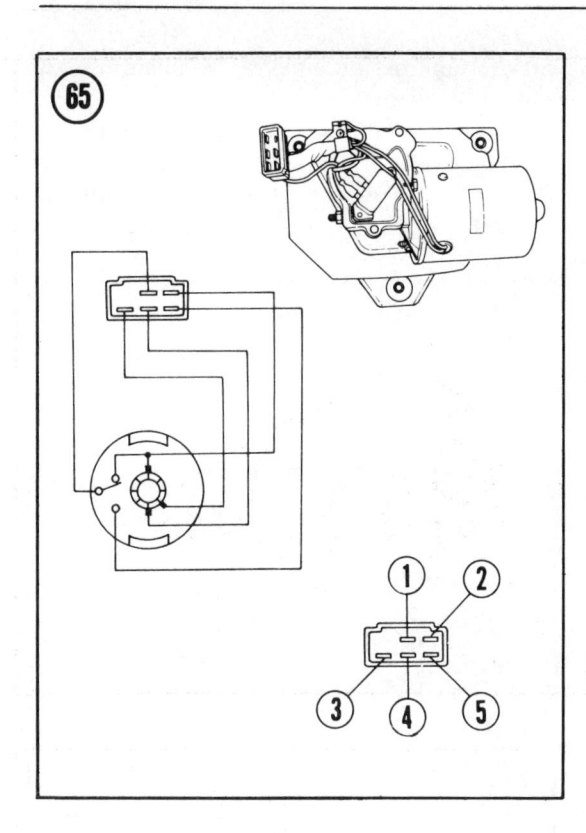

## Wiper Motor Test (F10)

Refer to **Figure 65** for this procedure.

1. There should be continuity between terminals 1 and 2, 2 and 3, and 2 and 4. An ohmmeter connected between these terminals should indicate continuity, or a self-powered test lamp should light.

2. Connect the positive terminal of a 12-volt battery to terminal 2. Connect the negative terminal to terminal 3. The motor should run.

3. Leave the battery positive terminal connected to terminal 2. Connect the negative terminal to terminal 4. The motor should run again.

4. With the motor running, check continuity between terminals 1 and 5. Continuity should appear and disappear (test lamp should turn on and off) periodically.

## Wiper Motor Removal/Installation (310)

1. Disconnect negative cable from battery.
2. Remove motor mounting bolts (**Figure 66**).

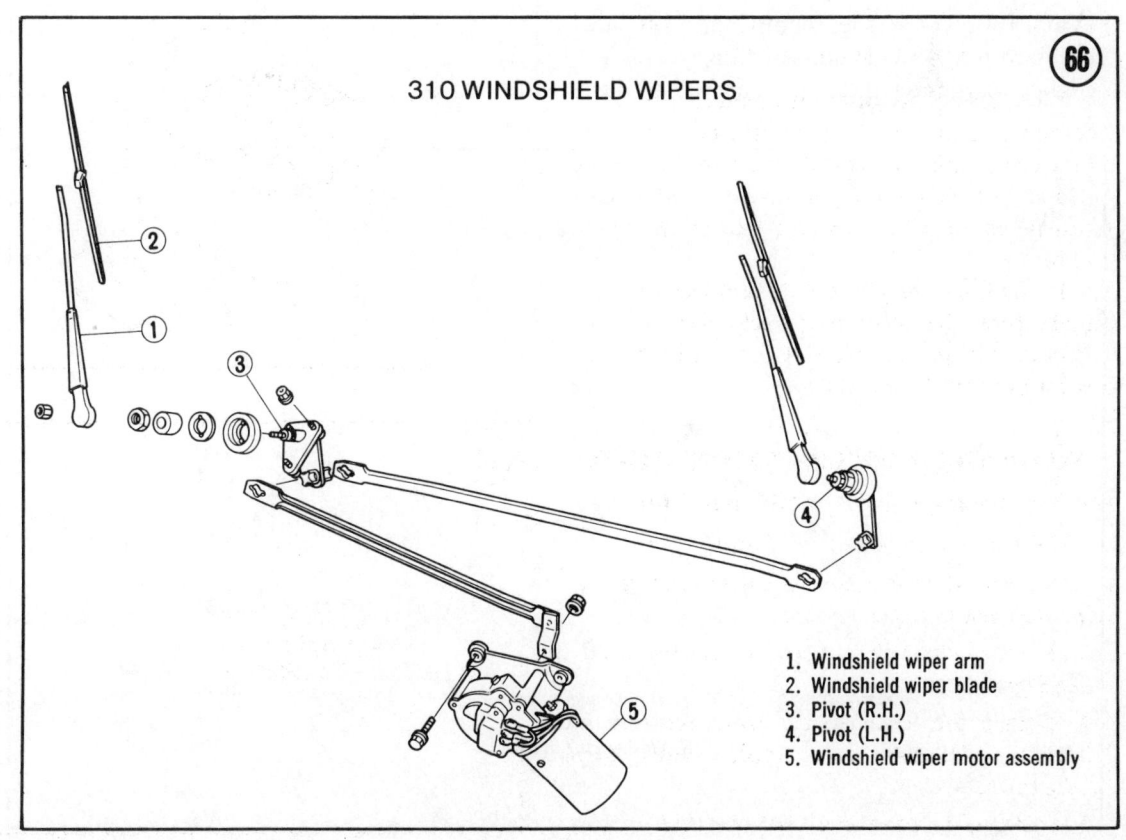

**310 WINDSHIELD WIPERS**

1. Windshield wiper arm
2. Windshield wiper blade
3. Pivot (R.H.)
4. Pivot (L.H.)
5. Windshield wiper motor assembly

3. Pull the motor partway out. Remove the nut securing the motor to the linkage, then remove the motor.

4. Installation is the reverse of removal.

**Wiper Motor Test (310)**

Refer to **Figure 67** for this procedure.

1. There should be continuity between terminals 1 and 4, and 1 and 5. An ohmmeter connected between these terminals should show continuity, or a self-powered test lamp should light.
2. Connect the positive terminal of a 12-volt battery to terminal 1. Connect the negative terminal to terminal 5. The motor should run.
3. Switch the negative terminal to terminal 4. The motor should run.
4. With the motor running, check continuity between terminals 1 and 2, and 2 and 3. Continuity should appear and disappear (test lamp should go on and off) periodically.

**Washer Pump and Tank Replacement**

On F10's, the washer pump and tank are combined into a single unit. See **Figure 68**. To remove, disconnect the hoses and wires, then lift the tank off its bracket. Installation is the reverse of removal.

On 310's, the tank and motors are separate. See **Figure 69**. To remove, disconnect the hoses (and wires on motors) and remove from the bracket. Installation is the reverse of removal.

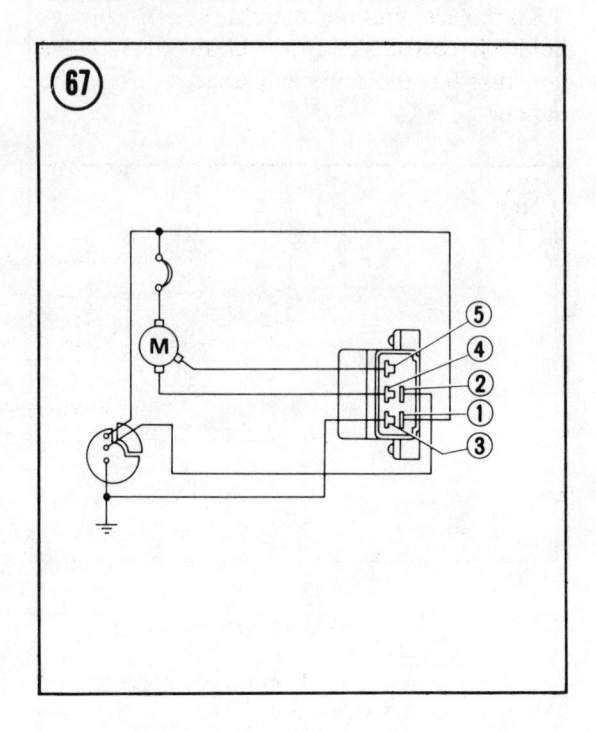

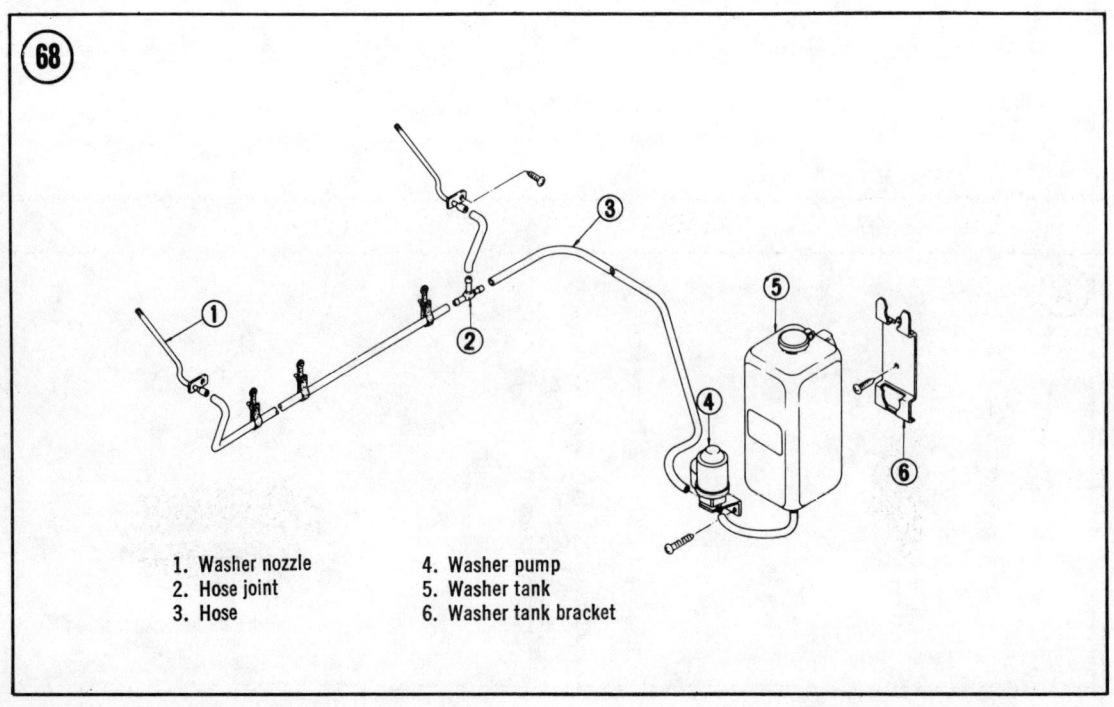

1. Washer nozzle
2. Hose joint
3. Hose
4. Washer pump
5. Washer tank
6. Washer tank bracket

# ELECTRICAL SYSTEM

## FUSES AND FUSIBLE LINKS

The fuse block is located under the instrument panel on the left side. See **Figure 70** (F10) or **Figure 71** (310). Fuse specifications are printed on the fuse block cover.

To replace, pull out the old fuse. If the fuse holder is corroded, sand it clean. Be sure the new fuse has the same amperage rating as the old one.

Whenever a fuse blows, find out the cause before replacing it. Usually the trouble is a short circuit in the wiring. This may be caused by worn-through insulation, or by a wire that works its way loose and shorts to ground. Carry several spare fuses in the glove compartment.

CAUTION
*Never substitute tinfoil or wire for a fuse. An overload could cause a fire and complete loss of the car.*

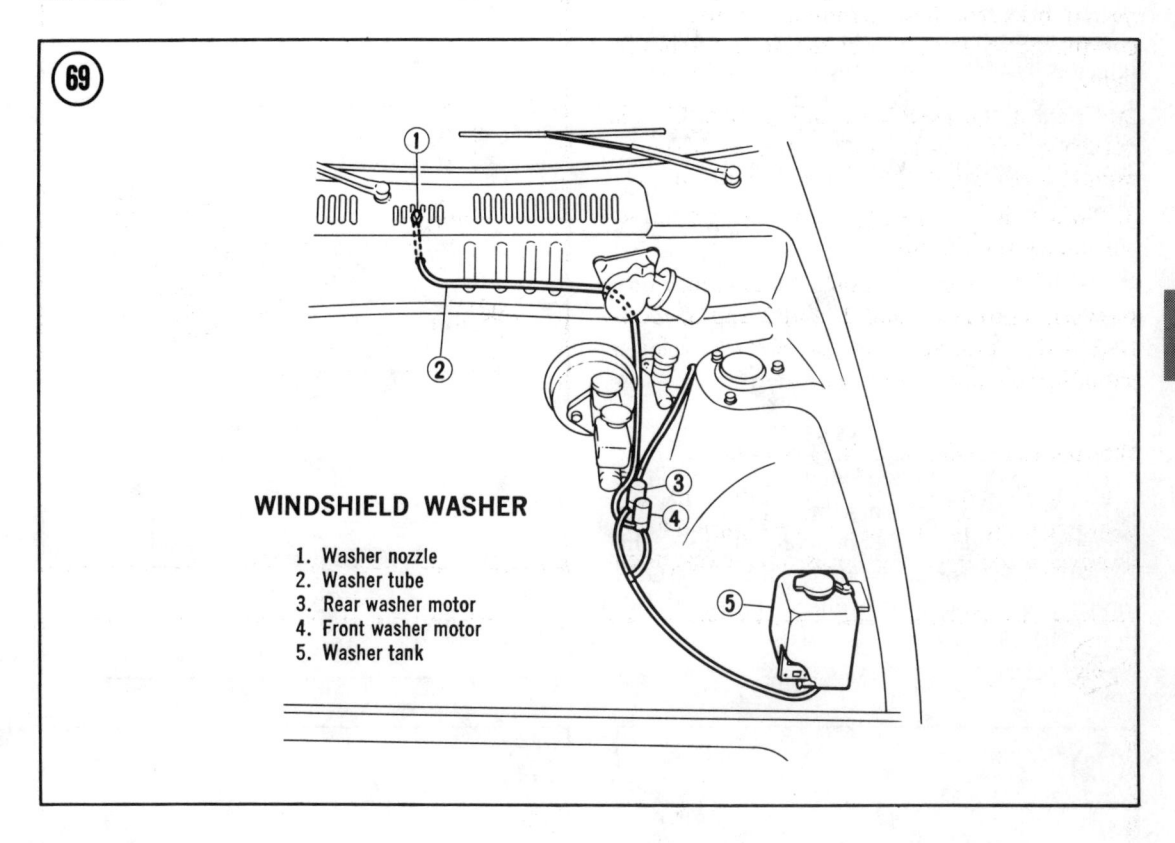

**WINDSHIELD WASHER**

1. Washer nozzle
2. Washer tube
3. Rear washer motor
4. Front washer motor
5. Washer tank

Fuse block

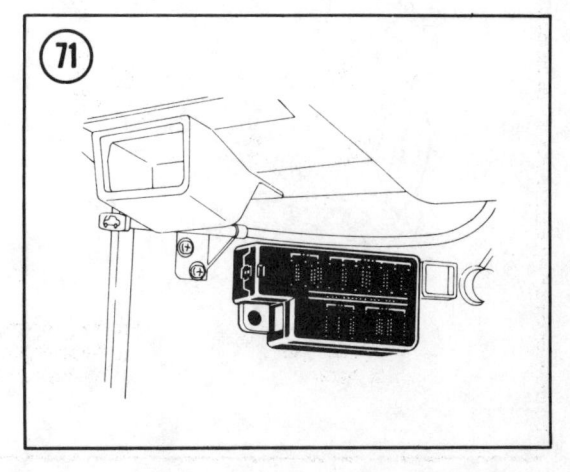

CHAPTER SEVEN

## Fusible Link Replacement

Fusible links are short sections of thin wire in a thicker wire. They are intended to burn out if an overload occurs, thus protecting the wiring harnesses.

The F10 fusible links are located in the wiring harnesses. See **Figure 72** (with air conditioning) or **Figure 73** (without air conditioning). The 310 fusible links are located in a holder near the ignition coil (**Figure 74**).

Burned-out fusible links can usually be detected by melted or burned insulation. They may give off smoke before burning out. Suspect links with no apparent damage can be tested for continuity with an ohmmeter or a test lamp like the one shown in **Figure 18**. If a link burns out, unplug it and plug in a new one.

> **CAUTION**
> *Never replace a fusible link with a link of larger capacity. Never wrap fusible links with tape.*

## IGNITION SYSTEM

There are three types of ignition systems. The 1976-1977 non-California models, and 1978 Canadian models, use breaker points. The 1976-1977 California models, and 1978 U.S. models, use a pulse-controlled transistor ignition. The 1979 cars use a pulse-controlled integrated circuit ignition system.

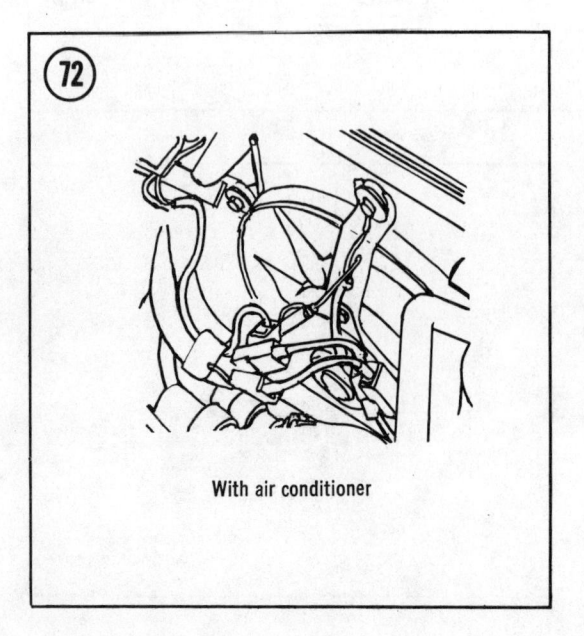

The breaker point system and integrated circuit system can easily be tested. The transistor system, however, requires special equipment, and it can easily be damaged by incorrect equipment hookups. This system should be tested by a Datsun dealer or competent automotive electrical shop.

### Testing (Breaker Point Ignition)

Refer to **Figure 75** for this procedure.

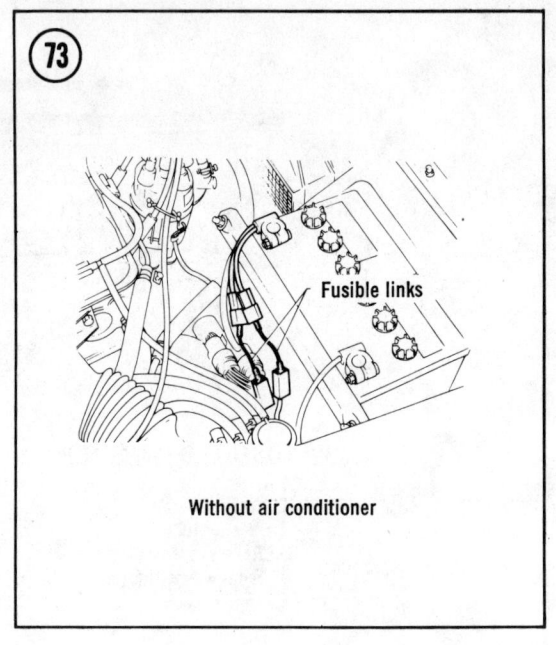

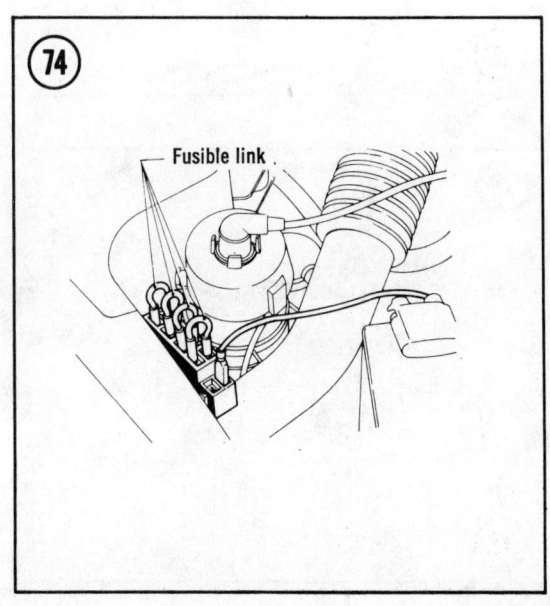

**⑦③**
Fusible links
Without air conditioner

**⑦②**
With air conditioner

**⑦④**
Fusible link

# ELECTRICAL SYSTEM

**(75)**

**WEAK OR NO SPARK AT ALL**

**IGNITION COIL TEST**
Disconnect the coil wire from the center of the distributor cap. Position the end of wire about ½ in. from any ground by propping it or tying it in place.

**CRANK ENGINE**

**WEAK SPARK OCCURS**
Check:
- Rotor
- Point gap
- Distributor cap
- Worn distributor lobes

**NO SPARK**

Check for opens (breaks) in the secondary (high voltage) wire.

**WIRE DEFECTIVE**
Replace it.

**WIRE GOOD**
Crank engine until the breaker points are at maximum open position. Check point condition and gap.

**POINTS BAD**
Clean or replace.

**POINTS GOOD**
Check voltage from coil negative terminal (black wire) to ground.

**VOLTAGE PRESENT**
Defective coil. Replace it.

**VOLTAGE NOT PRESENT**
Check wiring connections to coil and distributor.

**CONNECTIONS BAD**
Repair or replace.

**CONNECTIONS GOOD**
Disconnect coil negative wire (thin, black). Measure voltage from coil negative terminal to ground.

**VOLTAGE PRESENT**
The distributor is shorted. Have it repaired or replace it.

**VOLTAGE NOT PRESENT**
Indicates a break or bad connection between positive terminal and battery. Check wiring and ignition switch.

**VOLTAGE NOT PRESENT**
Measure voltage from coil positive terminal to ground.

**VOLTAGE PRESENT**
Coil probably defective. Have it checked or replace it.

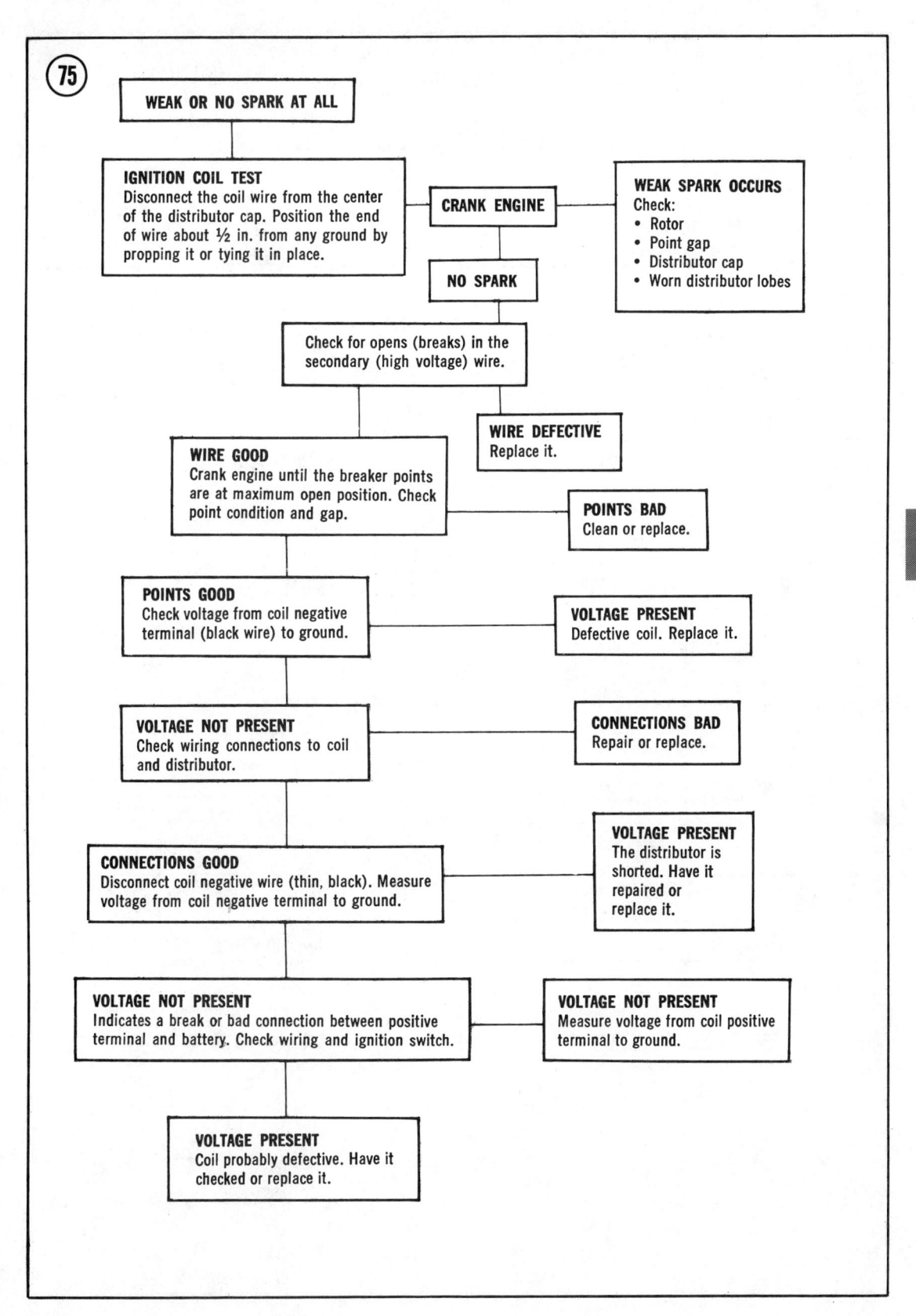

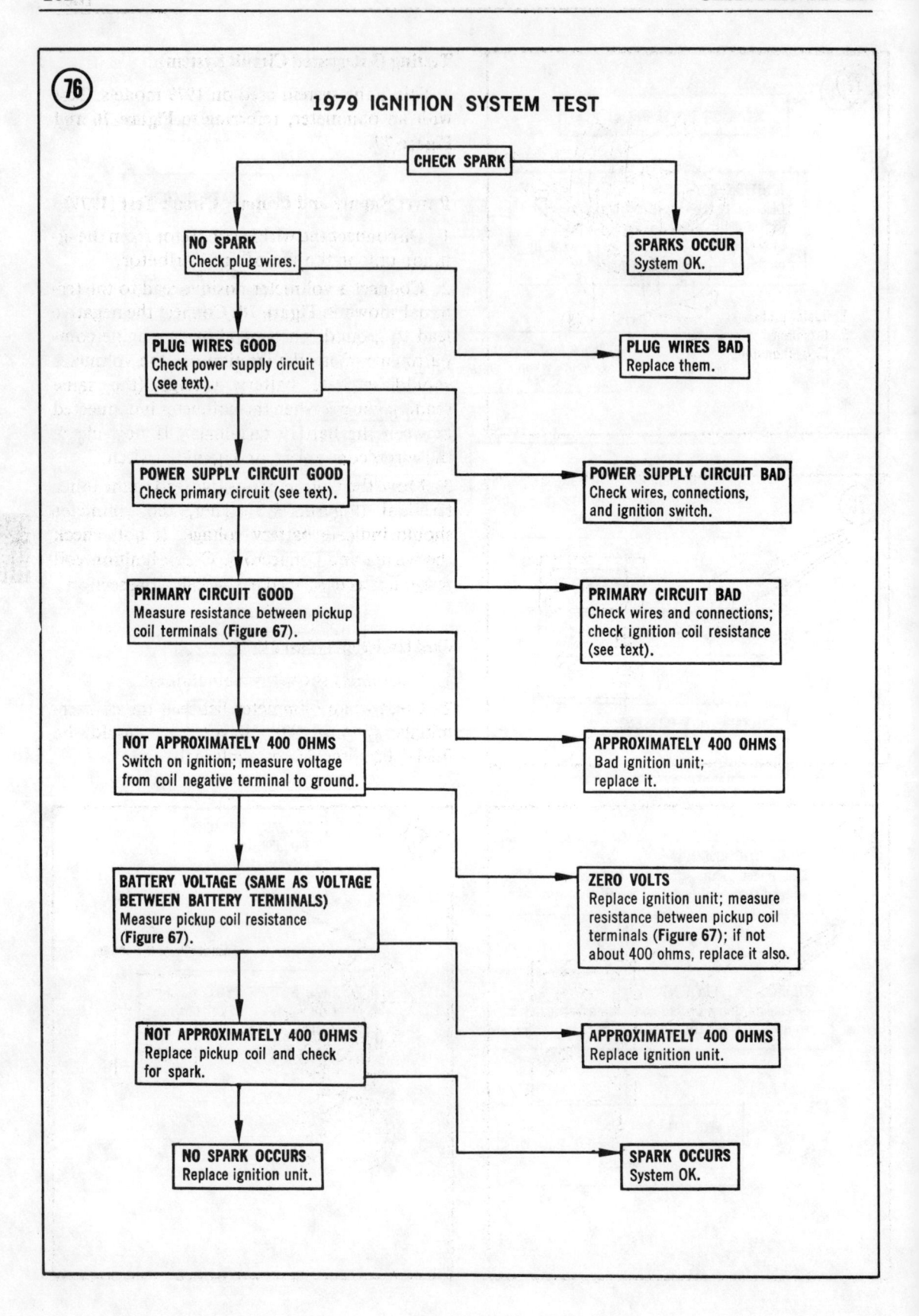

## 1979 IGNITION SYSTEM TEST

(76)

**CHECK SPARK**
- **SPARKS OCCUR** — System OK.
- **NO SPARK** — Check plug wires.
  - **PLUG WIRES BAD** — Replace them.
  - **PLUG WIRES GOOD** — Check power supply circuit (see text).
    - **POWER SUPPLY CIRCUIT BAD** — Check wires, connections, and ignition switch.
    - **POWER SUPPLY CIRCUIT GOOD** — Check primary circuit (see text).
      - **PRIMARY CIRCUIT BAD** — Check wires and connections; check ignition coil resistance (see text).
      - **PRIMARY CIRCUIT GOOD** — Measure resistance between pickup coil terminals (**Figure 67**).
        - **APPROXIMATELY 400 OHMS** — Bad ignition unit; replace it.
        - **NOT APPROXIMATELY 400 OHMS** — Switch on ignition; measure voltage from coil negative terminal to ground.
          - **ZERO VOLTS** — Replace ignition unit; measure resistance between pickup coil terminals (**Figure 67**); if not about 400 ohms, replace it also.
          - **BATTERY VOLTAGE (SAME AS VOLTAGE BETWEEN BATTERY TERMINALS)** — Measure pickup coil resistance (**Figure 67**).
            - **APPROXIMATELY 400 OHMS** — Replace ignition unit.
            - **NOT APPROXIMATELY 400 OHMS** — Replace pickup coil and check for spark.
              - **SPARK OCCURS** — System OK.
              - **NO SPARK OCCURS** — Replace ignition unit.

# ELECTRICAL SYSTEM

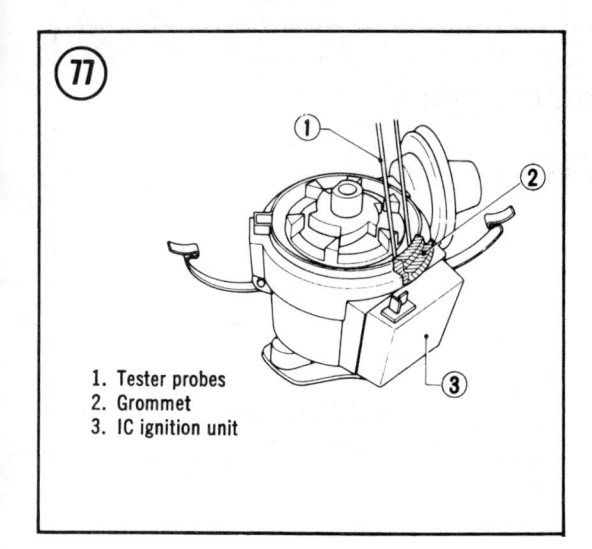

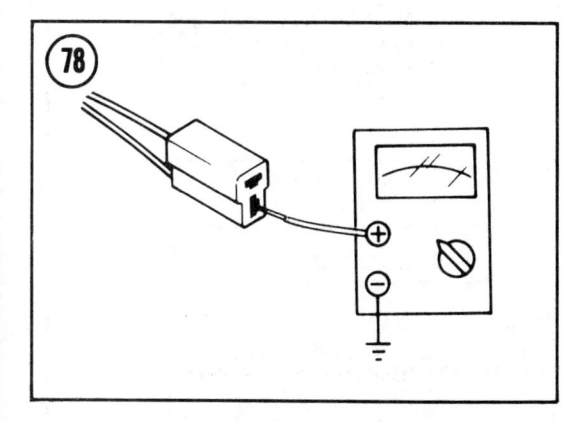

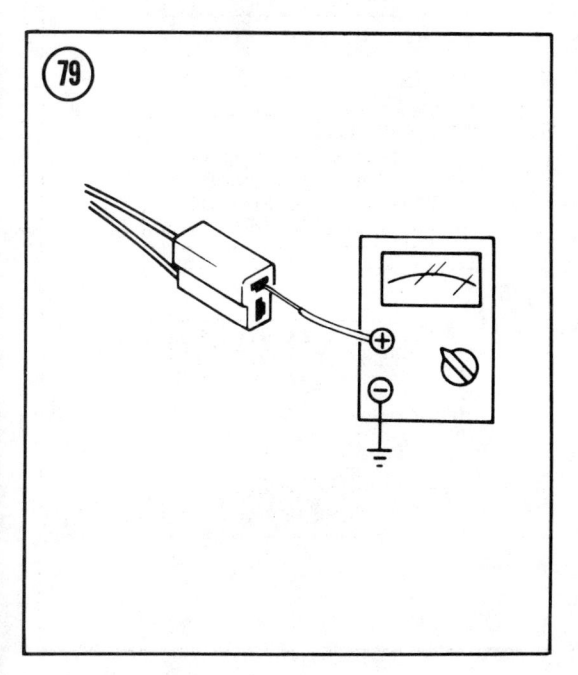

**Figure 77**
1. Tester probes
2. Grommet
3. IC ignition unit

## Testing (Integrated Circuit System)

This is the system used on 1979 models. Test with an ohmmeter, referring to **Figure 76 and Figure 77**.

### Power Supply and Primary Circuit Test (1979)

1. Disconnect the wiring connector from the ignition unit on the side of the distributor.

2. Connect a voltmeter positive lead to the terminal shown in **Figure 78**. Connect the negative lead to ground (bare metal in the engine compartment). With the ignition on, the voltmeter should indicate battery voltage (the same reading you get when the voltmeter is connected between the battery terminals). If not, check the wires, connectors, and ignition switch.

3. Move the voltmeter positive lead to the other terminal (**Figure 79**). Again, the voltmeter should indicate battery voltage. If not, check the wiring and connectors. Check ignition coil resistance as described in the following section.

### Coil Resistance Test (1979)

1. Disconnect the wires from the coil.

2. Connect an ohmmeter between the coil terminals (**Figure 80**). Resistance should be 0.84-1.02 ohms. If not, replace the coil.

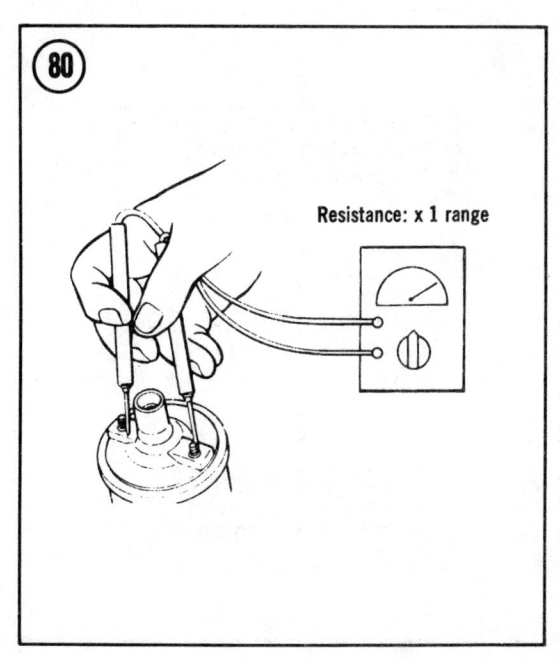

Figure 78

Figure 79

Figure 80 — Resistance: x 1 range

## Ignition Coil Replacement

The ignition coil is located at the rear of the engine compartment. To replace, disconnect the coil wires and take the coil out. Install in the reverse order. Be sure the coil cap (**Figure 81**) is positioned properly.

## Ignition Unit Removal (1979)

1. Disconnect negative cable from battery.
2. Remove the distributor cap and rotor.
3. Disconnect the wiring connector from the ignition unit. Remove its screws (**Figure 82**) and take it off the distributor.
4. Disconnect the pickup coil wires from the ignition unit (**Figure 83**).

#### CAUTION
*Be sure to use needle-nosed pliers as shown to prevent pulling on the wire.*

## Ignition Unit Installation (1979)

Installation is the reverse of removal, plus the following.

1. Be sure the mating surfaces of distributor and ignition unit are clean and dry.
2. Carefully push the pickup coil wires on with fingers (**Figure 84**). Connect the wires according to the color marks on the upper side of the grommet.

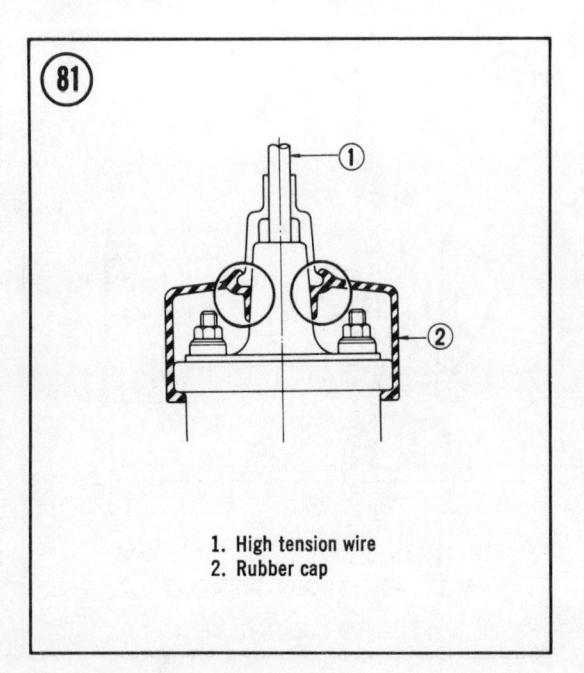

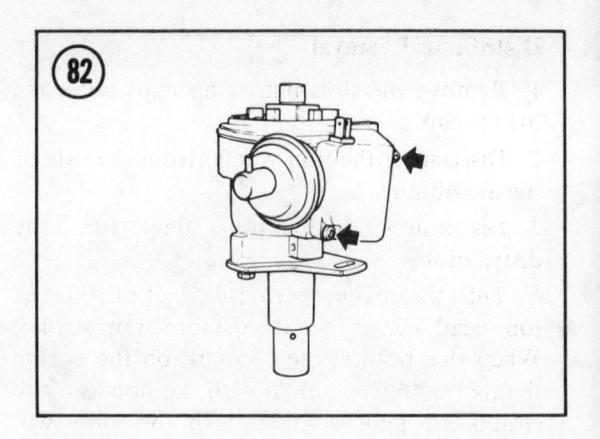

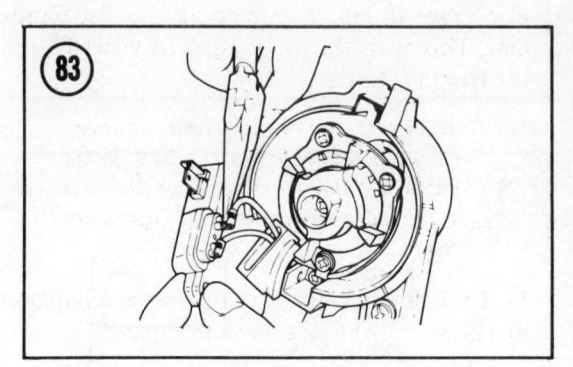

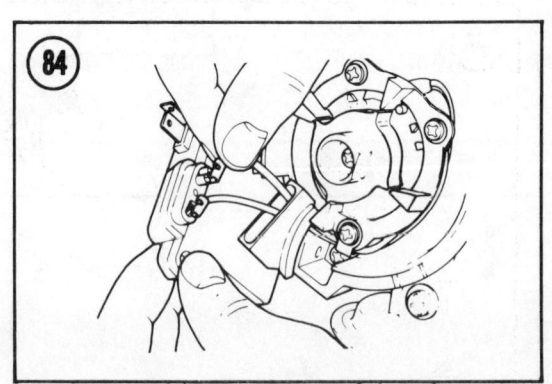

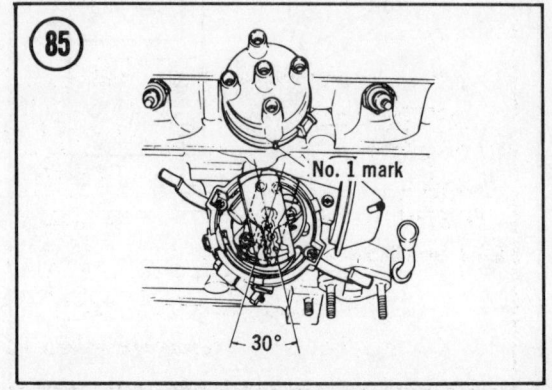

1. High tension wire
2. Rubber cap

# ELECTRICAL SYSTEM

## Distributor Removal

1. Remove the distributor cap clips and take off the cap.
2. Disconnect the thin wire(s) from the side of the distributor.
3. Disconnect the vacuum line from the distributor.
4. Turn the engine over until No. 1 piston is at top dead center on its compression stroke. When this occurs, the 0° mark on the engine timing pointer will align with the notch in the crankshaft pulley. The distributor rotor will also point to No. 1 terminal in the distributor cap. This is to the lower right as viewed from the front of the car.

> NOTE: *Be sure to check rotor position, as well as the timing marks. The timing marks also line up when No. 4 piston is at top dead center on its compression stroke.*

5. To simplify installation, make alignment marks on distributor body and engine.
6. Remove the distributor body setscrew. Lift the distributor out of the engine.

## Distributor Installation

1. If the engine was turned with the distributor out, place No. 1 piston at TDC on its compression stroke. See *Removal*, Step 4.
2. Insert the distributor in the engine. Start with the rotor 30° from its final position **(Figure 85)**. When the distributor is all the way in, the rotor should point to No. 1 terminal in the distributor cap as shown.
3. Install the body setscrew and distributor cap. Connect the vacuum line and thin wire(s) to the distributor.
4. Adjust ignition timing. See *Tune-Up*, Chapter Three.

## SPARK TIMING CONTROL SYSTEM

This system cuts off vacuum advance in all but fourth gear (and fifth gear, if so equipped). **Figure 86** shows the system used on all 1976

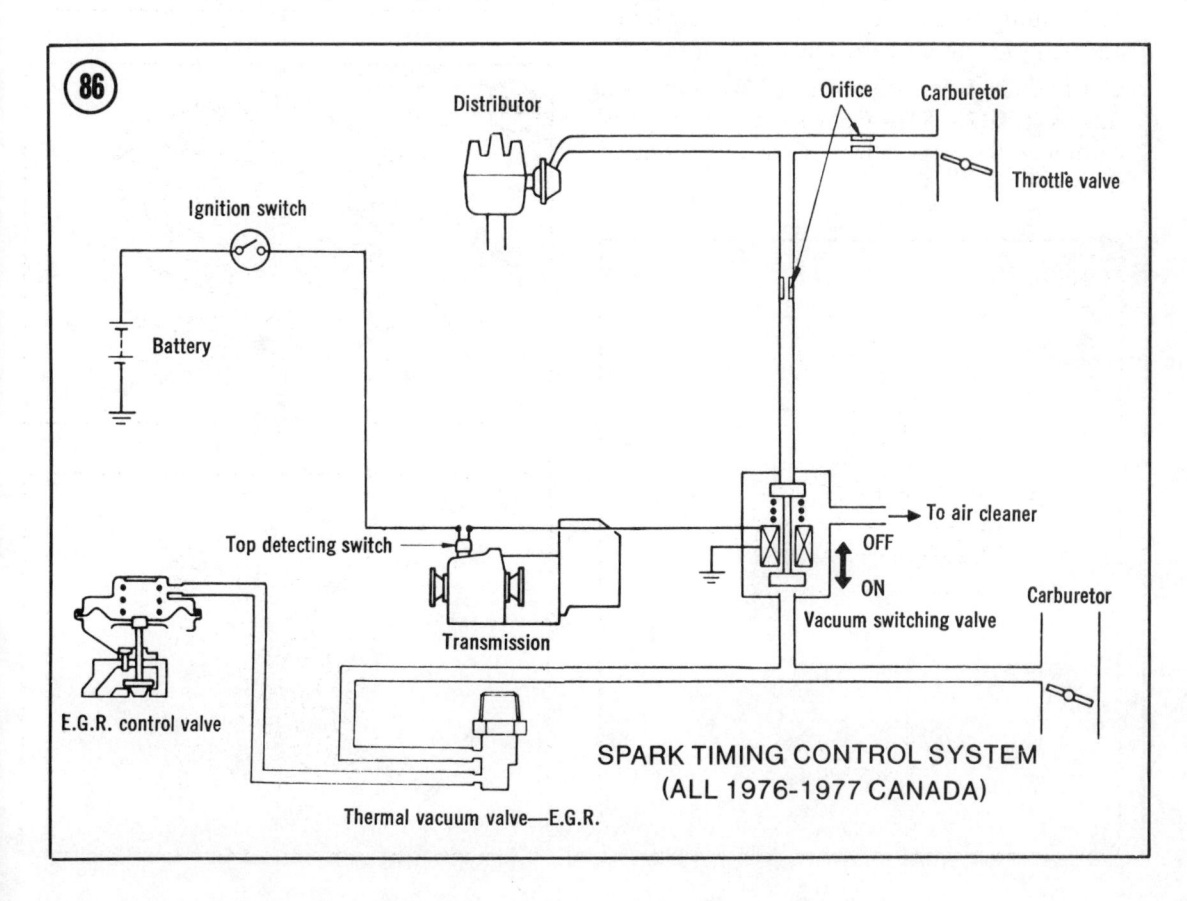

**86** SPARK TIMING CONTROL SYSTEM (ALL 1976-1977 CANADA)

**SPARK TIMING CONTROL SYSTEM (1977 U.S. MODELS)**

Top position: ON

1. Battery
2. Ignition key
3. Top detecting switch
4. Transmission
5. Vacuum switching valve
6. Distributor
7. Orifice
8. Thermal vacuum valve—T.C.S.
9. Carburetor

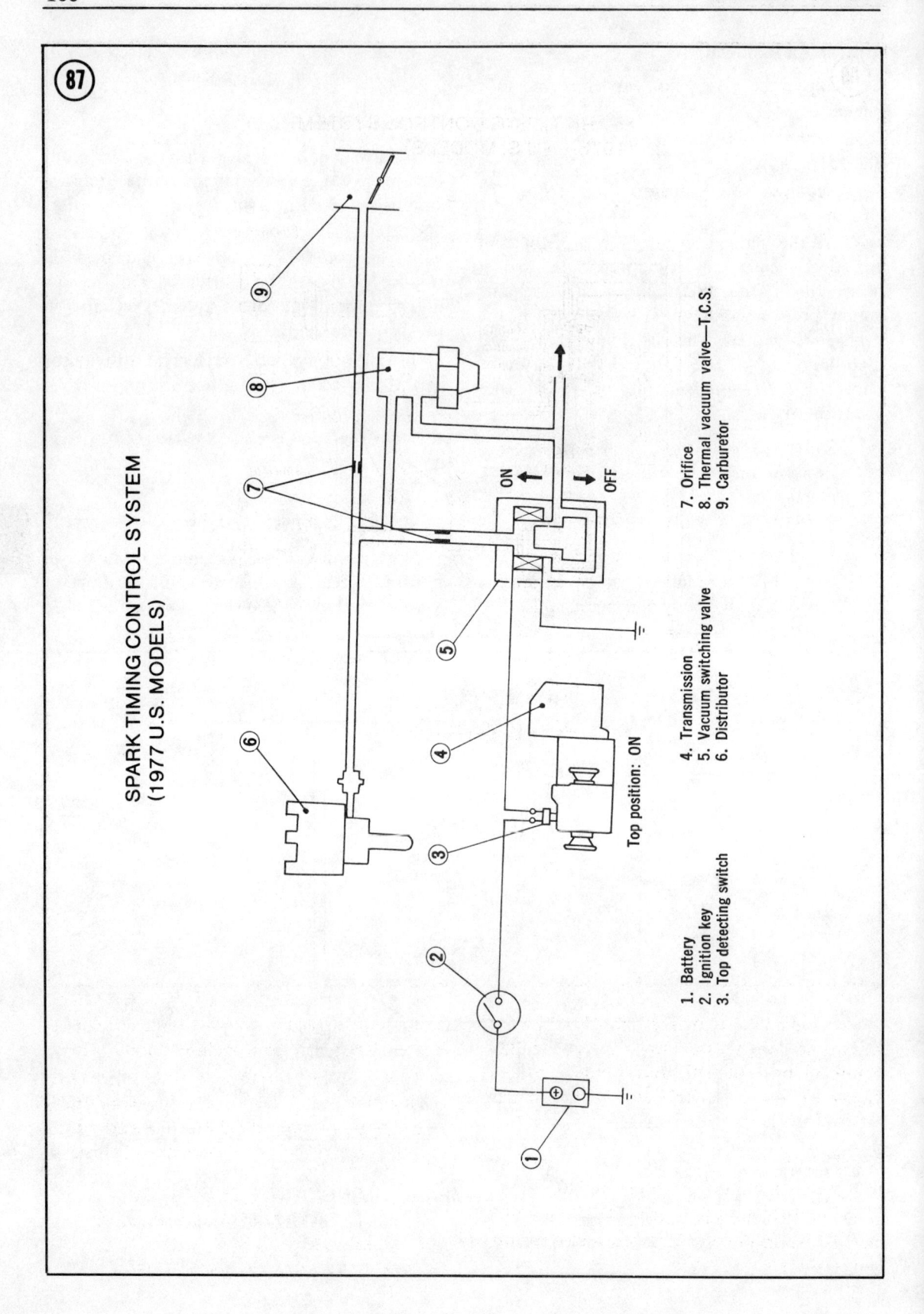

ELECTRICAL SYSTEM 167

**⑧⑧**

SPARK TIMING CONTROL SYSTEM
(1978-on U.S. MODELS)

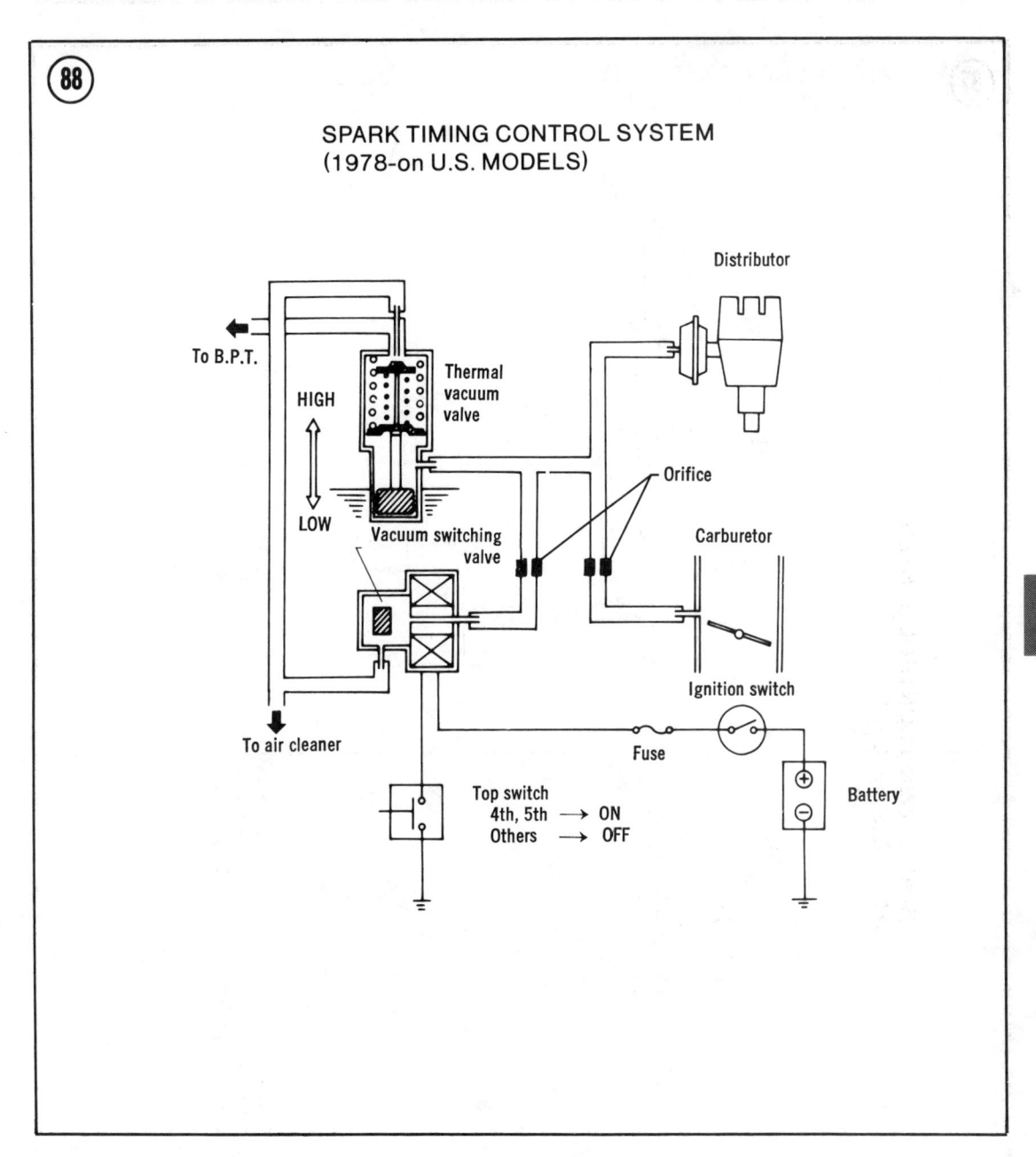

- To B.P.T.
- Thermal vacuum valve
- HIGH
- LOW
- Vacuum switching valve
- To air cleaner
- Distributor
- Orifice
- Carburetor
- Ignition switch
- Fuse
- Battery
- Top switch
  4th, 5th → ON
  Others → OFF

models, and 1977 Canadian models. **Figure 87** shows the design used on 1977 U.S. models. **Figure 88** shows the 1978 and later U.S. system; **Figure 89** shows the 1978 and later Canadian system.

**System Test**

1. Connect a timing light to the engine.
2. On all 1976 models, and 1977 and later U.S. models, warm the engine to normal operating temperature.

3. Have an assistant run the engine at approximately 1,600 rpm. Note ignition timing.

4. Have the assistant move the shift lever through the gears, keeping the engine at 1,600 rpm. Ignition timing should be more advanced in fourth and fifth gears than in the other gears.

*WARNING*
*Do not stand directly in front of the car. If the assistant's clutch foot slips, the car could jump forward and hit you.*

**CHAPTER SEVEN**

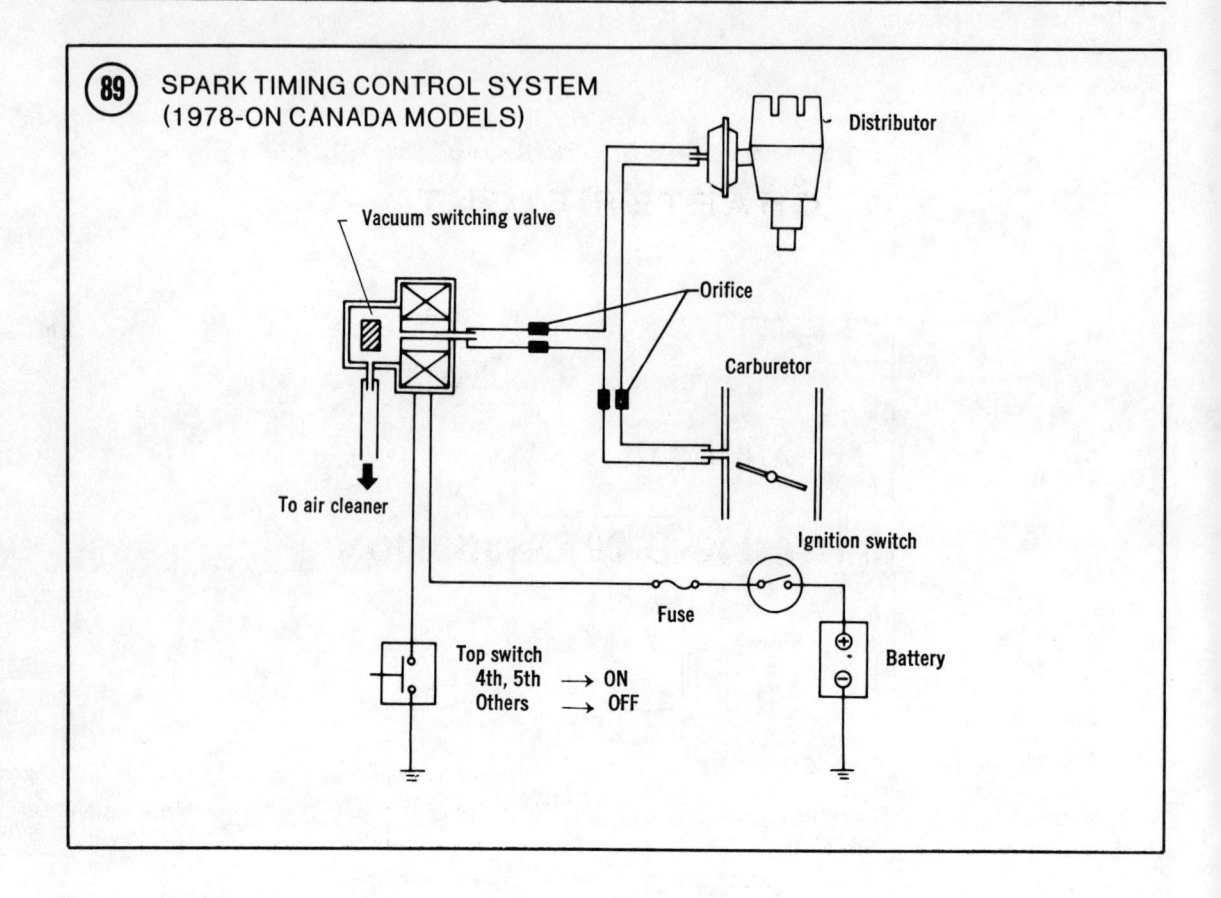

**89 SPARK TIMING CONTROL SYSTEM (1978-ON CANADA MODELS)**

# CHAPTER EIGHT

# CLUTCH AND TRANSMISSION

All models use a single dry plate clutch with diaphragm spring. The clutch is controlled by a hydraulic linkage. A four-speed manual transmission is standard, with a five-speed manual optional.

This chapter provides complete service procedures for the clutch and both transmissions.

## CLUTCH

The clutch differs from the basic design used in most front-engine, rear-drive cars. **Figure 1** is an assembled view; **Figure 2** shows the separate parts.

> NOTE: *Mechanics often use the terms "clutch cover" and "pressure plate" interchangeably, since they are usually part of the same assembly. On the 310 and F10, however, they are separate pieces as shown.*

### Pedal Adjustment

1. Measure pedal height from the floor (H, **Figure 3**). Compare with **Table 1**.

2. If pedal height is incorrect, loosen the pedal stopper locknut. Turn the pedal stopper to adjust the pedal height, then tighten the locknut.

3. Push the clutch pedal by hand. Note pedal free play (A, **Figure 3**). When free play ends, pedal resistance will increase abruptly.

4. If free play is incorrect, loosen the pushrod locknut. Turn the pushrod as needed to change free play. Then tighten the locknut.

## CLUTCH BLEEDING

The clutch hydraulic system must be bled whenever air enters it. Unlike hydraulic fluid, air is compressible. This means that instead of transmitting pedal pressure through the hydraulic line, the master cylinder will just compress the air in the line. The result is a mushy feeling pedal, incomplete clutch disengagement, and hard shifting.

> NOTE: *This procedure requires two people, one to operate the clutch pedal, the other to open and close the bleed valve.*

1. Clean the bleed valve on the operating cylinder (4, **Figure 1**).

2. Connect a plastic tube to the bleed valve. The tube should be a snug fit. Immerse the other end of the tube in a clear glass jar containing several inches of clean brake fluid.

170     CHAPTER EIGHT

① 

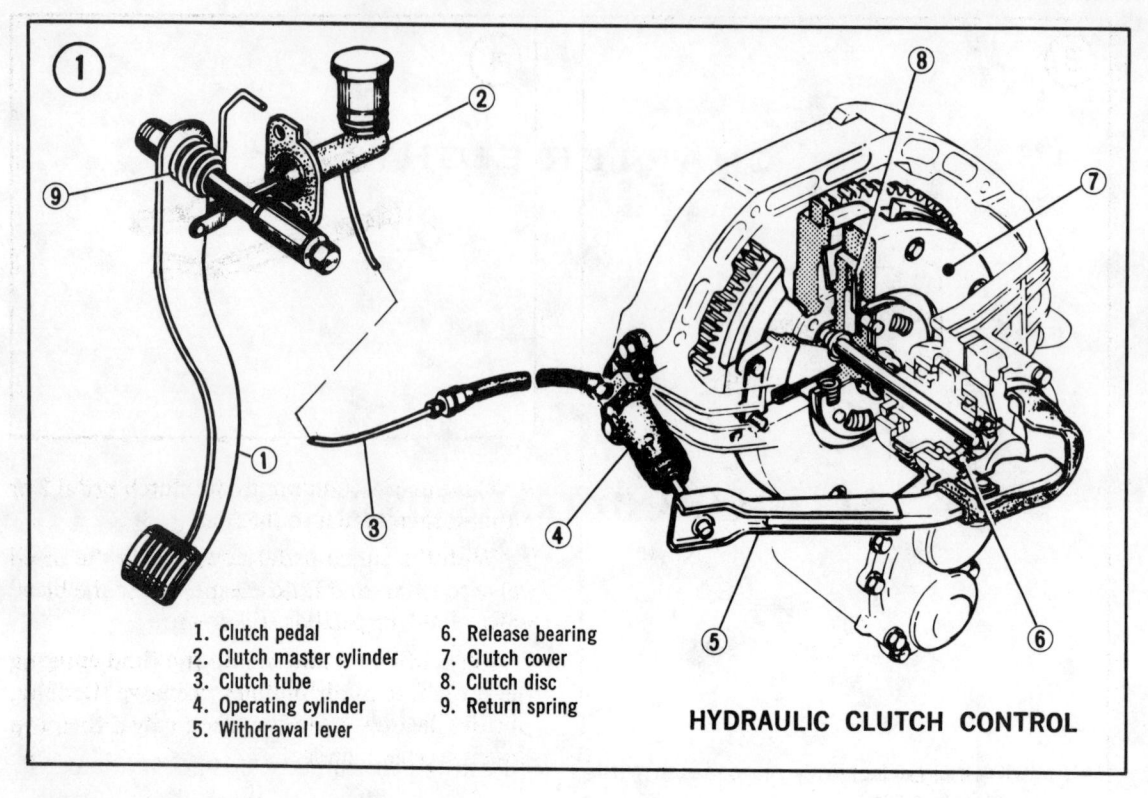

1. Clutch pedal
2. Clutch master cylinder
3. Clutch tube
4. Operating cylinder
5. Withdrawal lever
6. Release bearing
7. Clutch cover
8. Clutch disc
9. Return spring

**HYDRAULIC CLUTCH CONTROL**

②

**CLUTCH UNIT**

1. Flywheel
2. Diaphragm spring
3. Pressure plate
4. Clutch disc
5. Clutch cover
6. Push rod
7. Release bearing
8. Push rod A piece
9. O-ring
10. Bearing housing packing
11. Bearing housing
12. Withdrawal lever pin
13. Withdrawal lever
14. Dust cover

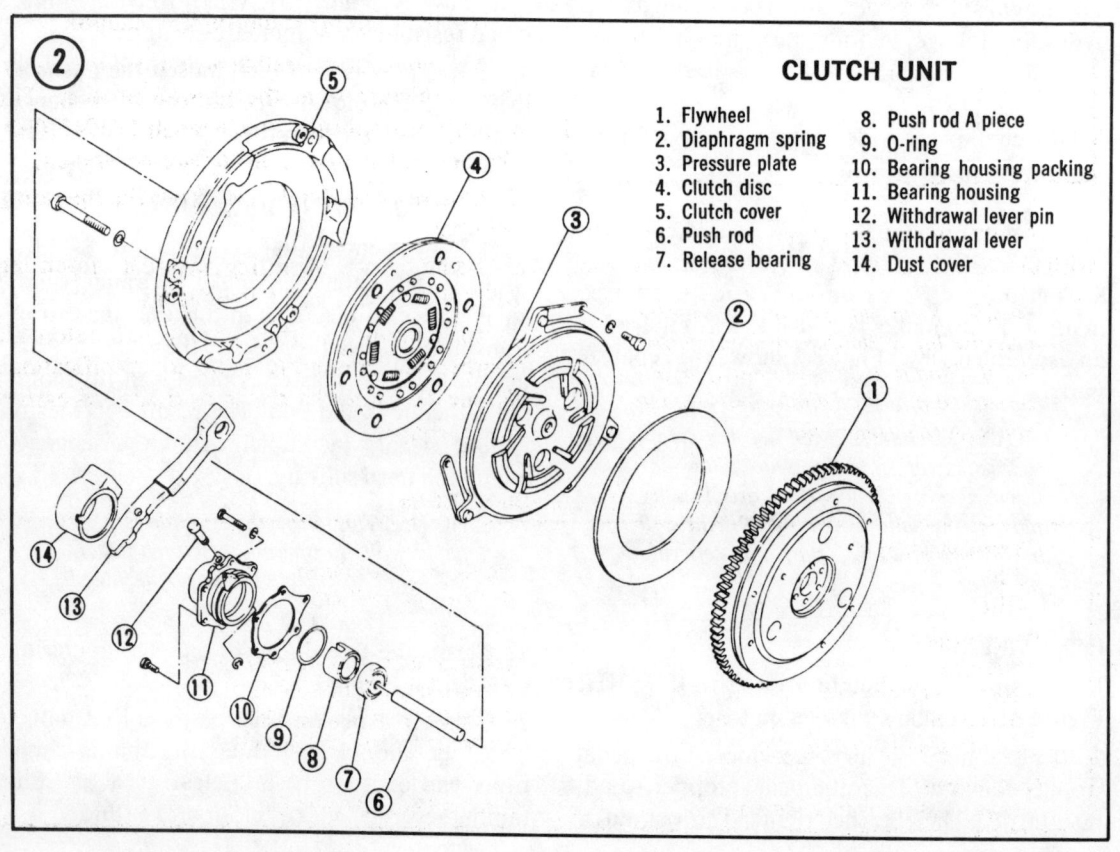

# CLUTCH AND TRANSMISSION

**3**

1. Pedal stopper lock nut
2. Push rod lock nut

**4**

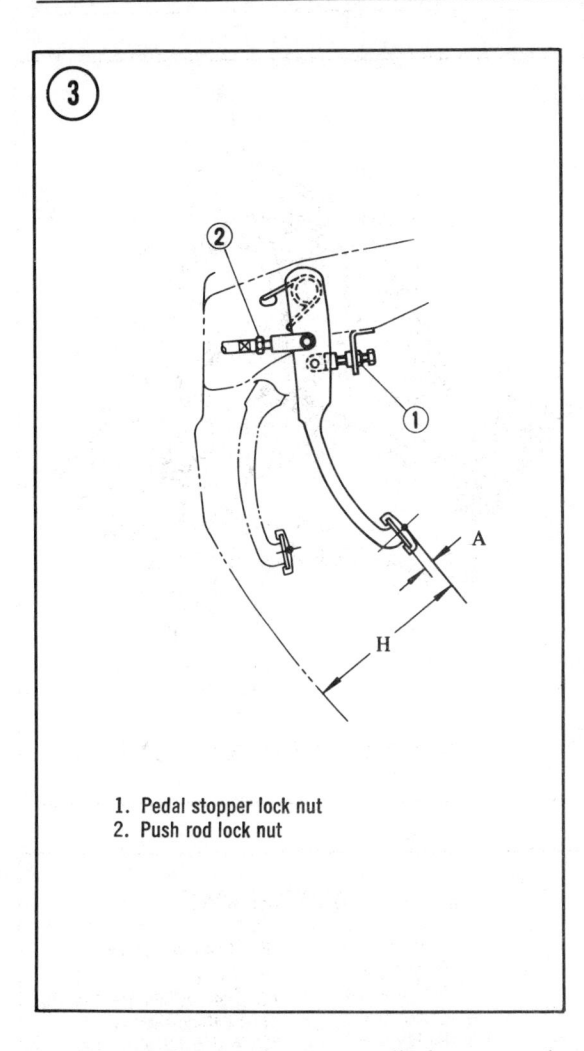

3. Top up the clutch master cylinder reservoir with brake fluid meeting DOT 3 specifications.

> NOTE: *Keep an eye on the master cylinder brake fluid level during bleeding. If fluid level drops too low, air will be sucked into the cylinder and bleeding will have to be repeated.*

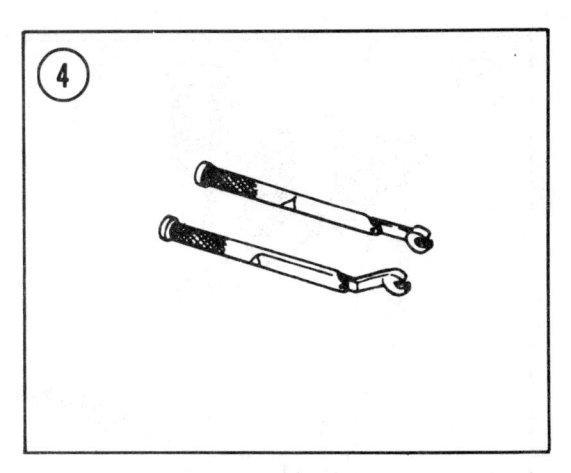

4. Have an assistant pump the clutch pedal 2 or 3 times, then hold it to the floor.

5. With the clutch pedal down, open the bleed valve to let air and fluid escape. Close the bleed valve while the pedal is still down.

6. Repeat Steps 4 and 5 until the fluid entering the jar is free of air bubbles. Remove the tube, put the dust cover on the bleed valve, then top up the master cylinder.

### Operating Cylinder Removal/Installation

1. Disconnect the flexible hose from the clutch line (3, **Figure 1**) at the firewall bracket. The factory recommends tube wrench GG94310000 (**Figure 4**). Do not use an adjustable wrench.

2. Unscrew the clutch hose from the operating cylinder.

3. Remove the operating cylinder mounting bolts. Lift it off the clutch housing.

4. Installation is the reverse of removal. Tigthen the mounting bolts to specifications (**Table 2**). Bleed the clutch as described earlier in this chapter.

Table 1    SPECIFICATIONS

| | |
|---|---|
| **Clutch pedal height** | |
| F10 | $6\frac{5}{8}$-$6\frac{7}{8}$ in. (169-175mm) |
| 310 | 7-$7\frac{1}{4}$ in. (179-185mm) |
| **Gear end play** | |
| First gear | 0.008-0.012 in. (0.2-0.3mm) |
| Second and third gear (four-speed) | 0.008-0.012 in. (0.2-0.3mm) |
| Second and third gear (five-speed) | 0.008-0.014 in. (0.20-0.35mm) |
| Fourth gear (five-speed only) | 0.008-0.012 in. (0.2-0.3mm) |
| **Counter gear** | |
| Reverse idler gear (five-speed only) | 0.0-0.-15 in. (0.0-0.37mm) |

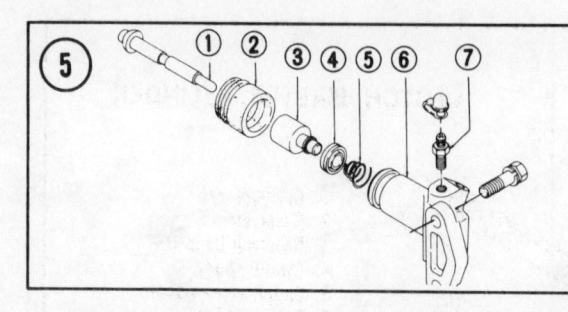

1. Push rod
2. Dust cover
3. Piston
4. Piston cup
5. Piston spring
6. Cylinder body
7. Bleeder screw

Table 2  TIGHTENING TORQUES

| Fastener | Ft.-lb. | Mkg |
| --- | --- | --- |
| Clutch cover to flywheel | 5-7 | 0.7-1.0 |
| Strap plate bolts | 5-6 | 0.7-0.8 |
| Master cylinder nuts | 5-8 | 0.8-1.1 |
| Slave cylinder bolts | | |
| F10 | 12-15 | 1.6-2.1 |
| 310 | 22-30 | 3.1-4.1 |
| Bleed valve | 5-6½ | 0.7-0.9 |
| Metal fluid line | 11-13 | 1.5-1.8 |
| Fluid hose | 12-14 | 1.6-2.0 |
| Clutch housing-to-engine | 10-13 | 1.4-1.8 |
| Clutch housing-to-transmission case | 12-17 | 1.6-2.3 |
| Starter bolts | 20-27 | 2.7-3.7 |
| Transmission bracket-to-engine | 43-58 | 6-8 |
| Bottom cover bolts | 4½-7 | 0.6-1.0 |
| Mainshaft locknut | 36-43 | 5-6 |
| Main drive shaft locknut | 43-58 | 6-8 |
| Check ball plugs | 8-22 | 1.1-3.0 |
| Reverse idler gear shaft locknut | 72-87 | 10-12 |
| Bearing retainer bracket | 12-17 | 1.6-2.3 |
| Transmission case cover | 12-17 | 1.6-2.3 |
| Main drive gear mounting nut | 43-58 | 6-8 |
| Primary gear cover | | |
| Small bolts | 4½-7 | 0.6-1.0 |
| Large bolts | 12-17 | 1.6-2.3 |
| Primary gear bearing housing-to-primary gear cover | 4½-7 | 0.6-1.0 |
| Speedometer pinion | 4½-7 | 0.6-1.0 |
| Mainshaft bearing retainer (four-speed) | 12-16 | 1.6-2.2 |
| Reverse fork lever bracket (four-speed) | 12-16 | 1.6-2.2 |
| Counter gear shaft retaining plate (five-speed) | 12-17 | 1.6-2.2 |
| Final gear side flange nut | 87-101 | 12-14 |
| Final gear to differential case | 43-58 | 6-8 |

# CLUTCH AND TRANSMISSION

**CLUTCH MASTER CYLINDER**

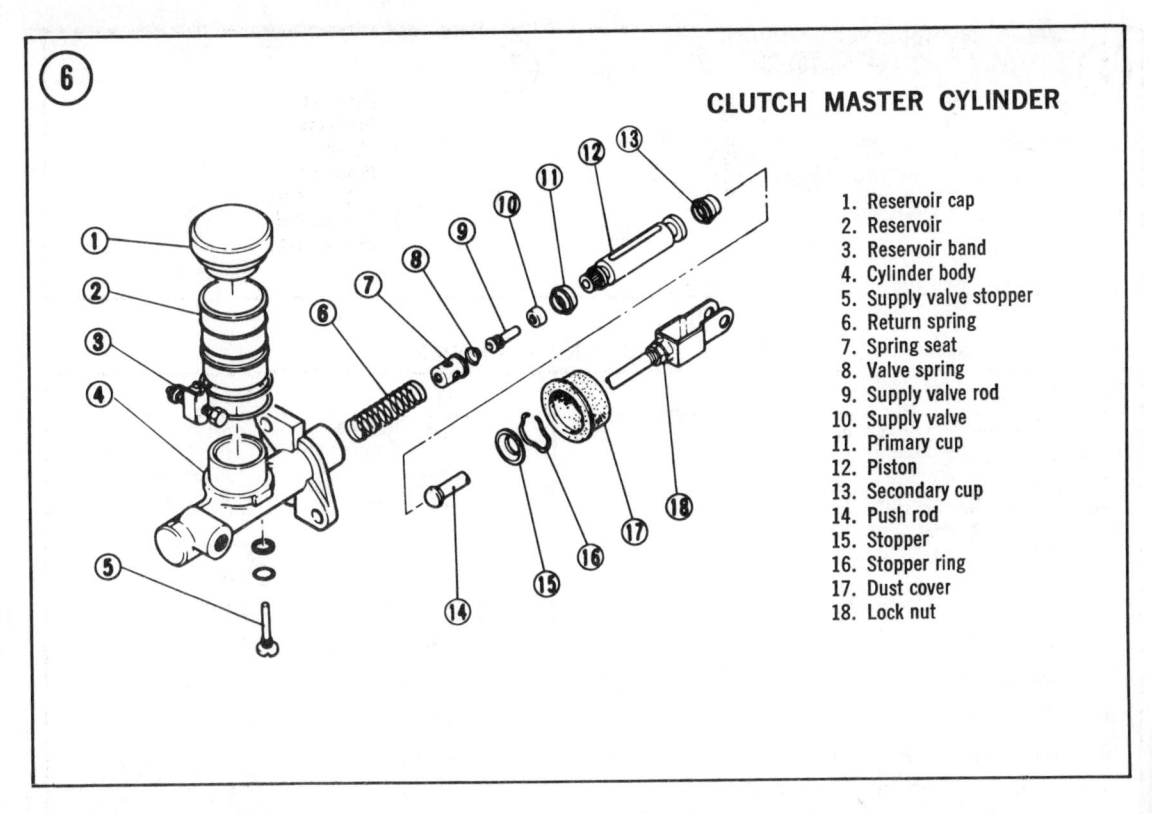

1. Reservoir cap
2. Reservoir
3. Reservoir band
4. Cylinder body
5. Supply valve stopper
6. Return spring
7. Spring seat
8. Valve spring
9. Supply valve rod
10. Supply valve
11. Primary cup
12. Piston
13. Secondary cup
14. Push rod
15. Stopper
16. Stopper ring
17. Dust cover
18. Lock nut

## Operating Cylinder Overhaul

1. Remove the pushrod and dust cover (**Figure 5**). Take out the piston, together with the piston cup, then remove the return spring.

2. Remove the bleed valve dust cap. Unscrew the bleed valve.

3. Remove the piston cup from the piston, then throw the cup away. Piston cups must not be reused, no matter how good they look.

4. Throughly clean all parts with clean brake fluid. Do not clean with gasoline, kerosene, or solvent.

5. Check the cylinder bore and the piston for scratches, rust, or pitting. Replace if these conditions can be seen.

6. Check the piston and cylinder bore for wear. If piston-to-bore clearance exceeds 0.006 in. (0.15mm), replace the operating cylinder.

7. Make sure the bleed valve and its passage in the cylinder body are clean.

8. Dip a new piston cup in brake fluid. Install it on the piston.

   NOTE: *The lip (wide side) of the piston cup faces into the cylinder.*

9. Install the return spring and piston. Be careful not to damage the lip of the piston cup.

10. Install a new dust cover and the pushrod.

11. Install the bleed valve and dust cover.

## Master Cylinder Removal/Installation

1. Remove the clip and clevis pin that secure the master cylinder pushrod to the clutch pedal arm.

2. Disconnect the fluid line from the master cylinder.

3. Remove the cylinder mounting nuts. Lift the cylinder out.

4. Installation is the reverse of removal. Tighten the mounting nuts to specifications (end of chapter). Bleed the clutch as described earlier in this chapter.

## Master Cylinder Overhaul

Refer to **Figure 6** for this procedure.

1. Pull back the dust cover.

2. Push the pushrod into the cylinder. Remove the snap ring with needle-nosed pliers, then take the pushrod out.

3. Unscrew the supply valve stopper.

4. Remove the piston, spring seat, and return spring.

5. Take the piston cups off the piston. Throw away the piston cups and dust cover.

6. Clean all parts with new brake fluid. Do not use gasoline, kerosene, or solvent.

7. Check the cylinder bore and the piston for scratches, rust, or pitting. Replace if these conditions can be seen.

8. Check the cylinder bore and piston for wear. If piston-to-bore clearance exceeds 0.006 in. (0.15mm), replace the master cylinder.

9. Make sure all internal passages are clear.

10. Dip new piston cups in clean brake fluid. Install them on the piston.

NOTE: *The lip (wide end) of each cup faces into the cylinder.*

11. Coat the cylinder bore and piston with clean brake fluid.

12. Assemble the master cylinder by reversing disassembly. **Figure 7** is an assembled view.

## CLUTCH REMOVAL

The clutch can be removed without removing the engine or transmission from the car.

1. Disconnect negative cable from battery.

2. Disconnect the air cleaner fresh air duct.

3. Disconnect the thick wire running from coil to distributor.

4. On 310's, remove the fuel filter as described in Chapter Five. Remove the clutch operating cylinder as described earlier in this chapter.

5. On F10's, disconnect the wiring harness connectors on the clutch housing. Also disconnect the evaporative emission canister hoses.

6. Turn the steering wheel all the way to the right. Remove the access cover (**Figure 8**) from the wheel well.

7. Remove the inspection cover from the upper part of the clutch housing.

8. Working through the access hole in the wheel well, remove the primary drive gear assembly. See **Figure 9**.

9. Unbolt the clutch cover from the flywheel (**Figure 10**). Turn the flywheel with a prying tool as shown to expose the bolts.

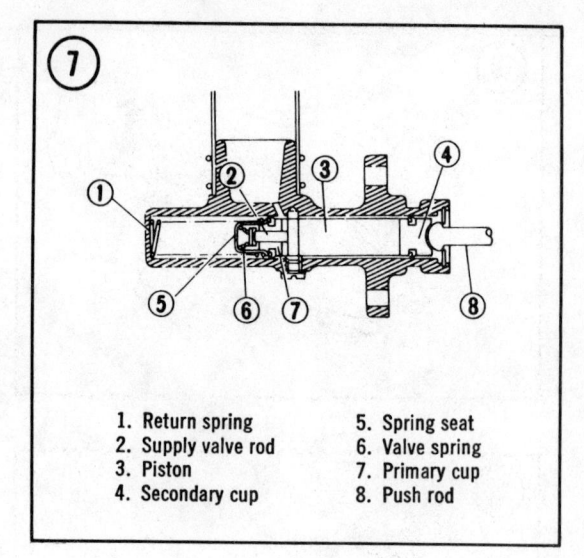

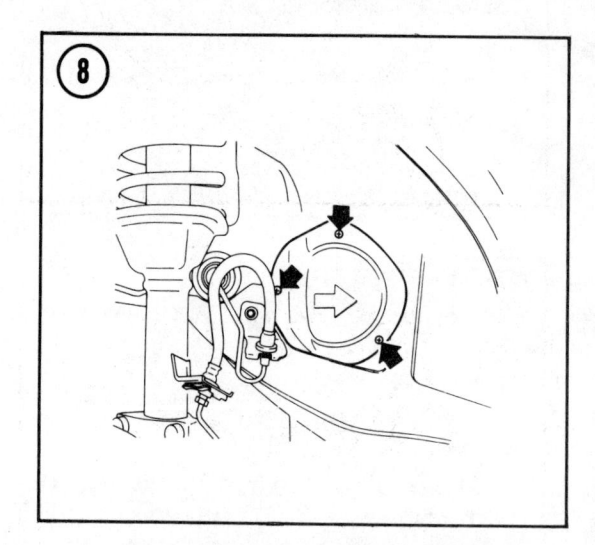

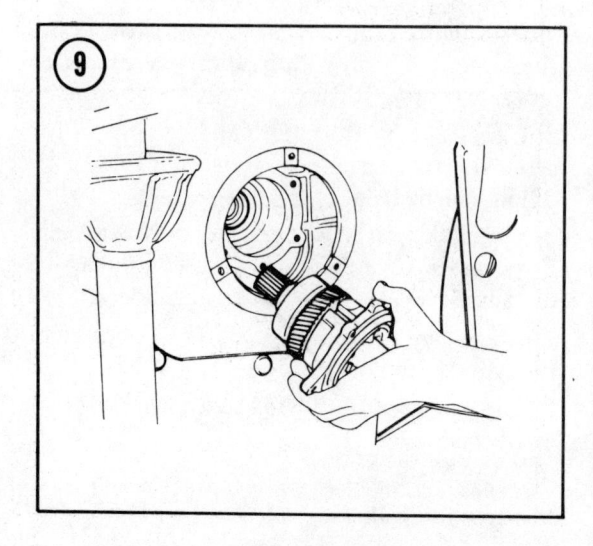

⑦

1. Return spring
2. Supply valve rod
3. Piston
4. Secondary cup
5. Spring seat
6. Valve spring
7. Primary cup
8. Push rod

⑧

⑨

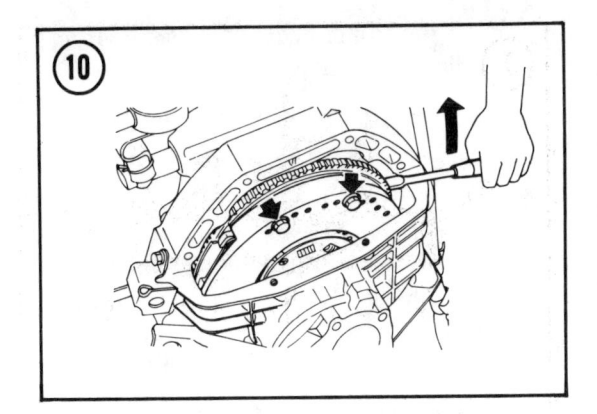

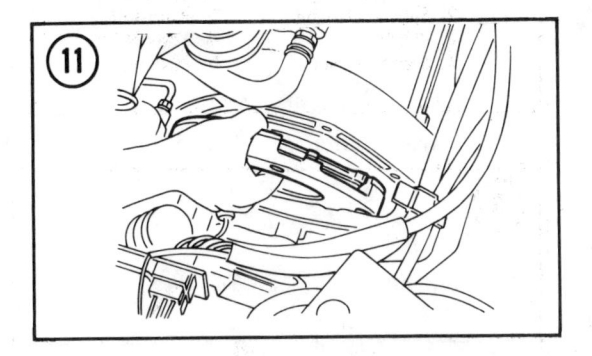

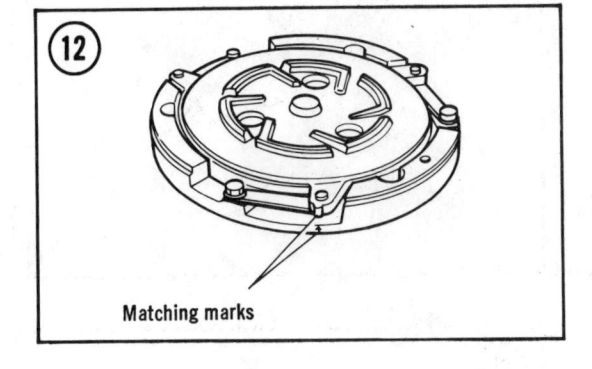

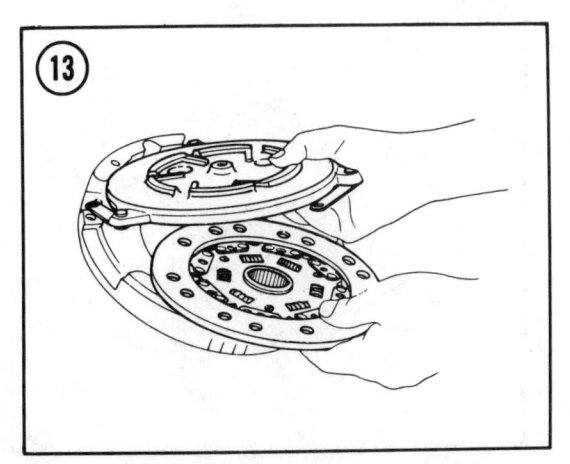

# CLUTCH AND TRANSMISSION

### CAUTION
*Loosen the bolts in several stages, a little at a time, to prevent warping the clutch cover.*

10. Lift the clutch out **(Figure 11)**.

11. Check for match marks on clutch cover and pressure plate **(Figure 12)**. If these are not visible, make your own marks. The clutch cover and pressure plate must be reassembled in their original positions.

12. Remove 3 bolts securing the clutch cover straps to the pressure plate. Separate the pressure plate, clutch cover, and disc **(Figure 13)**.

## CLUTCH INSPECTION

1. Check the diaphragm spring for wear or damage. Since spring weakness is hard to judge, the spring should be replaced if there is any doubt about its condition.

2. Check the clutch disc for the following:
   a. Oil or grease on the facings
   b. Glazed facings
   c. Warped facings
   d. Loose or missing rivets
   e. Broken springs
   f. Loose fit or rough movement on the primary gear splines

3. Measure the depth of the facing material with a vernier caliper **(Figure 14)**. If within 0.012 in. (0.3mm) of any rivet, replace the clutch disc.

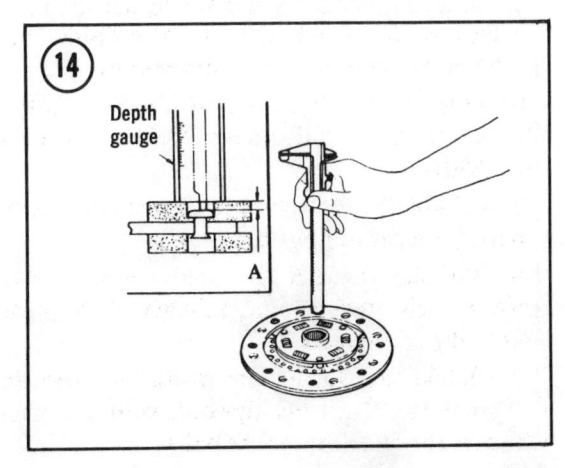

4. Check the pressure plate for scoring, cracks, or burn marks (blue-tinted areas). Replace the pressure plate if these can be seen.

## CLUTCH INSTALLATION

1. Be sure your hands are clean.

2. Bolt the clutch cover to the pressure plate, with the disc between them. Be sure the match marks (**Figure 12**) are lined up. Tighten the bolts to 7-9 ft.-lb. (1.0-1.3 mkg).

3. Position the clutch cover assembly on the 2 flywheel dowels. Install the clutch cover bolts and tighten them slightly.

4. Insert the primary gear into the clutch disc to align the disc. Tighten the primary gear bolts to 4½-7 ft.-lb. (0.6-1.0 mkg).

5. Tighten the clutch cover bolts to 5-7 ft.-lb. (0.7-1.0 mkg).

6. Install the inspection cover on the upper part of the clutch housing. Tighten its bolts to 4½-7 ft.-lb. (0.6-1.0 mkg).

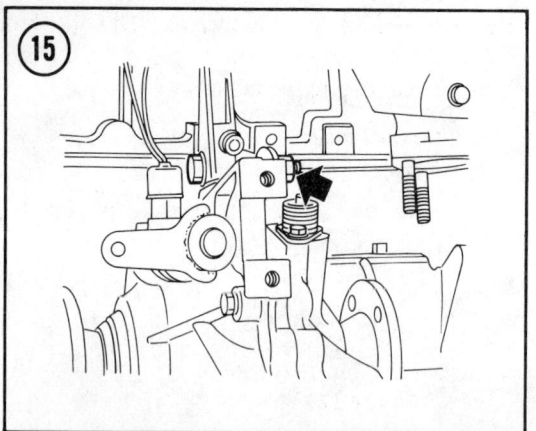

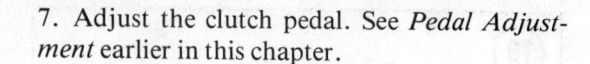

7. Adjust the clutch pedal. See *Pedal Adjustment* earlier in this chapter.

## RELEASE BEARING REPLACEMENT

1. Remove the clutch operating cylinder as described earlier in this chapter.

2. Remove the withdrawal lever pin (12, **Figure 2**). Then take out the withdrawal lever.

3. Remove the bearing housing assembly. Take the O-ring, pushrod "A" piece, and release bearing out of the housing. See **Figure 2**.

4. To inspect the bearing, hold the inner race in one hand and turn the outer race with the other. If the bearing is rough or noisy, replace it.

## TRANSMISSION

### Removal/Installation

1. Remove the engine and transmission from the car. See *Engine Removal/Installation*, Chapter Four.

2. Remove the starter (Chapter Seven).

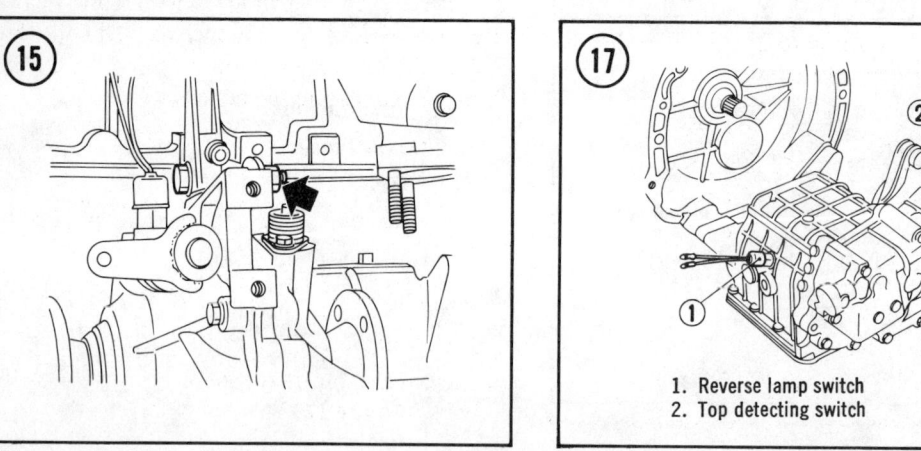

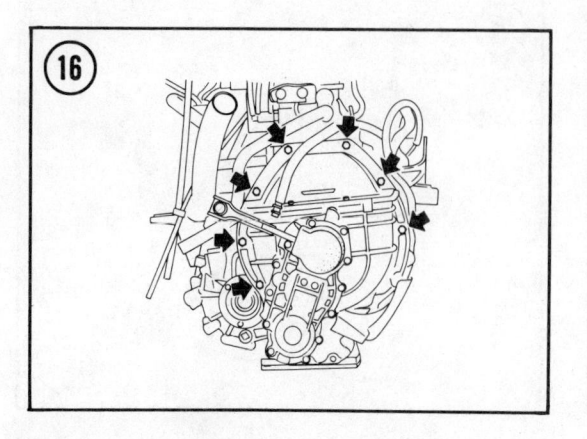

1. Reverse lamp switch
2. Top detecting switch

# CLUTCH AND TRANSMISSION

3. Unbolt the transmission mounting bracket from the transmission (**Figure 15**).

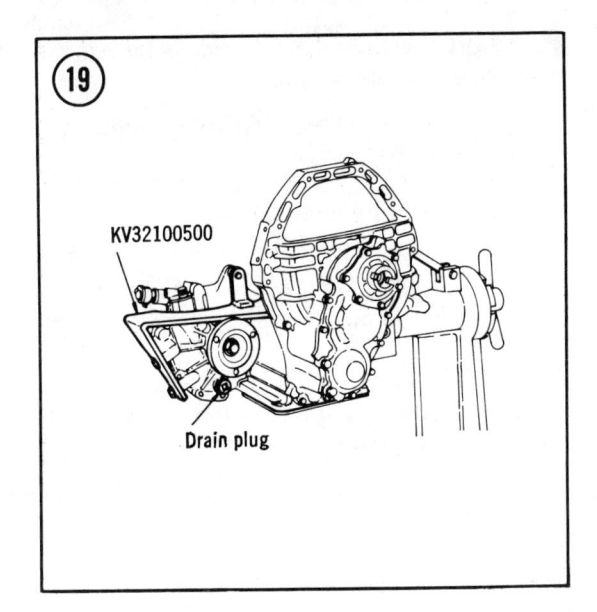

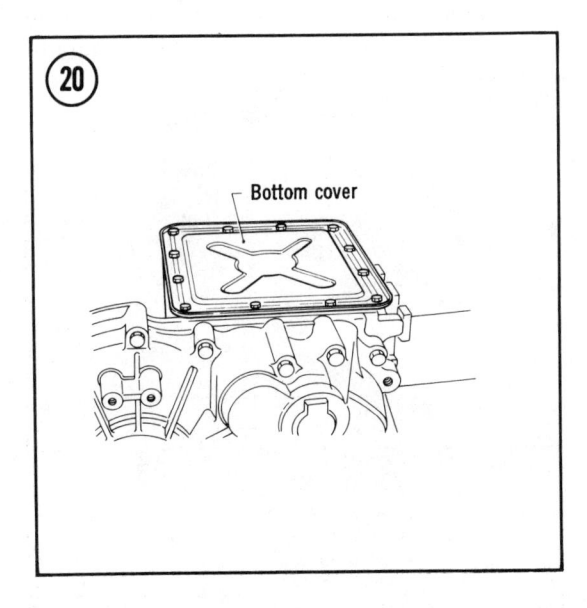

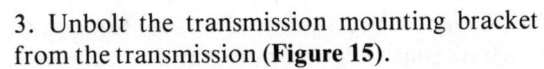

4. Unbolt the clutch housing from the engine (**Figure 16**). Separate the transmission from the engine.

5. Installation is the reverse of removal. Tighten the lower 3 clutch housing bolts before tightening the other 4 bolts.

**Disassembly (Four-Speed)**

1. Thoroughly clean the outside of the transmission with solvent.

2. Unscrew the transmission electrical switches (**Figure 17**).

3. Remove the speedometer pinion (**Figure 18**).

4. If available, place the transmission on a stand as shown in **Figure 19**.

5. Remove the drain plug (**Figure 19**). Drain the transmission oil.

6. Unbolt the bottom cover from the transmission (**Figure 20**).

7. Secure each side flange with a tool such as the one shown in **Figure 21** (Datsun No. KV38103701; Kent-Moore No. J26085). Remove the side flange nuts as shown.

> NOTE: *The side flange nuts are tightened to 87-101 ft.-lb. (12-14 mkg). If you don't have the correct socket, have the nuts removed by a Datsun dealer.*

8. Once the nuts are removed, take off the side flanges. If this is difficult, use a gear puller. These are available from rental dealers.

9. Remove the primary drive gear and bearing housing (**Figure 22**).

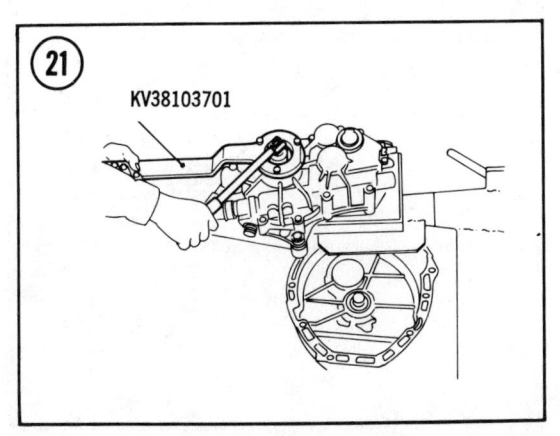

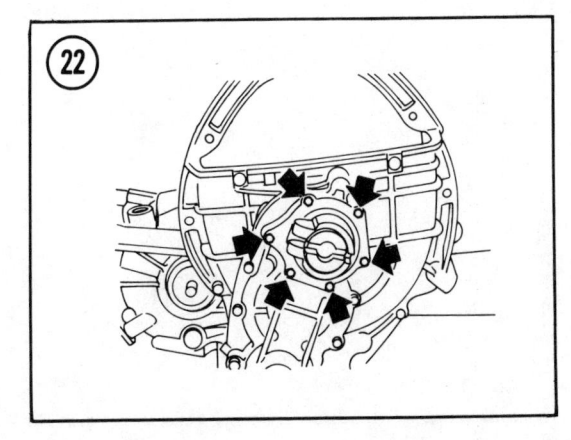

10. Remove the primary gear cover (**Figure 23**). Note that the bolts are different lengths. Label them so they can be reinstalled in the right places. In addition, bolt No. 3 is coated with hard resin to prevent oil leaks. This bolt must not be reused.

11. Lock the transmission by engaging 2 gears at once. To do this, push the reverse shift fork (**Figure 24**) toward the transmission case wall to engage reverse gear. Push the first-second shift fork in either direction until the synchronizer sleeve covers the clutch teeth (small teeth) on the gear next to it.

12. With the transmission locked, loosen the input gear locknut (**Figure 25**).

13. Remove the main drive input gear and bearing (**Figure 26**).

14. Remove the primary idler gear (**Figure 27**). If it is difficult to remove, use a puller.

15. Remove the clutch housing (**Figure 28**). Note that the bolts are different lengths. Label them so they can be reinstalled correctly.

NOTE: *The No. 1 bolts are coated with hard resin to prevent oil leaks. These cannot be reused.*

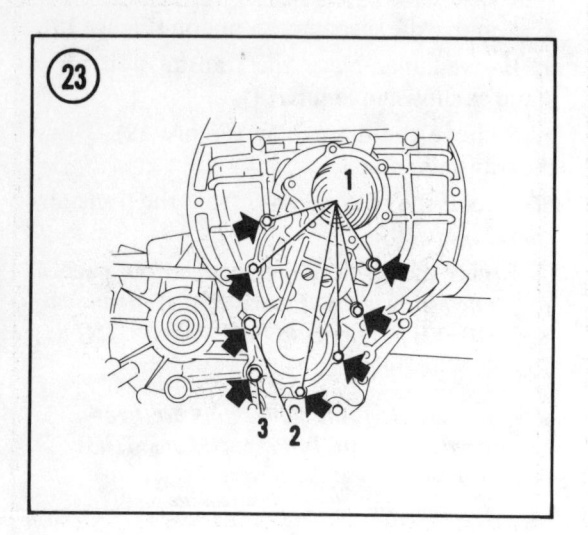

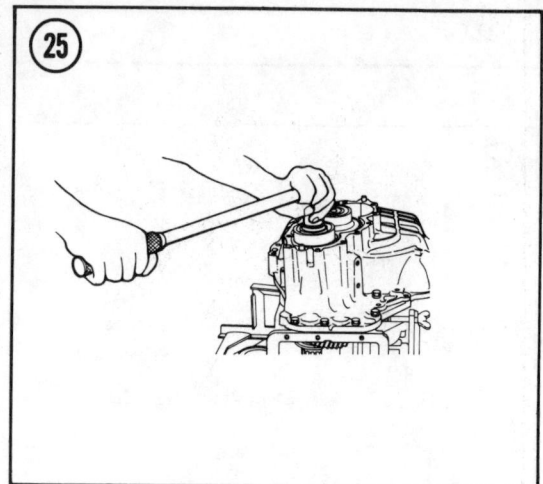

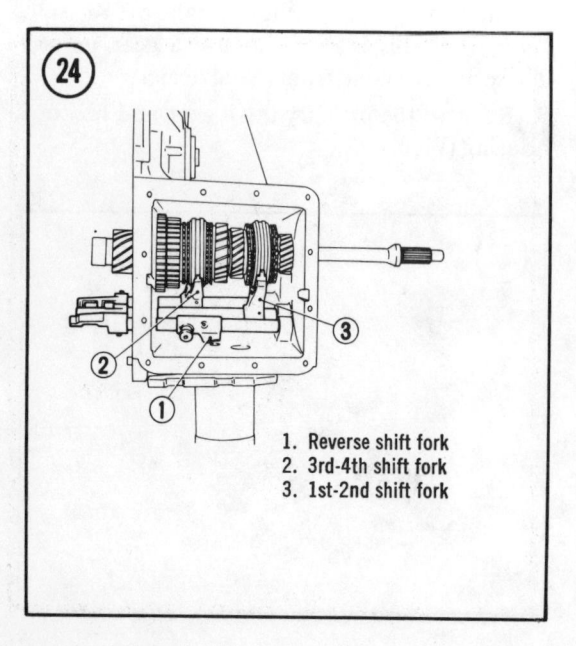

23

25

24

1. Reverse shift fork
2. 3rd-4th shift fork
3. 1st-2nd shift fork

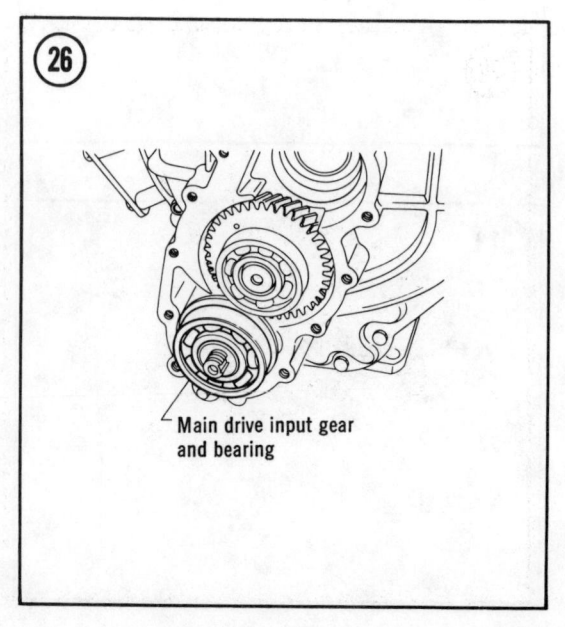

26

Main drive input gear and bearing

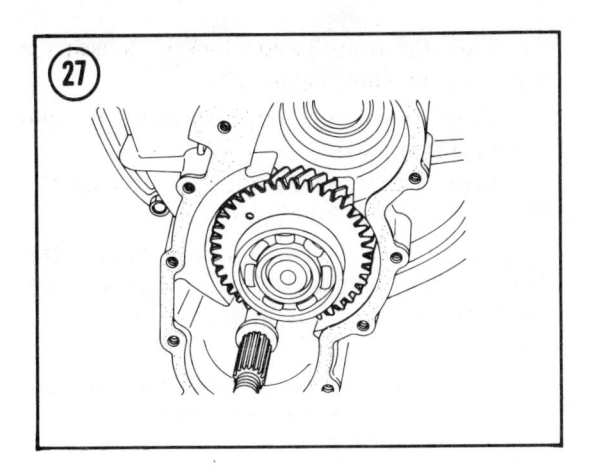

# CLUTCH AND TRANSMISSION

16. Remove the transmission case cover **(Figure 29)**. Note that the bolts are different lengths. Label them so they can be reinstalled in the correct holes. Note that bolts 5, 6, 7, and 8 are coated with hard resin to prevent oil leaks. These cannot be reused.

17. Remove the check ball plugs **(Figure 30)**. Take out the locking springs and steel detent balls.

18. Remove the transmission case taper thread plug with an inpact screwdriver. See **Figure 31**.

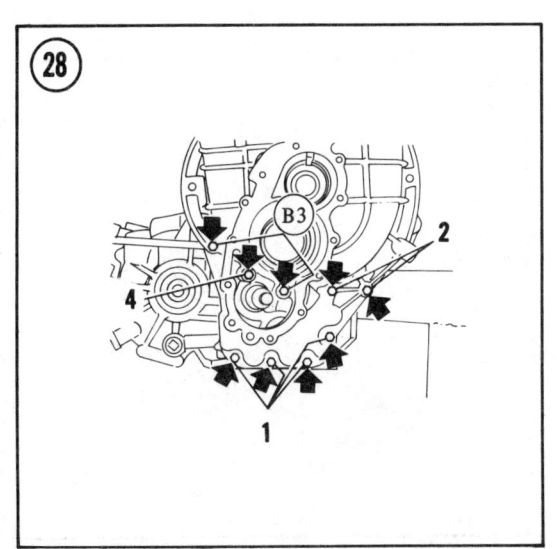

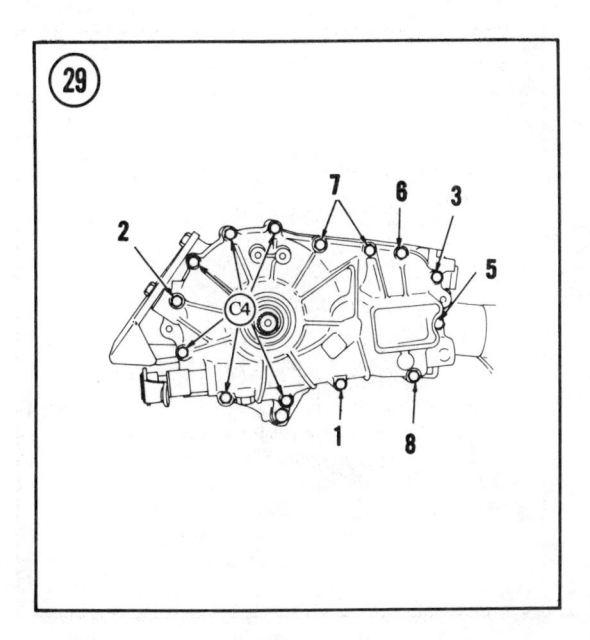

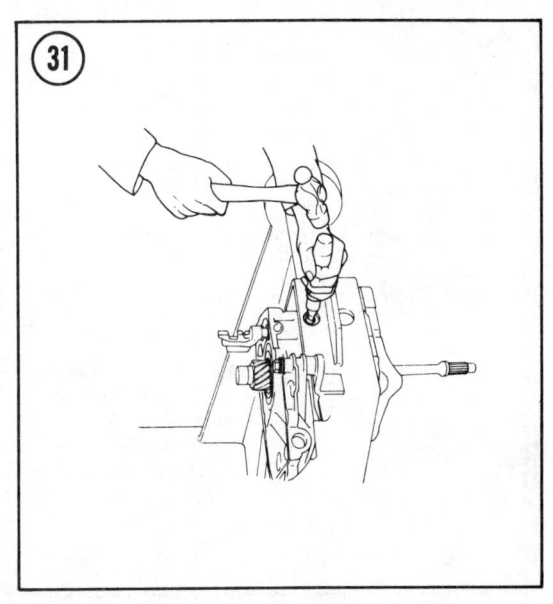

**CHAPTER EIGHT**

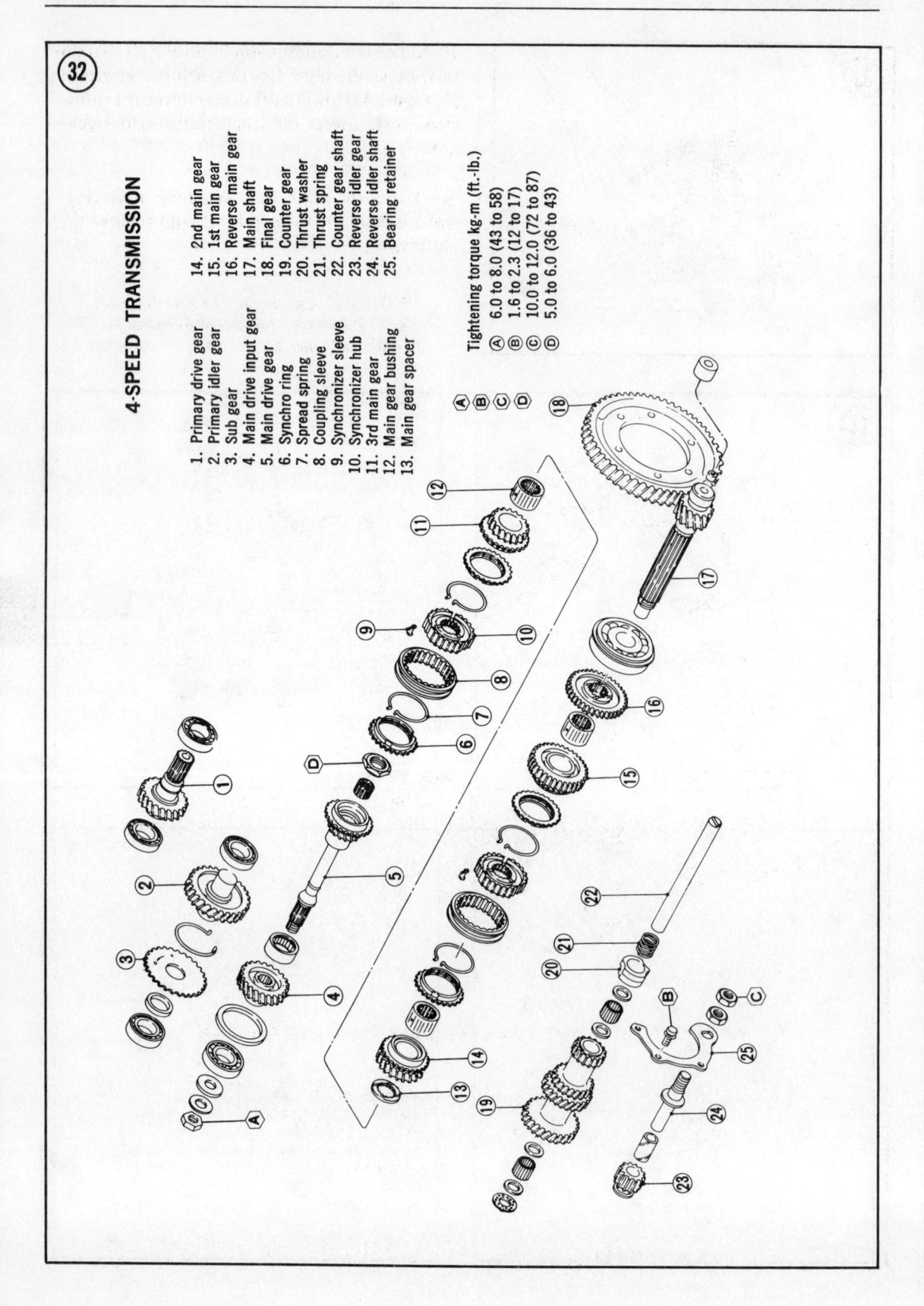

**④ 4-SPEED TRANSMISSION**

1. Primary drive gear
2. Primary idler gear
3. Sub gear
4. Main drive input gear
5. Main drive gear
6. Synchro ring
7. Spread spring
8. Coupling sleeve
9. Synchronizer sleeve
10. Synchronizer hub
11. 3rd main gear
12. Main gear bushing
13. Main gear spacer
14. 2nd main gear
15. 1st main gear
16. Reverse main gear
17. Main shaft
18. Final gear
19. Counter gear
20. Thrust washer
21. Thrust spring
22. Counter gear shaft
23. Reverse idler gear
24. Reverse idler shaft
25. Bearing retainer

Tightening torque kg-m (ft.-lb.)
Ⓐ 6.0 to 8.0 (43 to 58)
Ⓑ 1.6 to 2.3 (12 to 17)
Ⓒ 10.0 to 12.0 (72 to 87)
Ⓓ 5.0 to 6.0 (36 to 43)

# CLUTCH AND TRANSMISSION

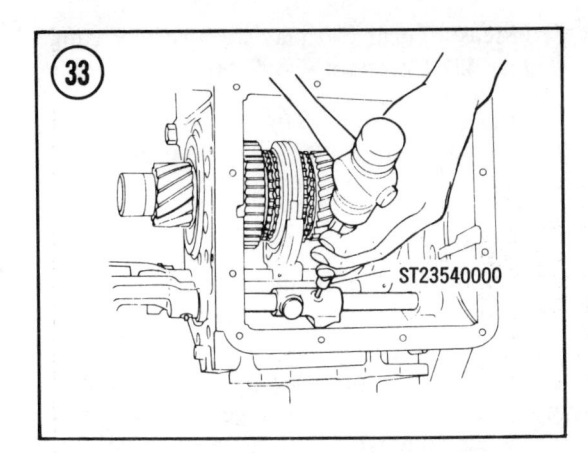

19. Place the transmission in third gear. To do this, move the third-fourth synchronizer sleeve (8, **Figure 32**) toward the center of the transmission, so it covers the small teeth on the gear next to it.

20. Drive the shift fork retaining pins (**Figure 33**). Pull the shift rods out and remove the shift forks.

NOTE: *Be careful not to lose the interlock plungers, located in the gearbox wall. See* **Figure 34**.

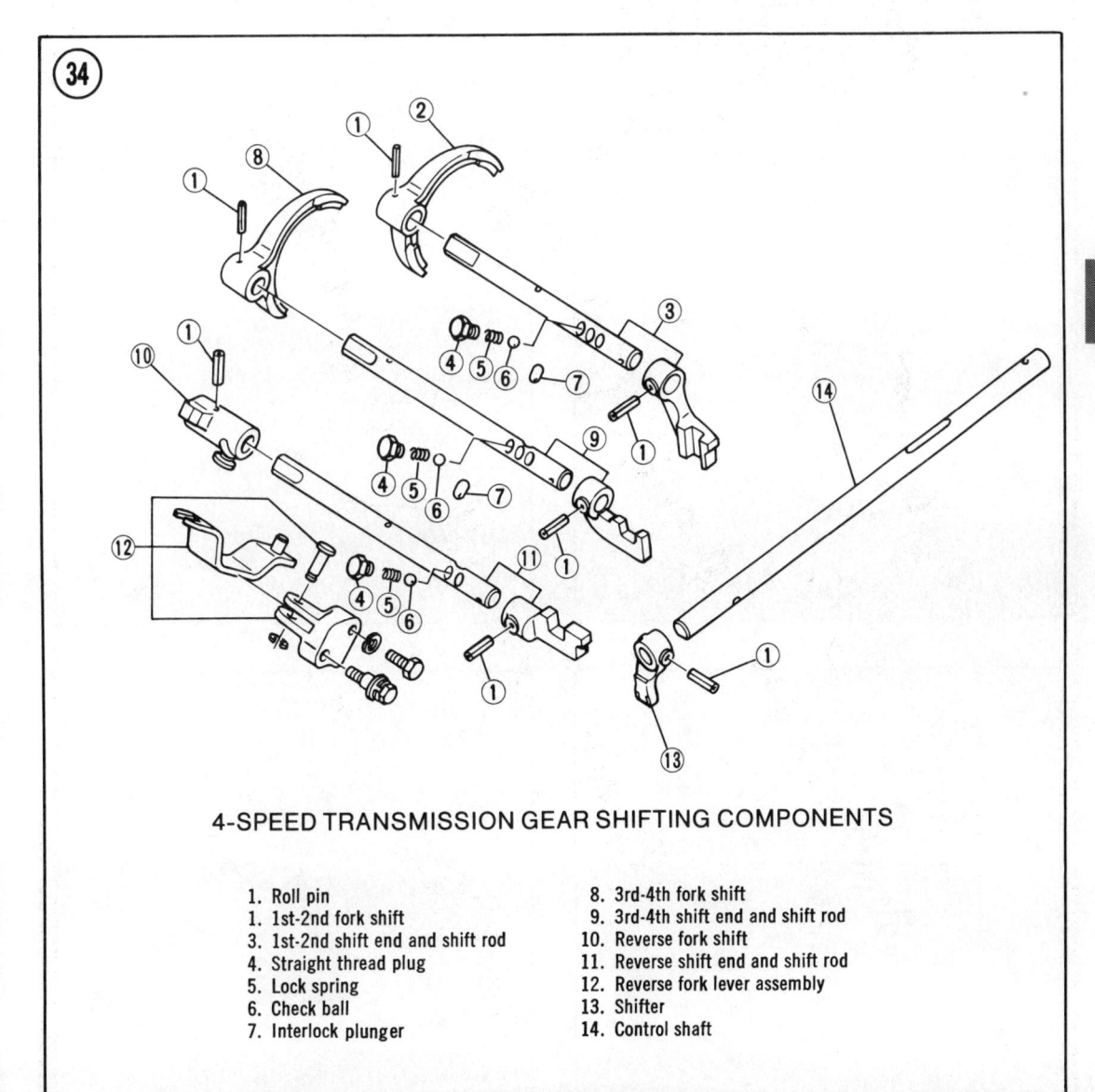

**4-SPEED TRANSMISSION GEAR SHIFTING COMPONENTS**

1. Roll pin
1. 1st-2nd fork shift
3. 1st-2nd shift end and shift rod
4. Straight thread plug
5. Lock spring
6. Check ball
7. Interlock plunger
8. 3rd-4th fork shift
9. 3rd-4th shift end and shift rod
10. Reverse fork shift
11. Reverse shift end and shift rod
12. Reverse fork lever assembly
13. Shifter
14. Control shaft

21. Remove final drive assembly (**Figure 35**).

22. Remove the reverse fork lever and bracket (**Figures 36 and 37**).

23. Remove the main shaft bearing retainer (**Figure 38**).

24. Remove the reverse idler gear and shaft (**Figure 39**).

25. Push out the counter gear shaft with a tool such as ST23100000 (Kent-Moore No. J25685). **Figure 40** shows the tool in use; **Figure 41** gives dimensions.

CAUTION
*The needle roller bearings at either end of the countershaft are not caged. Do not let them fall out, or they may be lost.*

26. Remove the main shaft gear assembly and main drive gear from the transmission case. See **Figure 42**.

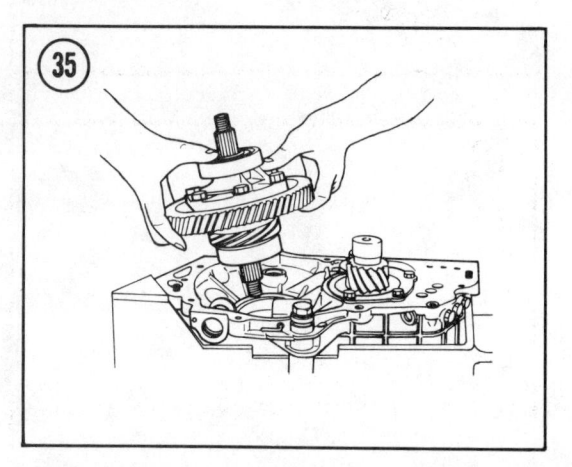

27. Measure gear end play with a feeler gauge (**Figure 43**). Compare with **Table 1**.

28. Place the main shaft in a soft-jawed vise (**Figure 44**). Remove the main shaft locknut as shown.

NOTE: *The locknut is punched into a groove on the main shaft. File this part away, or bend it up with a sharp punch.*

29. Remove the following parts in order from the main shaft: third-fourth gear synchronizer, third gear, main gear bushing, gear spacer, second gear, main gear bushing, first-second gear synchronizer, first gear, and gear bushing. If the bearing is worn or damaged, have it pressed off by a machine shop. See **Figure 45**.

30. Disassemble the synchronizers. See **Figure 46**.

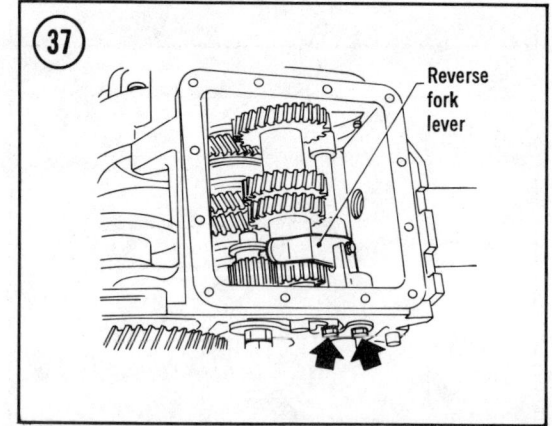

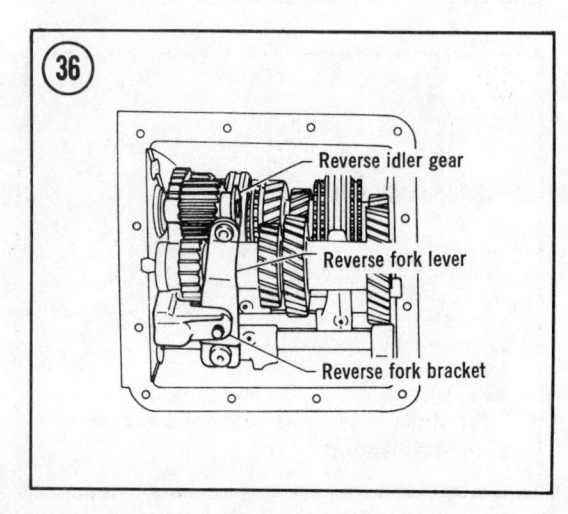

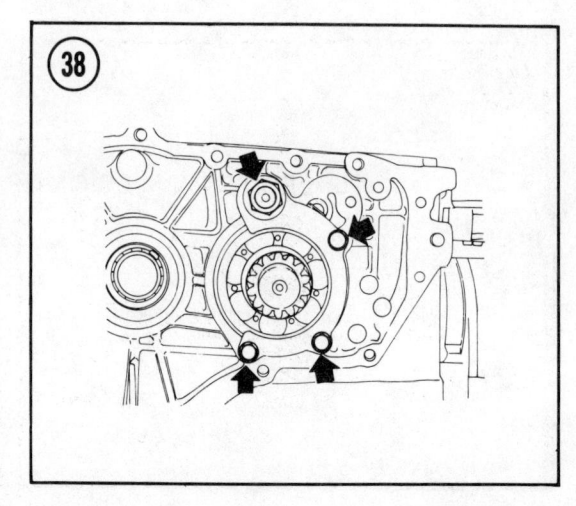

# CLUTCH AND TRANSMISSION

183

39

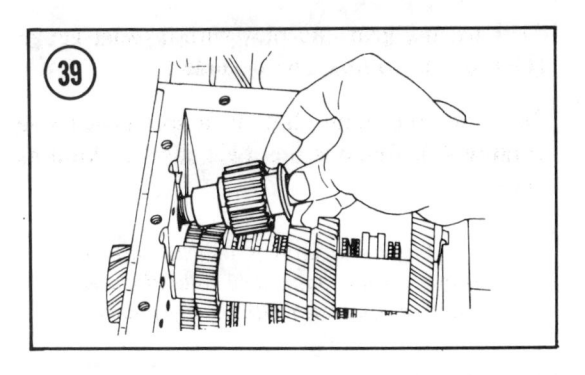

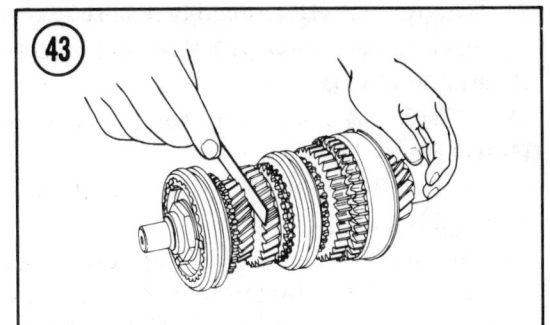

40 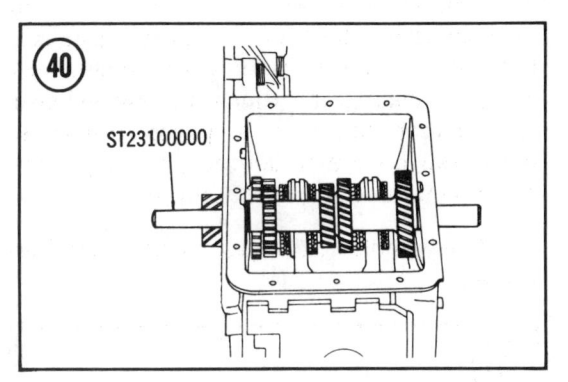 ST23100000

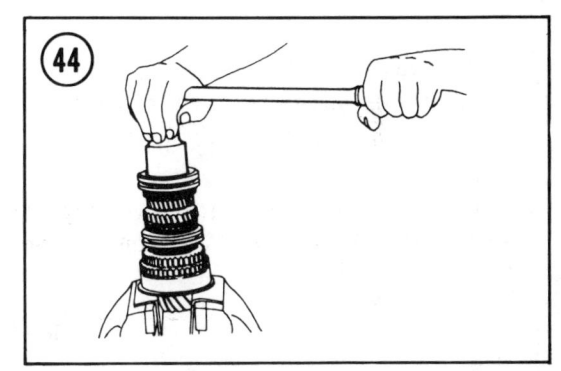

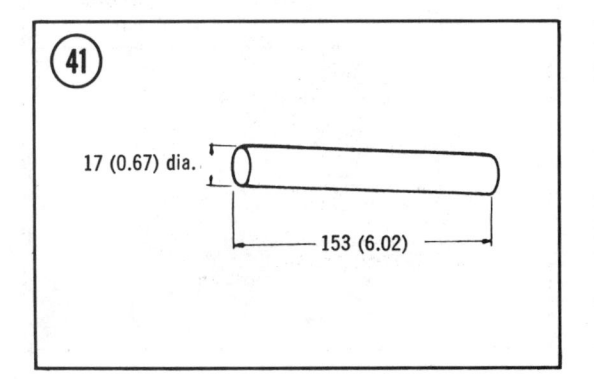

41 17 (0.67) dia. — 153 (6.02)

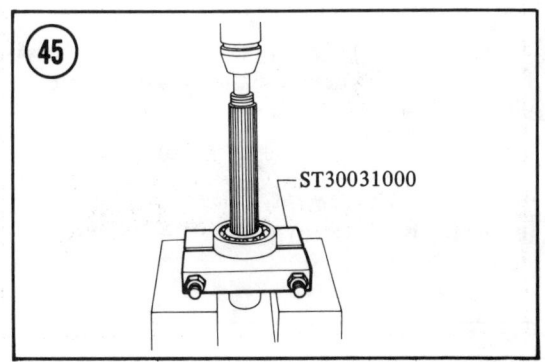

42

43

44

45 ST30031000

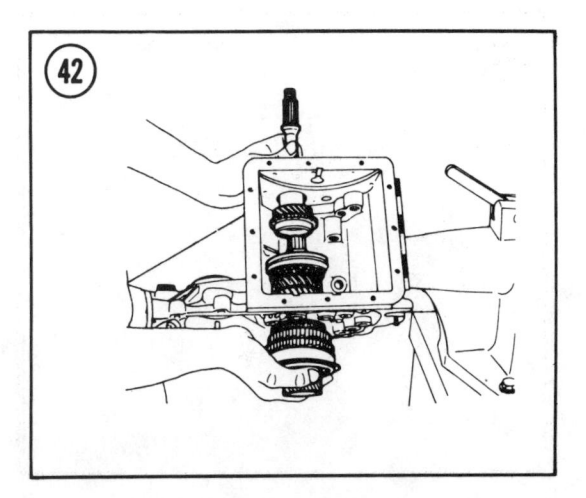

46
1. Baulk ring 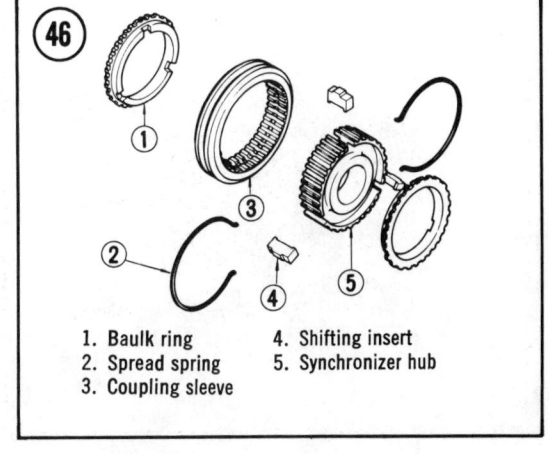
2. Spread spring
3. Coupling sleeve
4. Shifting insert
5. Synchronizer hub

8

31. Place the final gear assembly in a soft-jawed vise (**Figure 47**). Pull off the side bearings with a puller and adapter as shown (Datsun No. KV38103800; Kent-Moore No. J22888). If you don't have these tools, have the bearings pressed off by a machine shop.

32. Unbolt the differential case from the final gear (**Figure 48**).

33. Take off the differential case. Take out the pinion shaft, side gears, and pinion mates. See **Figure 49**.

**Inspection (Four-Speed)**

1. Thoroughly clean all parts in solvent. Blow dry with compressed air.

2. Check the castings (**Figure 50**) for cracks or damaged gasket surfaces. Small burrs can be removed from gasket surfaces with an oilstone. Small dents can be filled with gasket sealer. Replace any parts that are more seriously damaged.

3. Rotate the ball bearings (**Figure 51**). Check for roughness, looseness, or excessive noise. If any bearings are in doubtful condition, have them pressed off by a machine shop.

4. Check gears for wear, cracks, chips, or missing teeth. Replace any gears with these conditions.

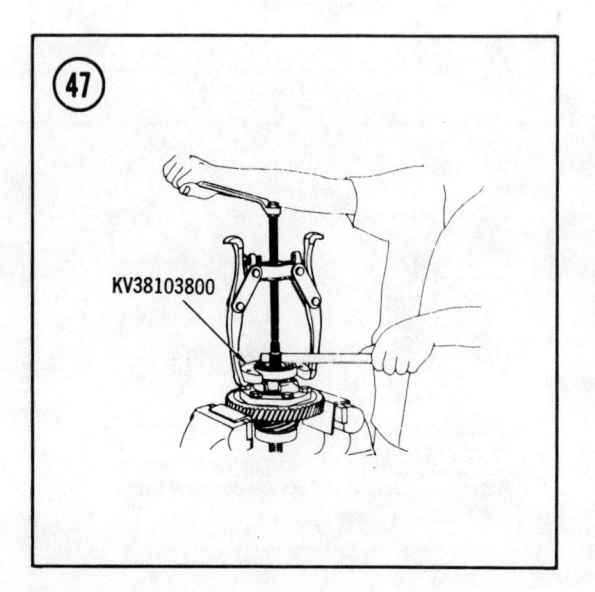

5. Check shafts for bending, cracks, wear, or damaged splines. **Figure 52** shows examples of damaged splines.

6. Check the speedometer pinion for stripping or damage. Replace as needed.

7. Check the synchronizer rings for visible wear or damage. Replace if these can be seen. Place each synchro ring on a gear cone.

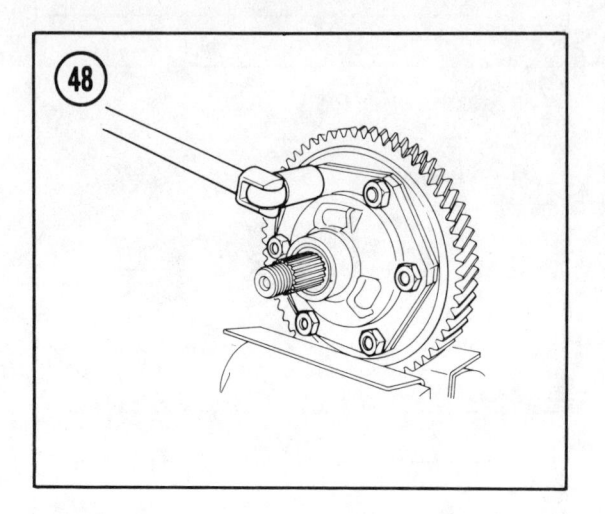

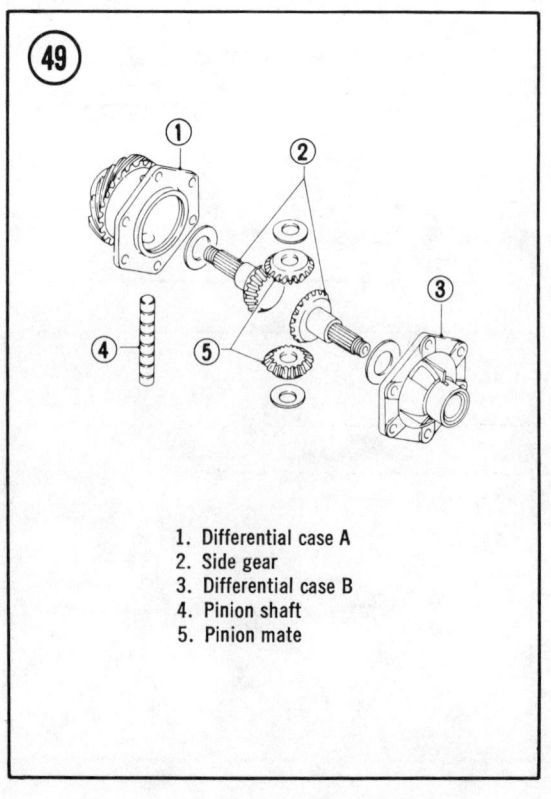

1. Differential case A
2. Side gear
3. Differential case B
4. Pinion shaft
5. Pinion mate

# CLUTCH AND TRANSMISSION

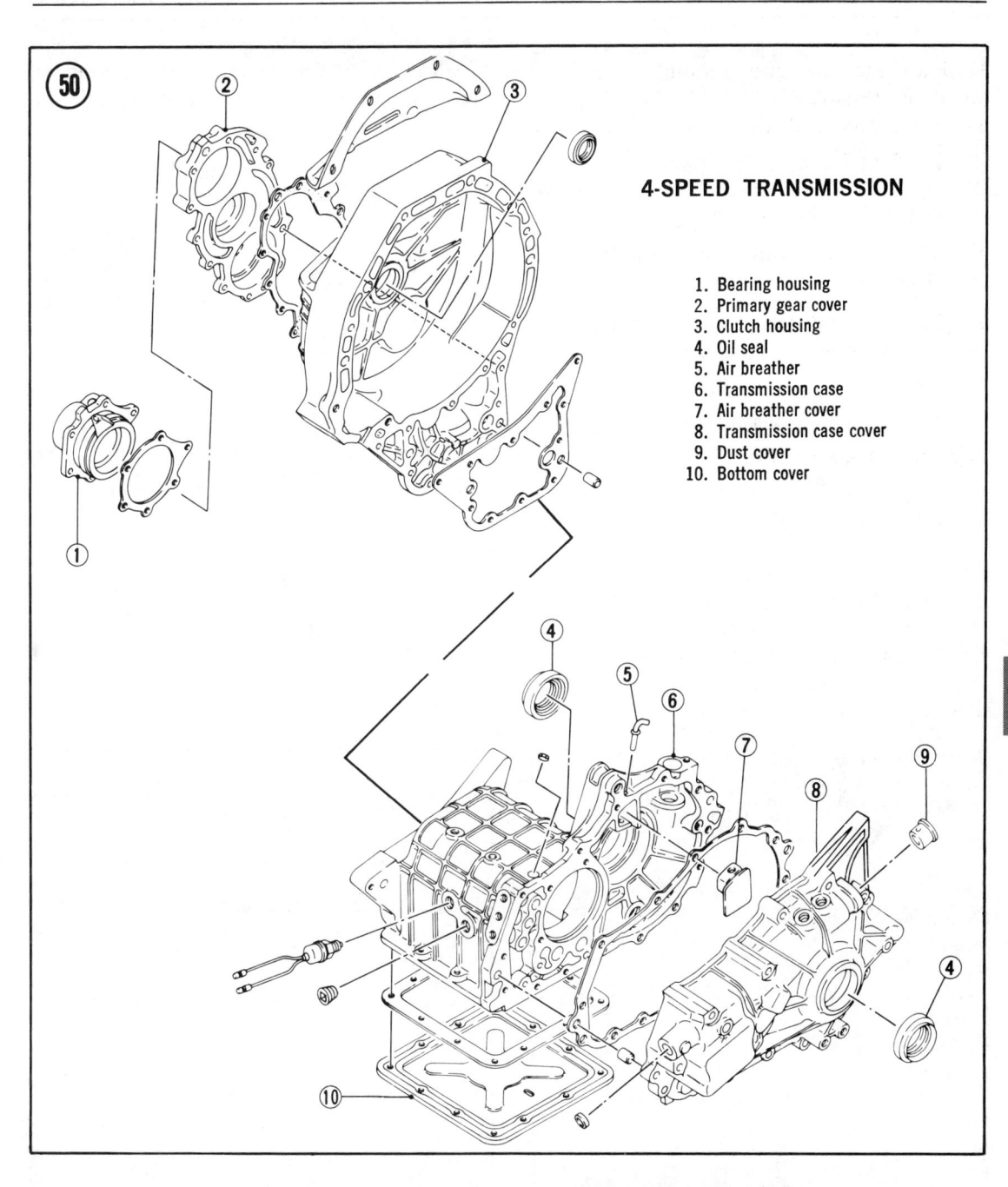

**4-SPEED TRANSMISSION**

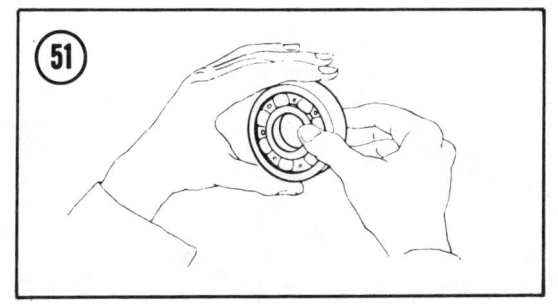

1. Bearing housing
2. Primary gear cover
3. Clutch housing
4. Oil seal
5. Air breather
6. Transmission case
7. Air breather cover
8. Transmission case cover
9. Dust cover
10. Bottom cover

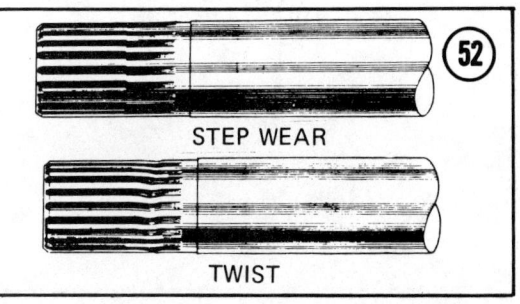

STEP WEAR

TWIST

Measure the gap between gear cone and synchro ring with a feeler gauge (**Figure 53**). If it exceeds 0.020 in. (0.5mm), replace the synchro ring.

8. Since needle bearing wear can be difficult to see, the needle bearings should be replaced whenever the transmission is overhauled.

9. Discard all oil seals, O-rings, and gaskets.

**Assembly (Four-Speed)**

1. During assembly, coat all gears, bearings, and shafts with clean gear oil.

2. Assemble the synchronizers (**Figure 46**). Position the springs so their gaps are not directly opposite each other. See **Figure 54**.

3. Install reverse gear, first gear, and the main gear bushing on the main shaft. See **Figure 55**.

> NOTE: *Be sure the bushing oil hole aligns with the main shaft oil hole.*

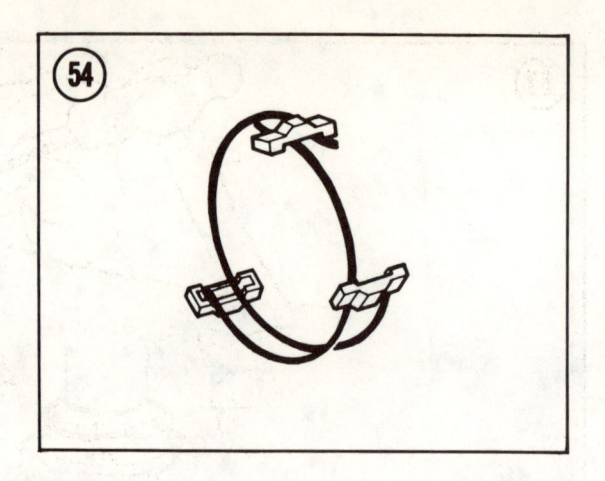

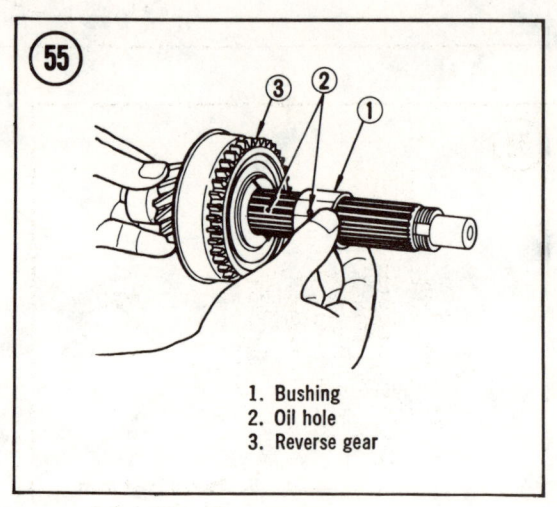

4. Install the following parts in order on the main shaft: synchro ring, synchronizer assembly, main gear bushing, synchro ring, second gear, and main gear spacer.

> NOTE: *Be sure the second main gear bushing's oil hole aligns with the main shaft oil hole.*

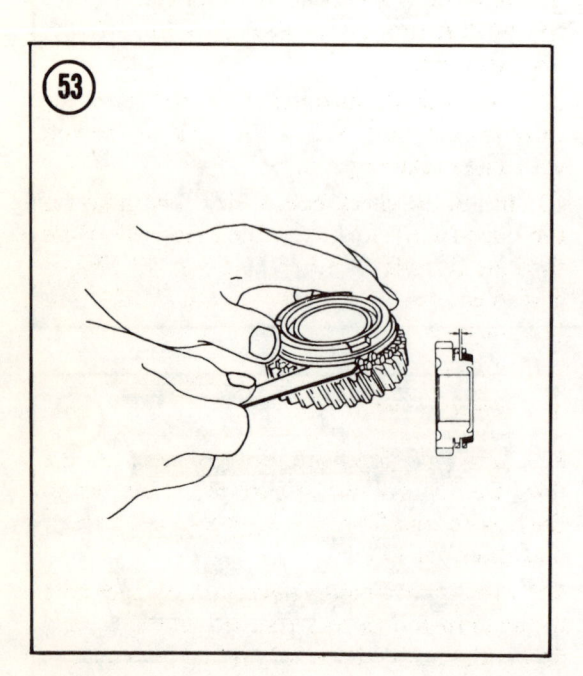

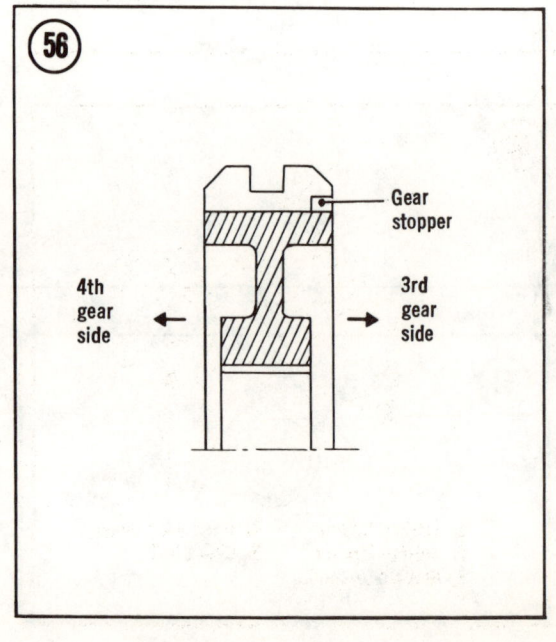

# CLUTCH AND TRANSMISSION

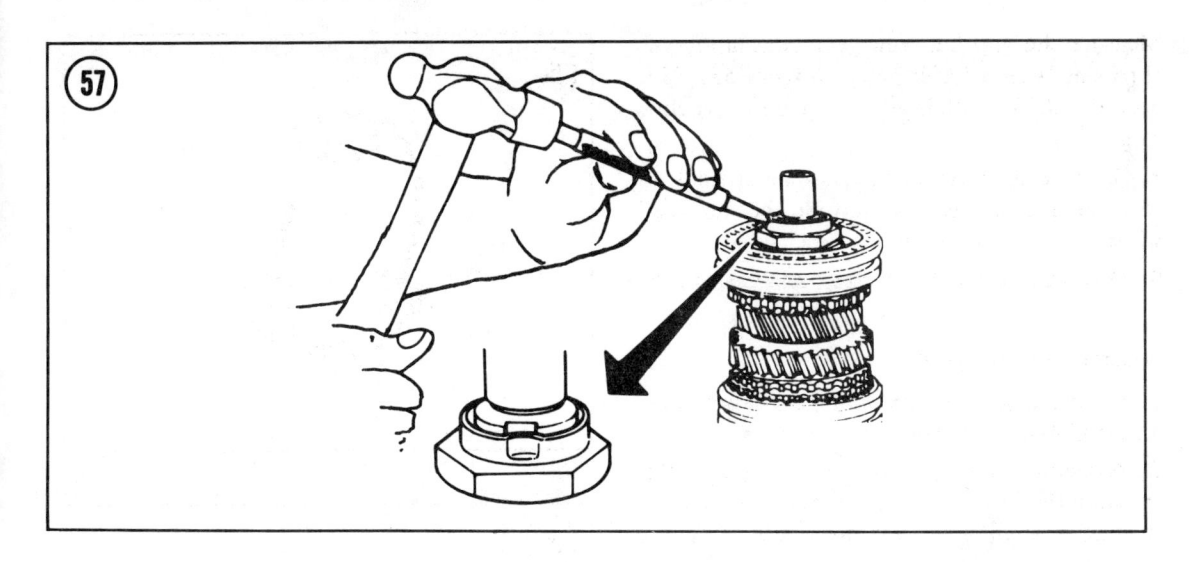

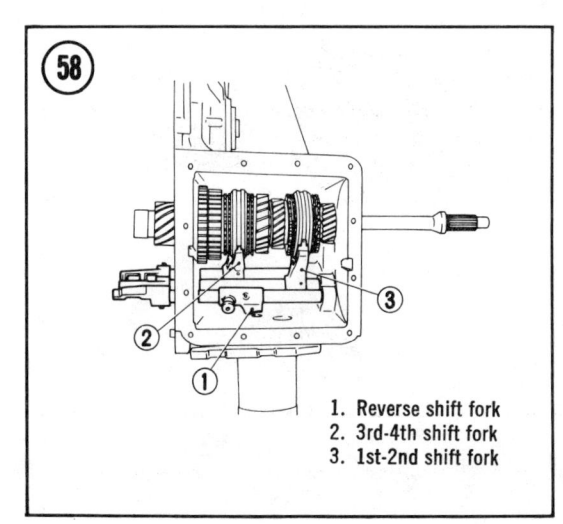

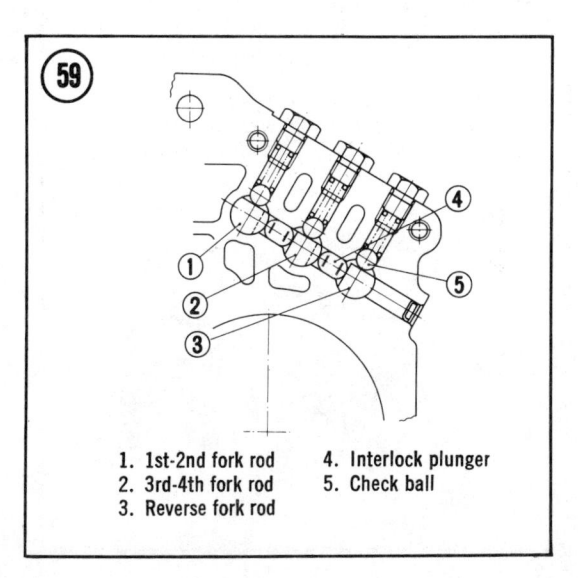

**(57)**

**(58)**

1. Reverse shift fork
2. 3rd-4th shift fork
3. 1st-2nd shift fork

**(59)**

1. 1st-2nd fork rod
2. 3rd-4th fork rod
3. Reverse fork rod
4. Interlock plunger
5. Check ball

5. Install the third main gear bushing, third gear, synchro ring, synchronizer, and locknut.

> NOTE: *Be sure the third-fourth gear synchronizer is installed in the right direction. See* **Figure 56**.

6. Place the main shaft in a soft-jawed vise. Tighten the main shaft locknut to 36-43 ft.-lb. (5-6 mkg). After tightening, punch the locknut lip into the main shaft groove (**Figure 57**).

7. Install the main drive gear and main shaft in the transmission case.

8. Place the first-second shift fork and third-fourth shift fork on their synchronizer sleeves. See **Figure 58**.

9. Insert the third-fourth fork rod into the transmission case. Secure the fork to the rod with a new retaining pin.

10. Install the check ball, spring, and plug for the third-fourth fork rod. See **Figure 59**. Coat the plug threads with Loctite Lock 'N' Seal or equivalent, then tighten the plug to 18-22 ft.-lb. (1.1-3.0 mkg).

> NOTE: *Align the fork rod notch with the check ball.*

11. Install an interlock plunger next to the third-fourth fork rod (**Figure 59**). Then install the first-second fork rod, its check ball, spring, and check ball plug.

12. Install the remaining interlock plunger next to the third-fourth fork rod. Install the reverse fork rod, then its check ball, spring, and plug.

13. Lubricate the countershaft needle bearings with multipurpose grease. Install the spacers and needle bearings in the counter gear (**Figure 60**).

14. Insert the dummy countershaft into the counter gear to hold the bearings in place. See **Figure 40**.

15. Install the thrust washers and spring. Install the counter gear in the transmission case (**Figure 61**).

NOTE: *Position the large thrust washer as shown in* **Figure 62**. *Position the flat thrust washer so its tab fits into the notch in the transmission case wall.*

16. Assemble the final drive assembly. See **Figure 49**.

17. Check side gear end play with a dial indicator. See **Figure 63**.

18. Install the reverse idler shaft, reverse idler gear, and bearing retainer.

19. Install the reverse fork lever and fork bracket.

NOTE: *Be sure the cutout in the reverse idler shaft lines up with the bearing retainer.*

20. Install the transmission case cover (8, **Figure 50**) on the transmission case.

NOTE: *Be sure the shifter (13,* **Figure 34**) *aligns properly with the shift fork ends (3,* **Figure 64**).

21. Install the transmission case cover bolts. Refer to **Figure 65** and **Table 3** for correct bolt locations.

NOTE: *Use new seal bolts.*

22. Install the differential side flanges (**Figure 66**). Use Loctite Lock 'N' Seal or the equivalent on the threads. Tighten the flange nuts to 87-101 ft.-lb. (12-14 mkg).

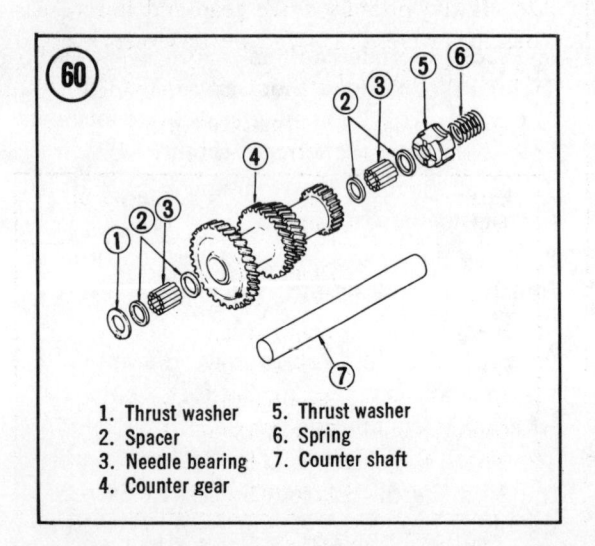

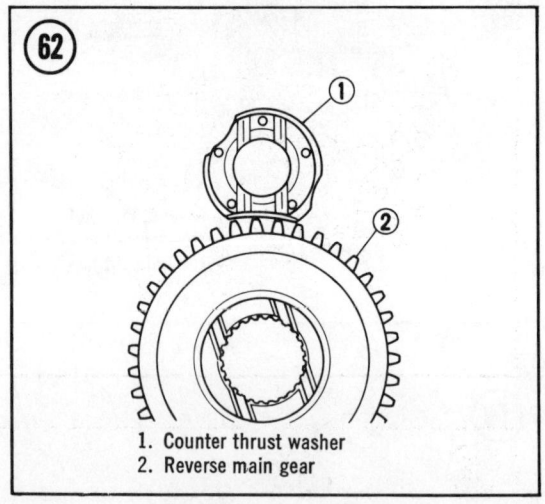

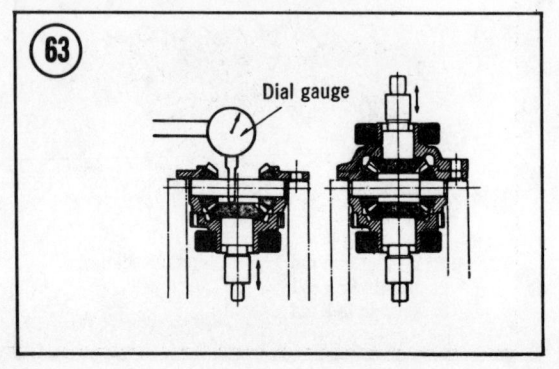

**60**
1. Thrust washer
2. Spacer
3. Needle bearing
4. Counter gear
5. Thrust washer
6. Spring
7. Counter shaft

**62**
1. Counter thrust washer
2. Reverse main gear

**61**

**63**
Dial gauge

# CLUTCH AND TRANSMISSION

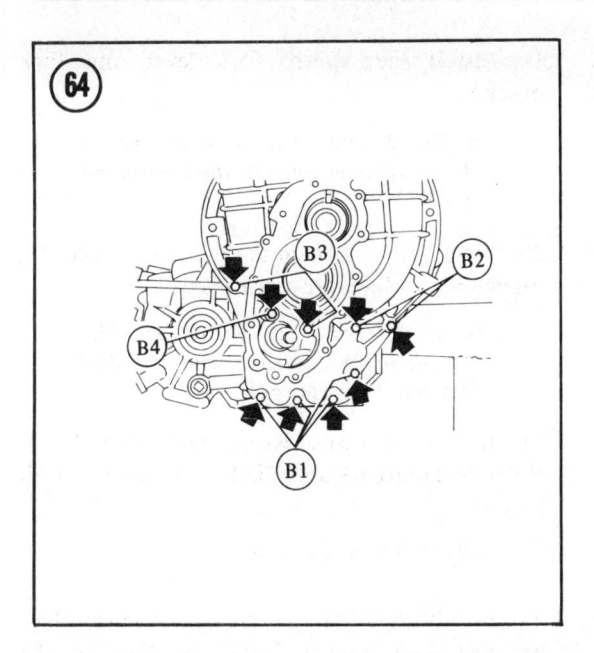

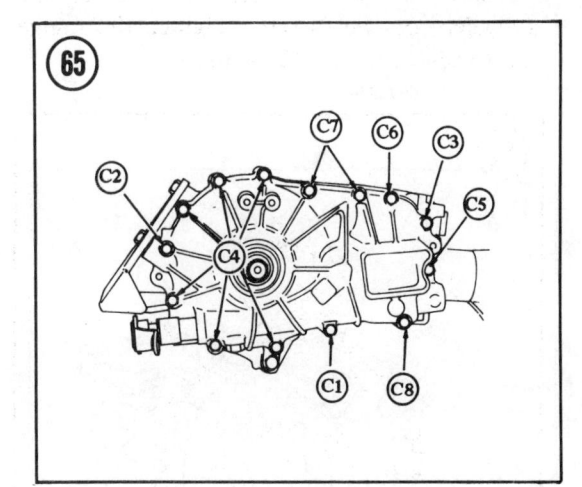

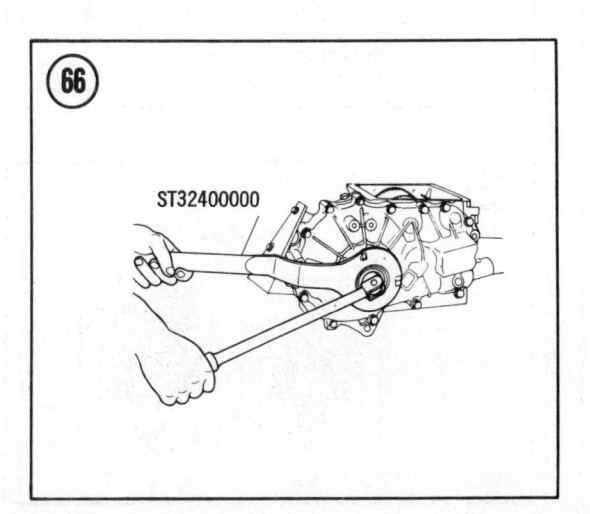

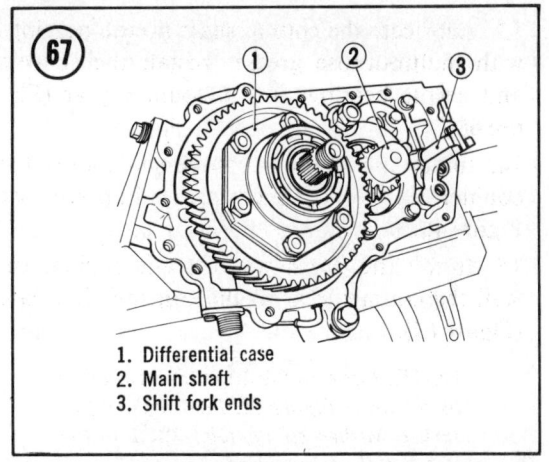

**64**

**65**

**66**
ST32400000

**67**
1. Differential case
2. Main shaft
3. Shift fork ends

NOTE: *Use new flange nuts.*

23. Install the clutch housing on the transmission case. Refer to **Figure 67** and **Table 4** for correct bolt locations.

NOTE: *Use new seal bolts.*

24. If the primary drive gear and input gear bearings were removed, have new ones pressed on by a machine shop.

Table 3    TRANSMISSION CASE COVER BOLTS

| Bolt number | Length | Number of bolts |
|---|---|---|
| 1 | 3.27 in. (83mm) | 1 |
| 2 | 2.76 in. (70mm) | 1 |
| 3 | 1.18 in. (30mm) | 1 |
| 4 | 1.77 in. (45mm) | 6 |
| 5* | 2.36 in. (60mm) | 1 |
| 6* | 1.38 in. (35mm) | 1 |
| 7* | 1.77 in. (45mm) | 2 |
| 8* | 3.27 in. (83mm) | 1 |

*Seal bolts. Do not reuse.

Table 4    CLUTCH HOUSING BOLTS

| Bolt number | Length | Number of bolts |
|---|---|---|
| 1* | 0.98 in. (25mm) | 4 |
| 2 | 1.18 in. (30mm) | 2 |
| 3 | 2.76 in. (70mm) | 2 |
| 4 | 4.33 in. (110mm) | 1 |

*Seal bolt. Do not reuse.

25. Install the ring spring, subgear, and spacer on the primary idler gear. See **Figure 68**. Have the bearing pressed on by a machine shop.

26. Check subgear end play. It should be 0.004 in. (0.11mm) or less. If incorrect, change spacer thickness.

27. Install the primary idler gear in the clutch housing.

28. Install main drive input gear (**Figure 69**). Position the primary idler gear as needed by inserting a probe through the subgear.

29. Install the thrust washer, lockwasher, and main drive gear locknut. The chamfered side of the nut faces the lockwasher.

30. Tighten the main drive gear locknut to 43-58 ft.-lb. (6-8 mkg). To keep the transmission from turning, move the reverse fork rod to shift the transmission into reverse. Then pull the first-second fork rod so the transmissoin is in second gear. Engaging 2 gears at once locks the transmission.

> NOTE: *Once the locknut is tightened, place the transmission in neutral.*

31. Install the primary gear cover. Tighten the small bolts to 4½-7 ft.-lb. (0.6-1.0 mkg). Tighten the large bolts to 12-17 ft.-lb. (1.6-2.3 mkg).

> NOTE: *The bolts are different lengths. Refer to **Figure 70** and **Table 5** for locations. Use a new seal bolt.*

32. Install the bearing housing and primary drive gear.

33. Install the bottom cover.

34. Install the speedometer pinion.

35. Install the electrical switches.

36. Install the drain plug and service hole plug. Use gasket sealer on the plug threads.

Table 5   PRIMARY GEAR COVER BOLTS

| Bolt number | Length | Number of bolts |
|---|---|---|
| 1 | 0.98 in. (25mm) | 6 |
| 2 | 6.93 in. (176mm) | 1 |
| 3* | 6.93 in. (176mm) | 1 |

*Seal bolt. Do not reuse.

## Disassembly (Five-Speed)

The five-speed transmission is similar to the four-speed, and overhaul procedures are much the same. Differences are noted as they occur.

**Figure 71** shows the gears and shafts. **Figure 72** shows the shift rods and forks.

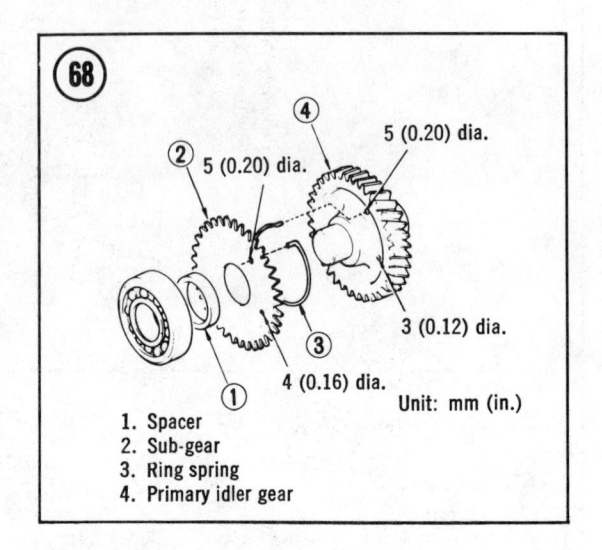

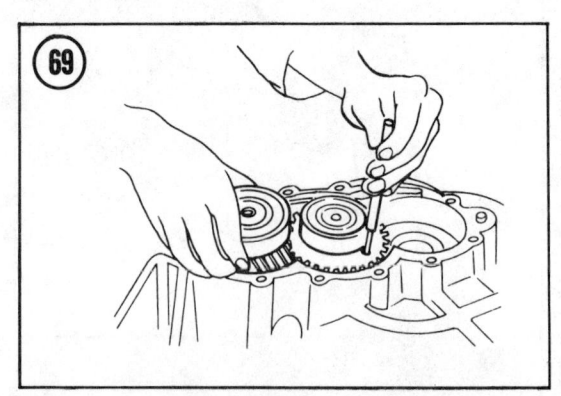

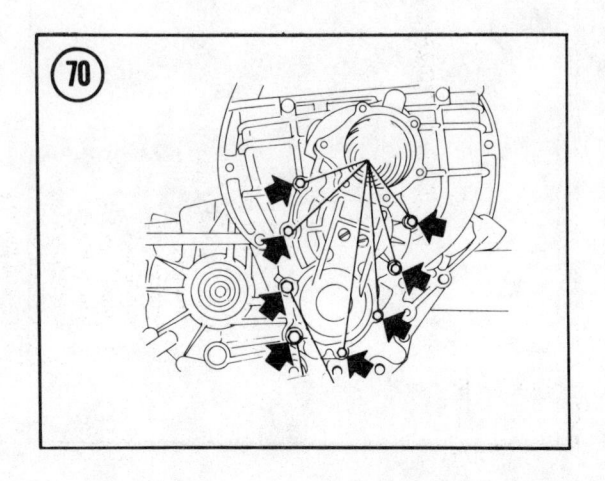

**68**

5 (0.20) dia.
2  5 (0.20) dia.
3 (0.12) dia.
4 (0.16) dia.
Unit: mm (in.)

1. Spacer
2. Sub-gear
3. Ring spring
4. Primary idler gear

**69**

**70**

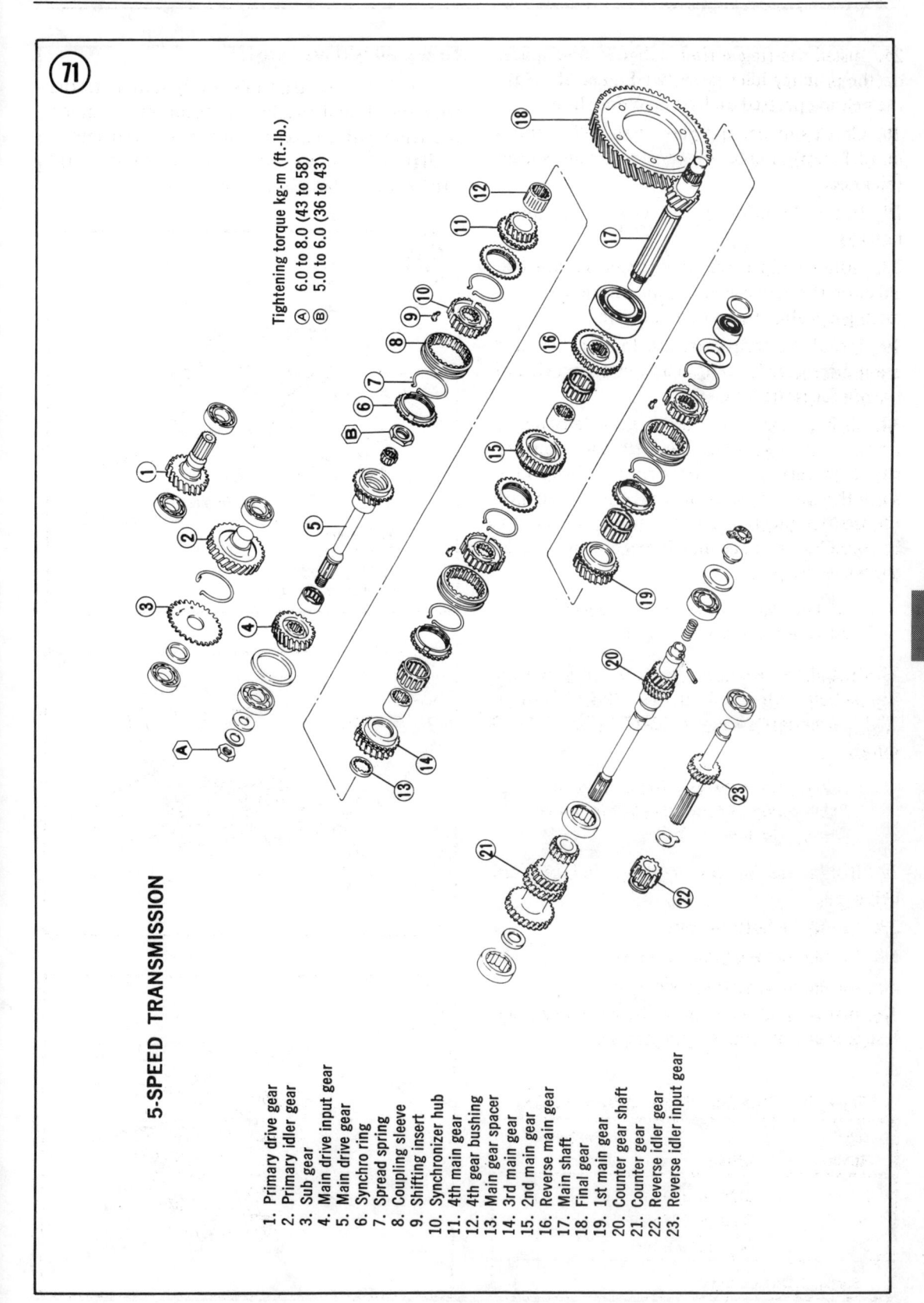

# CLUTCH AND TRANSMISSION

## 5-SPEED TRANSMISSION

Tightening torque kg-m (ft.-lb.)
Ⓐ 6.0 to 8.0 (43 to 58)
Ⓑ 5.0 to 6.0 (36 to 43)

1. Primary drive gear
2. Primary idler gear
3. Sub gear
4. Main drive input gear
5. Main drive gear
6. Synchro ring
7. Spread spring
8. Coupling sleeve
9. Shifting insert
10. Synchronizer hub
11. 4th main gear
12. 4th gear bushing
13. Main gear spacer
14. 3rd main gear
15. 2nd main gear
16. Reverse main gear
17. Main shaft
18. Final gear
19. 1st main gear
20. Counter gear shaft
21. Counter gear
22. Reverse idler gear
23. Reverse idler input gear

# CHAPTER EIGHT

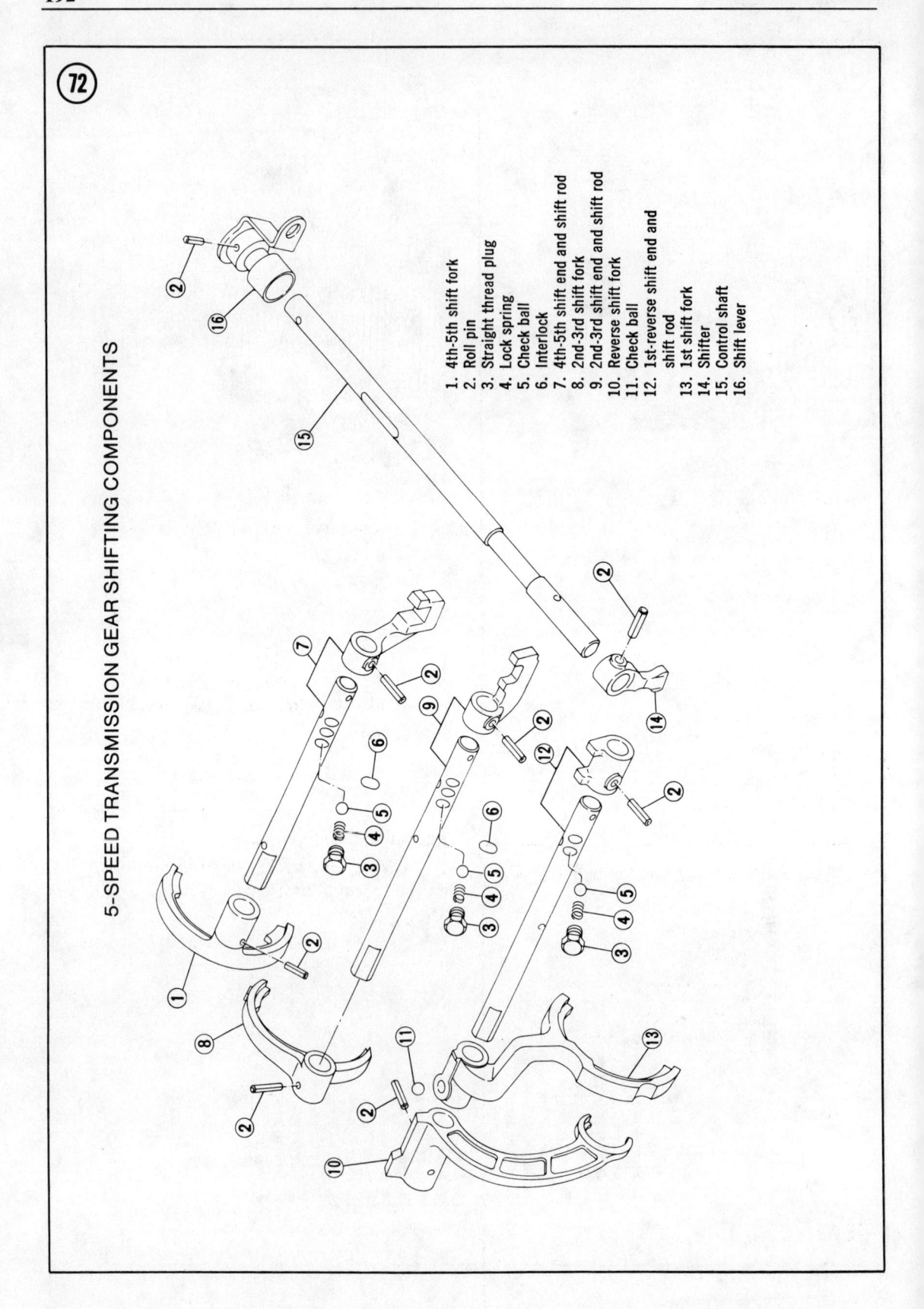

**5-SPEED TRANSMISSION GEAR SHIFTING COMPONENTS**

1. 4th-5th shift fork
2. Roll pin
3. Straight thread plug
4. Lock spring
5. Check ball
6. Interlock
7. 4th-5th shift end and shift rod
8. 2nd-3rd shift fork
9. 2nd-3rd shift end and shift rod
10. Reverse shift fork
11. Check ball
12. 1st-reverse shift end and shift rod
13. 1st shift fork
14. Shifter
15. Control shaft
16. Shift lever

# CLUTCH AND TRANSMISSION

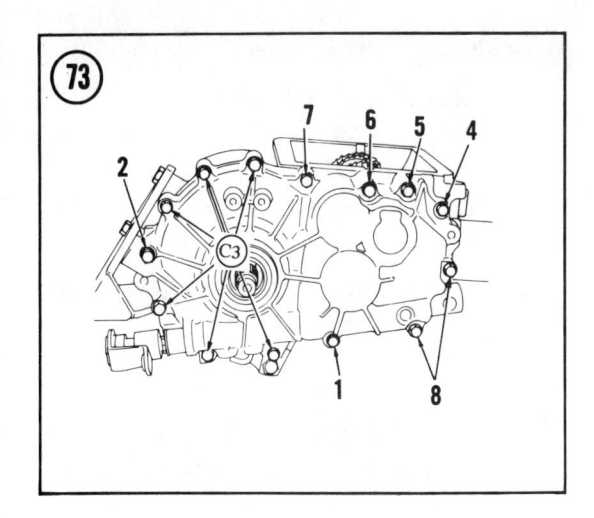

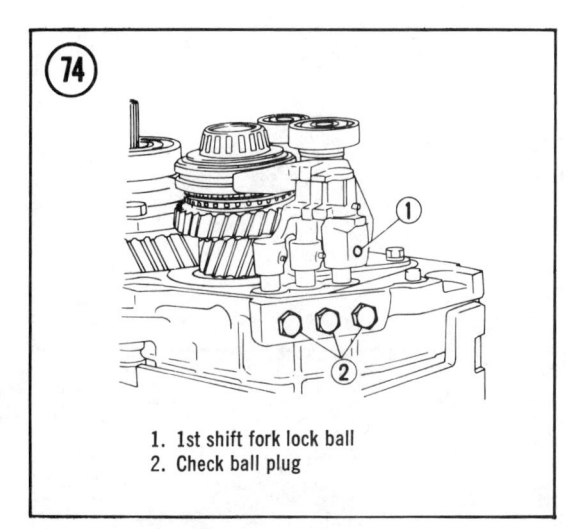

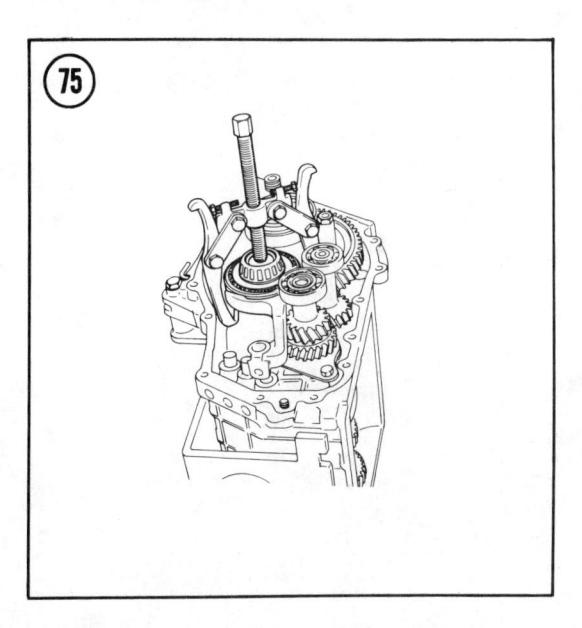

1. Remove the side flanges.

2. Remove the transmission case cover (**Figure 73**). Note that the bolts are different lengths. Label them so they can be reinstalled in the correct locations.

3. Remove the counter thrust bushing from the counter gear.

4. Remove the first shift fork lock ball (1, **Figure 74**). Remove the check ball plugs (2), then take out the springs and check balls.

5. Remove the shift fork retaining pins. Take out the shift rods.

6. Remove the tapered roller bearing with a gear puller (**Figure 75**).

7. At the same time, remove the retainer washer, first gear synchronizer, first gear, needle bearing, first gear shift fork, reverse idler input gear, reverse idler gear, and first-reverse counter gear.

8. Remove the differential.

9. Remove the bearing retainer (**Figure 76**).

10. Tap the main shaft assembly and main drive gear out of the transmission. See **Figure 77**.

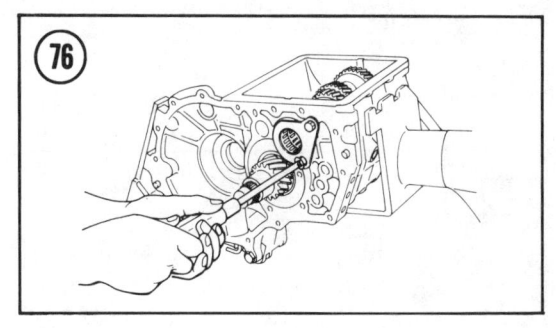

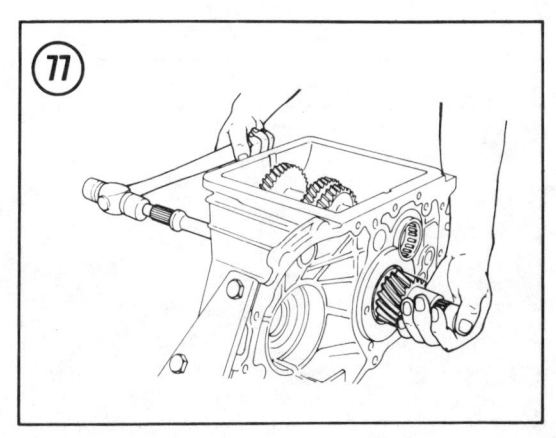

1. 1st shift fork lock ball
2. Check ball plug

11. Remove the counter gear and shift forks.

12. Place the main shaft in a soft-jawed vise and remove the locknut.

> NOTE: *The locknut lip is punched into a groove on the main shaft. It is not necessary to bend this part back. Simply undo the nut.*

13. Disassemble the main shaft, referring to **Figure 71**. The tapered roller bearing must be pressed off **(Figure 78)**.

### Inspection (Five-Speed)

This is the same as for four-speed transmissions, described earlier in this chapter.

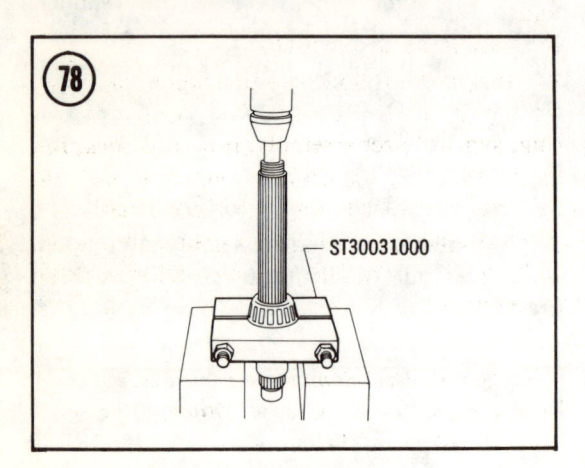

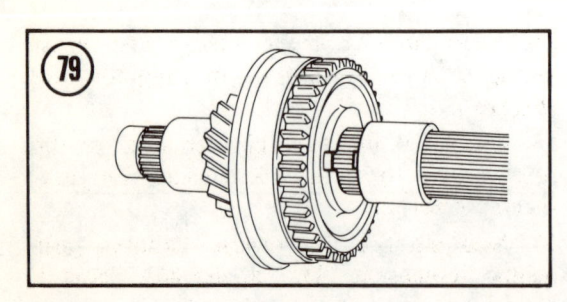

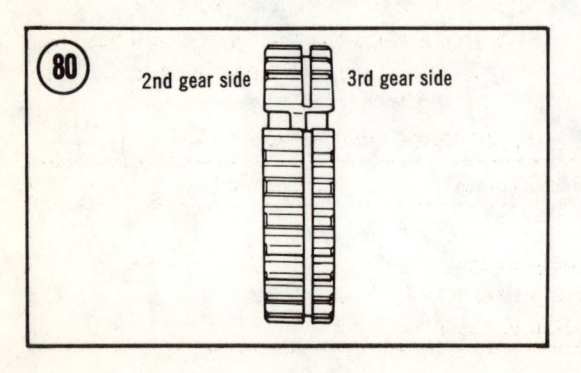

### Assembly (Five-Speed)

1. When assembling the main shaft, align the second gear bushing tab with the notch in reverse gear **(Figure 79)**.

2. Install the second-third synchronizer with its thinner side toward third gear. See **Figure 80**.

3. Align the third gear bushing tabs **(Figure 81)** with the main gear spacer **(Figure 82)**. The notched side of the main gear spacer faces away from third gear.

4. Install the differential side flange oil seals.

5. Install the main drive gear and main shaft assembly in the transmission case.

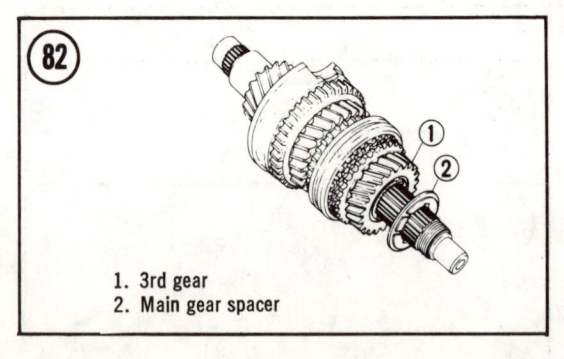

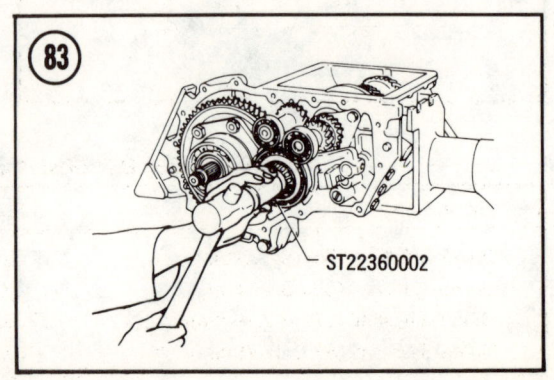

---

⑦⑧ ST30031000

⑦⑨

⑧⓪ 2nd gear side    3rd gear side

⑧① Tabs

⑧② 
1. 3rd gear
2. Main gear spacer

⑧③ ST22360002

# CLUTCH AND TRANSMISSION

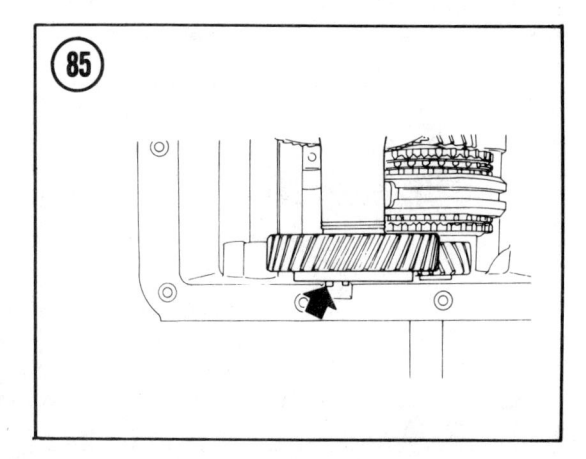

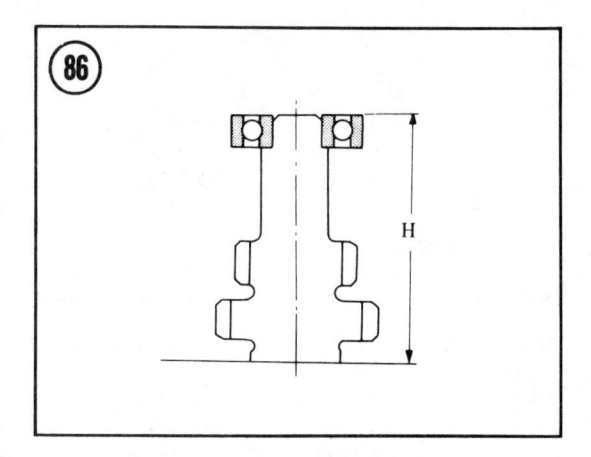

6. Install the second-third shift fork, fourth-fifth shift fork, and counter gear with thrust washer.

7. Install the countershaft bearing retainer plate.

8. Install the counter gear shaft.

9. Install the second-third fork rod. Secure it to the fork with a new retaining pin.

10. Install the second-third check ball, spring, and plug. Use gasket sealer on the plug threads.

11. Install an interlock plunger next to the second-third fork rod, then install the fourth-fifth fork rod.

12. Install the counter needle bearing and retainer.

13. Install the differential.

14. Install the following parts simultaneously: First gear needle bearing, first gear, synchro ring, synchronizer assembly, first shift fork, retainer washer, first-reverse counter gear, ball bearing, reverse idler input gear (with ball bearing and washer), reverse idler gear, and reverse shift fork. Then drive the tapered roller bearing on (**Figure 83**).

NOTE: *The tab on the reverse idler input gear washer faces the bottom of the transmission (Figure 84).*

15. Align the tabs on the counter gear thrust washer with the notch in the transmission case (**Figure 85**).

16. Install an interlock plunger next to the second-third fork rod, then install the first-reverse fork rod.

17. Measure the first-reverse counter gear height (dimension "H," **Figure 86**). Select a shim or shims, referring to **Table 6**.

18. Select a shim for the main shaft bearing. Place the outer race on the bearing and measure

Table 6    FIRST-REVERSE COUNTER GEAR BEARING SHIMS

| Dimension "H" | Shim thickness | Number of shims |
|---|---|---|
| 3.472-3.480 in. (88.19-88.20mm) | 0.0 | — |
| 3.465-3.472 in. (88-88.19mm) | 0.008 in. (0.2mm) | 1 |
| 3.457-3.464 in. (87.8-87.99mm) | 0.016 in. (0.4mm) | 2 |
| 3.449-3.456 in. (87.6-87.79mm) | 0.024 in. (0.6mm) | 3 |

its height (dimension "B," **Figure 87**). Then measure the depth of the bearing's recess in the case cover (dimension "A," **Figure 88**). Subtract "A" from "B," then add 0.020 in. (0.5mm) to the result. This is the needed shim thickness.

19. Once the shim has been selected, install it in the case cover. Press the outer race into the case cover on top of the shim.

20. Install the thrust bushing on the first-reverse counter gear (**Figure 89**).

21. Using multipurpose grease, stick the shims (selected in Step 17) and thrust washer to the transmission case cover. See **Figure 90**.

22. Install the case cover. Be sure the shifter fits into the fork rod ends. Refer to **Figure 91** and **Table 7** for bolt tightening.

> NOTE: *Be sure to install the first fork lock ball (Figure 74). If it is left out, the transmission won't shift into first gear.*

## SHIFT LINKAGE

All models use an external shift linkage. Refer to **Figure 92** (F10) or **Figure 93** (310).

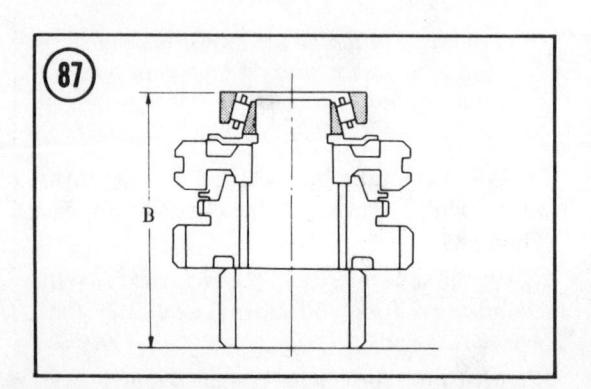

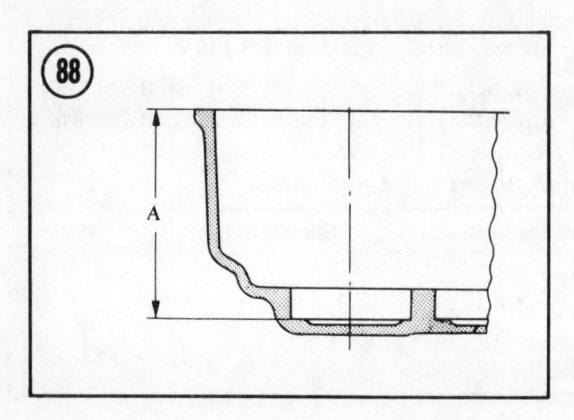

Table 7  FIVE-SPEED CASE COVER BOLTS

| Bolt number | Length | Number of bolts |
|---|---|---|
| 1 | 3.27 in. (83mm) | 1 |
| 2 | 2.76 in. (70mm) | 1 |
| 3 | 1.77 in. (45mm) | 6 |
| 4 | 1.18 in. (30mm) | 1 |
| 5* | 2.76 in. (70mm) | 1 |
| 6* | 3.27 in. (83mm) | 1 |
| 7* | 1.77 in. (45mm) | 1 |
| 8* | 3.27 in. (83mm) | 2 |

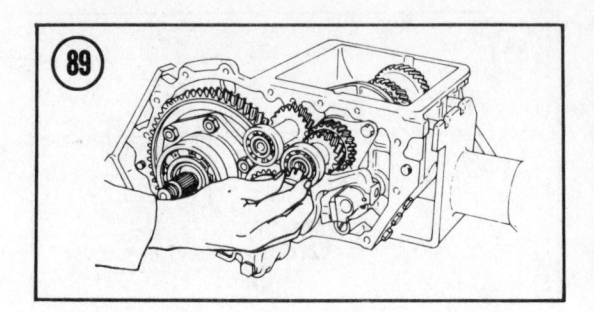

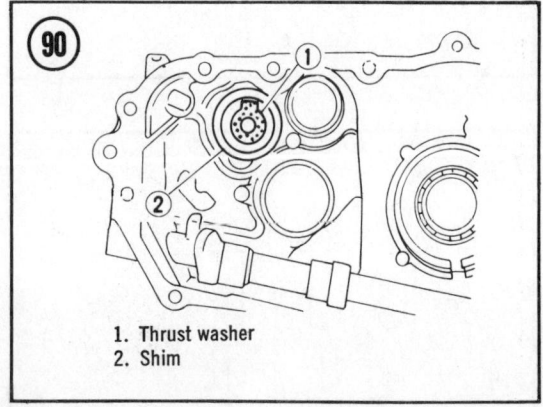

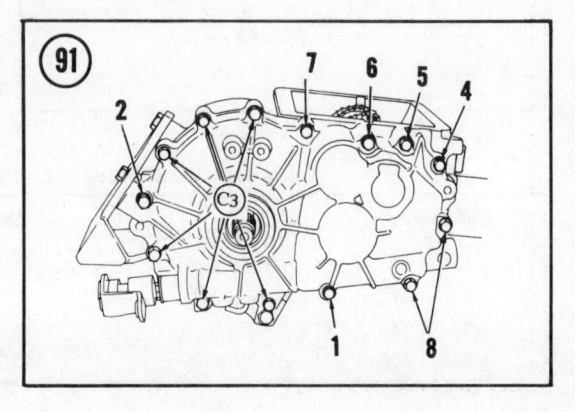

*Seal bolts. Do not reuse.

**CLUTCH AND TRANSMISSION** 197

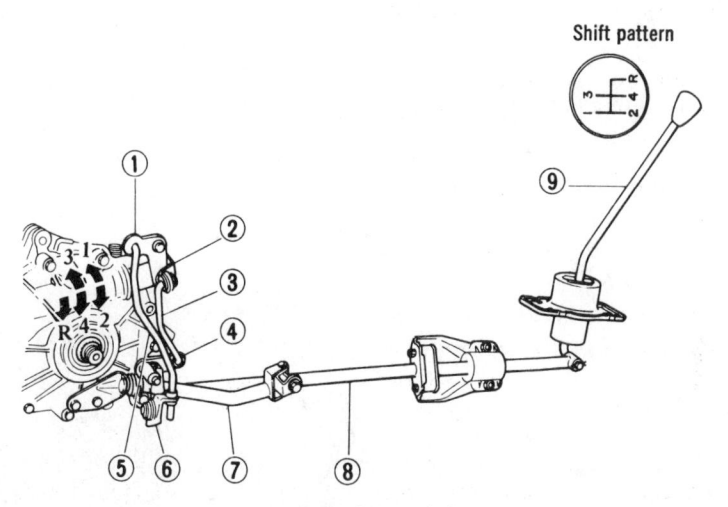

⑨²

For 4-speed transmission

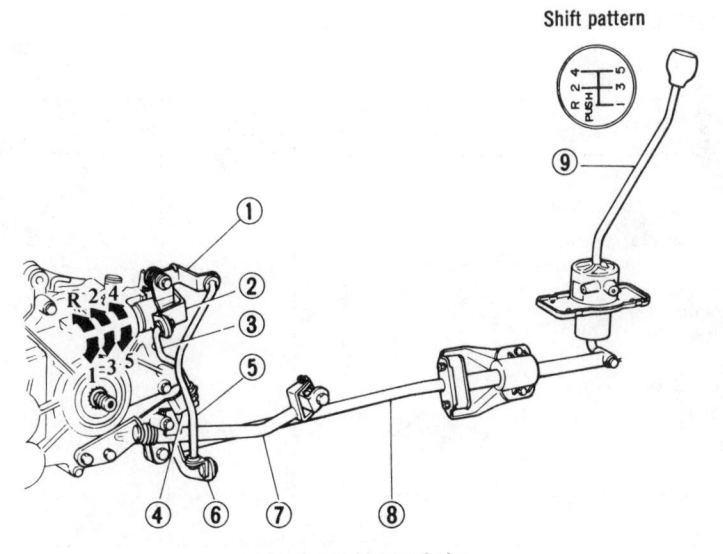

For 5-speed transmission

### F10 SHIFT LINKAGE

1. Select lever E
2. Shift lever E
3. Shift rod
4. Shift lever
5. Select rod
6. Select lever
7. Radius link
8. Control rod
9. Hand lever

198                                                                                           CHAPTER EIGHT

**310 SHIFT LINKAGE**

1. Rubber boot
2. Dust cover holder
3. Dust cover
4. Control rod base bracket
5. Support rod
6. Shift rod
7. Select rod
8. Select lever
9. Transmission mounting bracket
10. Support bracket
11. Bushing
12. O-ring
13. Bushing
14. Spring washer
15. Bushing
16. Trunnion
17. Control rod
18. Control rod bracket mounting
19. O-ring
20. Bushing
21. Lower control lever
22. Control lever pin
23. Bushing
24. Upper control lever
25. Control knob

Tightening torque kg-m (ft.-lb.)
- Ⓐ 0.8 to 1.1 (5.8 to 8.0)
- Ⓑ 1.6 to 2.2 (12 to 16)
- Ⓒ 0.33 to 0.44 (2.4 to 3.2)

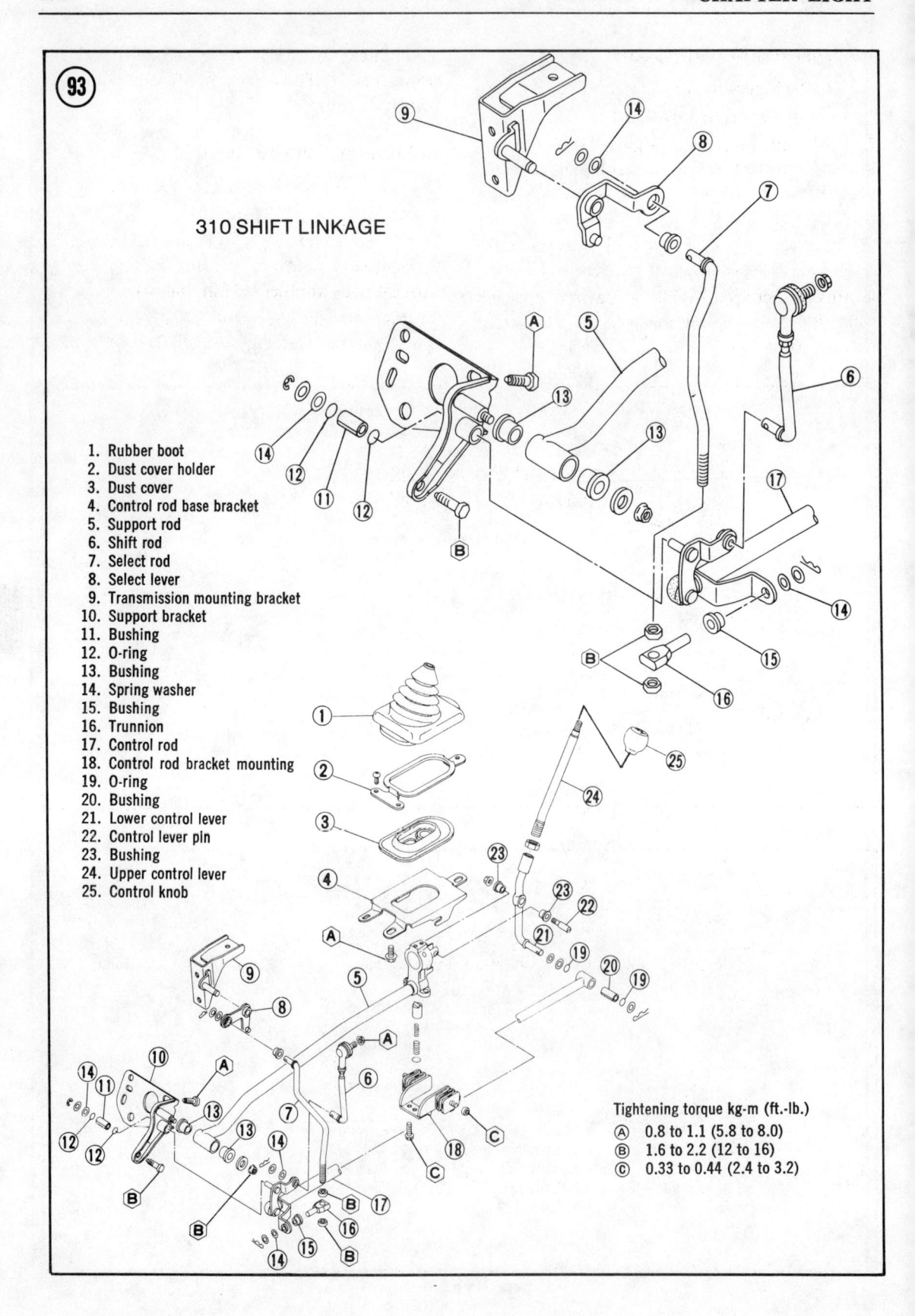

# CLUTCH AND TRANSMISSION

## Adjustment (F10 Four-Speed)

Refer to **Figure 94** for this procedure.

1. Loosen adjusting nuts 1 and 2.
2. Push shift lever E (3, **Figure 94**) all the way into the transmission case cover and measure its protrusion as shown. Then adjust the lever so it protrudes another 0.31 in. (8mm).
3. Place the shift lever inside the car in fourth gear. This moves shift lever E all the way down.
4. Push select lever (4) all the way up. This will cause hand lever guide plate (5) to contact detent pin (6).
5. Tighten adjusting nut (1) until it touches trunnion (7). Then back it off one turn and secure with nut (2).

## Adjustment (F10 Five-Speed)

Refer to **Figure 95** for this procedure.

1. Loosen nuts, (1, 2, 3, and 4).
2. Move shift lever E (5) all the way toward the transmission case and measure its protrusion. Move it back another 0.31 in. (8mm).
3. Place the shift lever inside the car in third gear. This moves shift lever E all the way down.

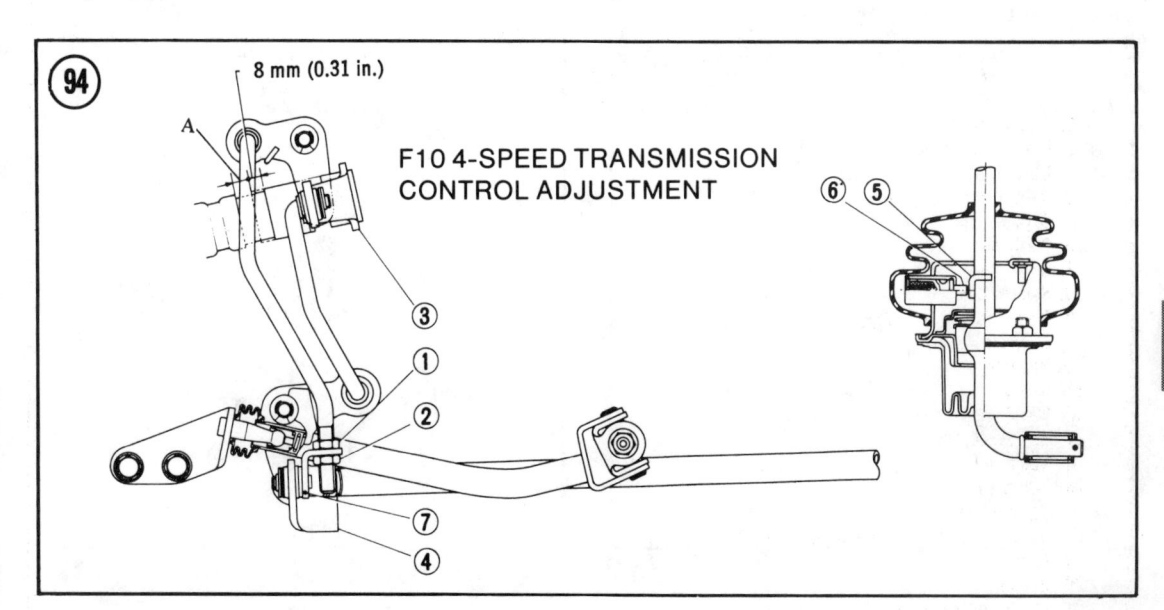

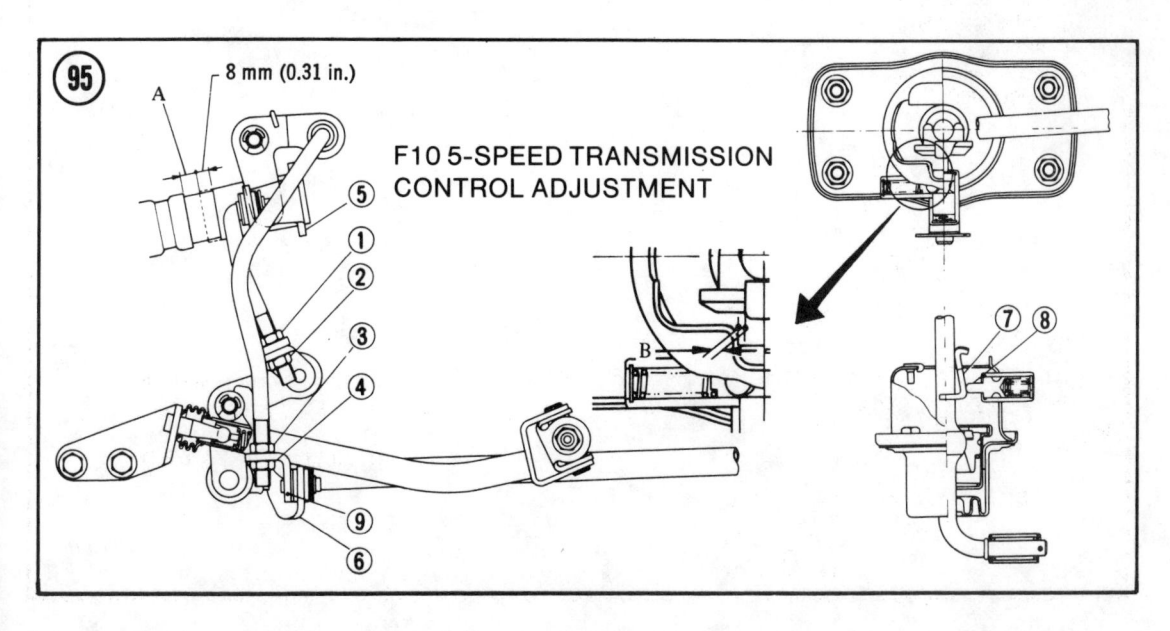

4. Push select lever (6) all the way down. This brings hand lever guide plate (7) into contact with detent pin (8). Turn nut (3) until it contacts trunnion (9). Turn it one or 2 more turns, then secure with nut (4).

5. Remove the boot from the shift lever in the passenger compartment.

6. Adjust dimension "B" at the base of the lever to 0.039-0.079 in. (1-2mm). Then tighten nuts (1 and 2).

**Adjustment (310)**

1. Loosen both adjusting nuts (**Figure 96**).
2. Set the shift lever inside the car in NEUTRAL.
3. Push the select lever as far as it will go in direction P1 (**Figure 97**). Then back it up 0.31 in. (8mm) on 4-speeds; or 0.45 in. (11.5mm) on 5-speeds.
4. Hold the select lever in position. Push the shift lever in direction P2. This engages third gear on 4-speeds, and second gear on 5-speeds.
5. Push the control rod select lever as far as it will go in direction P3. Turn the adjusting nut against the trunnion, then ¼ turn more. Secure with the locknut.

96
1. Select lever
2. Select rod
3. Adjusting nut

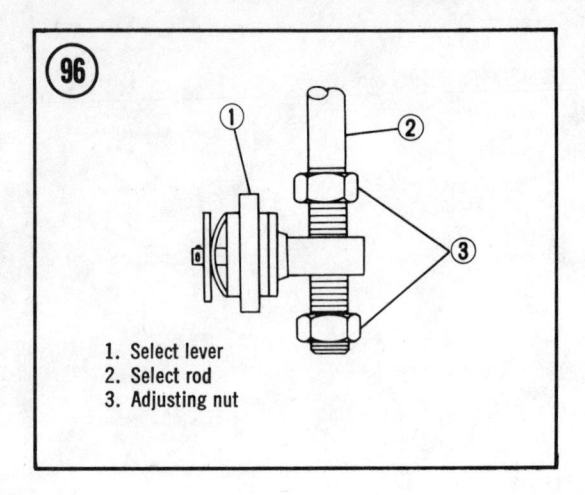

97 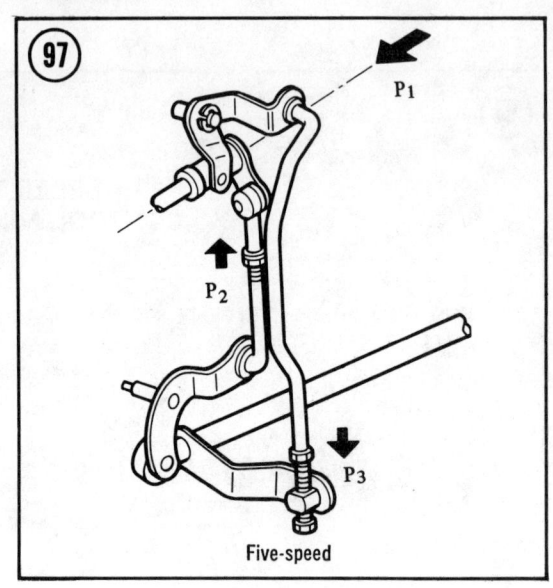
Five-speed

NOTE: If you own a 1980 or 1981 model, first check the Supplement at the back of the book for any new service information.

# CHAPTER NINE

# BRAKES

All models use disc brakes on the front and drum brakes on the rear. The front brakes are single-piston floating caliper types. The rear brakes use dual-piston wheel cylinders.

A vacuum-operated brake booster (power brakes) is used on all models. The handbrake is a mechanical type, operating the rear brakes through a cable linkage.

## FRONT BRAKES

**Figure 1** shows the front brake assembly for one side of the car.

**Pad Replacement**

1. Set the handbrake and place the transmission in gear.
2. Loosen the front wheel nuts, jack up the front of the car, place it on jackstands, and remove the front wheels.
3. Remove the retaining clip (12, **Figure 1**) from the retaining pins. Pull out the pins (13).

> NOTE: *There is a coil spring on the upper pin. Hold it while removing the pin so it won't pop out and be lost.*

4. Take off the hanger springs, then pull the pads out with pliers. See **Figure 2**.

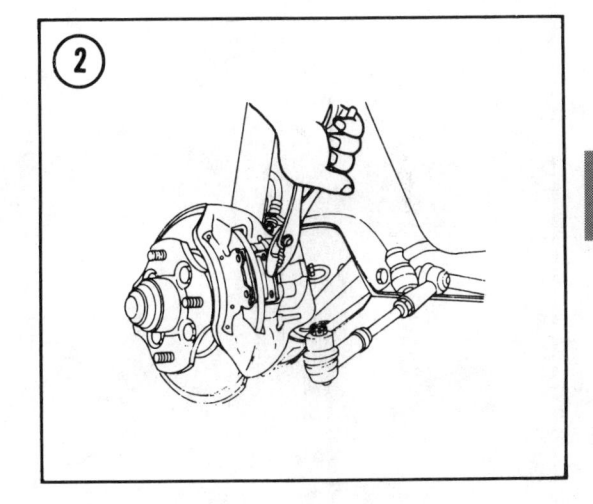

CAUTION
*Do not press the brake pedal with the pads removed, or the piston will fall out of the cylinder. The caliper will then have to be overhauled.*

5. Check the pads for wear. If the friction material is worn thinner than 0.08 in. (2mm), replace the pads. Always replace the pads in full sets of 4.
6. If the pads are only slightly oily or greasy, they can be cleaned with sand paper or a commercial brake cleaner. *Do not* clean with gasoline, kerosene, solvent, or brake fluid. These

202　　　　　　　　　　　　　　　　　　　　　　　　　　　　　　　　　　　　　CHAPTER NINE

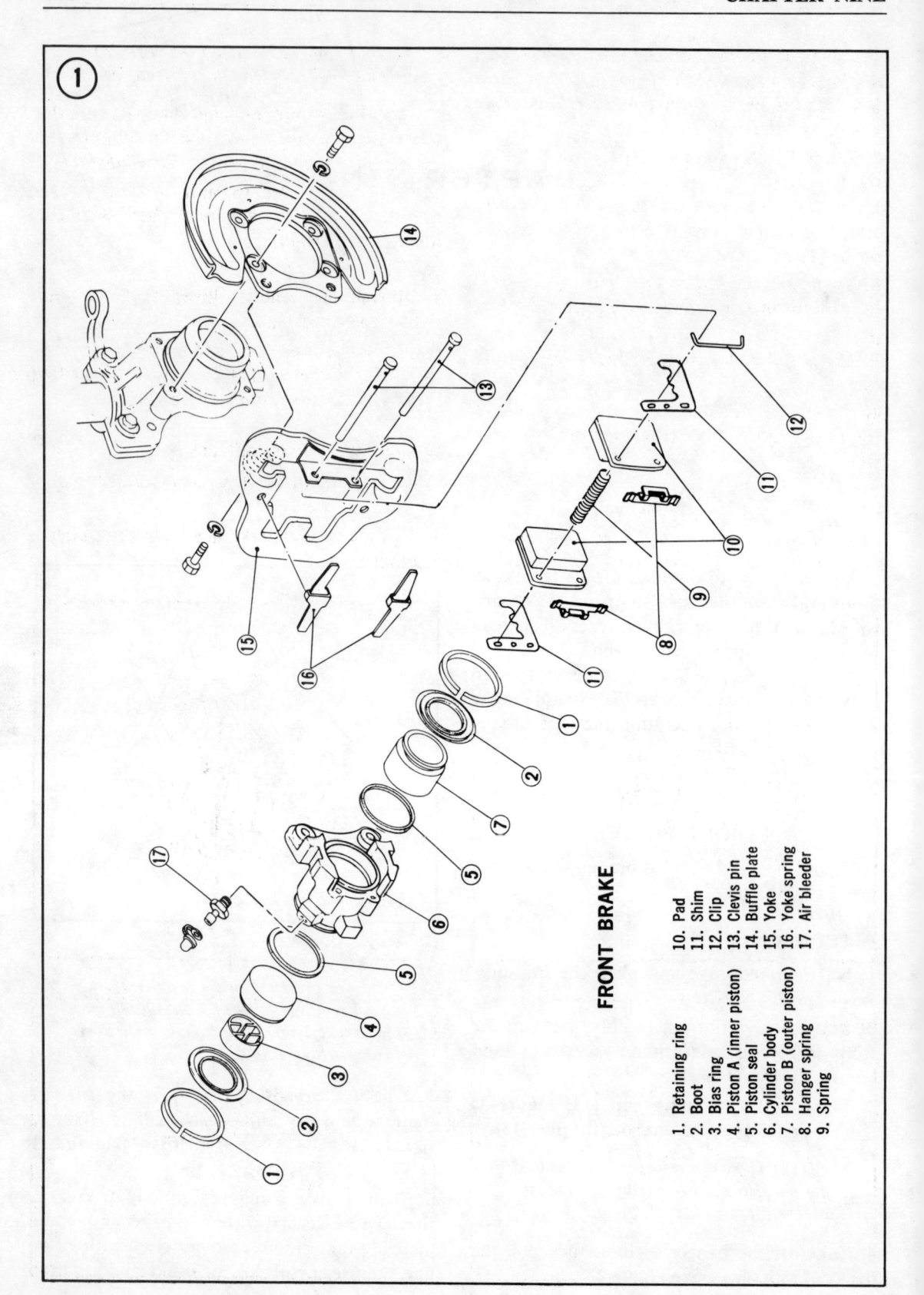

**FRONT BRAKE**

1. Retaining ring
2. Boot
3. Bias ring
4. Piston A (inner piston)
5. Piston seal
6. Cylinder body
7. Piston B (outer piston)
8. Hanger spring
9. Spring
10. Pad
11. Shim
12. Clip
13. Clevis pin
14. Buffle plate
15. Yoke
16. Yoke spring
17. Air bleeder

# BRAKES

can damage the friction material. If the pads are heavily stained with grease or oil, or show any sign of brake fluid contamination, they should be replaced.

7. Carefully clean out the space which holds the brake pads. Inspect the cylinder body. If brake fluid has been leaking past the piston seal, the caliper must be overhauled as described later.

8. To install new brake pads, first place rags beneath the master cylinder to protect the body paint.

9. Open the bleed valve to let excess brake fluid escape.

10. Push the outer piston into the cylinder (**Figure 3**) until it is flush with the retaining ring.

### CAUTION
*Do not push the piston in past the retaining ring. If pushed too far, the piston will hang up on the seal. The caliper will then have to be overhauled.*

11. Install the outer pad and shim. Be sure the arrow mark on the shim points in the forward rotating direction of the wheel. See **Figure 4**.

12. Pull on the yoke to push the inner piston into the cylinder. See **Figure 5**. Then install the inner pad and shim.

13. Tighten the bleed valve.

14. Install the wheels, lower the car, and tighten the wheel nuts.

15. Press the brake pedal several times to seat the pads. If the pedal feels mushy, bleed the brakes. See *Brake Bleeding* in this chapter.

## Caliper Removal/Installation

1. Remove the brake pads as described in the preceding section.

2. Disconnect the brake line from the caliper. See **Figure 6**.

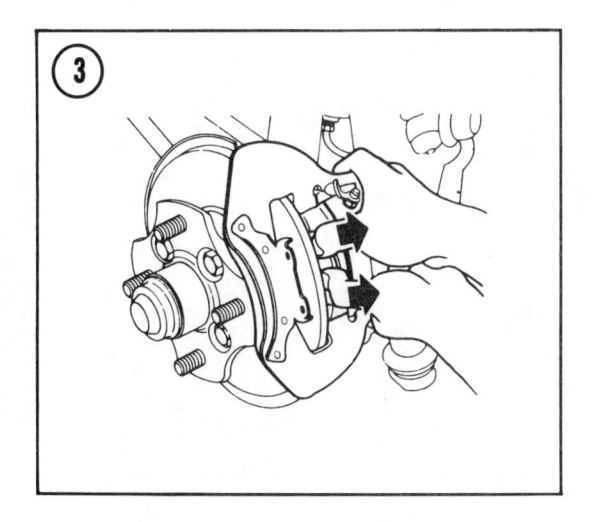

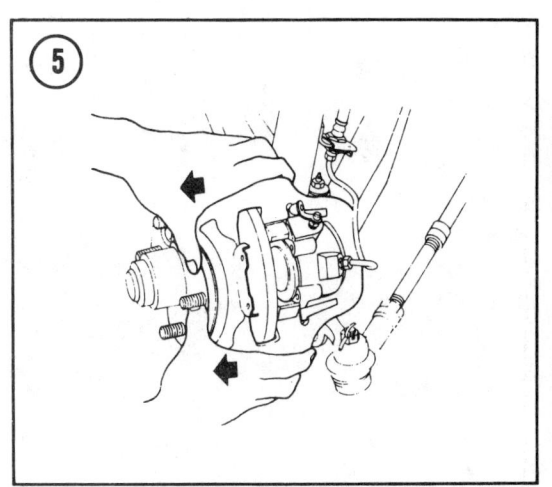

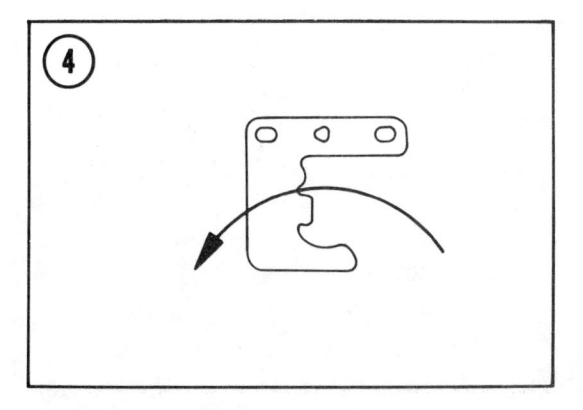

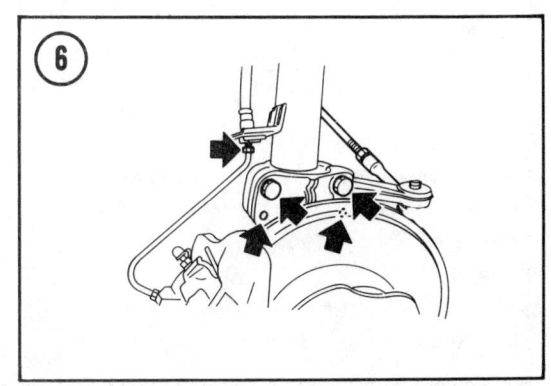

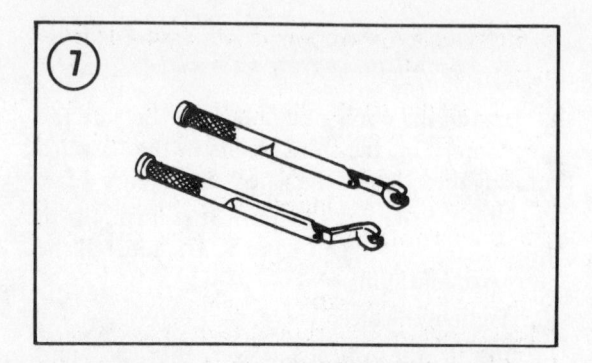

NOTE: *The factory recommends that brake line tube wrench GG94310000 (Figure 7) be used instead of an open-end wrench. Do not use an adjustable wrench.*

3. Remove the strut-to-knuckle arm bolts (**Figure 6**). Separate the strut from the knuckle arm.

4. Remove the caliper mounting bolts. Take the caliper off.

5. Installation is the reverse of removal. Tighten the caliper mounting bolts to 40-47 ft.-lb. (5.5-6.5 mkg). Tighten the strut-to-knuckle arm bolts to 24-33 ft.-lb. (3.3-4.5 mkg). After installation, bleed the brakes as described in this chapter.

**Caliper Overhaul**

1. Clean the caliper thoroughly before taking it apart. Clean with rubbing alcohol, brake fluid, or commercial brake cleaner. Do not use gasoline, kerosene, or solvent.

CAUTION
*If cleaning with brake fluid, do not let it touch the friction material on the pads.*

2. Drain the brake fluid through the bleed valve hole.

3. Place the caliper yoke in a vise and tap gently with a hammer (**Figure 8**). This will free the cylinder body so it can be removed.

4. Remove the bias ring from piston A. See **Figure 9**. Then remove the retaining rings and boots from both pistons.

5. Remove the pistons from the cylinder. If necessary, blow them out with compressed air. Use a service station air hose if you don't have a compressor.

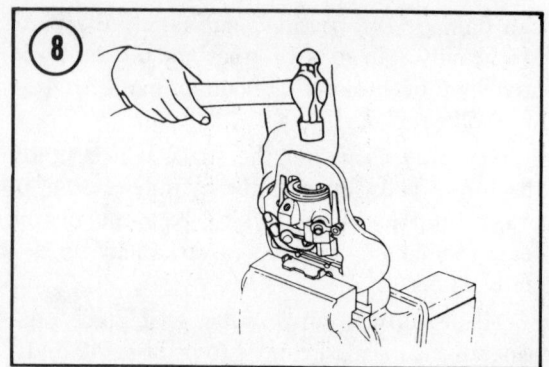

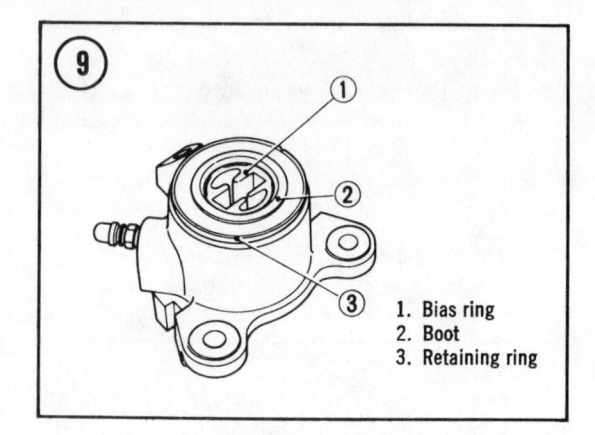

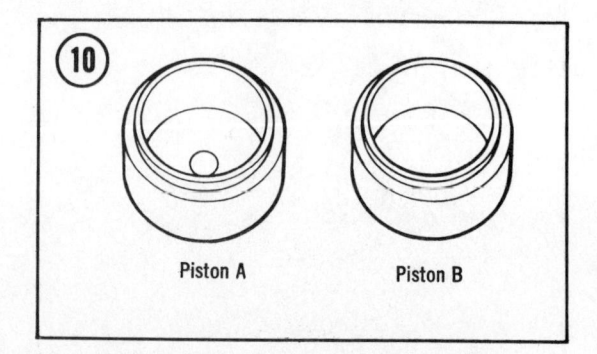

1. Bias ring
2. Boot
3. Retaining ring

Piston A    Piston B

WARNING
*The pistons may shoot out like bullets if air is applied too suddenly. Hold the caliper inside a sturdy box, and apply air pressure gradually.*

6. Note which bore each piston was removed from. Piston A (inner piston) has a dimple in the bottom. Piston B (outer piston) does not. See **Figure 10**.

7. Carefully remove the piston seals. Use fingers only so the cylinder bore won't be scratched.

# BRAKES

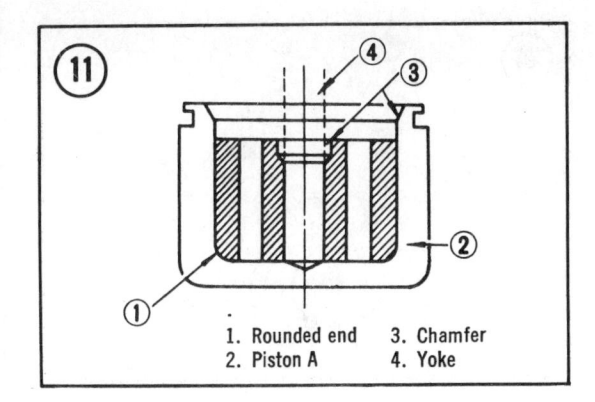

**Figure 11**
1. Rounded end
2. Piston A
3. Chamfer
4. Yoke

**Figure 12**
Yoke spring — Front face 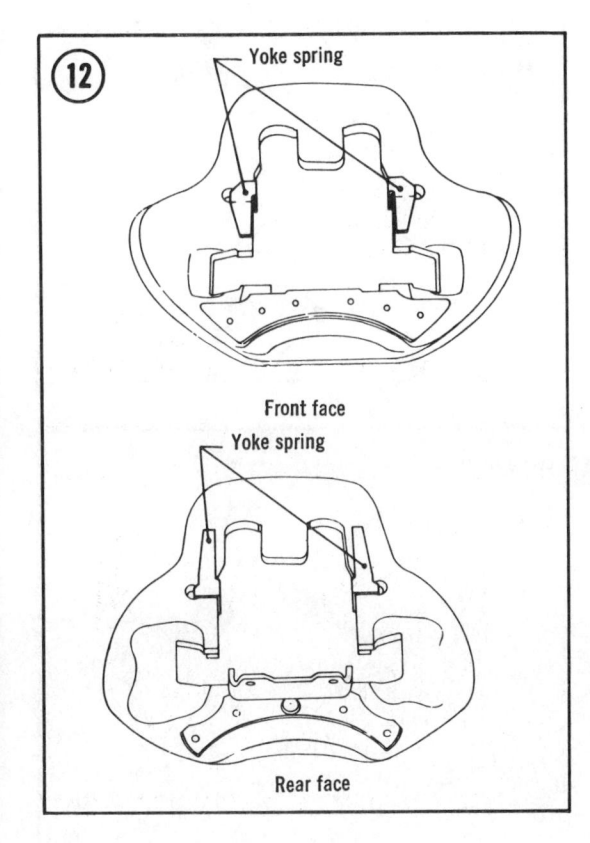
Yoke spring — Rear face

**Figure 13**

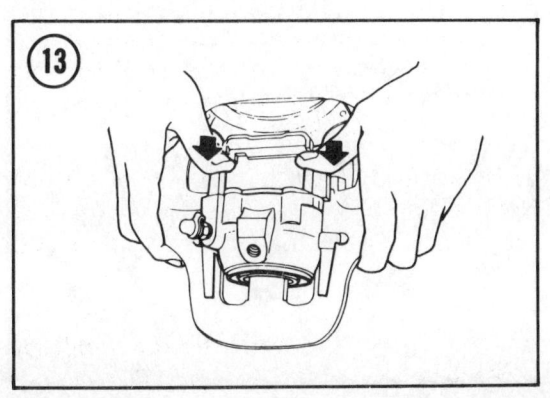

8. Remove the yoke springs (16, **Figure 1**) from the caliper yoke.

9. Throughly clean all parts with alcohol, brake fluid, or commercial brake cleaner. Do not use gasoline, kerosene, or solvent.

10. Inspect the cylinder walls. If they are scored or worn, replace the cylinder assembly. Small amounts of dirt or corrosion may be removed with fine emery paper as long as the cylinder walls are not scratched in the process.

11. Check the brake pads as described under *Pad Replacement*.

12. Inspect the pistons. Since they are plated, they cannot be cleaned with emery paper. If the pistons cannot be cleaned with a cloth and brake fluid, replace them. They must also be replaced if scored or worn.

13. The piston seals must be replaced whenever the cylinder is disassembled. In addition to stopping leaks, the piston seals retract the pistons when the brake pedal is let up. They are also the key part to the automatic adjustment of the front brakes. Therefore, seal replacement is very important.

14. Coat the cylinder bores with brake fluid and install the piston seals.

15. Place the bias ring in piston A (the piston with the dimple in the bottom). The rounded end of the bias ring goes in first (**Figure 11**).

16. Dip the pistons in brake fluid, then insert them into the cylinder. Be sure the pistons are placed in the correct ends of the cylinder. **Figure 10** identifies pistons A and B. **Figure 1** shows which ends of the cylinder they belong in.

> CAUTION
> *Do not push the pistons in too far, or they will hang up on the piston seals. If this happens, the pistons will have to be removed and new piston seals installed.*

17. After the pistons are installed, turn the bias ring so its groove will align with the caliper yoke when the cylinder body is installed.

18. Referring to **Figure 9**, install the boots and retaining rings.

19. Install yoke springs as shown in **Figure 12**.

20. Assemble cylinder and yoke (**Figure 13**).

21. Install the bleed valve.

## Disc Inspection

1. Set the handbrake and place the transmission in gear.

2. Loosen the front wheel nuts, jack up the front of the car, place it on jackstands, and remove the front wheel(s).

3. Check the disc for scoring, cracks, chips, or rust. Have the disc turned by a machine shop if these are found.

4. Check the disc for excessive runout. To do this, connect a dial gauge as shown in **Figure 14**. Rotate the disc one full turn and note the reading. If it exceeds specifications (**Table 1**), the disc can be turned by a machine shop.

5. Measure disc thickness (**Figure 15**). If thinner than the minimum (**Table 1**), the disc must be replaced. Variation in disc thickness must also be within specified limits.

## Disc Removal/Installation

The disc is removed together with the hub and the wheel bearings. See *Wheel Bearings*, Chapter Eleven.

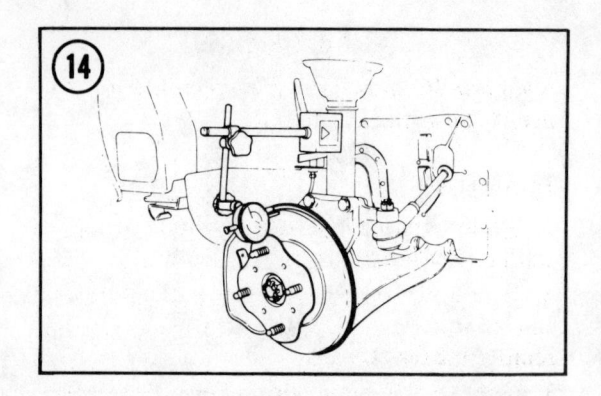

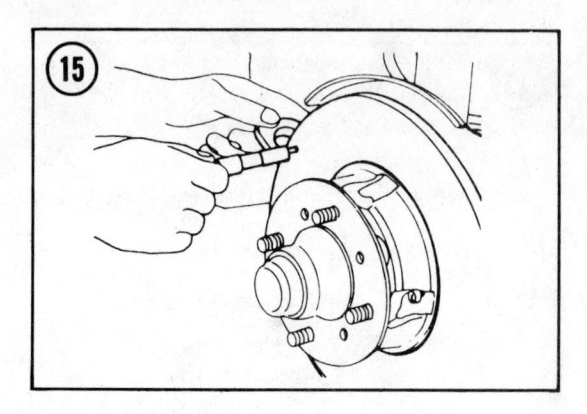

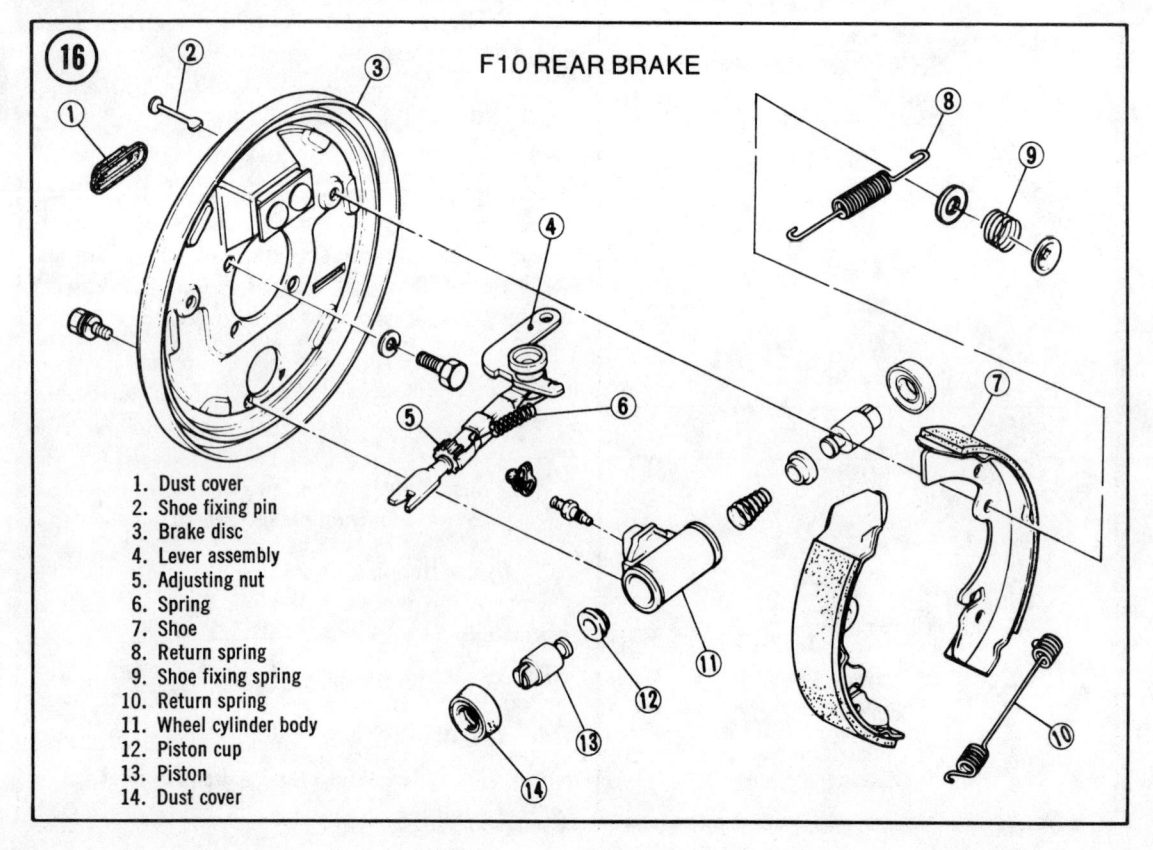

**F10 REAR BRAKE**

1. Dust cover
2. Shoe fixing pin
3. Brake disc
4. Lever assembly
5. Adjusting nut
6. Spring
7. Shoe
8. Return spring
9. Shoe fixing spring
10. Return spring
11. Wheel cylinder body
12. Piston cup
13. Piston
14. Dust cover

# BRAKES

## REAR BRAKES

**Figure 16** shows the F10 rear brakes. **Figure 17** shows the 310 design.

### Removal

1. Securely block both front wheels so the car can't roll in either direction.

2. Loosen the rear wheel nuts, jack up the rear end of the car, place it on jackstands, and remove the rear wheels.

3. Remove the brake drum. See *Rear Wheel Bearings*, Chapter Ten.

4. Note that the friction material on the brake shoes is not centered. Carefully note how the shoes are installed, so they can be put back the same way.

5. Grasp the shoe fixing pins with pliers and turn them ¼ turn. Remove the pins, retaining collars, springs, and spring washers.

6. Remove the return springs. Lift the brake shoes off.

7. Detach the adjuster assembly from the handbrake operating lever. Take the adjuster assembly out.

8. Pull back the dust covers on the ends of the wheel cylinder. If there is brake fluid in the covers, the cylinder must be overhauled. Disconnect the brake line, remove the mounting bolts, and take the cylinder out.

NOTE: *The factory recommends brake line torque wrench GG94310000 (Figure 7) instead of an open-end wrench. Do not use an adjustable wrench to disconnect the brake line.*

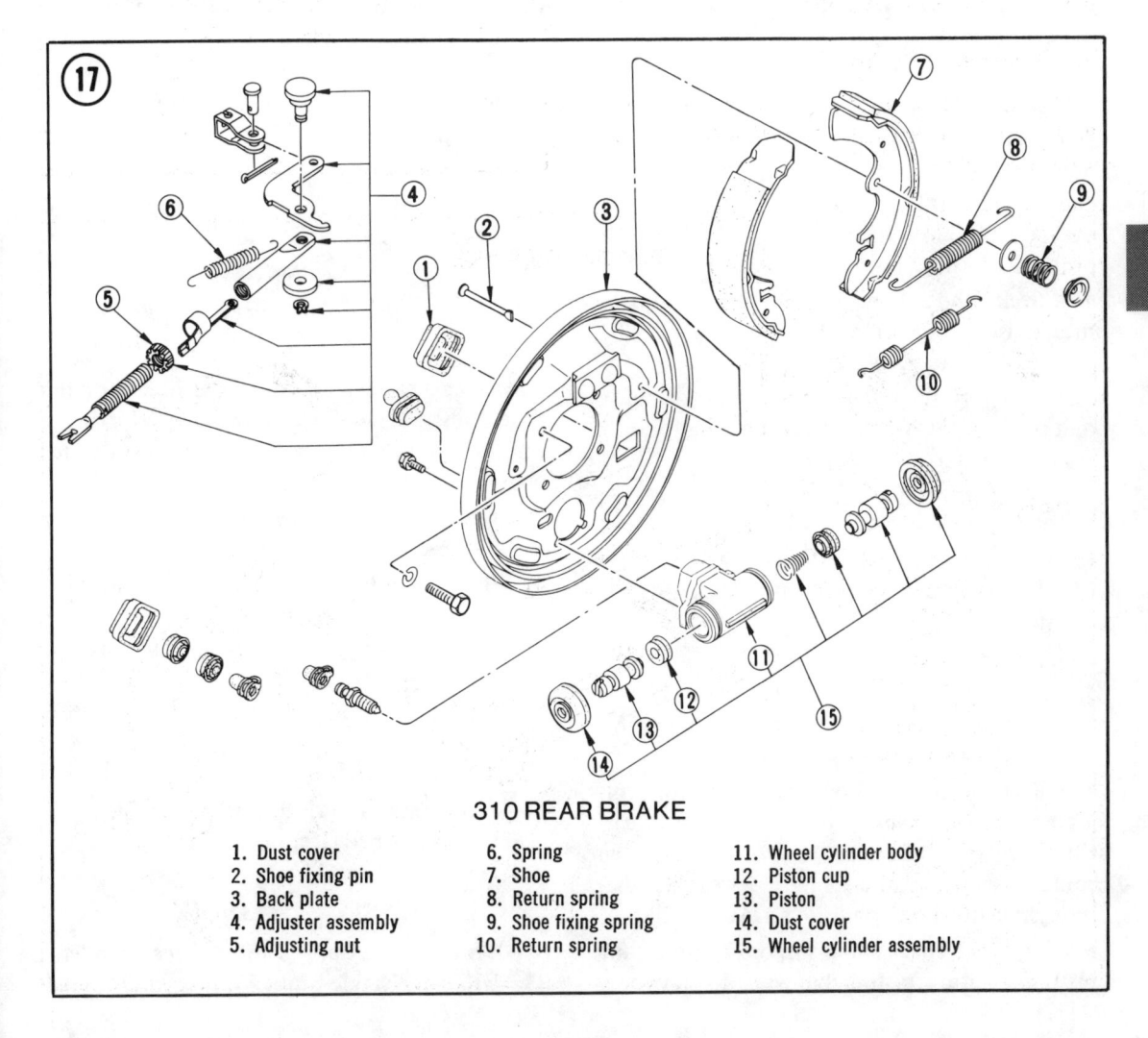

**310 REAR BRAKE**

1. Dust cover
2. Shoe fixing pin
3. Back plate
4. Adjuster assembly
5. Adjusting nut
6. Spring
7. Shoe
8. Return spring
9. Shoe fixing spring
10. Return spring
11. Wheel cylinder body
12. Piston cup
13. Piston
14. Dust cover
15. Wheel cylinder assembly

## Inspection

1. Clean all parts with alcohol, brake fluid, or commerical brake cleaner. Do not use gasoline, kerosene, or solvent.

### CAUTION
*If cleaning with brake fluid, keep it off the linings. Brake fluid will ruin the linings, and they will have to be replaced.*

2. Check drums for visible scoring, excessive or uneven wear, and corrosion. If you have precision measuring equipment, check the drum for wear and out-of-roundness. If you don't have the equipment, this measurement can be done by a dealer or machine shop. Maximum wear and out-of-roundness are listed in **Table 1**. The machine shop can turn the drums to correct scoring or out-of-roundness. However, they must not be turned beyond the maximum diameter listed in **Table 1**.

3. Inspect the lining material on the brake shoes. Make sure it is not cracked, unevenly worn, or separated from the shoes. Light surface stains may be removed with sandpaper or commercial brake cleaner. If the linings are soaked with oil or grease or contaminated with brake fluid, they must be replaced.

4. Check the shoe fixing pins, adjuster mechanism, and handbrake operating lever for wear or damage. Replace as needed.

5. Check return springs for weakness or deformation. Replace if their condition is in doubt.

## Wheel Cylinder Overhaul

1. Remove the dust covers from the ends of the wheel cylinders. Take out the pistons, cups, and return springs.

2. Unscrew the bleed valve.

3. Clean the pistons and cylinder in alcohol, brake fluid, or commercial brake cleaner. Do not use gasoline, kerosene, or solvent.

4. Check the cylinder bore and piston for scoring, cracks, corrosion, dirt, or excessive wear. Check the ends of the pistons for wear at their contact points with the brake shoes. Replace the cylinder if these conditions are detected.

5. As a final check on a suspect cylinder and piston, measure piston diameter and cylinder bore. If there is more than 0.006 in. (0.15mm) clearance between pistons and bore, replace the cylinder.

6. Coat *new* piston cups with rubber grease and install them on the piston. Do not reuse piston cups.

NOTE: *Nabco and Tokico brand wheel cylinders are used in production. Repair kits for the two brands are not interchangeable. Be sure to get the same brand as your wheel cylinder.*

7. Coat the piston cups with rubber grease and install them on the pistons. The wide end of each piston cup faces into the cylinder.

8. Coat the cylinder bore with brake fluid.

9. Install the return spring and pistons.

10. Pack the dust covers with rubber grease. Install them on the cylinder.

## Installation

Installation is the reverse of removal, plus the following.

1. Apply brake grease to the metal-to-metal friction points shown in **Figure 18**.

### CAUTION
*Keep grease off the brake shoe lining material.*

2. Put gasket sealer around the holes for the shoe fixing pins and wheel cylinder bolts (if the wheel cylinder was removed). This keeps water out of the drum.

3. Tighten the wheel cylinder bolts to 4½-6 ft.-lb. (0.6-0.8 mkg).

4. When installing brake shoes, the shoe with the low lining goes toward the front of the car. The shoe with the high lining goes toward the rear.

5. When installing the brake drum, adjust wheel bearing preload. See *Rear Wheel Bearings*, Chapter Ten.

6. After installation, adjust and bleed the brakes as described later in this chapter.

## MASTER CYLINDER

All models use dual-piston master cylinders. The F10 has a single brake fluid reservoir, con-

# BRAKES

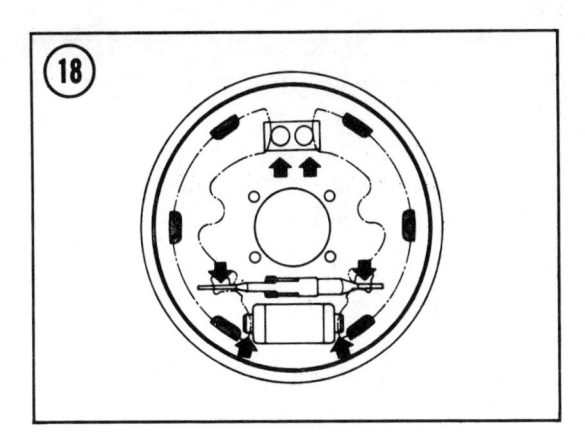

**(18)**

nected to the cylinder body by tubes (**Figure 19**). The 310 uses separate reservoirs, mounted on the cylinder body. See **Figure 20**.

### Removal/Installation

1. Disconnect the warning light wires from the master cylinder.

2. Place rags beneath the master cylinder to protect the paint from brake fluid.

3. Disconnect the brake lines from the master cylinder. On F10's, disconnect the fluid hoses as well.

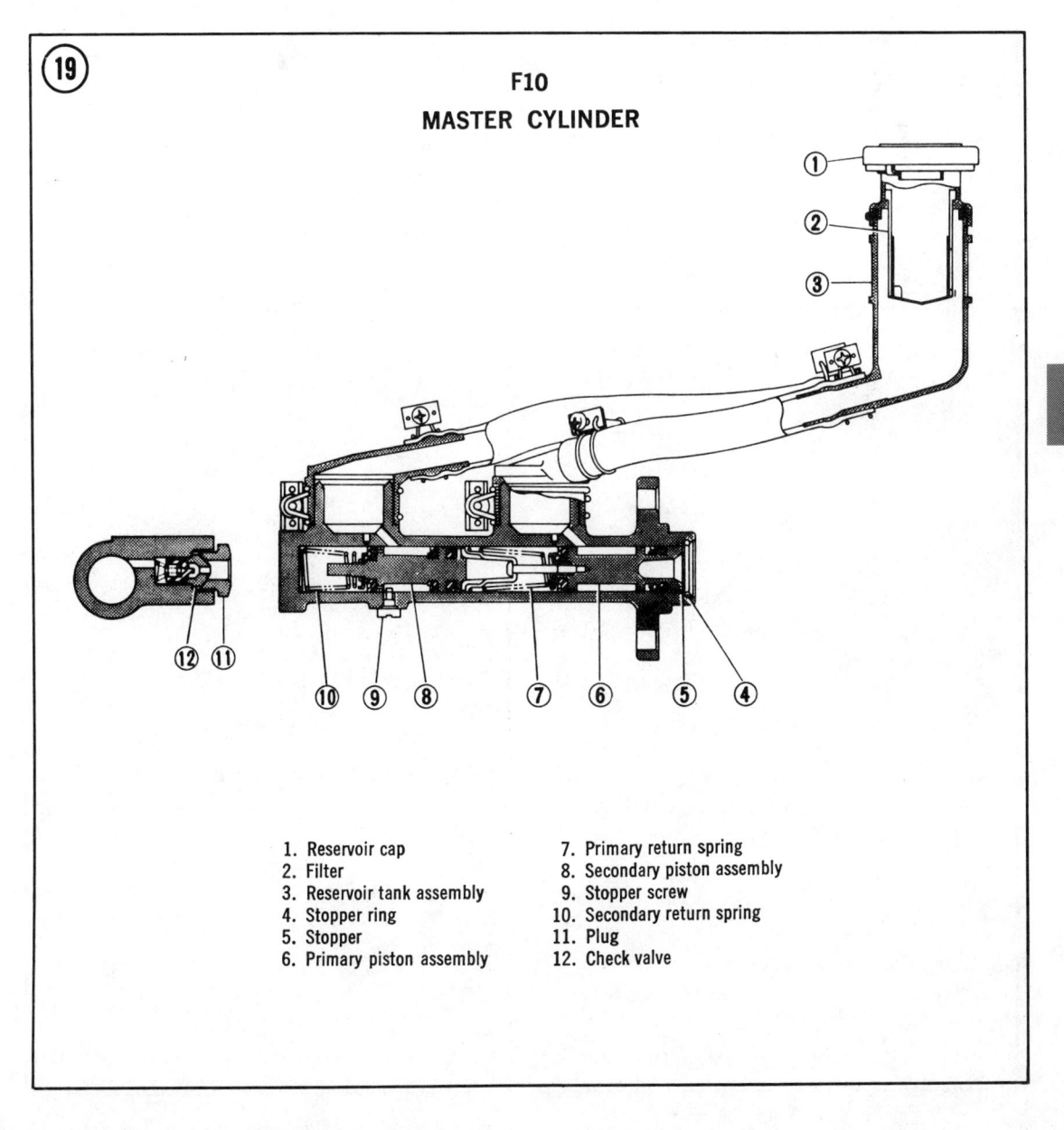

**(19) F10 MASTER CYLINDER**

1. Reservoir cap
2. Filter
3. Reservoir tank assembly
4. Stopper ring
5. Stopper
6. Primary piston assembly
7. Primary return spring
8. Secondary piston assembly
9. Stopper screw
10. Secondary return spring
11. Plug
12. Check valve

# CHAPTER NINE

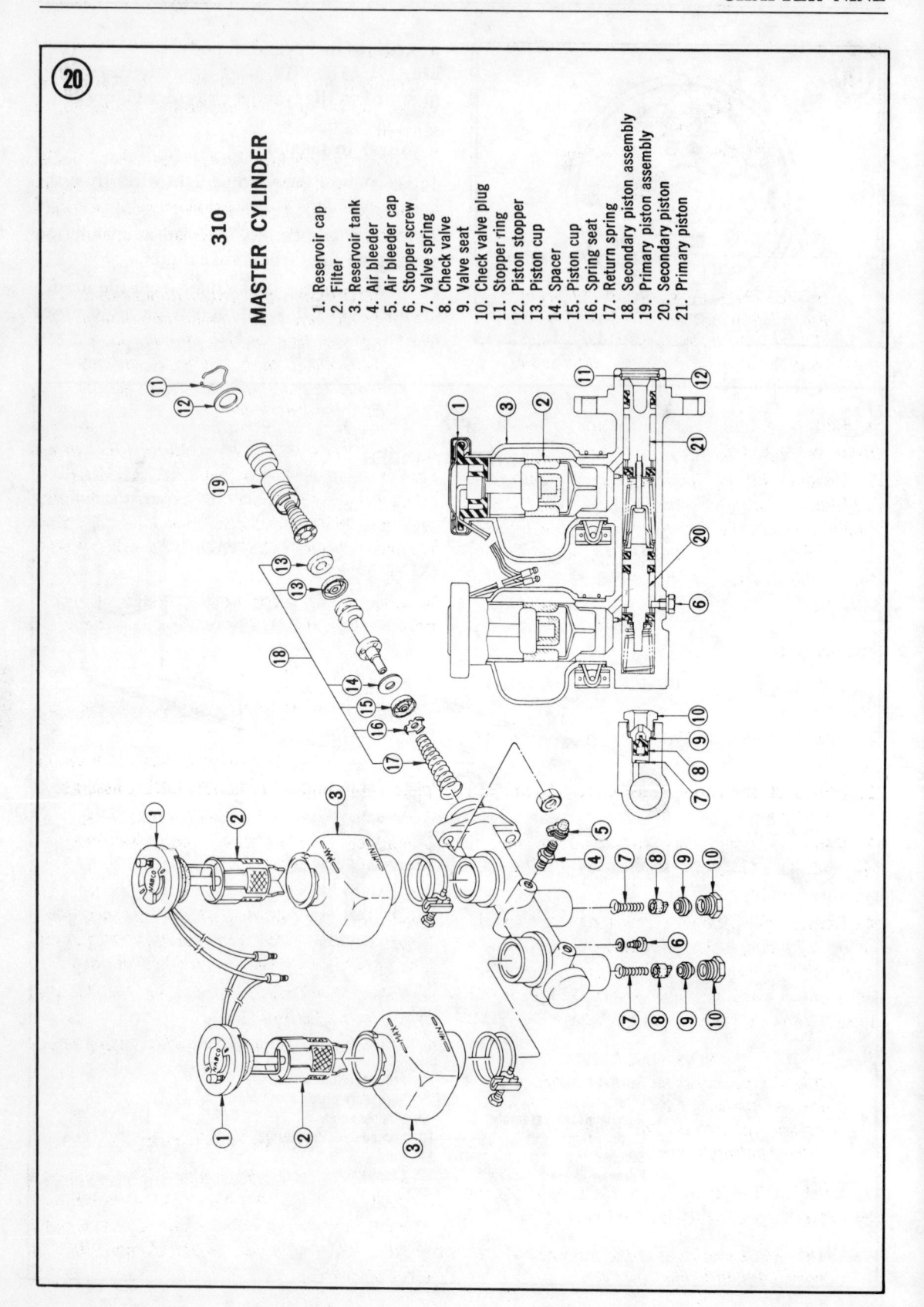

**310 MASTER CYLINDER**

1. Reservoir cap
2. Filter
3. Reservoir tank
4. Air bleeder
5. Air bleeder cap
6. Stopper screw
7. Valve spring
8. Check valve
9. Valve seat
10. Check valve plug
11. Stopper ring
12. Piston stopper
13. Piston cup
14. Spacer
15. Piston cup
16. Spring seat
17. Return spring
18. Secondary piston assembly
19. Primary piston assembly
20. Secondary piston
21. Primary piston

# BRAKES

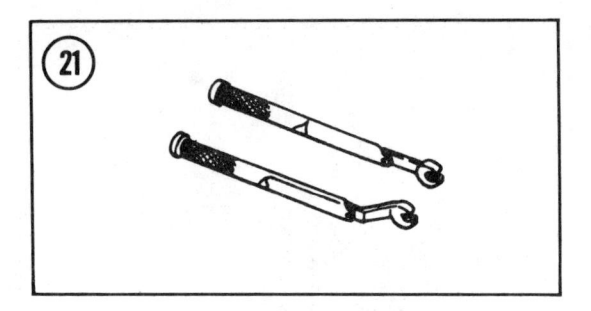

NOTE: *The factory recommends using brake line torque wrench GG94310000 (Figure 21) instead of an open-end wrench. Do not use an adjustable wrench on brake line nuts.*

4. Remove 2 nuts securing the master cylinder to the brake booster. Lift the cylinder out.

5. Installation is the reverse of removal. Tighten the cylinder securing nuts to 6-8 ft.-lb. (0.8-1.1 mkg). Tighten the brake line nuts to 11-13 ft.-lb. (1.5-1.8 mkg). After installation, bleed the brakes as described later in this chapter.

## Disassembly

Refer to the appropriate illustration for this chapter.

1. Pour the brake fluid out of the master cylinder.
2. Pull back the rubber dust cover on the end of the cylinder.
3. Remove the snap ring from inside the end of the cylinder. Take out the washer and primary piston assembly.
4. Remove the stopper screw from the bottom of the cylinder. Take out the secondary piston assembly. If necessary, blow the pistons out with compressed air. Use a service station air hose if you don't have a compressor.

### WARNING
*The pistons may shoot out like bullets if air is applied suddenly. Point the cylinder at a block of wood on the ground and apply pressure slowly.*

5. Unscrew the fittings from the bottom of the cylinder. Take out the check valves.

NOTE: *Do not remove the fluid reservoirs unless installing new ones.*

## Inspection

1. Thoroughly clean all parts in alcohol or brake fluid. Do not use gasoline, kerosene, or solvent.

2. Discard the piston cups, check valves, packing rings, and rubber dust cover. If the repair kit comes with pre-assembled pistons, don't bother to take the old ones apart; just throw them away.

NOTE: *Nabco and Tokico brand master cylinders are used in production. Repair kits for these brands are not interchangeable. Check the name on the master cylinder, and be sure to get the same brand of repair kit.*

3. Check the cylinder bore and pistons for excessive or uneven wear, scoring, or corrosion. Wear is excessive if clearance between cylinder wall and pistons exceeds 0.006 in. (0.15mm). Replace cylinder and pistons if these conditions can be seen.

4. Check springs for wear, damage, or weakness. Replace if these conditions are evident.

## Assembly

Assembly is the reverse of disassembly, plus the following.

1. Apply rubber grease or brake fluid to the piston cups and cylinder bore before assembly.
2. If the repair kit didn't come with pre-assembled pistons, be sure to install the new piston cups on the pistons exactly as the old ones were. The lip (wide side) of all piston cups faces into the cylinder, *except* the secondary piston cup closest to the cylinder bore opening. The lip of this cup faces out of the cylinder.
3. After installation, bleed the brakes and check for brake fluid leaks.
4. Check brake pedal height. See *Adjustments* in this chapter.

## BRAKE BOOSTER

The F10 uses a rebuildable 4.5 in. diameter brake booster. The 310 uses a non-rebuildable 6 in. unit. The brake booster and its check valve should be tested at intervals specified in Chapter Three.

## Check Valve Test

1. Remove the check valve from its clip on the firewall **(Figure 22)**.

2. Disconnect the hose from the brake booster side of the valve. Connect a vacuum gauge in its place.

3. Run the engine at a fast idle. Turn the engine off when vacuum reaches approximately 20 in. (500mm).

4. Watch the vacuum gauge for 15 seconds. On F10's, there should be no vacuum drop at all. On 310's, there should be no more than 1 in. (25mm).

5. If vacuum dropped more than specified, the vacuum line or check valve is defective. If the vacuum line is tightly connected and in good condition, replace the check valve.

## Airtightness Test (No Load)

1. Using a T-fitting, connect a vacuum gauge into the line between check valve and brake booster. See **Figure 23** (F10) or **Figure 24** (310).

2. Run the engine at a fast idle. Turn it off when vacuum reaches 20 in. (500mm).

3. With the engine off, watch the vacuum gauge for 15 seconds. There should be no vacuum drop on F10's; no more than 1 in. (25mm) on 310's. Too much vacuum drop means a defective vacuum line or brake booster. If the vacuum lines are tightly connected and in good condition, the brake booster must be overhauled (F10) or replaced (310).

## Airtightness Test (Under Load)

1. Connect a vacuum gauge as described in Step 1 of the previous procedure. Place the gauge where it can be seen from the driver's seat, or have an assistant watch it for you.

2. With the engine running, press the brake pedal as far as it will go.

3. When vacuum reaches approximately 20 in. (500mm), shut the engine off. Keep the brake pedal down. Watch the vacuum gauge for 15 seconds. There should be no pressure drop (F10) or less than 1 in. (25mm) on 310's. If pressure drops too much, the brake booster must be overhauled or replaced.

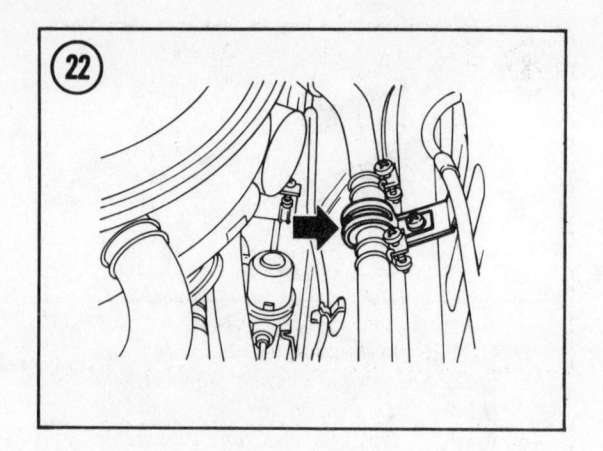

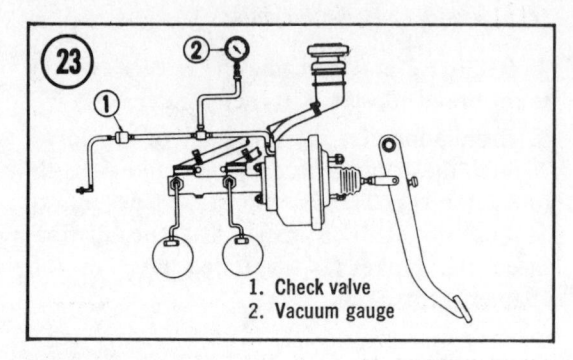

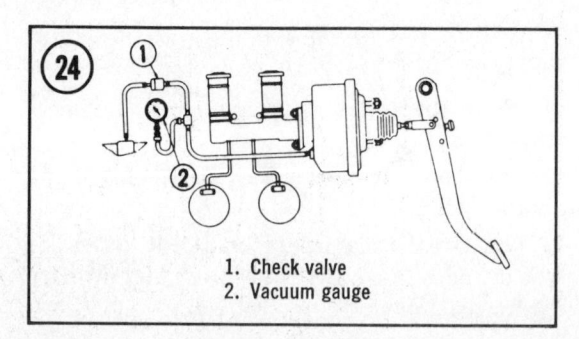

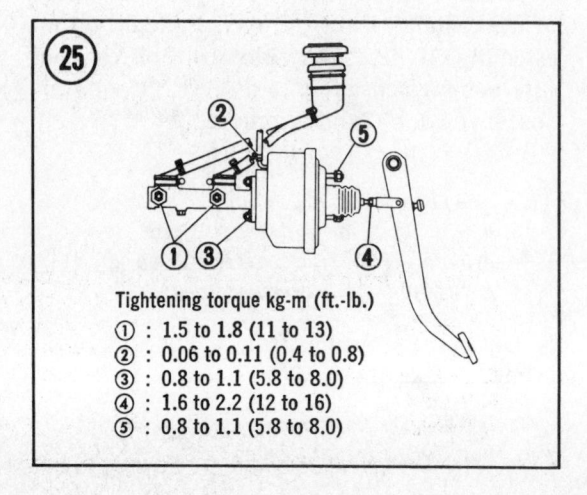

1. Check valve
2. Vacuum gauge

1. Check valve
2. Vacuum gauge

Tightening torque kg-m (ft.-lb.)
① : 1.5 to 1.8 (11 to 13)
② : 0.06 to 0.11 (0.4 to 0.8)
③ : 0.8 to 1.1 (5.8 to 8.0)
④ : 1.6 to 2.2 (12 to 16)
⑤ : 0.8 to 1.1 (5.8 to 8.0)

# BRAKES

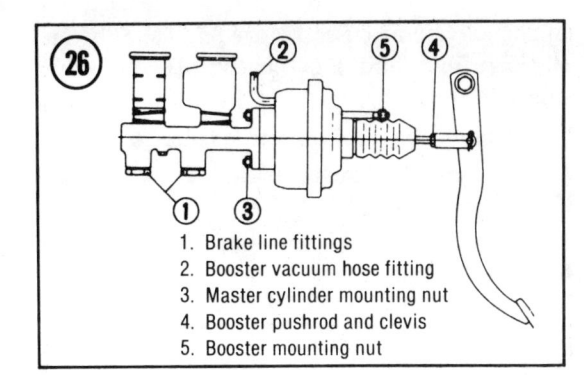

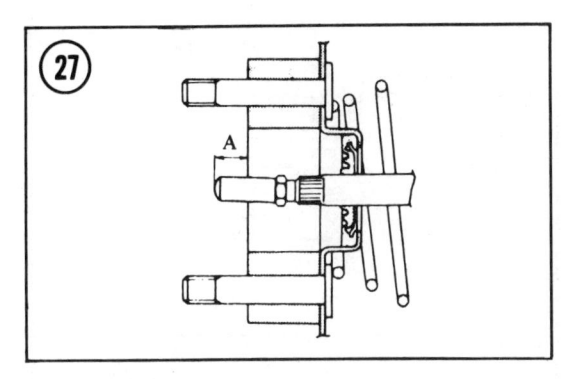

**26**

1. Brake line fittings
2. Booster vacuum hose fitting
3. Master cylinder mounting nut
4. Booster pushrod and clevis
5. Booster mounting nut

**27**

## Booster Removal/Installation

Refer to **Figure 25** (F10) or **Figure 26** (310).

1. Remove the master cylinder. See *Master Cylinder* earlier in this chapter.

2. Disconnect the brake booster pushrod from the brake pedal arm.

3. Remove 4 nuts securing the brake booster to the firewall. These are accessible from inside the car.

4. Lift the brake booster out.

5. Installation is the reverse of removal. If installing a new or rebuilt unit, check operating rod length (dimension "A," **Figure 27**). It should be 0.384-0.394 in. (9.75-10.0mm). After installation, bleed the brakes and check brake pedal height. See *Brake Bleeding* and *Adjustments* in this chapter.

## Disassembly

The following sections apply to the F10's only. The 310 brake booster is not rebuildable.

Refer to **Figure 28** for this procedure.

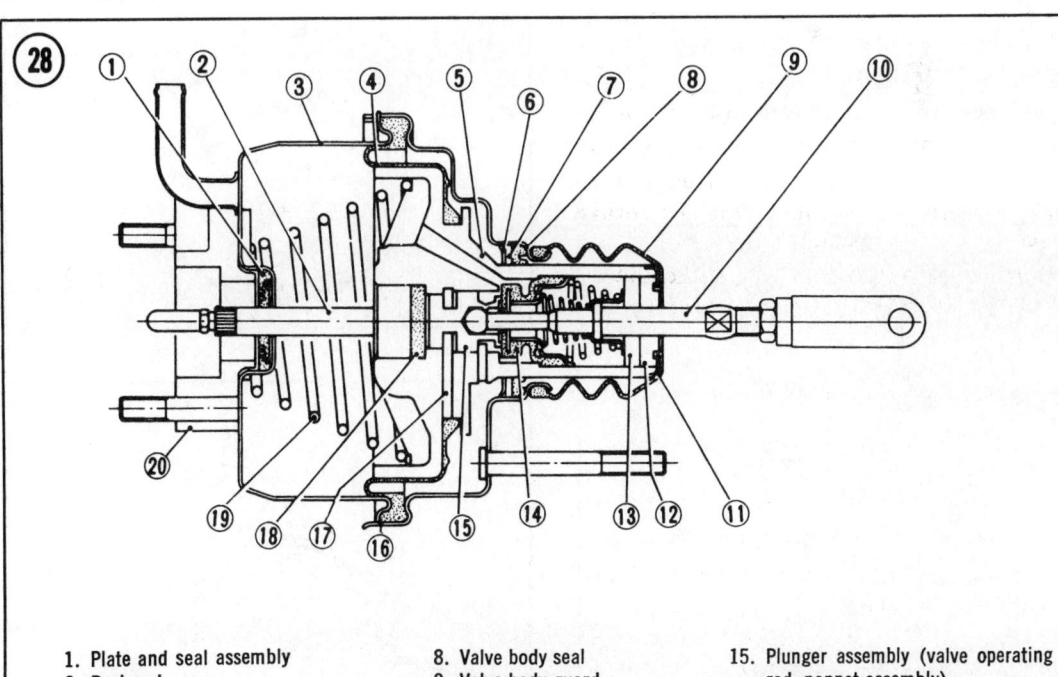

**28**

1. Plate and seal assembly
2. Push rod
3. Front shell
4. Diaphragm
5. Diaphragm plate and valve body
6. Retainer
7. Bearing
8. Valve body seal
9. Valve body guard
10. Valve operating rod
11. Silencer retainer
12. Silencer filter (felt)
13. Silencer (rubber)
14. Poppet assembly
15. Plunger assembly (valve operating rod, poppet assembly)
16. Rear shell
17. Valve plunger stop key
18. Reaction disc
19. Diaphragm return spring
20. Flange

1. Thoroughly clean the outside of the booster before disassembly. Be sure your work area is *clean*.

2. Paint or scribe mating marks on the front and rear shells. This ensures that the parts will be reassembled in their original positions.

3. Place the booster in a soft-jawed vise as shown in **Figure 29**. Remove the pushrod, locknut, and valve body guard.

4. Separate the front and rear shells. Use the Datsun special tool shown in **Figure 30** or improvise a substitute.

5. Carefully pry the seal retainer loose with a screwdriver (**Figure 31**). Remove the seal and discard it.

6. Remove the diaphragm from the diaphragm plate (**Figure 32**). Throw the diaphragm away.

7. Carefully pry the air silencer retainer loose from the valve body (**Figure 33**).

### CAUTION
*Don't hit the screwdriver with a hammer, or the valve body may be damaged.*

8. Press in the valve operating rod and take out the stop key. See **Figure 34**.

9. Pull the valve plunger assembly, together with the air silencer filter and air silencer, out of the valve body. See **Figure 35**. Take the air silencer filter and air silencer off the valve operating rod.

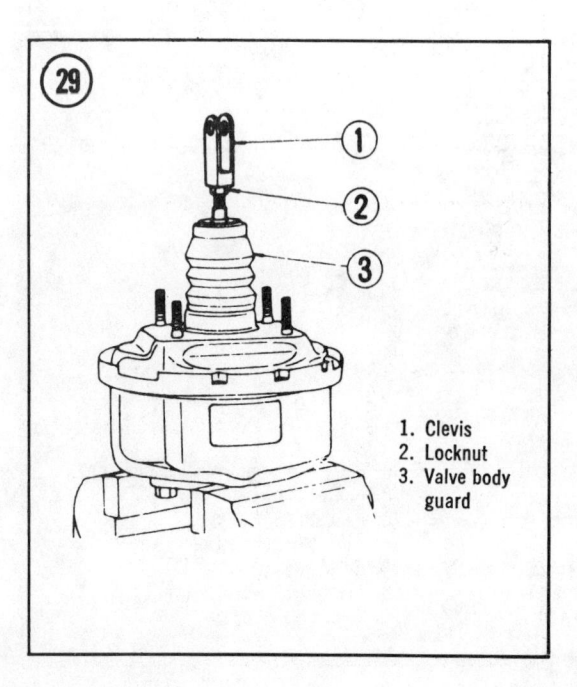

10. Remove the reaction disc (**Figure 36**).

11. Detach the flange from the front shell (**Figure 37**).

12. Remove the plate and seal assembly from beneath the flange.

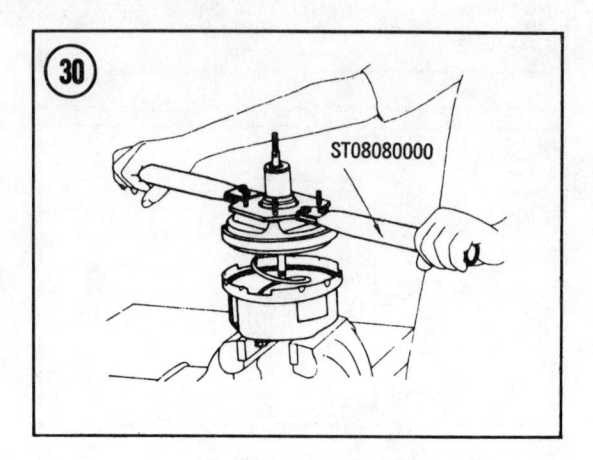

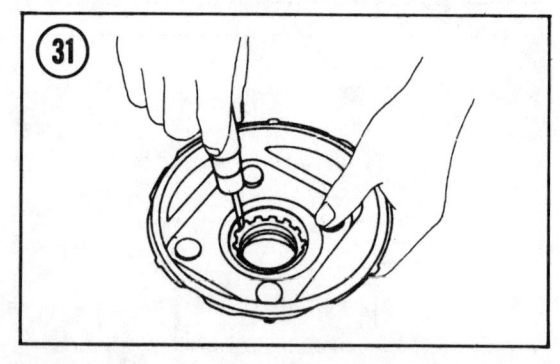

1. Clevis
2. Locknut
3. Valve body guard

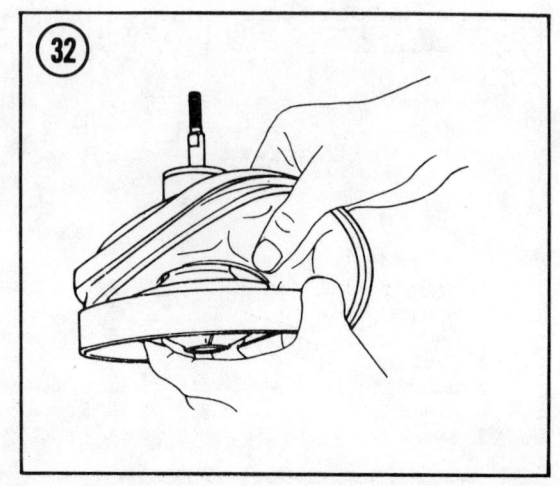

# BRAKES

## Inspection

1. Discard the following parts:

   a. Plunger assembly
   b. Air silencer, filter, and retainer
   c. Diaphragm
   d. Reaction disc

2. Throughly clean all remaining parts with alcohol.

3. Check front and rear shells for wear or damage. Replace if these conditions are found.

4. Examine stud threads. Repair lightly damaged threads with a die. Replace parts with severely damaged threads. Check welds at bases of the studs for cracks. Replace cracked parts.

5. Check for corrosion in the areas marked "D" in **Figure 38**. Clean off light corrosion with emery paper. Replace heavily corroded parts.

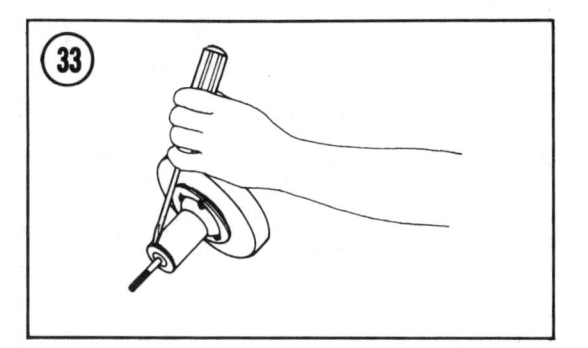

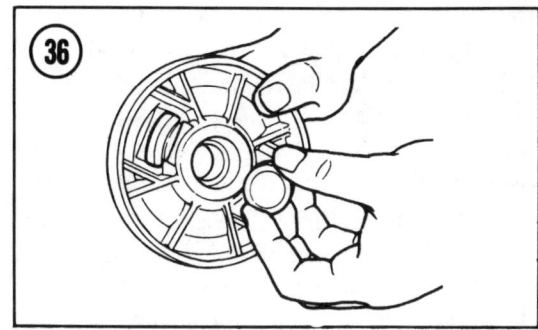

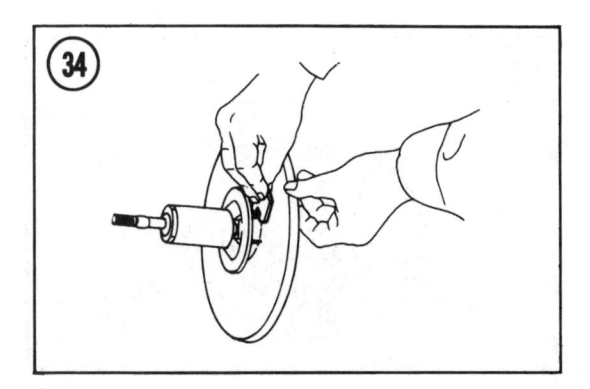

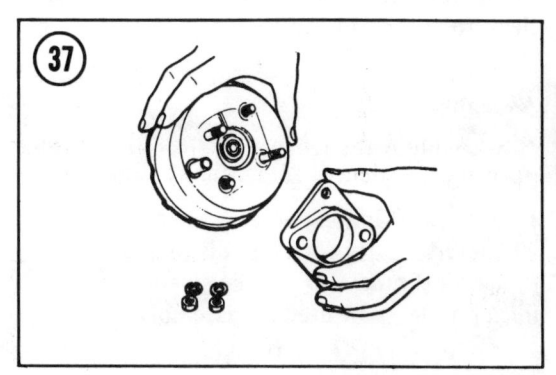

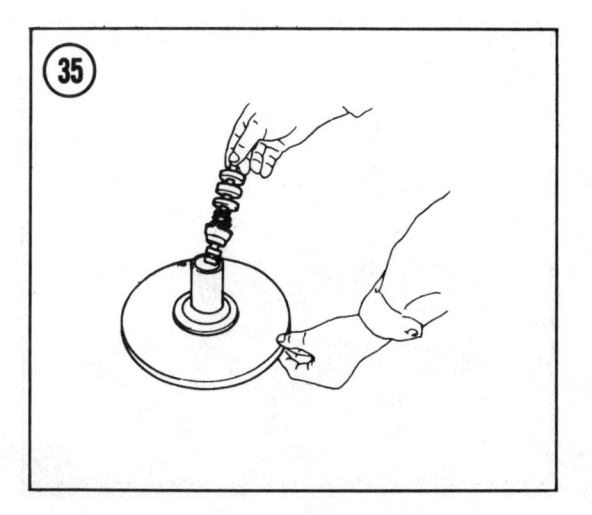

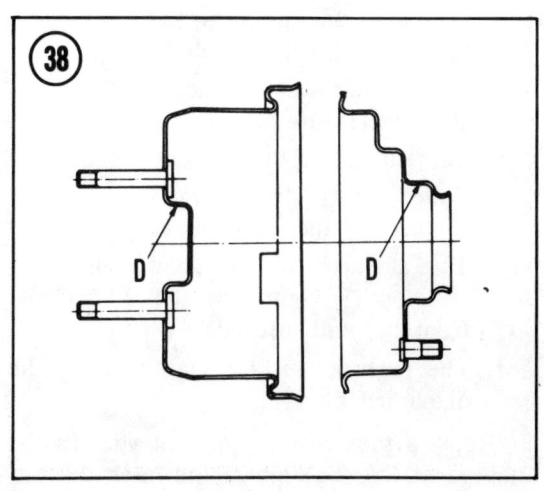

6. Inspect the outer surface of the valve body (E, **Figure 39**). If any visible damage can be seen, including slight scratches, replace the valve body.

7. Check bearing movement on the valve body. Replace the bearing if it doesn't move smoothly.

8. Carefully examine the groove in the diaphragm plate (F, **Figure 39**). Replace the diaphragm plate if wear or damage can be seen.

9. Check the diaphragm plate and valve body for cracks. Replace if cracks can be found.

10. Check the flange for cracks or rust. Replace if these can be found.

11. Check the diaphragm spring for weakness, rust, or deformation. Replace if these can be seen.

12. Check the pushrod for rust. Remove light rust with emery paper. Replace heavily rusted pushrods. The pushrod must also be replaced if the friction surface (**Figure 40**) is scored.

13. Check the stop key and pedal-to-booster clevis for wear and damage. Replace as needed.

**Assembly**

Assembly is the reverse of disassembly, plus the following.

1. Replace all parts contained in the repair kit whenever the booster is disassembled. The repair kit should also contain silicon grease and mica powder to be used for assembly.

2. Apply a light coat of silicon grease to the following:

   a. On the seal (8, **Figure 28**), the lip, and the face contacting the rear shell.
   b. Friction surfaces of the valve plunger assembly (**Figure 41**).
   c. Both surfaces of the reaction disc.
   d. The edge of the diaphragm where it makes contact with the front and rear shells.
   e. The surfaces on the plate and seal assembly (1, **Figure 28**), that contact the front shell and pushrod.
   f. The pushrod surface that contacts the diaphragm plate.

3. Apply a thin coat of mica powder to the diaphragm. Don't get any on the outer edge.

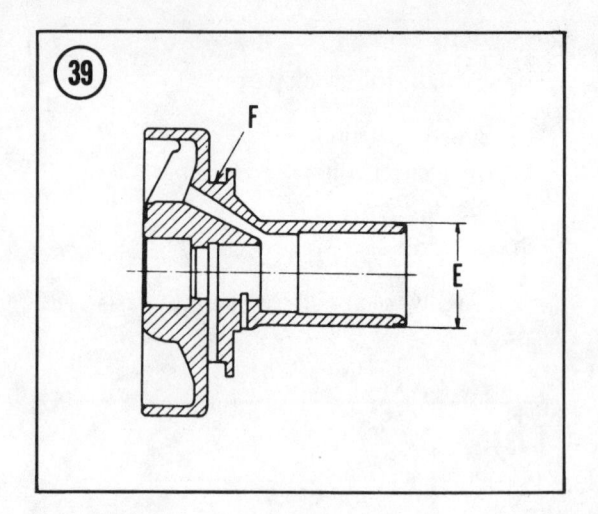

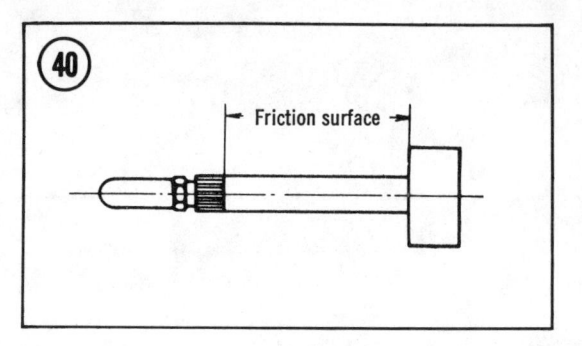

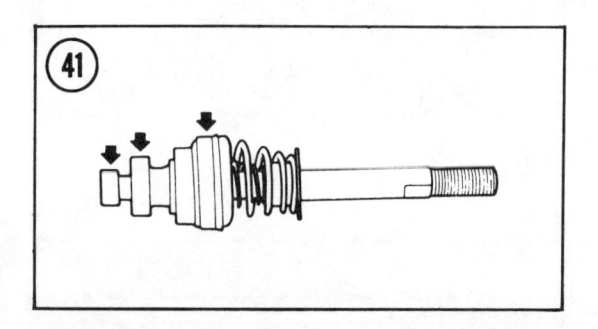

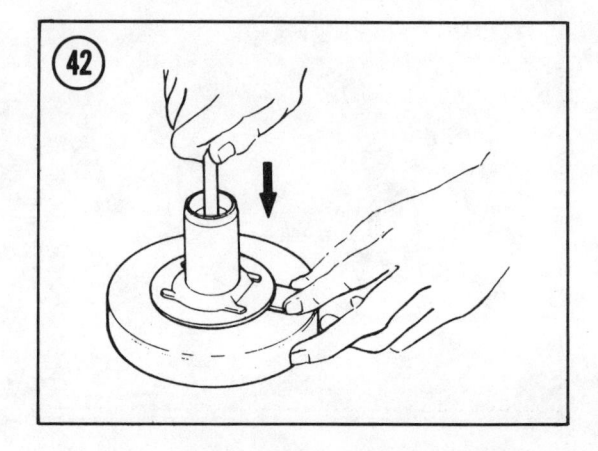

# BRAKES

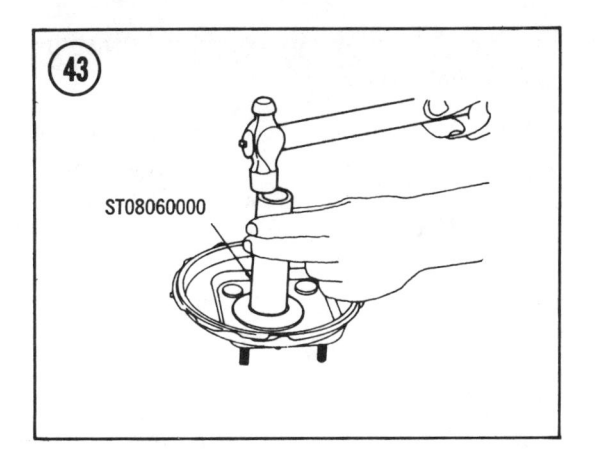

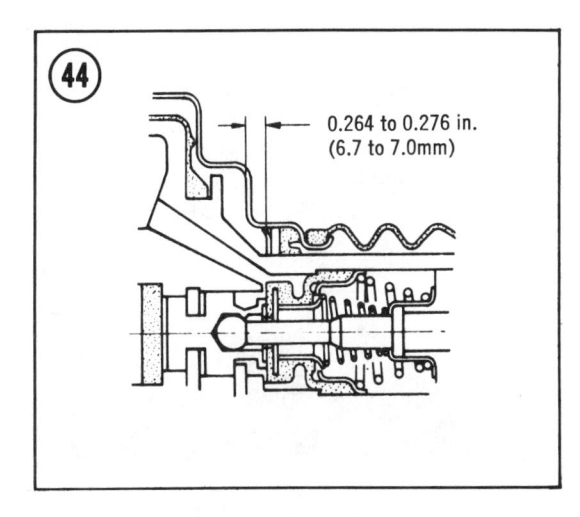

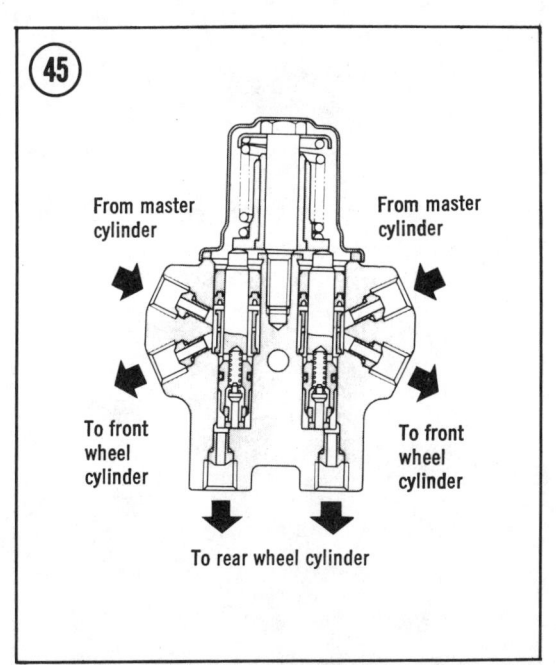

**43** ST08060000

**44** 0.264 to 0.276 in. (6.7 to 7.0mm)

**45** From master cylinder / From master cylinder / To front wheel cylinder / To front wheel cylinder / To rear wheel cylinder

4. When inserting the valve operating rod in the valve body, be sure the rod goes straight in and is not tilted to either side. When the rod is in, press it down against its spring and insert the stop key. See **Figure 42**.

5. When installing the bearing and seal retainer in the rear shell, use a drift such as ST08060000 (**Figure 43**). Tap the retainer in until the flange on the drift contacts the rear shell. If you can't get the special tool, tap the retainer in until it is 0.264-0.276 in. (6.7-7.0mm) deep in its recess. See **Figure 44**.

## PROPORTIONING VALVE

The proportioning valve (**Figure 45**) regulates hydraulic pressure to the front and rear brakes. This prevents premature rear wheel lockup.

### Testing

Test the proportioning valve at intervals specified in Chapter Three. To test, drive the car at a speed in excess of 31 mph (50 kph). Brake hard enough to lock the wheels *slightly*. If the front wheels lock before or at the same time as the rear wheels, the valve is good. If the rear wheels lock first, the valve is defective and must be replaced.

> **CAUTION**
> *Do not lock the wheels completely, or the tires will be flat-spotted.*

To remove the valve, disconnect the brake lines, noting carefully how they are connected. Remove the mounting bolt and take the valve out. Installation is the reverse of removal.

## BRAKE BLEEDING

The brakes should be bled whenever air enters the hydraulic system, reducing braking effectiveness. If the pedal feels spongy, or if pedal travel increases considerably, brake bleeding is usually called for. Bleeding is also necessary whenever a brake line is disconnected or the hydraulic system is repaired.

Brake fluid should be pumped out and replaced with new fluid at intervals listed in Chapter Three.

This procedure requires handling brake fluid. Be careful not to get any fluid on brake pads, shoes, discs, or drums. Clean all dirt from bleed valves before beginning. Two people are required, one to operate the brake pedal and the other to open and close the bleed valves.

Bleeding should be done in the following order: Master cylinder (310 only); right rear; left rear; right front; left front.

1. Clean away any dirt around the master cylinder. Top up the reservoir with brake fluid marked DOT 3.

2. Attach a plastic tube to the bleed valve. Dip the other end of the tube in a jar containing several inches of clean brake fluid. See **Figure 46**.

> NOTE: *Do not allow the end of the tube out of the brake fluid during bleeding. This could allow air into the hydraulic system, requiring that the bleeding procedure be done over.*

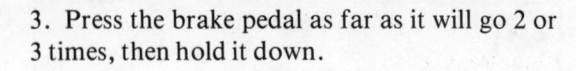

3. Press the brake pedal as far as it will go 2 or 3 times, then hold it down.

4. With the brake pedal down, open the bleed valve until the pedal goes to the floor, then close the bleed valve. Do not let the pedal up until the bleed valve is closed.

5. Let the pedal back up slowly.

6. Repeat Steps 3-5 until the fluid entering the jar is free of air bubbles.

7. Repeat the procedure for the other bleed valves.

> NOTE: *Keep an eye on the brake fluid level in the master cylinder during bleeding. If the reservoir level drops too low, air will be sucked into the system and the brakes will have to be bled again.*

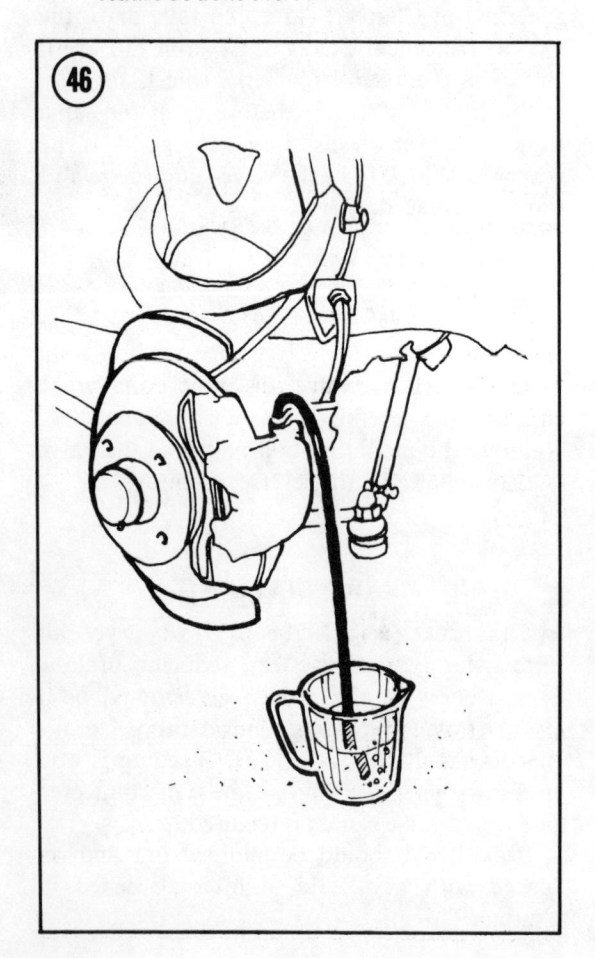

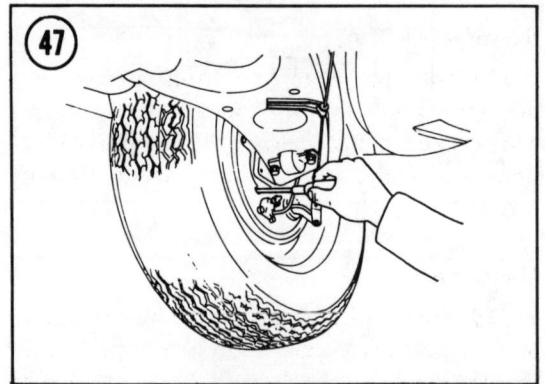

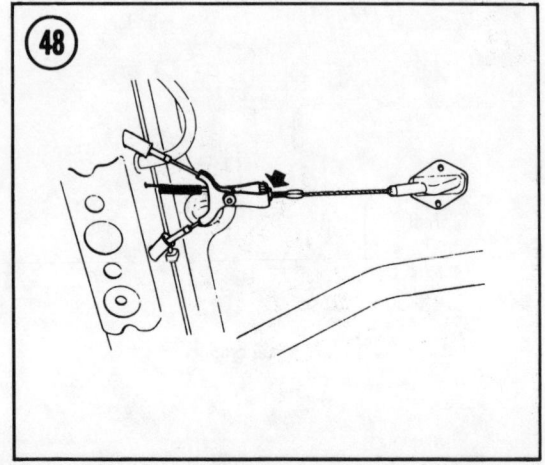

# BRAKES

## ADJUSTMENTS

### Front Brakes

The front brakes are adjusted automatically by the piston seals. Therefore, no means of manual adjustment is necessary or provided.

### Rear Brakes (F10)

1. Press the brake pedal several times to seat the shoes.
2. Securely block both front wheels so the car will not roll in either direction. Jack up the rear end and place it on jackstands.
3. Make sure the transmission is in neutral and the handbrake is off.
4. Insert a screwdriver or brake adjusting tool in the adjusting hole (**Figure 47**). Turn the adjuster wheel until the shoes lock the drum.
5. Back off the adjuster wheel 5 or 6 notches, until the drum just turns freely. If drag can still be felt, back off another one notch.

### Rear Brakes (310)

The 310 rear brakes are self-adjusting, and do not require regular adjustment. To adjust after brake repairs, operate the handbrake lever until no more clicks are heard from the rear wheels.

### Handbrake

*F10 Hatchback and Sedan* — The handbrake lever should rise 5 or 6 notches at a pull of 55 lb. (25 kg). If the handbrake is loose, check rear brake adjustment first. If the rear brakes are adjusted correctly, loosen the cable locknut (**Figure 48**). Turn the adjusting nut to take up slack in the cable, then tighten the locknut.

*F10 wagon* — The handbrake should rise 6 to 8 notches at a pull of 55 lb. (25 kg). If the handbrake is loose, check rear brake adjustment first. If the rear brakes are adjusted correctly, loosen the cable locknut (**Figure 49**). Turn the adjuster to take up slack in the cable, then tighten the locknut.

*310* — The handbrake lever should rise 7 or 8 notches at a pull of 44 lb. (20 kg). If the handbrake is loose, operate the handbrake lever several times and listen for clicks from the rear wheels. If no clicks can be heard, loosen the cable locknut (**Figure 50**). Turn the adjusting nut to take up slack in the cable, then tighten the locknut.

### Brake Pedal

1. Check pedal height from the floor (H, **Figure 51**). It should be $6\frac{7}{8}$-$7\frac{1}{16}$ in. (174-180mm) on F10's, and $7\frac{1}{16}$-$7\frac{5}{16}$ in. (180-186mm) on 310's.
2. If pedal height is incorrect, loosen the locknuts on pedal pushrod and stoplight switch. Turn the pushrod and switch to change pedal height, then tighten the locknuts.
3. Check pedal stroke (S, **Figure 51**). It should be 5-5 1/4 in. (127-133mm) on F10's, and 4 1/3-4 1/2 in. (110-116mm) on 310's. If the stroke is too long, the brakes should be bled. On F10's, check rear brake adjustment as well.

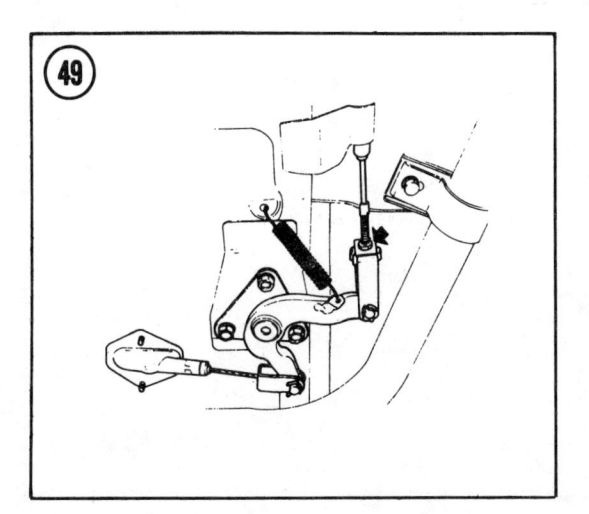

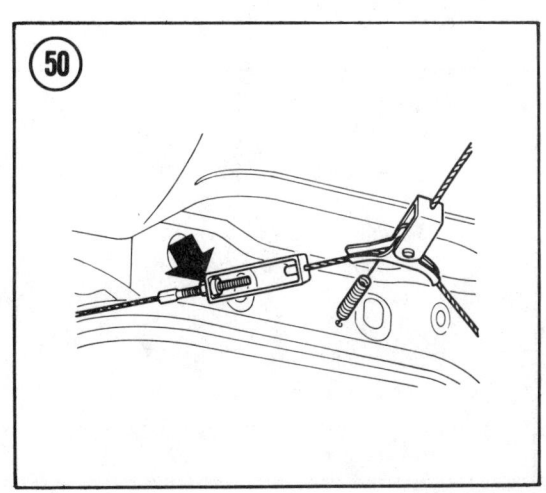

# CHAPTER NINE

⑤¹

**BRAKE PEDAL**

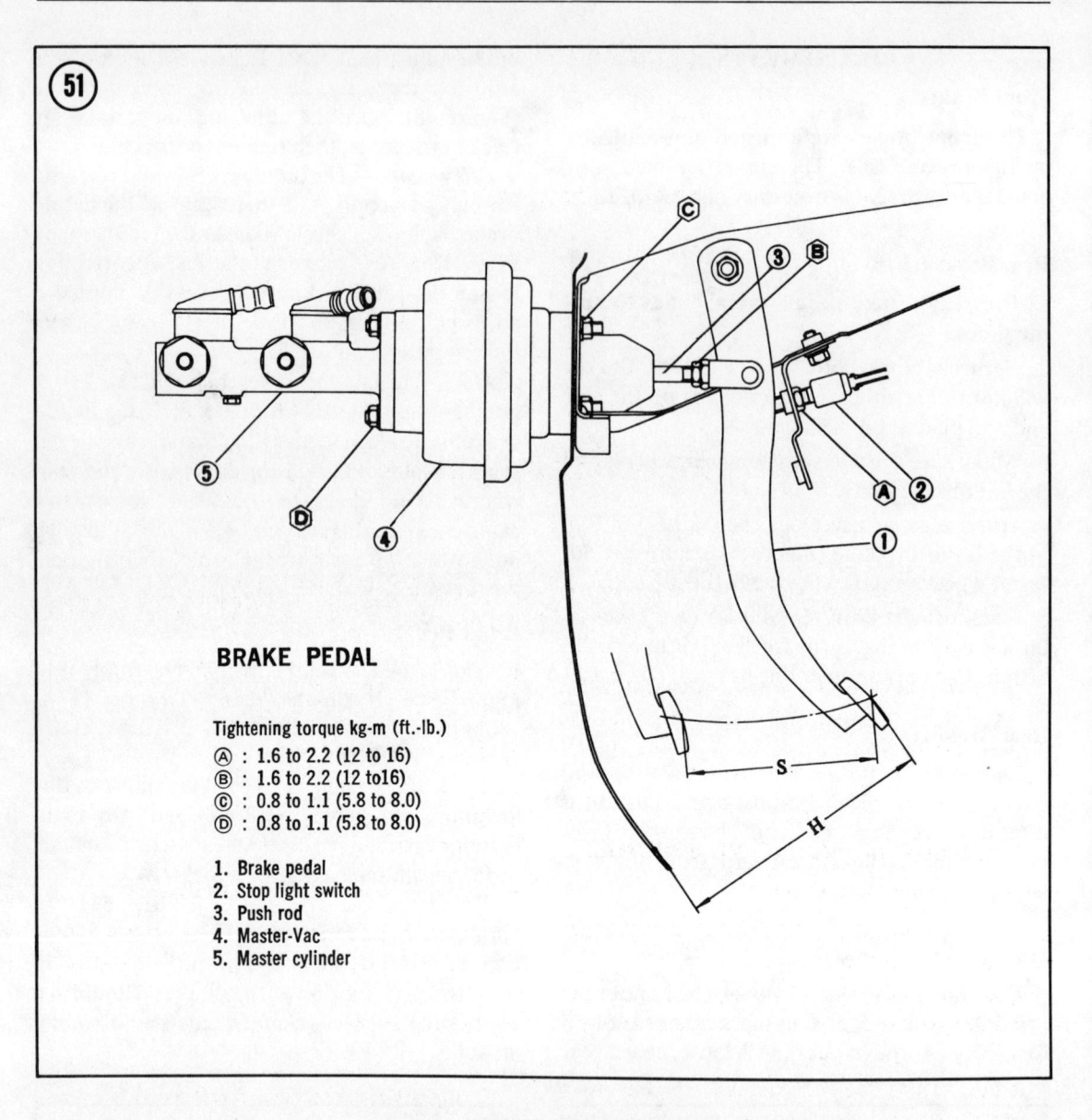

Tightening torque kg-m (ft.-lb.)
Ⓐ : 1.6 to 2.2 (12 to 16)
Ⓑ : 1.6 to 2.2 (12 to16)
Ⓒ : 0.8 to 1.1 (5.8 to 8.0)
Ⓓ : 0.8 to 1.1 (5.8 to 8.0)

1. Brake pedal
2. Stop light switch
3. Push rod
4. Master-Vac
5. Master cylinder

Table 1   BRAKE SPECIFICATIONS

| | |
|---|---|
| Brake disc | |
| Minimum thickness | 0.339 in. (8.6mm) |
| Maximum runout (F10) | 0.006 in. (0.15mm at 8.66 in. (220mm from center) |
| Maximum runout (310) | 0.005 in. (0.12mm) at 7.56 in. |
| Maximum thickness variation (parallelism) | 0.0012 in. (0.03mm) |
| | |
| Brake drum | |
| Maximum inner diameter | 8.05 in. (204.5mm) |
| Maximum out-of-round (F10) | 0.002 in. (0.05mm) |
| Maximum out-of-round (310) | 0.0008 in. (0.02mm) |

# CHAPTER TEN

# REAR SUSPENSION

The F10 sedan and hatchback, and all 310's, use an independent coil spring rear suspension (**Figure 1**). The F10 wagon uses a leaf spring with a rigid tube axle (**Figure 2**).

## REAR SUSPENSION (COIL SPRINGS)

### Shock Absorber Replacement

1. Securely block both front wheels so the car will not roll in either direction.
2. Loosen the rear wheel nuts, jack up the rear end of the car, place it on jackstands, and remove the rear wheels.
3. Place a jack beneath the suspension arm. Raise the arm to relieve tension on the shock absorber.
4. Remove the upper nut (**Figure 3**) and lower bolts (**Figure 4**). Take the shock absorber out.
5. Installation is the reverse of removal. Position the upper end bushings and washers as shown in **Figure 5**. Tighten nuts and bolts to specifications (**Table 1**).

### Coil Spring Removal/Installation

Refer to **Figure 6** for this procedure.

1. Remove the shock absorber as described earlier.
2. Slowly lower the jack beneath the suspension arm until all spring pressure is released.
3. Take out the spring and its rubber insulator.
4. Installation is the reverse of removal. Replace the rubber insulator if worn or deteriorated.

### Suspension Arm Removal/Installation

1. Remove the shock absorber and coil spring as described earlier.
2. Disconnect the brake line and handbrake cable from the brake backing plate.

> NOTE: *Place a container beneath the brake line to catch dripping brake fluid.*

3. Plug the brake line. If you don't have a threaded cap that will fit, wrap aluminum foil over the line and secure it with tape.
4. If the suspension arm is to be replaced, remove the wheel bearings and brake drum as described later in this chapter. Then remove the rear brake assembly (**Figure 7**).
5. Remove the suspension arm pivot bolts (**Figure 8**). Take the suspension arm out.
6. Check the suspension arm bushings for wear, damage, cracked rubber, or looseness. If any of these conditions are found, have the bushings replaced by a machine shop.

CHAPTER TEN

①

REAR SUSPENSION
(ALL 310; F10 SEDAN AND HATCHBACK)

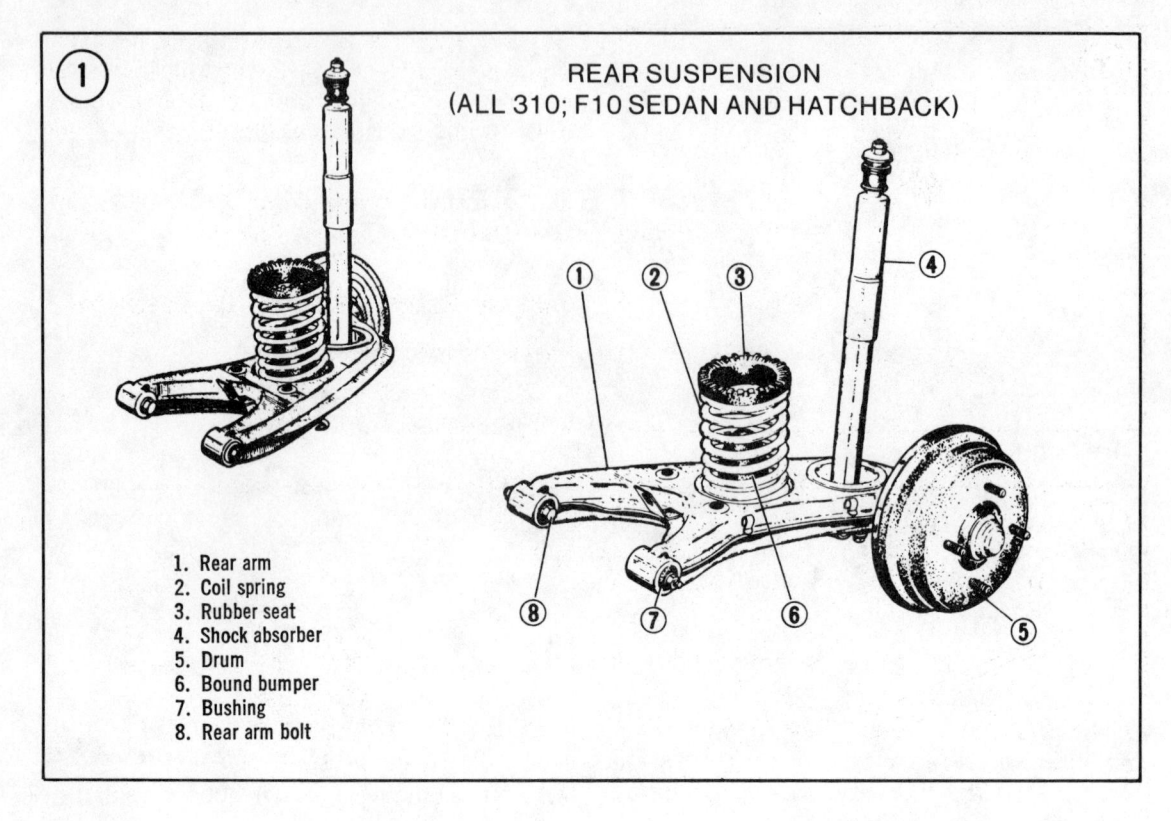

1. Rear arm
2. Coil spring
3. Rubber seat
4. Shock absorber
5. Drum
6. Bound bumper
7. Bushing
8. Rear arm bolt

②

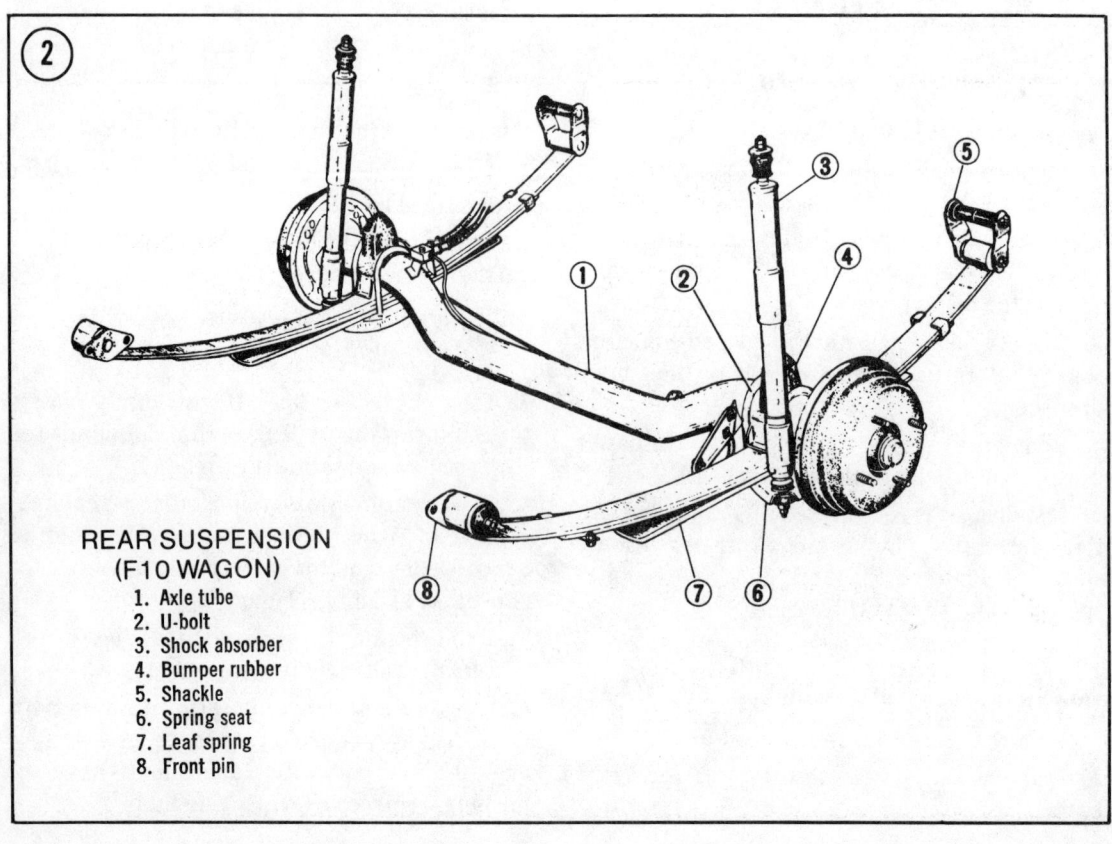

REAR SUSPENSION
(F10 WAGON)

1. Axle tube
2. U-bolt
3. Shock absorber
4. Bumper rubber
5. Shackle
6. Spring seat
7. Leaf spring
8. Front pin

# REAR SUSPENSION

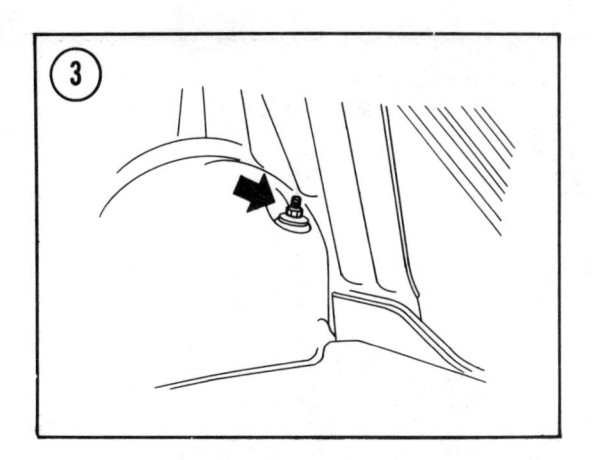

7. Installation is the reverse of removal. Tighten the pivot bolts slightly and lower the car. Tighten the pivot bolts to final specifications with the car's weight resting on the wheels.

8. After installation, bleed the brakes. See *Brake Bleeding*, Chapter Nine.

## REAR SUSPENSION (LEAF SPRINGS)

### Shock Absorber Replacement

1. Securely block both front wheels so the car will not roll in either direction.

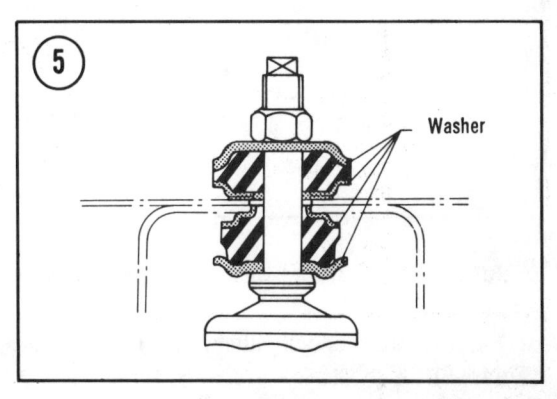

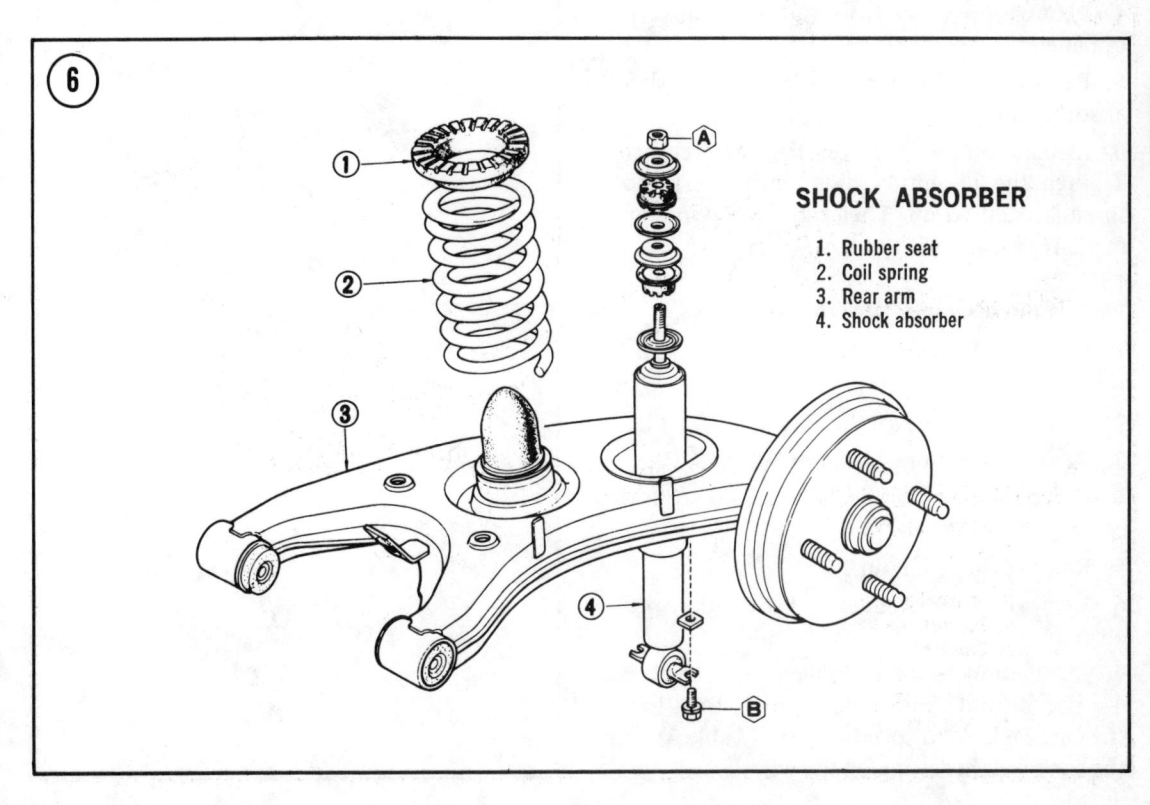

**SHOCK ABSORBER**

1. Rubber seat
2. Coil spring
3. Rear arm
4. Shock absorber

**7**

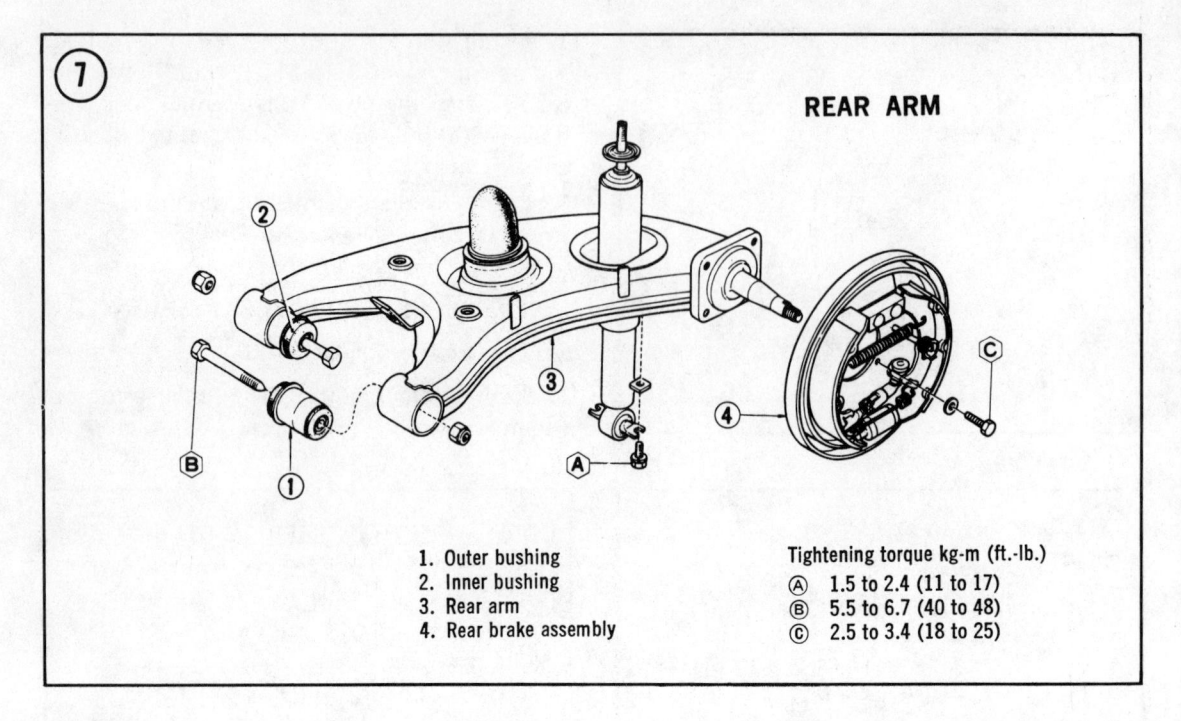

**REAR ARM**

1. Outer bushing
2. Inner bushing
3. Rear arm
4. Rear brake assembly

Tightening torque kg-m (ft.-lb.)
- Ⓐ 1.5 to 2.4 (11 to 17)
- Ⓑ 5.5 to 6.7 (40 to 48)
- Ⓒ 2.5 to 3.4 (18 to 25)

2. Jack up the rear end of the car, and place it on jackstands.

3. Remove the shock absorber lower nuts **(Figure 9)**.

4. Remove the cover from the shock absorber upper end.

5. Remove the upper nuts and take the shock absorber out.

6. Installation is the reverse of removal. Tighten the mounting nuts until the spindle threads are used up. Then tigthen the locknuts to 12-16 ft.-lb. (1.6-2.2 mkg).

**Leaf Spring Removal/Installation**

1. Remove the shock absorber as described earlier.

2. Place a jack beneath the axle to support it.

3. Remove the U-bolts **(Figure 10)**.

4. Detach the handbrake cable clamp from the spring.

5. Remove the front pin.

6. Remove the rear shackle. Lower the spring clear.

7. Installation is the reverse of removal. Tighten all nuts and bolts slightly, then lower the car. Tighten to specifications **(Table 1)** with the car's weight resting on the wheels.

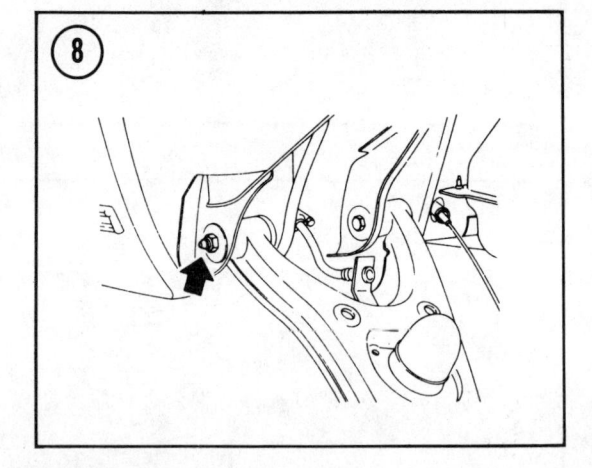

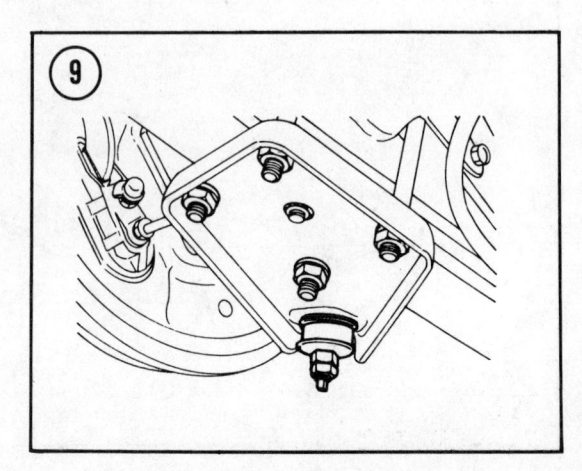

**8**

**9**

# REAR SUSPENSION

**⑩**

**LEAF SPRINGS**

1. Leaf spring
2. Bumper rubber
3. U-bolt
4. Front pin
5. Shackle
6. Axle tube

**⑪**

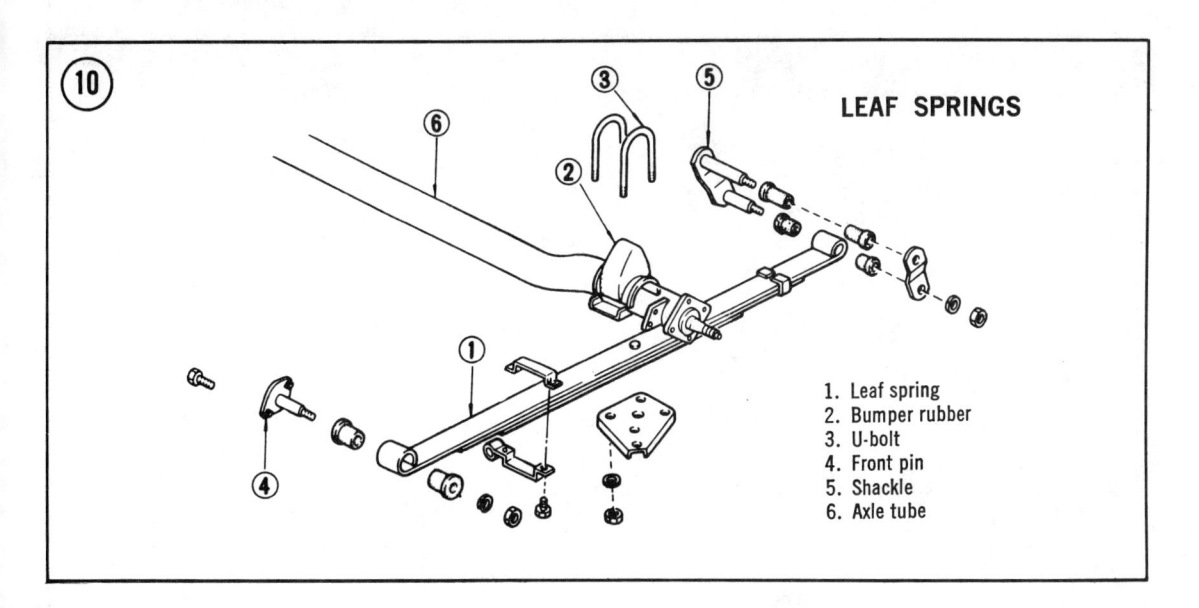

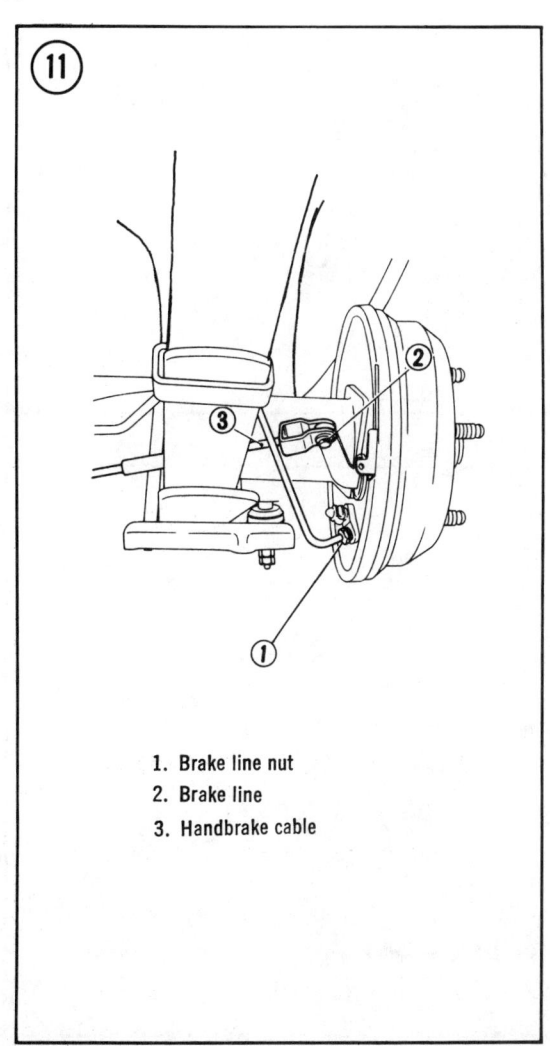

1. Brake line nut
2. Brake line
3. Handbrake cable

Table 1  TIGHTENING TORQUES

|  | Ft.-lb. | Mkg |
|---|---|---|
| Brake line and hose nuts | 11-13 | 1.5-1.8 |
| Brake backing plate bolts | 18-25 | 2.5-3.4 |
| Shock absorber lock nut (F10) | 12-16 | 1.6-2.2 |
| Shock absorber nut (310) | 6-9 | 0.8-1.2 |
| Leaf spring front pin bolt | 6½-10 | 0.9-1.4 |
| Leaf spring front pin nut | 12-16 | 1.6-2.2 |
| Leaf spring shackle bolts | 12-14 | 1.6-2.0 |
| Suspension arm pivot bolts | 40-48 | 5.5-6.7 |

### Axle Tube Removal/Installation

1. Securely block both front wheels so the car will not roll in either direction.

2. Loosen the rear wheel nuts, jack up the rear end of the car, place it on jackstands, and remove the rear wheels.

3. Disconnect the brake line and handbrake cable from the brake backing plate (**Figure 11**).

> NOTE: *Place a container beneath the brake line to catch dripping fluid.*

⑫

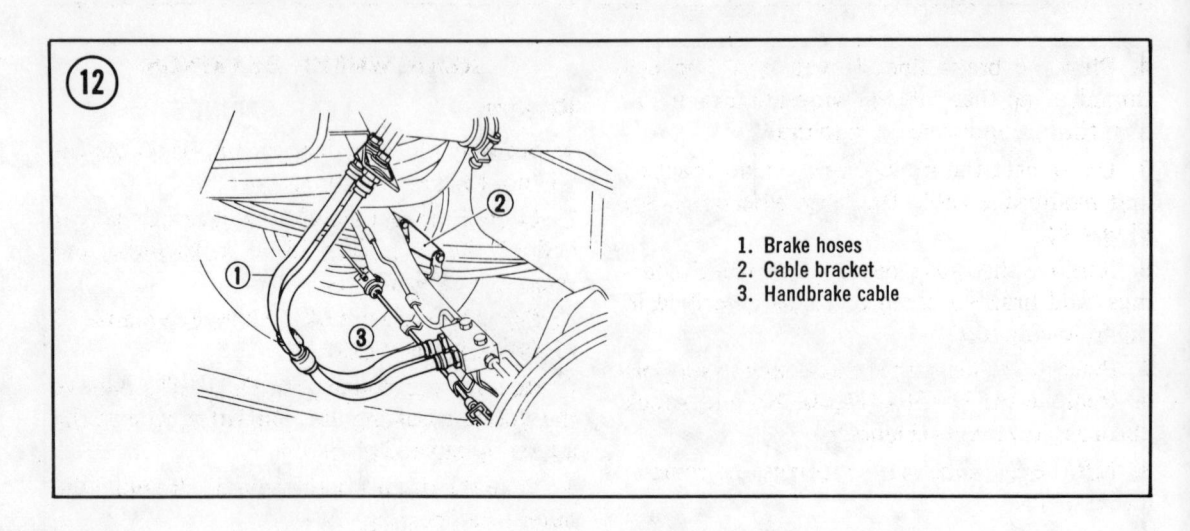

1. Brake hoses
2. Cable bracket
3. Handbrake cable

⑬

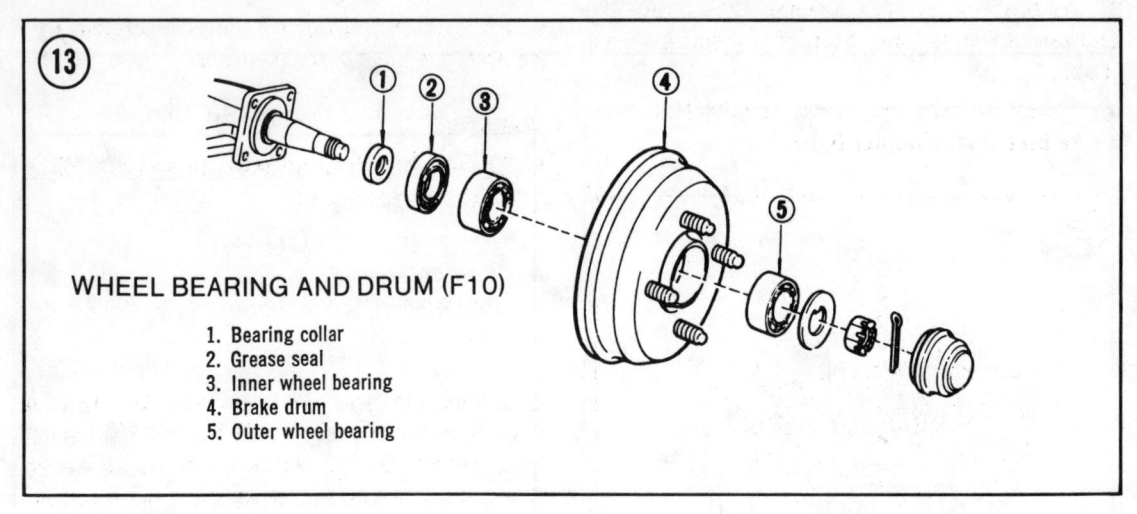

**WHEEL BEARING AND DRUM (F10)**

1. Bearing collar
2. Grease seal
3. Inner wheel bearing
4. Brake drum
5. Outer wheel bearing

⑭

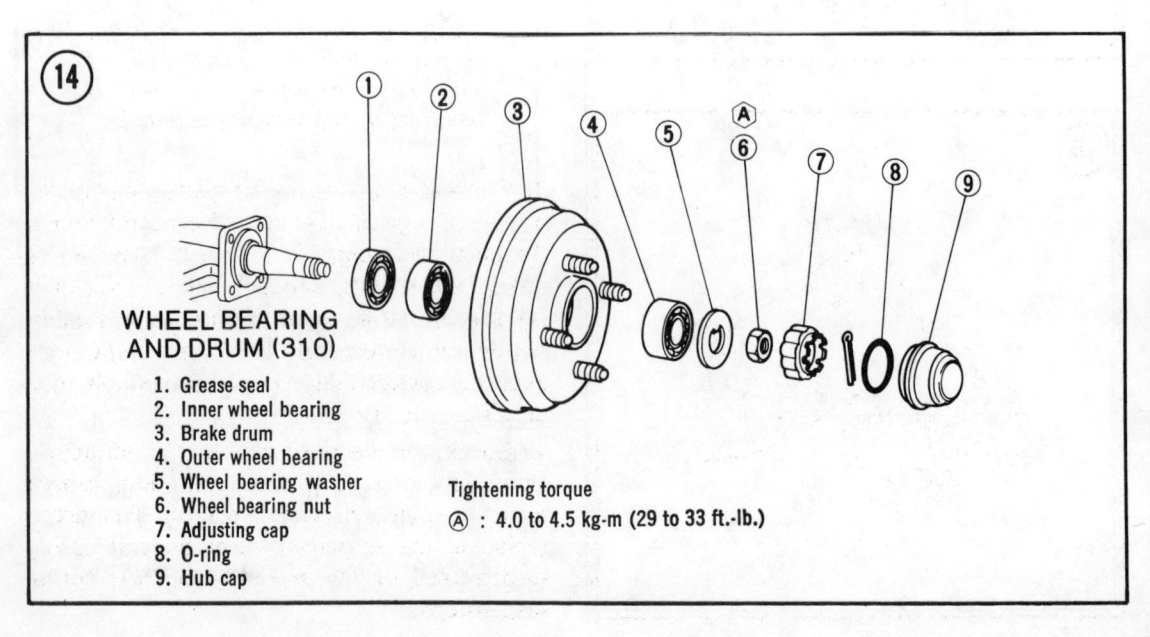

**WHEEL BEARING AND DRUM (310)**

1. Grease seal
2. Inner wheel bearing
3. Brake drum
4. Outer wheel bearing
5. Wheel bearing washer
6. Wheel bearing nut
7. Adjusting cap
8. O-ring
9. Hub cap

Tightening torque

Ⓐ : 4.0 to 4.5 kg-m (29 to 33 ft.-lb.)

# REAR SUSPENSION

4. Plug the brake line. If you don't have a threaded cap that will fit, wrap aluminum foil over the line and secure it with tape.

5. Disconnect the brake hoses, cable bracket, and handbrake cable from the axle tube. See **Figure 12**.

6. Remove the brake drums, rear wheel bearings, and brake backing plates as described in the following section.

7. Place a jack beneath the axle tube to support it. Remove the U-bolts **(Figure 10)** and shock absorber lower nuts **(Figure 9)**.

8. Lift the axle tube off the springs and remove it to one side.

9. Installation is the reverse of removal. Tighten all nuts and bolts to specifications **(Table 1)**.

10. After installation, bleed the brakes. See *Brake Bleeding*, Chapter Nine.

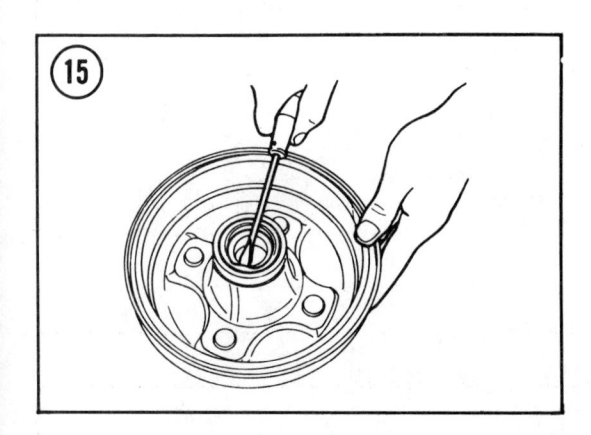

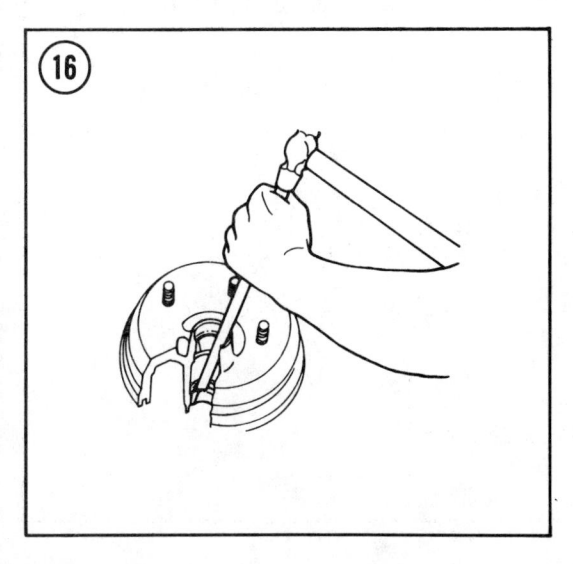

## REAR WHEEL BEARINGS

### Removal

1. Securely block both front wheels so the car will not roll in either direction.

2. Loosen the rear wheel nuts, jack up the rear end of the car, place it on jackstands, and remove the rear wheels.

3. Pry grease cap out of the hub. See **Figure 13** (F10) or **Figure 14** (310).

4. Remove the cotter pin. On F10's, remove the slotted adjusting nut. On 310's, remove the adjusting cap and plain nut.

5. Take the drum off the spindle. Remove the outer wheel bearing.

6. Pry grease seal out of the hub **(Figure 15)**. Remove the inner wheel bearing.

### Inspection

1. Thoroughly clean all parts in solvent. Blow dry with compressed air.

> **CAUTION**
> *Do not spin dry bearings with a compressed air hose. This can ruin the bearings.*

2. Check the bearings for wear, damage, or rust. Spin the bearings with fingers. Make sure they rotate freely, without excessive noise. Replace bearings if any defects can be found.

> NOTE: *If one bearing is replaced, replace the other bearing on that side of the car at the same time. This is necessary to prevent mixing of brands.*

3. Check the inner races for wear, pits, cracks, or rust. if any of these conditions can be seen, the inner races must be replaced. They can be driven out with a hammer and copper drift **(Figure 16)**. However, they must be pressed in, not driven. If you don't have a hydraulic press and the correct adapters, take this job to a dealer or machine shop.

4. Check the bearing spindle for scratches, chips, pits, or rust. If these conditions can be seen, the spindle must be replaced. This means replacing the suspension arm on coil spring suspensions, or the axle tube on leaf spring suspensions.

## Installation

1. If the outer races were removed, press them in with a press and adapters (**Figure 17**).

2. Pack the hub and grease cap with multipurpose lithium grease. Fill to the points shown in **Figure 18**.

3. Pack the wheel bearings with grease. Work as much grease as possible between the rollers.

4. Apply a thin coat of grease to the grease seal lip, spindle shaft and threads, wheel bearing nut, washer, and adjusting cap (310).

5. Install the wheel bearings in the hub. Carefully tap in the grease seal. Use a block of wood to spread the hammer's force, so the seal won't tip sideways and jam.

### CAUTION
*Keep grease off drum friction surface.*

6. Install the brake drum on the spindle. Install the washer, nut, and adjusting cap (310).

7. Adjust wheel bearings as described in the following section.

## Adjustment (F10)

1. Tighten the adjusting nut to 18-22 ft.-lb. (2.5-3.0 mkg). See **Figure 19**. Then turn the hub several turns each way to seat the bearings.

2. Back off the nut until it can be turned with fingers.

3. Tighten the nut firmly with a hand-held socket (**Figure 20**).

4. Install the cotter pin. If necessary, the nut may be tightened slightly to align the cotter pin holes.

5. Install the cotter pin and spread its ends. Install the grease cap.

6. Check wheel bearing preload. Rotate the hub with a spring scale (**Figure 21**) and measure the force necessary to turn the hub. With a used grease seal, it should be no more than 1.5 lb. (0.7 kg). With a new grease seal, it should be 1.1-2.6 lb. (0.5-1.2 kg).

## Adjustment (310)

1. Tighten the adjusting nut to 29-33 ft.-lb. (4.0-4.5 mkg). See **Figure 19**. Then turn the hub several turns in each direction to seat the bearings.

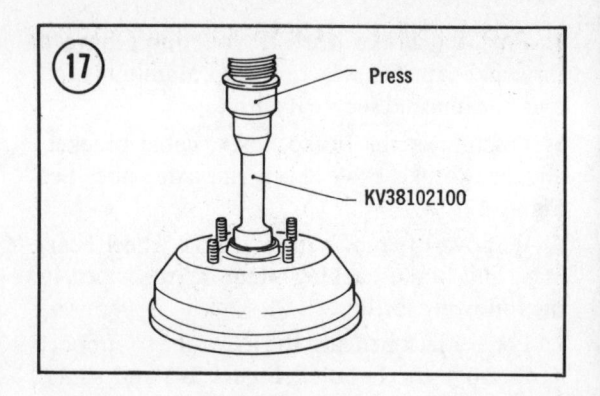

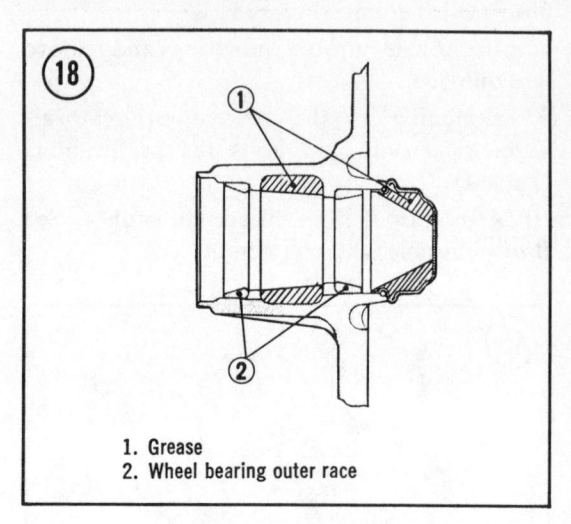

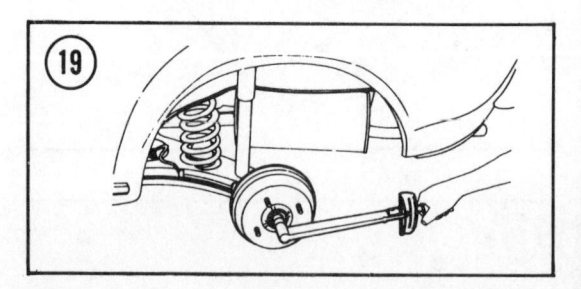

---

⑰ Press — KV38102100

⑱
1. Grease
2. Wheel bearing outer race

⑲

⑳

# REAR SUSPENSION

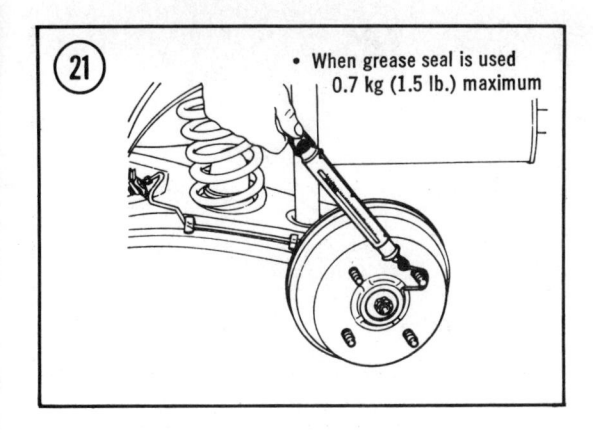

*When grease seal is used 0.7 kg (1.5 lb.) maximum*

2. Back off the wheel bearing nut ¼ turn. Install the adjusting cap. If the holes line up, install the cotter pin and spread its ends. If not, the nut may be tightened an additional 1/24 turn (15°).

3. Install the grease cap.

4. Check wheel bearing preload. Rotate the hub with a spring scale (**Figure 21**) and measure the force necessary to turn the hub. With a used grease seal, it should be 1.5 lb. (0.7 kg) or less. With a new grease seal, it should be 3.1 lb. (1.4 kg) or less.

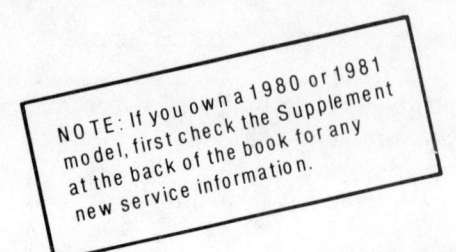

NOTE: If you own a 1980 or 1981 model, first check the Supplement at the back of the book for any new service information.

# CHAPTER ELEVEN

# FRONT SUSPENSION AND STEERING

All models use a MacPherson strut front suspension. The shock absorbers and springs are combined into a single unit. A knuckle bolted to the bottom of each strut carries the wheel hub, bearings, and brake disc. Rack-and-pinion steering is used. Power is transmitted to the wheel hubs by axle shafts with a constant velocity joint at each end. See **Figure 1**. Specifications and tightening torques are listed in **Tables 1 and 2** at the end of the chapter.

## WHEEL ALIGNMENT

Several suspension angles affect the running and steering of the front wheels. These angles must be properly aligned to prevent excessive tire wear, as well as to maintain directional stability and ease of steering. The angles are:

   a. Caster
   b. Camber
   c. Toe-in
   d. Steering axis inclination
   e. Steering lock angles

Caster, camber, and steering axis inclination are built in and cannot be adjusted. These angles are measured to check for bent suspension parts. Steering lock angles are adjusted automatically when toe-in is adjusted. These four angles cannot be checked without special alignment equipment. Toe-in, however, can be measured with a simple homemade tool.

### Pre-Alignment Check

Wheel alignment is affected by several factors. Perform the following steps before alignment is checked or adjusted.

1. Check tire pressure and wear. Pressures are listed on the *Quick Reference Page* at the front of the book. Wear patterns are illustrated under *Tire Wear Analysis*, Chapter Three.

2. Check front wheel bearings for looseness.

3. Check play in ball-joints.

4. Check for broken springs.

5. Remove any excessive load. This includes mud caked on the underside. Alignment should be checked with the gas tank, radiator, and crankcase full. The spare tire, jack, and floor mats should be in their correct positions.

6. Check shock absorbers.

7. Check the steering gear and the tie rods for looseness.

8. Check wheel balance.

9. Check *rear* suspension for looseness. Also check play in rear wheel bearings. See Chapter Ten.

# FRONT SUSPENSION AND STEERING

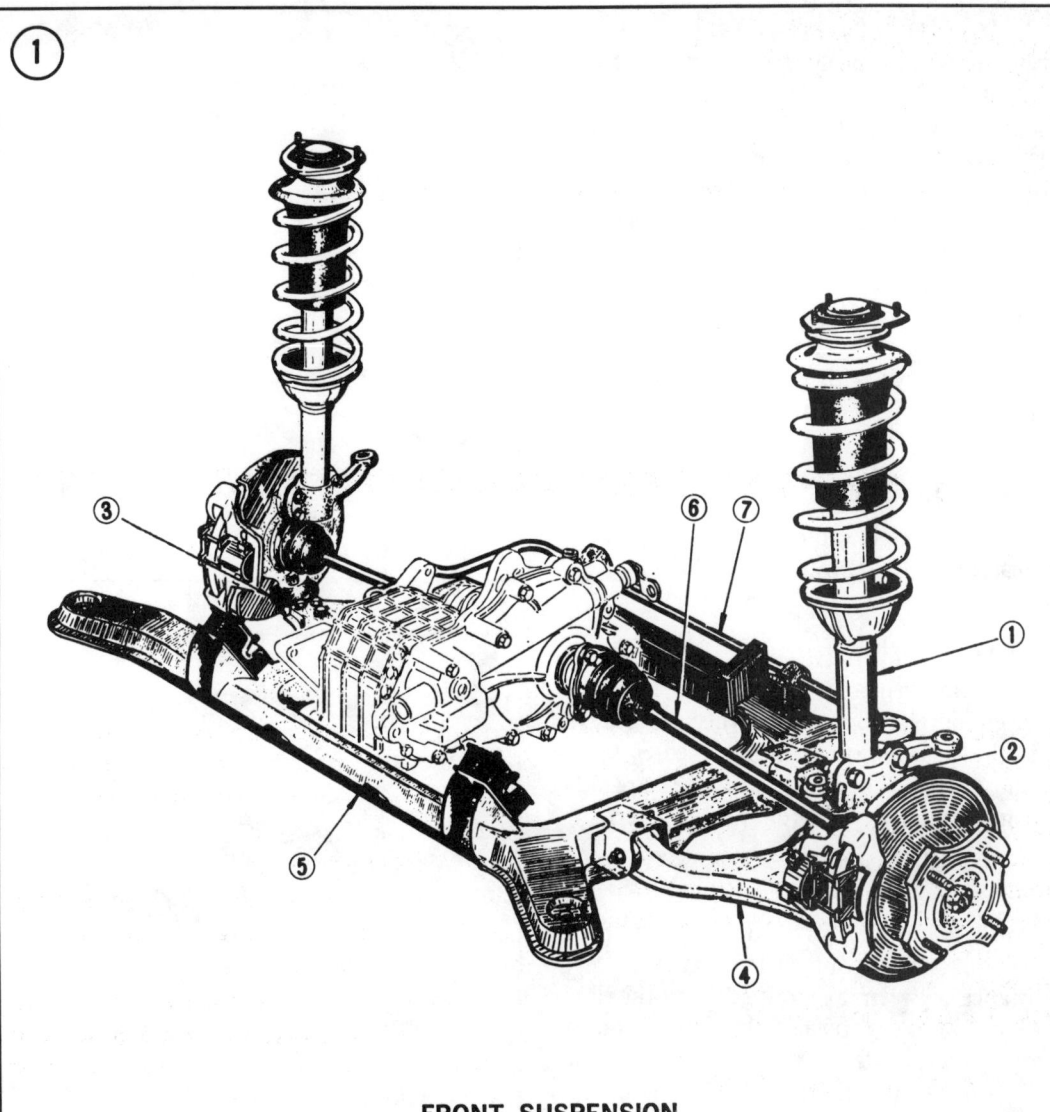

**FRONT SUSPENSION**

1. Strut assembly
2. Knuckle
3. Ball joint
4. Transverse link
5. Sub-frame
6. Drive shaft
7. Sway bar

## Caster and Camber

Caster is the inclination from vertical of the line through the ball-joints. Positive caster shifts the wheel forward; negative caster shifts the wheel rearward. Caster causes the wheels to return to the straight ahead position after a turn. It also prevents the wheels from wandering due to wind, potholes, or uneven road surfaces.

Camber is the inclination of the wheel from vertical. With positive camber, the top of the tire leans outward; with negative camber, the top of the tire leans inward.

## Steering Axis Inclination

Steering axis inclination is the inward or outward lean of the line through the ball-joints.

## Steering Lock Angles

When a car turns, the inside wheel makes a smaller circle than the outside wheels. Because of this, the inside wheel turns at a greater angle than the outside wheel. These angles are adjusted automatically when toe-in is adjusted.

## Toe-In

When a car moves forward, the front tires tend to point slightly outward. Because of this, the distance between the front edges of the tires (A, **Figure 2**) is slightly less than the distance between the rear edges (B) when the car is at rest. Toe-in is listed in **Table 1** at the end of the chapter.

To measure toe-in, use two pieces of telescoping aluminum tubing as follows:

1. Extend the tubes between the front edges of the tires. Scribe a mark on the inner tube where it enters the outer tube.

2. Extend the tubes between the rear edges of the tires. Again, scribe a mark on the inner tube where it enters the outer tube. The distance between the 2 scribe marks is toe-in.

3. If adjustment is necessary, loosen the tie rod locknuts (**Figure 3**). Turn the tie rods evenly to set toe-in, then tighten the locknuts.

> NOTE: *Tie rod lengths should be even after adjustment.*

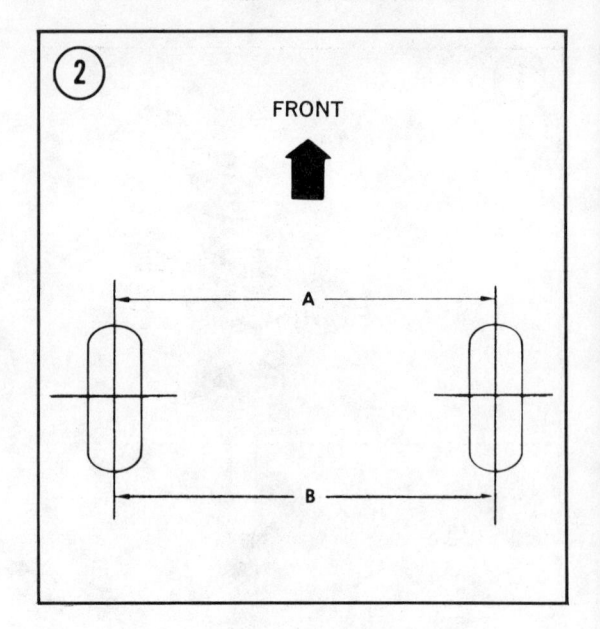

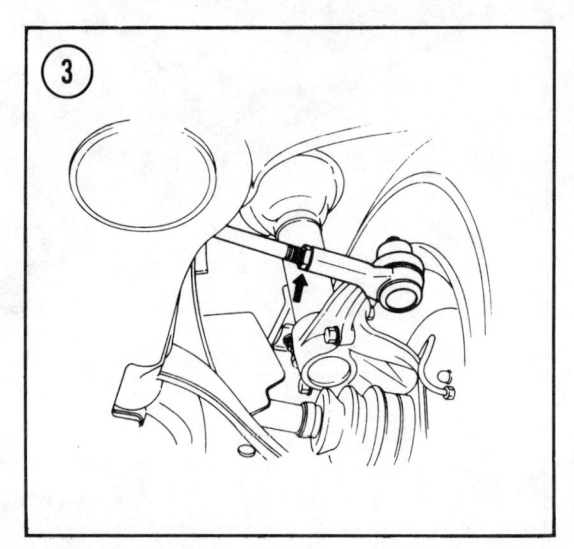

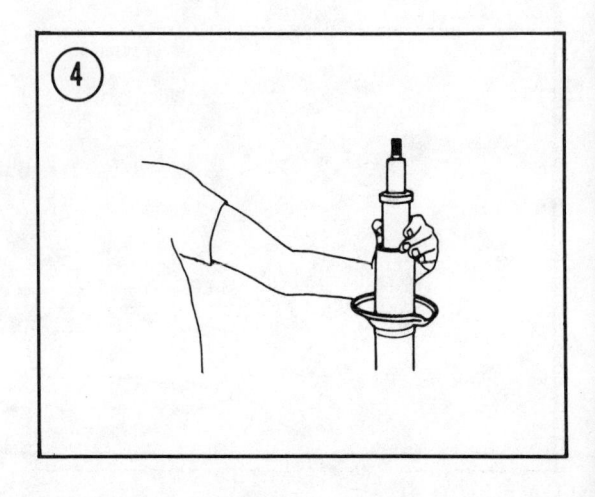

# FRONT SUSPENSION AND STEERING

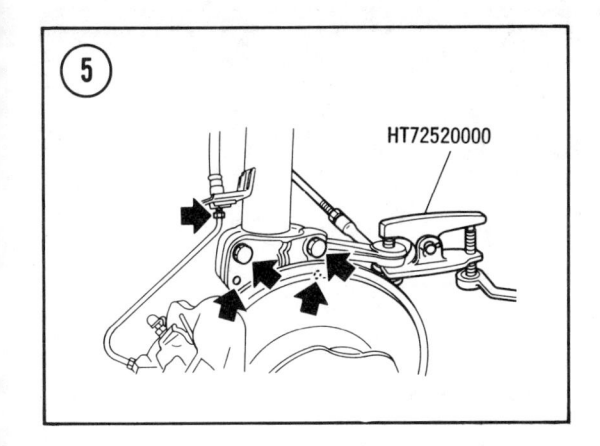

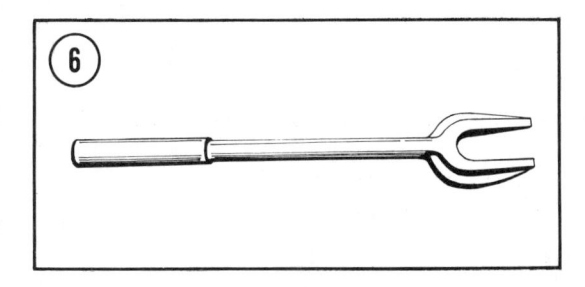

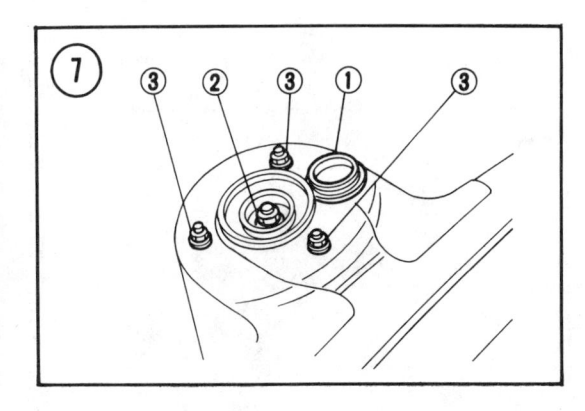

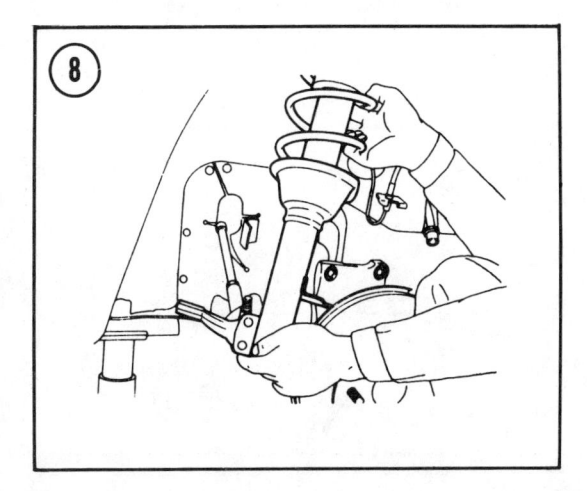

## SHOCK ABSORBER REPLACEMENT

The shock absorbers are located inside the suspension struts (**Figure 4**). Replacement requires several special tools, and should be left to a Datsun dealer. However, much of the cost can be saved by removing the strut assemblies yourself and taking them to a dealer for shock absorber replacement.

1. Securely block both rear wheels so the car will not roll in either direction. Loosen the front wheel nuts, jack up the front end, place it on jackstands, and remove the front wheels.

2. Remove the tie rod stud nut. Detach the tie rod from the knuckle arm with a puller such as Datsun tool HT72520000 (**Figure 5**) or a fork-type separator (**Figure 6**).

3. Disconnect the brake line (**Figure 5**). Remove the 4 bolts securing the strut to the knuckle.

4. Remove the strut cap (1, **Figure 7**). Remove the strut mounting nuts (3, **Figure 7**).

### WARNING
*Do not remove the center nut at the top of the strut (2, Figure 7). This could allow the coil spring to fly out and cause serious injury.*

5. Take the strut out (**Figure 8**).

6. Take the strut to a Datsun dealer for shock absorber replacement.

7. Install by reversing Steps 1-5.

## SPRING REPLACEMENT

The spring is part of the strut assembly. To replace the spring, follow the shock absorber procedure. Since the procedure requires special tools, it is best to remove the strut assembly yourself and let a dealer install the new spring. Then install the strut using the same procedure.

## SWAY BAR

### Removal/Installation (F10)

1. Securely block both rear wheels so the car will not roll in either direction.

2. Loosen the front wheel nuts, jack up the front end of the car, place it on jackstands, and remove the front wheels.

3. Remove the nuts connecting the sway bar to the transverse link. See **Figure 9**. Unbolt the sway bar brackets from the subframe and take the sway bar out.

4. Installation is the reverse of removal.

**Removal/Installation (310)**

1. Securely block both rear wheels so the car will not roll in either direction. Jack up the front end and place it on jackstands.

2. Support the subframe with a jack and heavy board such as a 4x4. See **Figure 10**.

3. Disconnect the front exhaust pipe from the manifold and body. The rear exhaust pipe need not be disconnected.

4. Disconnect the shift linkage from the transmission. See **Figure 11**.

5. Detach the sway bar from the transverse links (**Figure 9**).

6. Loosen, but do *not* remove, the subframe bolts. Lower the subframe just enough to provide access to the sway bar brackets (**Figure 12**).

7. Unbolt the sway bar brackets. Take the sway bar out.

8. Installation is the reverse of removal. Tighten all nuts and bolts to specifications (**Table 2**).

## TRANSVERSE LINKS

**Removal**

Refer to **Figure 13** for this procedure.

1. Securely block both rear wheels so the car will not roll in either direction.

2. Loosen the front wheel nuts, jack up the front end, place it on jackstands, and remove the front wheels.

3. Unbolt the ball-joint and stabilizer bar from the transverse link. See **Figure 14**.

4. Unbolt the transverse link from the subframe (**Figure 15**). Take the transverse link out.

**Inspection**

1. Check the transverse link for cracks or bending. Replace it if these are found.

2. Check the transverse link bushings for wear, cracks, or deterioration. If these are found, have the bushings pressed out, and new ones pressed in, by a machine shop.

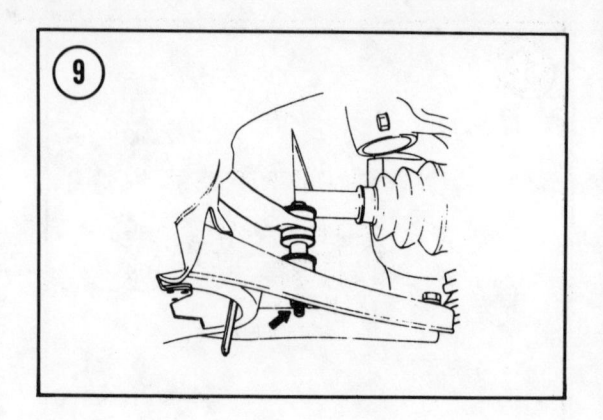

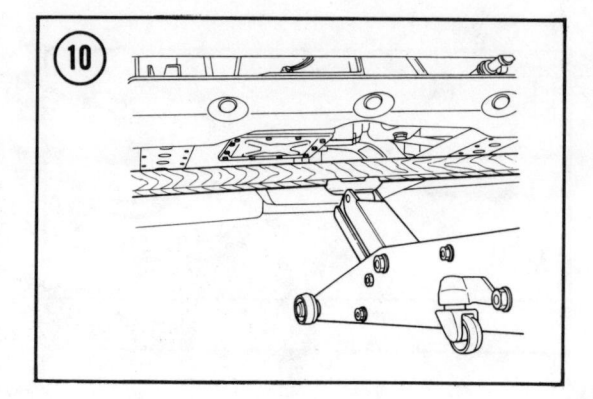

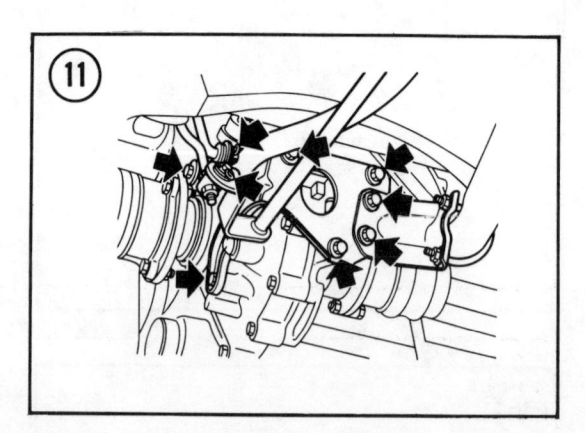

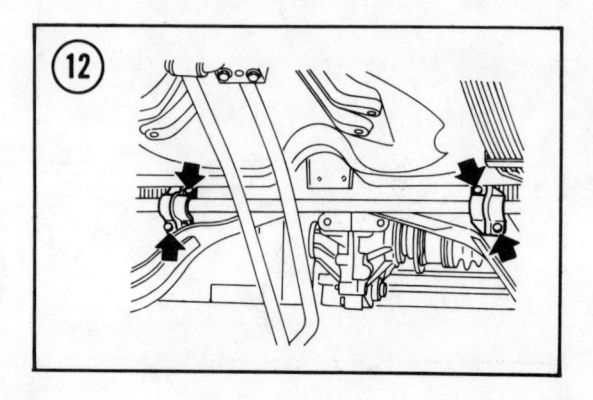

# FRONT SUSPENSION AND STEERING

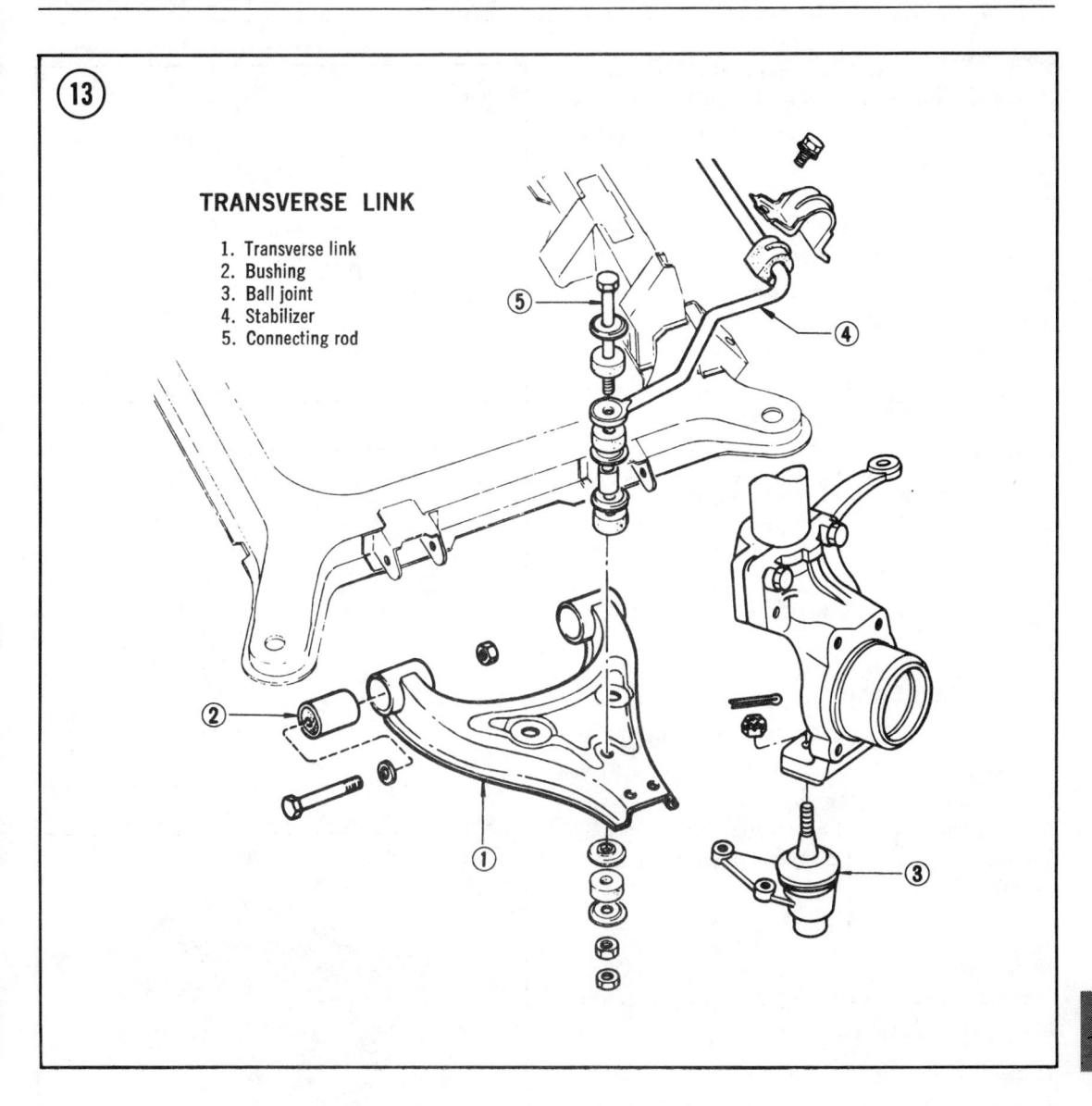

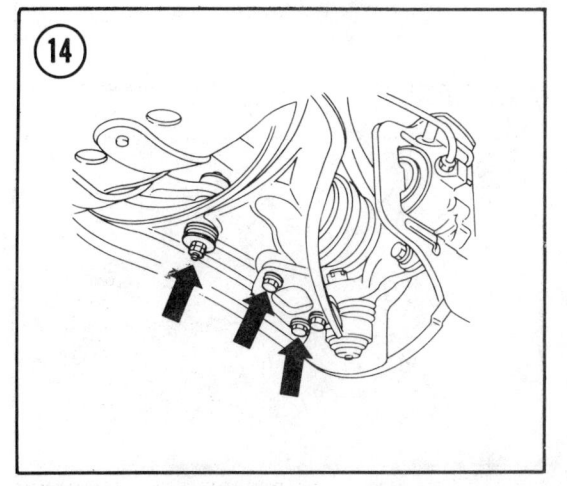

**TRANSVERSE LINK**

1. Transverse link
2. Bushing
3. Ball joint
4. Stabilizer
5. Connecting rod

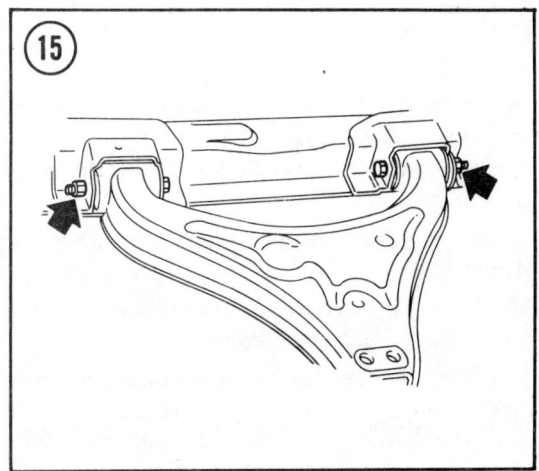

## Installation

Installation is the reverse of removal, plus the following.

1. Use new self-locking nuts on the transverse link pivot bolts.

2. Install the bolts with their heads toward the center of the transverse link.

3. Tighten the pivot bolts slightly, then lower the car. Tighten to final specifications with the car's weight resting on the wheels.

## BALL-JOINTS

### Removal/Installation

1. Securely block both rear wheels so the car will not roll in either direction.

2. Loosen the front wheel nuts, jack up the front end of the car, place it on jackstands, and remove the front wheels.

3. On 310's, remove the front axle shaft as described later in this chapter.

4. Remove the ball-joint stud nut. Separate the ball-joint from the knuckle arm with a fork-type separator (**Figure 16**). These are available from rental dealers.

5. Unbolt the ball-joint from the transverse link (**Figure 14**). Lift the ball-joint out.

6. Installation is the reverse of removal. Tighten all nuts and bolts to specifications (**Table 2**).

### Inspection

1. Check the rubber boot for cracks. Replace the ball-joint if these are found.

2. Thread the nut onto the ball-joint. Turn the stud with an inch-pound torque wrench. It should require at least 8.7 in.-lb. (10 kg-cm) to turn the stud. If it takes less, replace the ball-joint.

3. Set up a dial indicator with its pointer contacting the ball-joint stud. Pull the stud up and down and measure play. If more than 0.059 in. (1.5mm), replace the ball-joint.

## WHEEL BEARINGS

Refer to **Figure 17** for the following procedures.

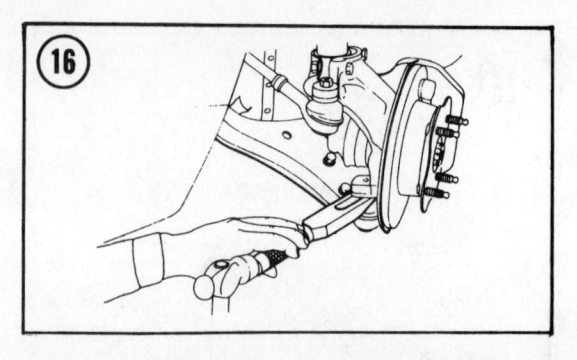

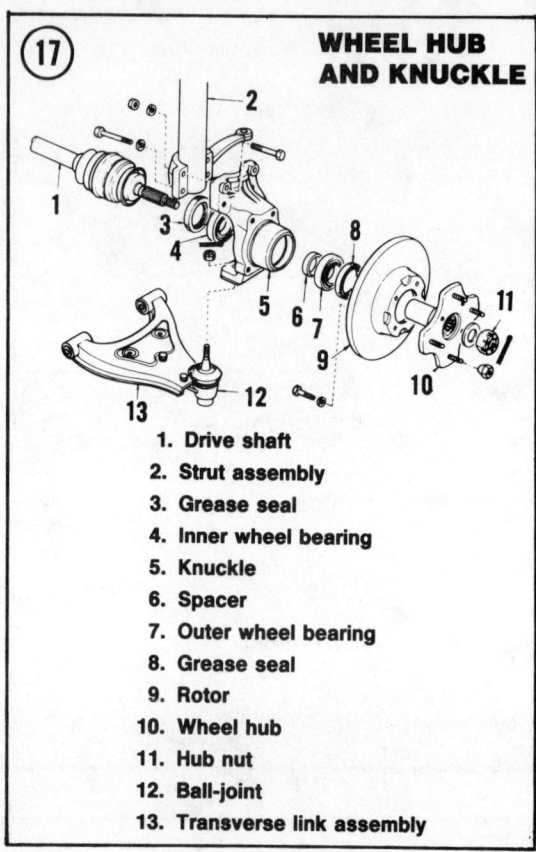

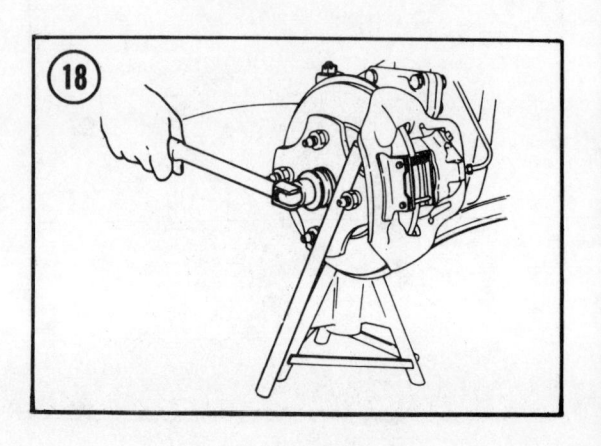

**WHEEL HUB AND KNUCKLE**

1. Drive shaft
2. Strut assembly
3. Grease seal
4. Inner wheel bearing
5. Knuckle
6. Spacer
7. Outer wheel bearing
8. Grease seal
9. Rotor
10. Wheel hub
11. Hub nut
12. Ball-joint
13. Transverse link assembly

# FRONT SUSPENSION AND STEERING

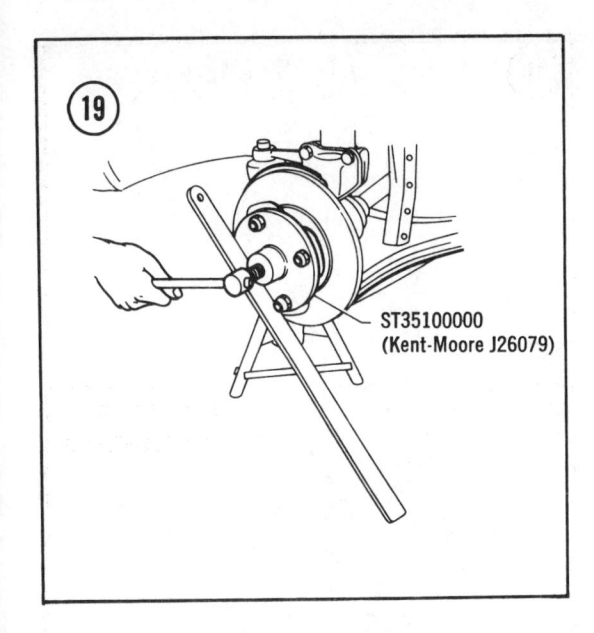

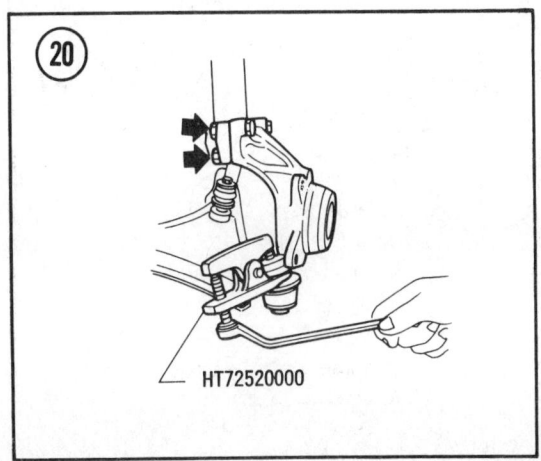

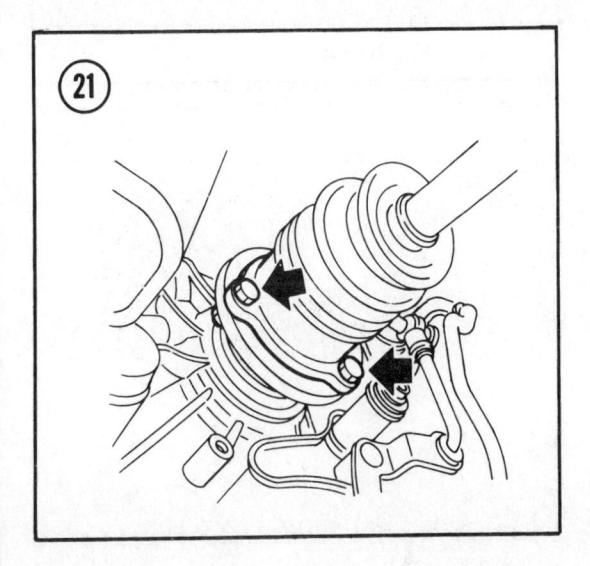

**Removal**

1. Securely block both rear wheels so the car will not roll in either direction.

2. Loosen the front wheel nuts, jack up the front end, place it on jackstands, and remove the front wheels.

3. Remove the brake caliper (Chapter Ten).

4. Remove the cotter pin from the hub nut.

5. Position a steel rod as shown in **Figure 18** to keep the hub from turning. Remove the hub nut.

CAUTION
*Thread the wheel nuts onto the studs so the threads won't be damaged.*

6. Pull the hub off with a tool like the one shown in **Figure 19**. If you don't have the tool, use a gear puller. These are available from tool rental dealers.

7. Disconnect the tie rod ball-joint and suspension ball-joint from the knuckle arm. Use a puller like the one shown in **Figure 20**, or a fork-type separator (**Figure 16**). These are available from rental dealers.

8. Remove 4 bolts securing the inner end of the axle shaft (**Figure 21**).

9. **Unbolt the knuckle from strut (Figure 17).**

10. Take out the knuckle and axle shaft.

**Wheel Bearing Inspection**

1. Throughly clean the wheel bearings, hub, and knuckle with solvent. Dry all parts with compressed air.

2. Check wheel bearings for wear, pitting, chips, or rust. If any of these conditions can be seen, the wheel bearings should be replaced. This requires a hydraulic press and adapters. Several special measuring tools are also required to determine correct spacer thickness. This job should be left to a dealer or machine shop.

3. Pack the wheel bearings with multipurpose grease.

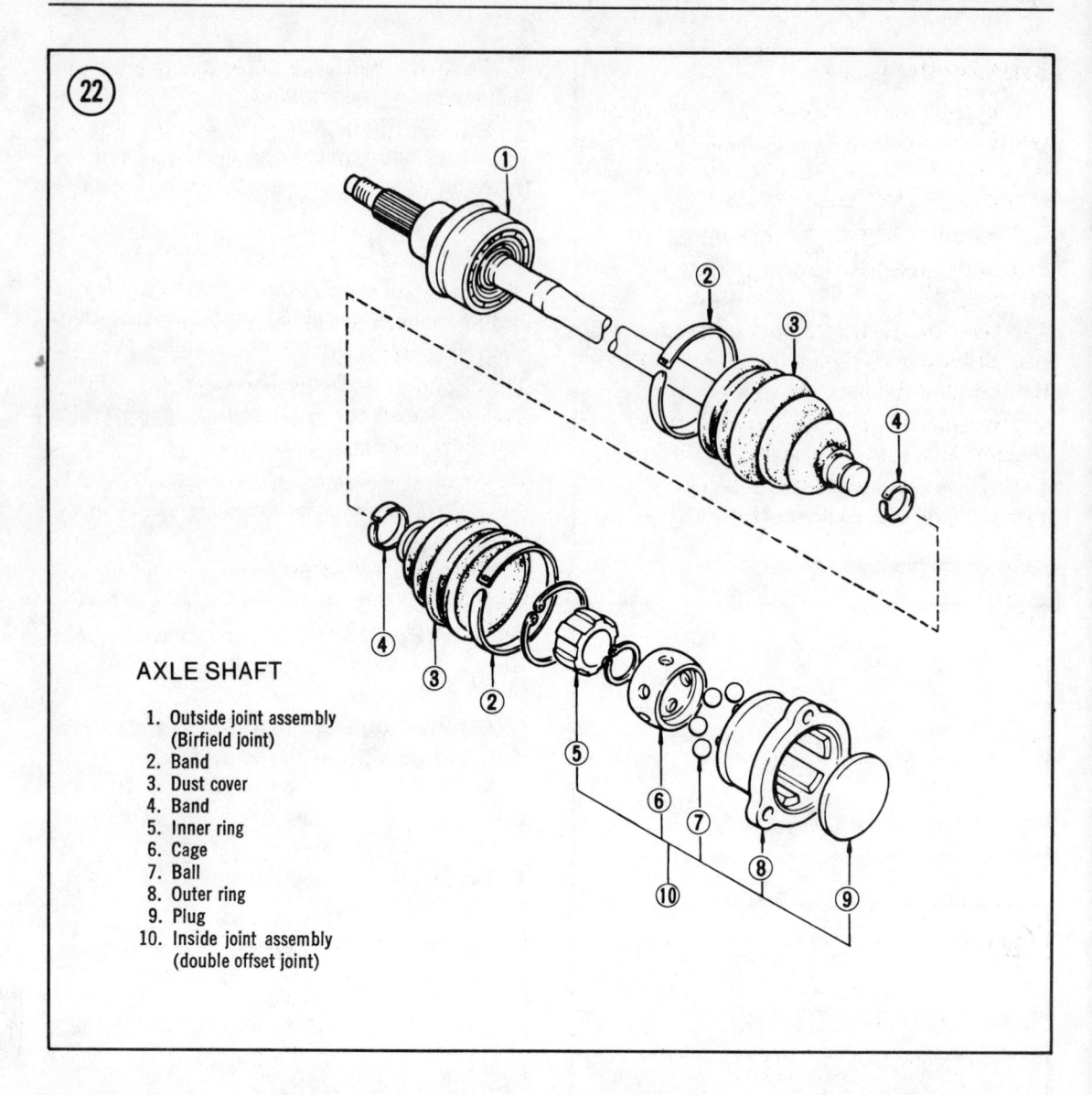

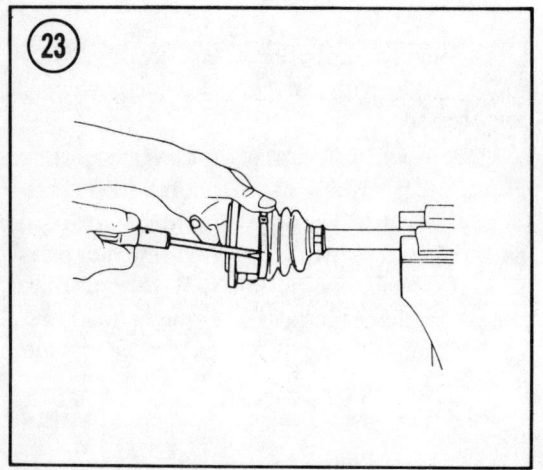

## AXLE SHAFT

1. Outside joint assembly (Birfield joint)
2. Band
3. Dust cover
4. Band
5. Inner ring
6. Cage
7. Ball
8. Outer ring
9. Plug
10. Inside joint assembly (double offset joint)

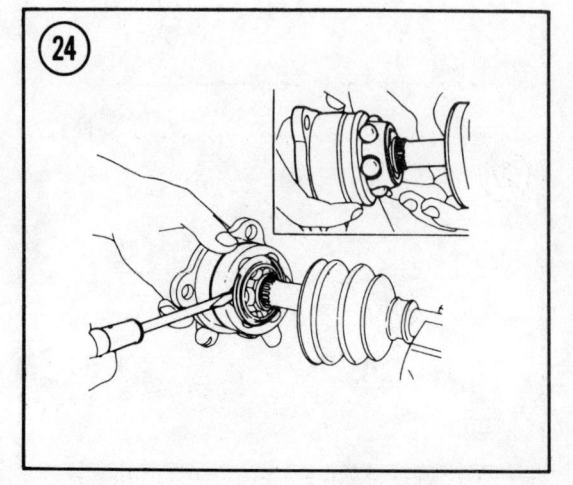

# FRONT SUSPENSION AND STEERING

## Axle Shaft Overhaul

The axle shafts (**Figure 22**) use constant velocity joints at each end. The inner joints are repairable double offset types. The outer joints are non-repairable Birfield types.

1. Place the axle shaft in a soft-jawed vise.
2. Remove the boot bands (**Figure 23**). Take off the dust boots.
3. Check the Birfield (outer) joint for wear, rust, or burn marks (blue-tinted areas). Replace the axle shaft if these can be seen.
4. Pry the clip off the double offset joint (**Figure 24**). Remove the outer ring as shown.
5. Wipe the grease off the ball cage. Push the balls out with a screwdriver (**Figure 25**).
6. Twist the ball cage approximately ½ turn and take it off the axle shaft.
7. Remove the snap ring (**Figure 26**). Tap the inner ring off the axle shaft with a mallet.
8. Thoroughly clean all parts in solvent.
9. Check the dust boots for cracks or deterioration. Replace if these are found.
10. Check all remaining parts for wear, chips, pitting, or burn marks. Replace parts with any of these defects.
11. Assemble by reversing Steps 1-7. During assembly, pack the joints with molybdenum disulphide-based grease.
12. Wrap each band twice around its dust boot. Tighten with a screwdriver and pliers (**Figure 27**) and bend the end 90°. Secure the end with a punch (**Figure 28**). Cut off the end leaving a length equal to the band's width. Bend the end over the punched part.

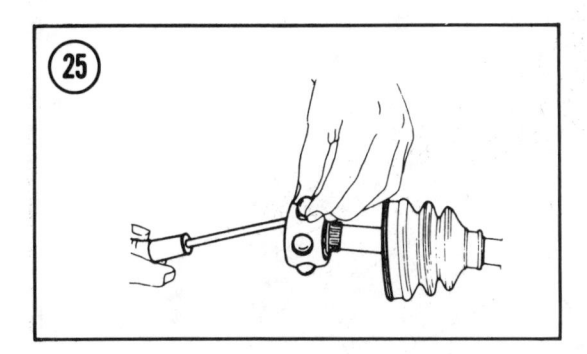

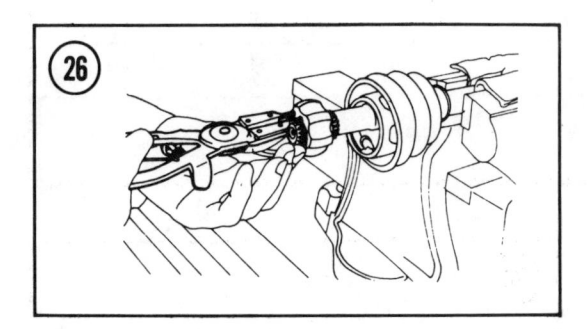

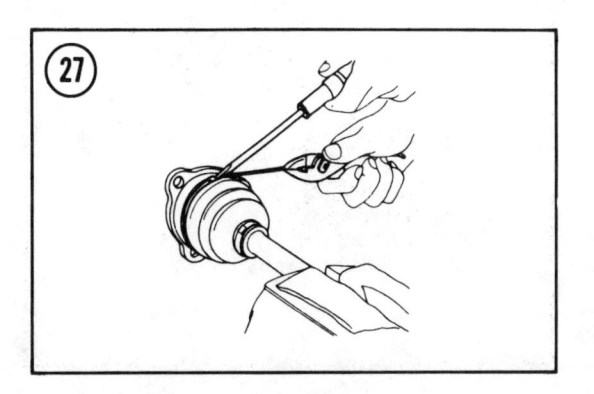

## Installation

1. Tap the hub and brake disc onto the axle shaft with a wooden mallet (**Figure 29**).

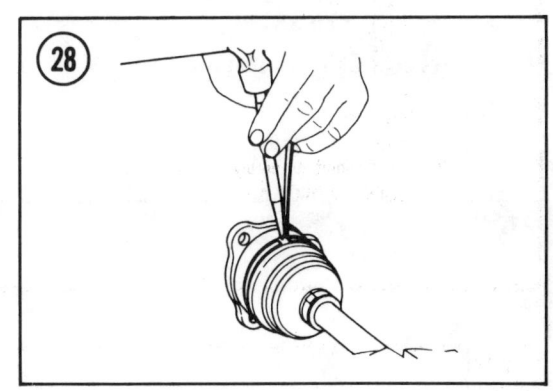

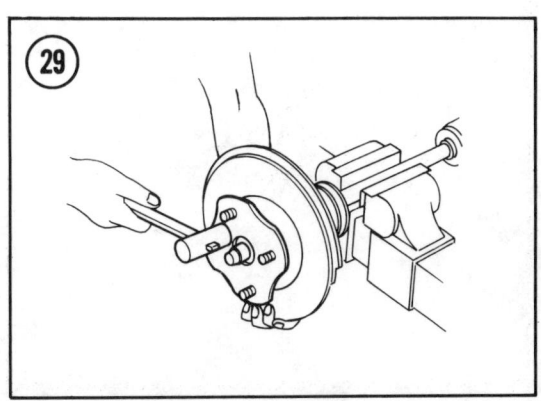

2. Hold the hub from turning with a steel bar (**Figure 30**). Tighten hub nut to 87-145 ft.-lb. (12-20 mkg).

3. Spin the hub several turns in both directions to make sure it turns freely.

4. Check bearing preload. Pull on a wheel stud with a spring scale as shown in **Figure 31**. It should take 3.1-10.8 lb. (1.4-4.9 kg) to turn the hub. If it takes less, the bearing spacer is too thin. Replace it with a thicker one. If it takes more, the spacer is too thick. Use a thinner one.

5. Install the axle shaft and hub assembly on the car. Tighten all nuts and bolts to specifications (**Table 2**).

   NOTE: *Use new self locking nuts on the axle shafts.*

6. Install the brake caliper. See *Caliper Removal/Installation*, Chapter Ten.

7. Install the wheels, lower the car, and tighten the wheel nuts.

8. Bleed the brakes. See *Brake Bleeding*, Chapter Ten.

## STEERING

### Steering Wheel Removal/Installation

1. Disconnect negative cable from battery.
2. Remove the horn pad. On bar-type pads,

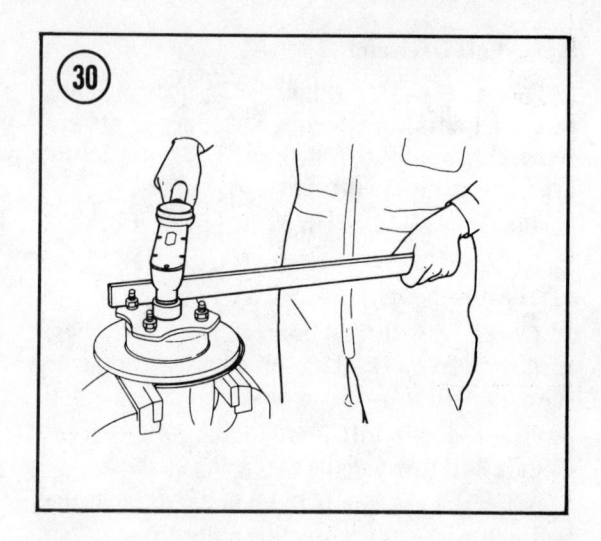

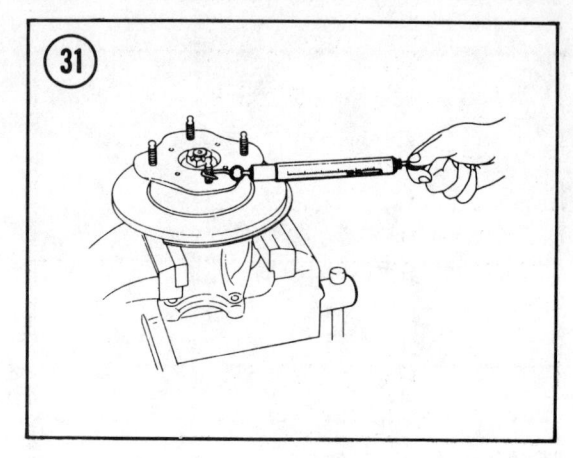

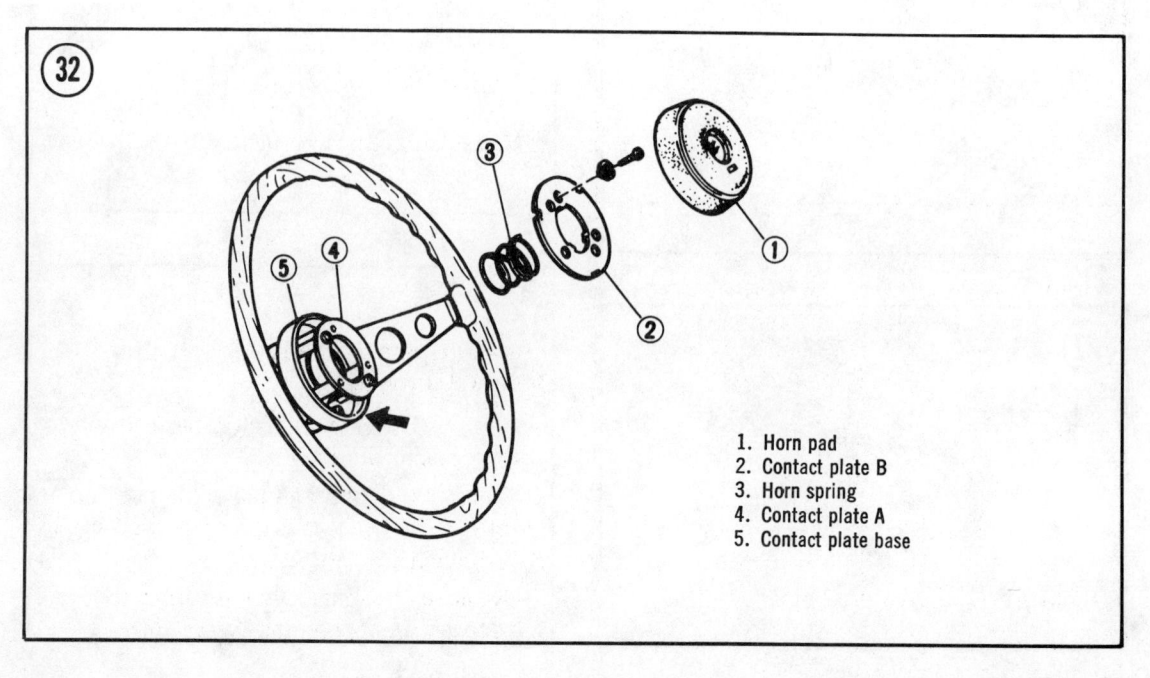

1. Horn pad
2. Contact plate B
3. Horn spring
4. Contact plate A
5. Contact plate base

# FRONT SUSPENSION AND STEERING

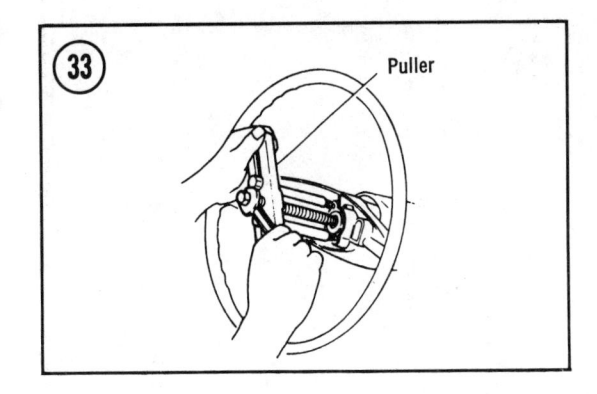

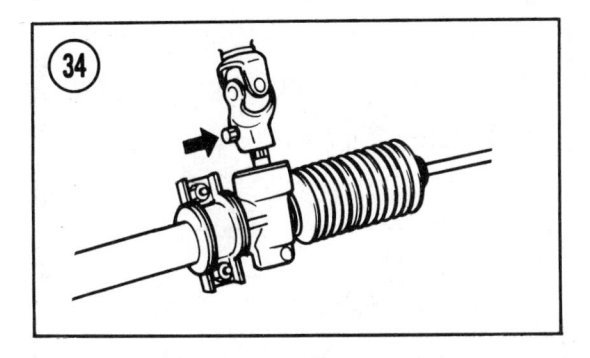

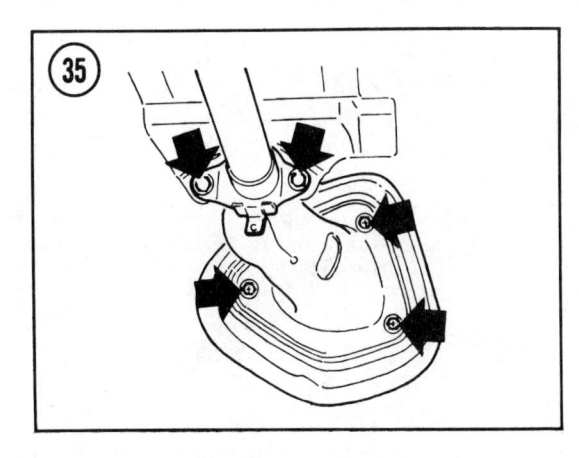

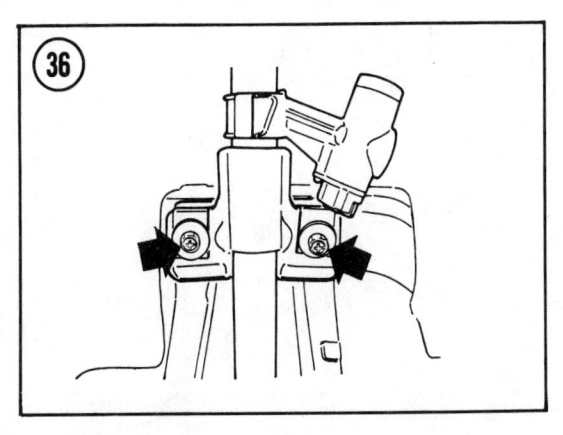

remove 2 screws from the back of the steering wheel. On button types, pull the pad off.

3. On F10's with a center horn button, remove the horn contact plates and base. See **Figure 32**.

4. Make sure the wheels are straight ahead. Make alignment marks on steering wheel and column.

5. Remove the steering wheel with a screw-type puller (**Figure 33**). These are available from tool rental dealers. Inexpensive pullers are also available at auto parts stores.

> CAUTION
> *Do not hit the steering wheel with a hammer. Do not use an impact puller.*

6. Installation is the reverse of removal. Tighten the steering wheel nut to 22-25 ft.-lb. (3.0-3.5 mkg).

## Steering Column Removal/Installation

1. Remove the steering wheel as described in the preceding section.

2. Remove the clamp bolt securing the lower joint to the steering pinion (**Figure 34**).

> NOTE: *The bolt fits through a groove in the pinion, so it must be removed, not just loosened.*

3. Remove the combination switch assembly. See *Switches*, Chapter Seven.

4. Remove the floor cover and lower bracket bolts (**Figure 35**).

5. Remove the upper bracket bolts (**Figure 36**). Remove the steering column into the passenger compartment.

6. Installation is the reverse of removal. Tighten all nuts and bolts to specifications (**Table 2**).

## Steering Gear and Linkage Removal/Installation

1. Securely block both rear wheels so the car will not roll in either direction.

2. Loosen the front wheel nuts, jack up the front end, place it on jackstands, and remove the front wheels.

3. Remove the clamp bolt securing the lower joint to the steering pinion (**Figure 34**).

NOTE: *The bolt fits through a groove on the pinion, so it must be removed, not just loosened.*

4. Detach the tie rod ends from the knuckle arms. If available, use a puller such as Datsun tool HT72520000 (**Figure 37**). If not, place a sledge hammer or heavy steel body tool against one side of the knuckle arm boss (the part which the tie rod stud goes through). Tap on the other side with a copper hammer. This will loosen the tie rod stud.

5. Remove the steering rack clamp bolts (**Figure 38**). Remove the steering rack to one side.

6. Installation is the reverse of removal. Tighten all nuts and bolts to specifications (**Table 2**). Check wheel alignment as described at the beginning of this chapter.

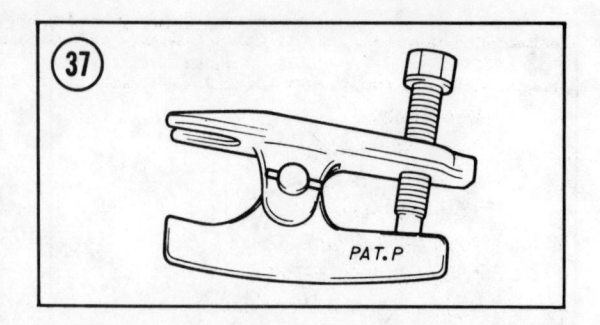

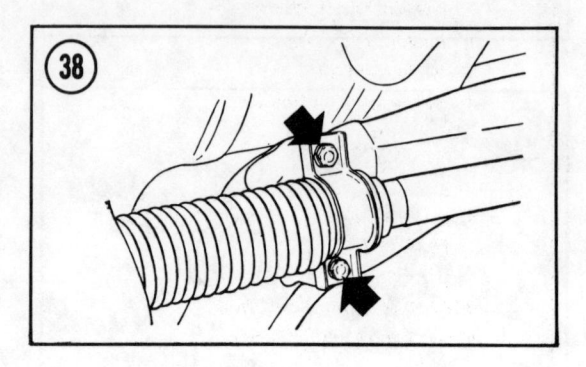

Table 1    SPECIFICATIONS

| | |
|---|---|
| **Caster** | |
| F10 | 20° to 1°50' |
| 310 | 25' to 1°55' |
| **Camber** | |
| F10 | 50' to 2°20' |
| **Toe-in** | |
| F10 (radial tires) | 0.0-1—0.08 in. (0-2mm) |
| F10 (bias ply tires, 1976-1977) | 0.20-0.28 in. (5-7mm) |
| F10 (bias ply tires, 1978) | 0.12-0.20 in. (3-5mm) |
| 310 | 0.0-1—0.08 in. (0-2mm) |
| **Steering axis inclination** | |
| F10 | 9°15' to 10°45' |
| 310 | 11°10' to 12°30' |
| **Steering lock angles** | |
| F10 (inside wheel) | 36°30' to 39°30' |
| F10 (outside wheel) | 31-34° |
| 310 (inside wheel) | 36°30' to 39°30' |
| 310 (outside wheel) | 29°30' to 32°30' |

# FRONT SUSPENSION AND STEERING

**Table 2   TIGHTENING TORQUES**

|  | Ft.-lb. | Mkg |
|---|---|---|
| Wheel hub nut | 87-145 | 12-20 |
| Axle shaft attaching bolts | 29-36 | 4-5 |
| Strut upper nuts | 11-17 | 1.5-2.4 |
| Strut lower bolts | 24-33 | 3.3-4.5 |
| Ball-joint stud nuts | 22-29 | 3-4 |
| Ball-joint attaching bolts | 40-47 | 5.5-6.5 |
| Brake caliper mounting bolts | 40-47 | 5.5-6.5 |
| Splash guard attaching bolts | 18-25 | 2.5-3.4 |
| Transverse link nut | 42-51 | 5.8-7.0 |
| Sway bar end nuts | 6-9 | 0.8-1.2 |
| Sway bar bracket bolts | 6-9 | 0.8-1.2 |
| Wheel nuts |  |  |
|   F10 | 58-65 | 8-9 |
|   310 | 58-72 | 8-10 |
| Steering wheel nut |  |  |
|   1976-1977 | 14-18 | 2.0-2.5 |
|   1978-on | 22-25 | 3.0-3.5 |
| Steering column clamp bolt | 6½-10 | 0.9-1.4 |
| Upper joint to column bolt |  |  |
|   F10 | 14-17 | 1.9-2.4 |
|   310 | 17-22 | 2.4-3.0 |
| Lower joint to pinion bolt |  |  |
|   F10 | 14-17 | 1.9-2.4 |
|   310 | 17-22 | 2.4-3.0 |
| Tie rod ball-joint stud nuts | 40-47 | 5.5-6.5 |
| Tie rod outer locknuts | 27-34 | 3.8-4.7 |
| Tie rod inner locknuts | 58-72 | 8-10 |
| Steering gear clamp bolts | 29-34 | 4-6 |

**11**

# CHAPTER TWELVE

# BODY

This chapter provides service procedures for the bumpers, front fenders, grille, hood, tailgate or trunk lid, doors, and door and rear quarter windows. Other body repairs require special skills and should be left to a dealer or body shop.

## BUMPERS

All models use shock-absorbing bumpers. To replace or disassemble, refer to the following illustrations:

**Figure 1** — F10 front bumper
**Figure 2** — F10 sedan and hatchback rear bumper
**Figure 3** — F10 wagon rear bumper
**Figure 4** — 310 front bumper
**Figure 5** — 310 rear bumper

## GRILLE
### Removal/Installation (F10)
1. Open the hood.
2. Remove the attaching screws (**Figure 6**) and take the grille out.
3. Installation is the reverse of removal.

### Removal/Installation (310)
1. Open the hood.
2. Remove grille attaching screws (**Figure 7**).
3. Installation is the reverse of removal. Be sure the plugs at the lower corners fit through the rubber spacers.

## FRONT FENDERS
### Removal/Installation (F10)
1. Disconnect negative cable from battery.
2. Remove fender mounting bolts (**Figure 8**). Lift the fender partway off, disconnect the side marker light wires, then remove the fender.

*CAUTION*
*Pull the fender away gradually so the sealing strip isn't damaged.*

3. Installation is the reverse of removal.

### Removal/Installation (310)
1. Disconnect negative cable from battery.
2. Disconnect the side marker light wires.
3. Remove fender mounting bolts (**Figure 9**). Lift the fender off.
4. Installation is the reverse of removal.

# BODY

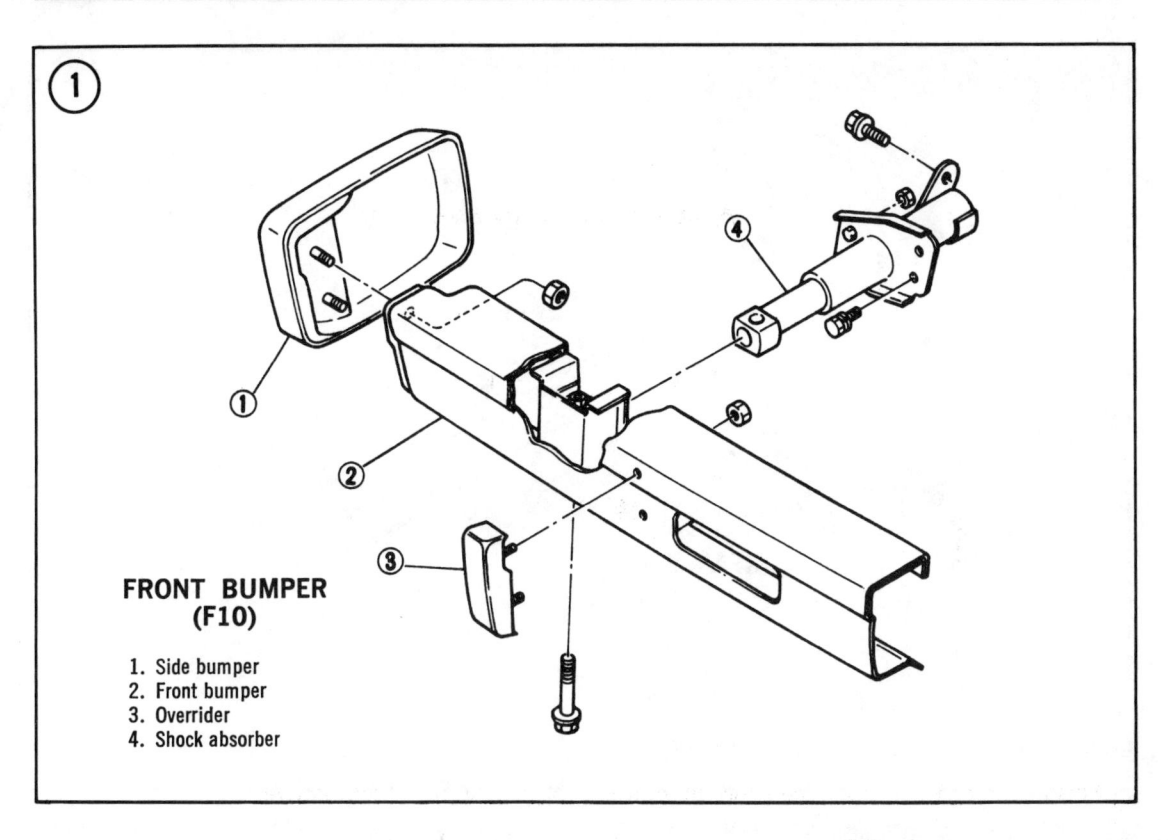

## FRONT BUMPER (F10)

1. Side bumper
2. Front bumper
3. Overrider
4. Shock absorber

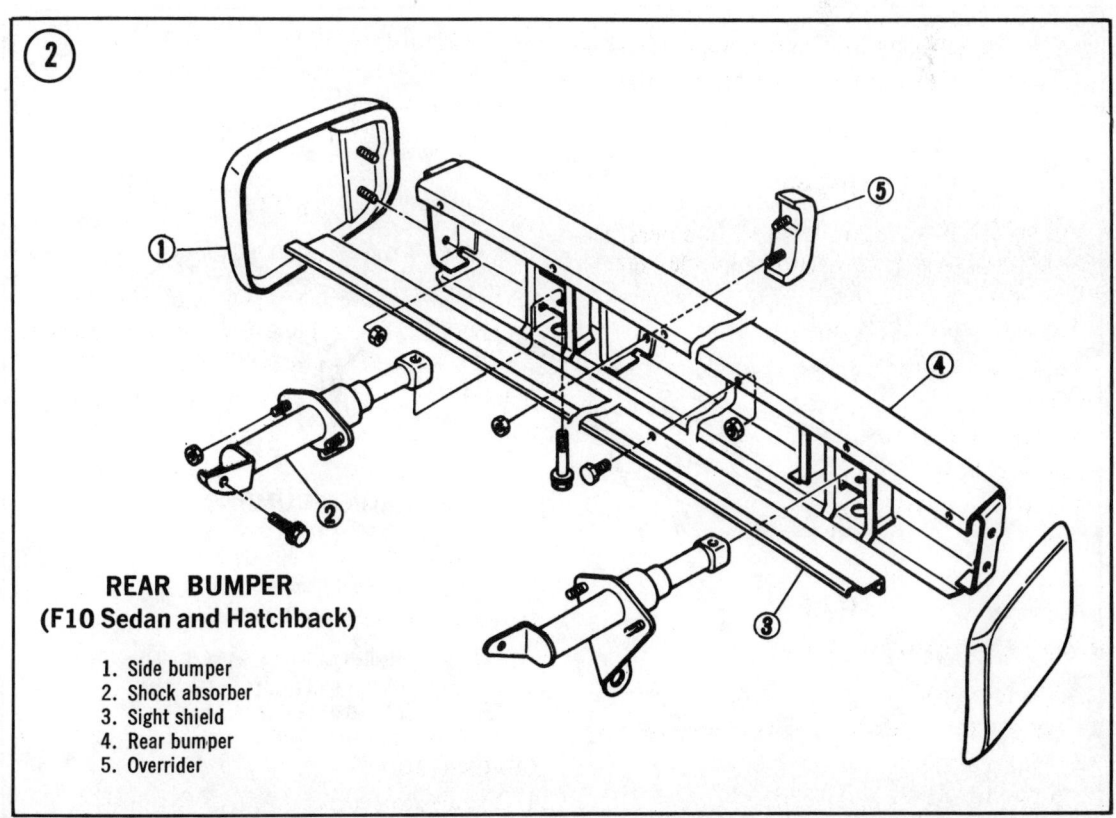

## REAR BUMPER (F10 Sedan and Hatchback)

1. Side bumper
2. Shock absorber
3. Sight shield
4. Rear bumper
5. Overrider

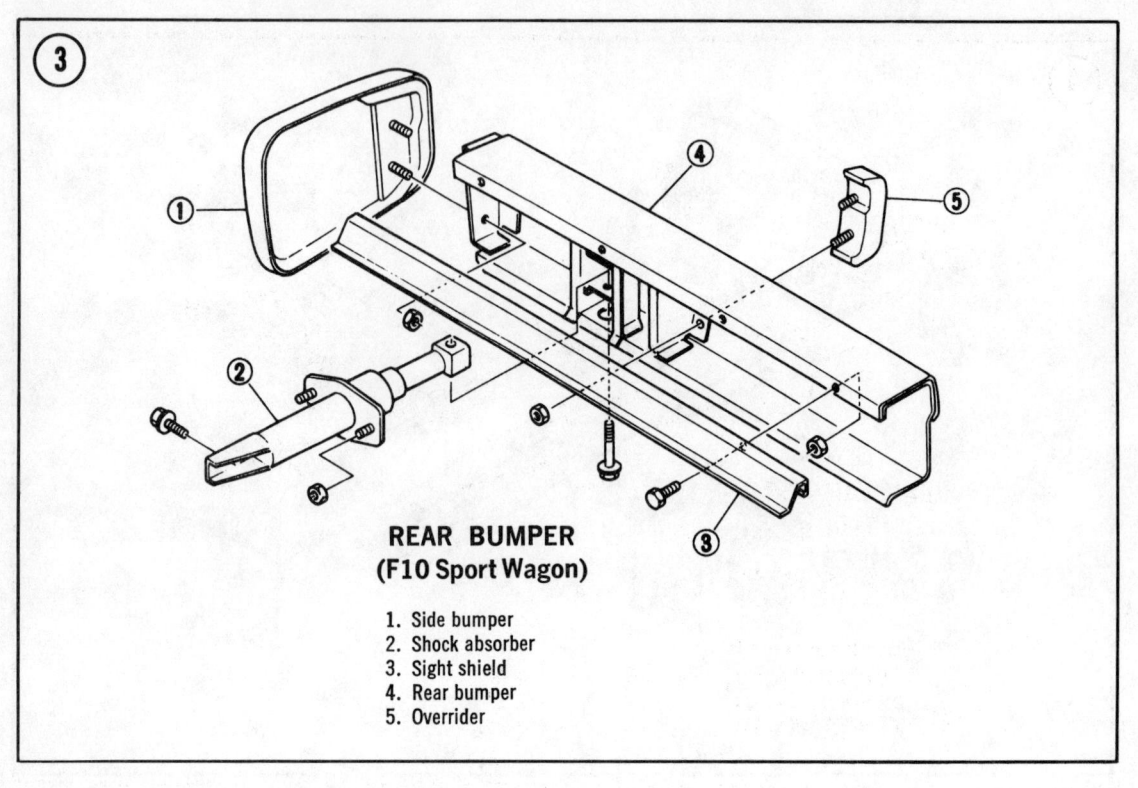

## REAR BUMPER
### (F10 Sport Wagon)

1. Side bumper
2. Shock absorber
3. Sight shield
4. Rear bumper
5. Overrider

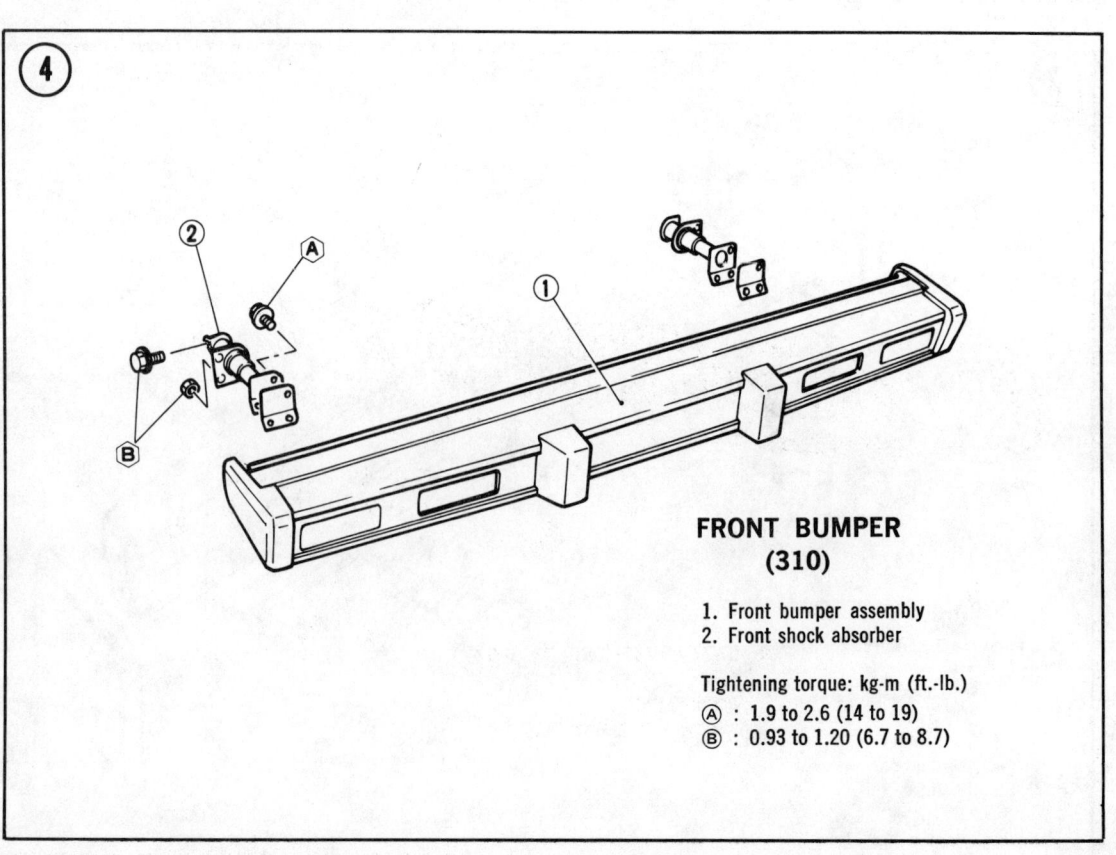

## FRONT BUMPER
### (310)

1. Front bumper assembly
2. Front shock absorber

Tightening torque: kg-m (ft.-lb.)
Ⓐ : 1.9 to 2.6 (14 to 19)
Ⓑ : 0.93 to 1.20 (6.7 to 8.7)

# BODY

### 5

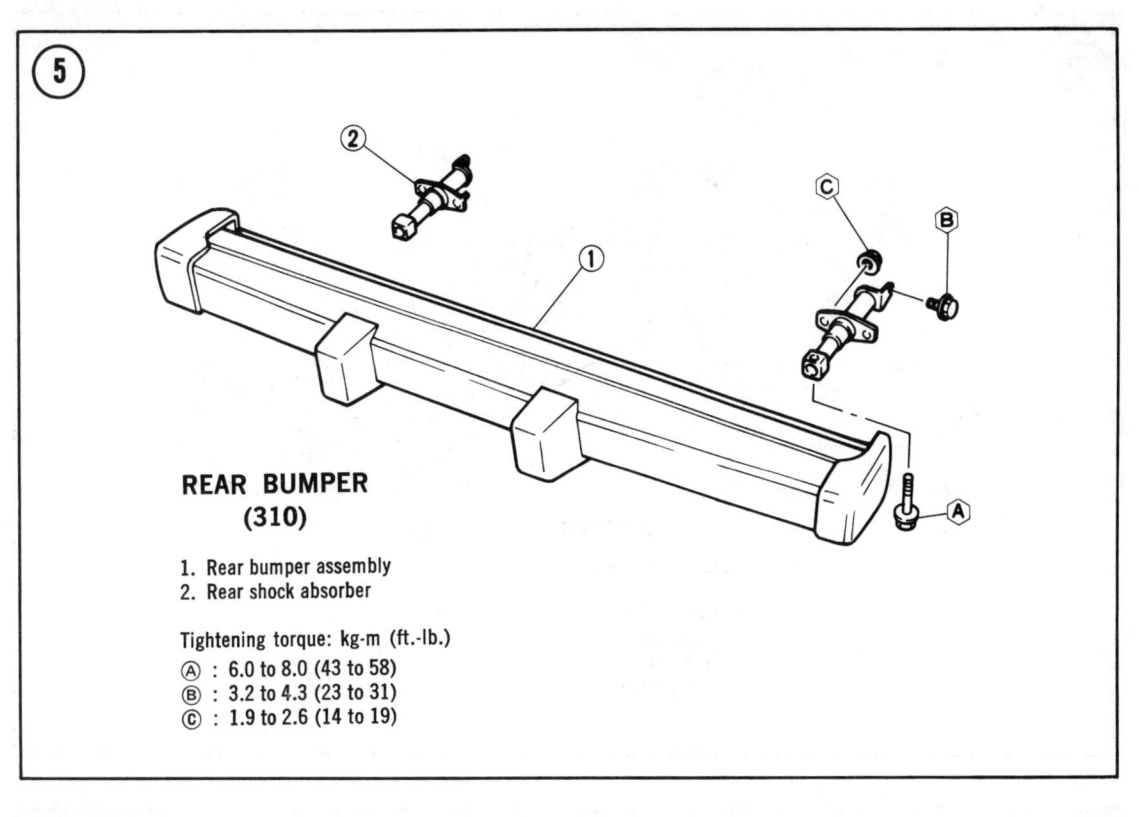

**REAR BUMPER (310)**

1. Rear bumper assembly
2. Rear shock absorber

Tightening torque: kg-m (ft.-lb.)
- Ⓐ : 6.0 to 8.0 (43 to 58)
- Ⓑ : 3.2 to 4.3 (23 to 31)
- Ⓒ : 1.9 to 2.6 (14 to 19)

### 6

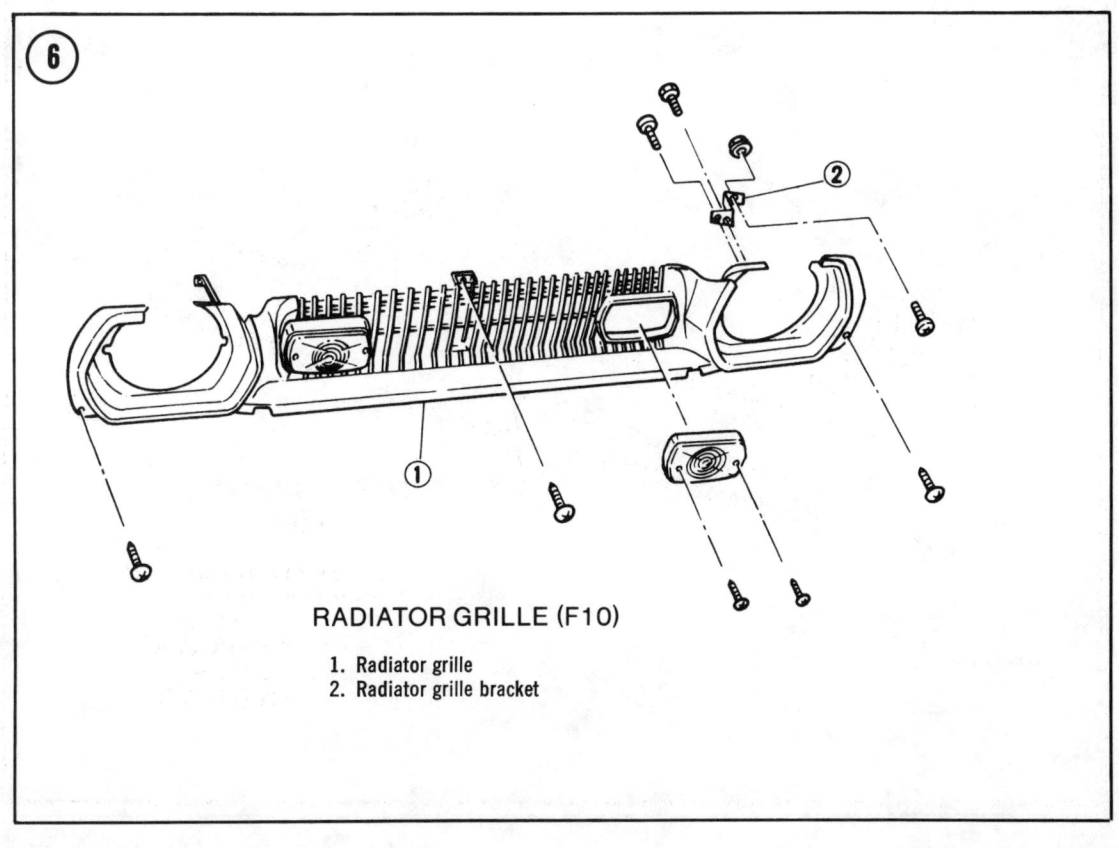

**RADIATOR GRILLE (F10)**

1. Radiator grille
2. Radiator grille bracket

CHAPTER TWELVE

⑦

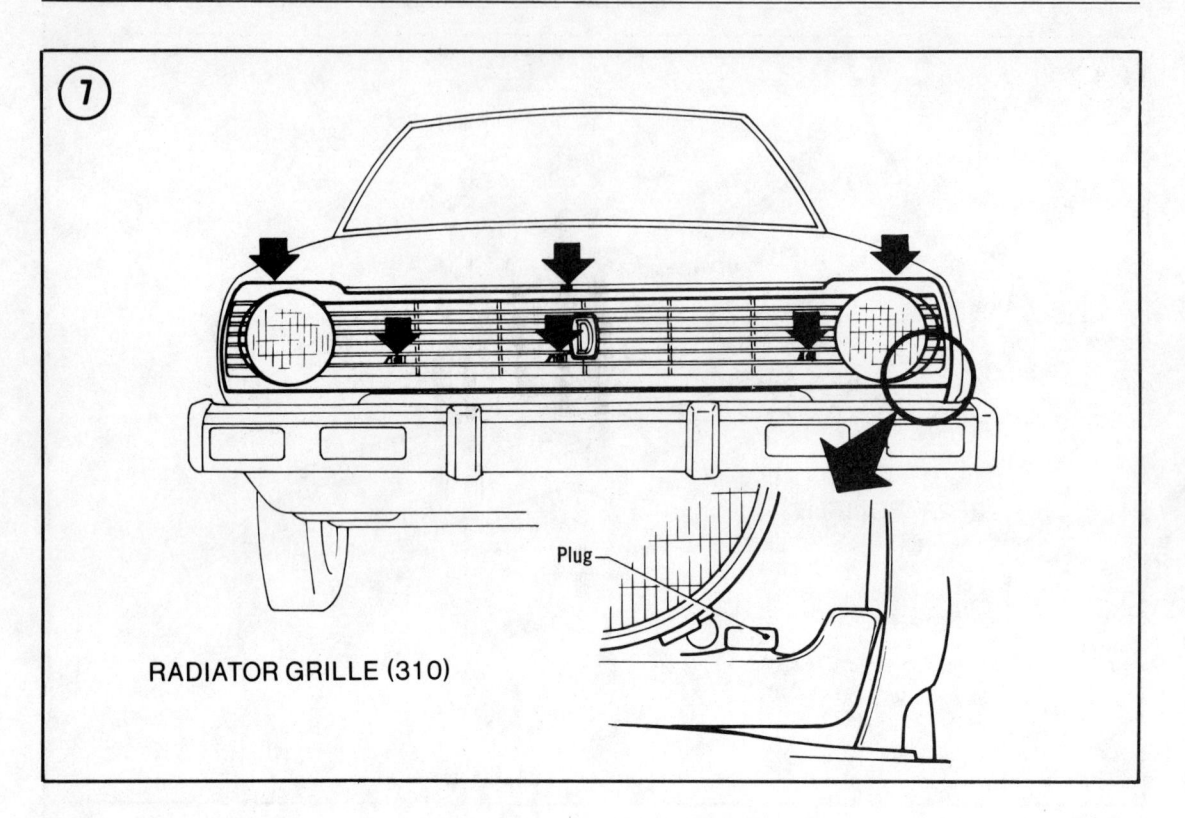

RADIATOR GRILLE (310)

Plug

⑧

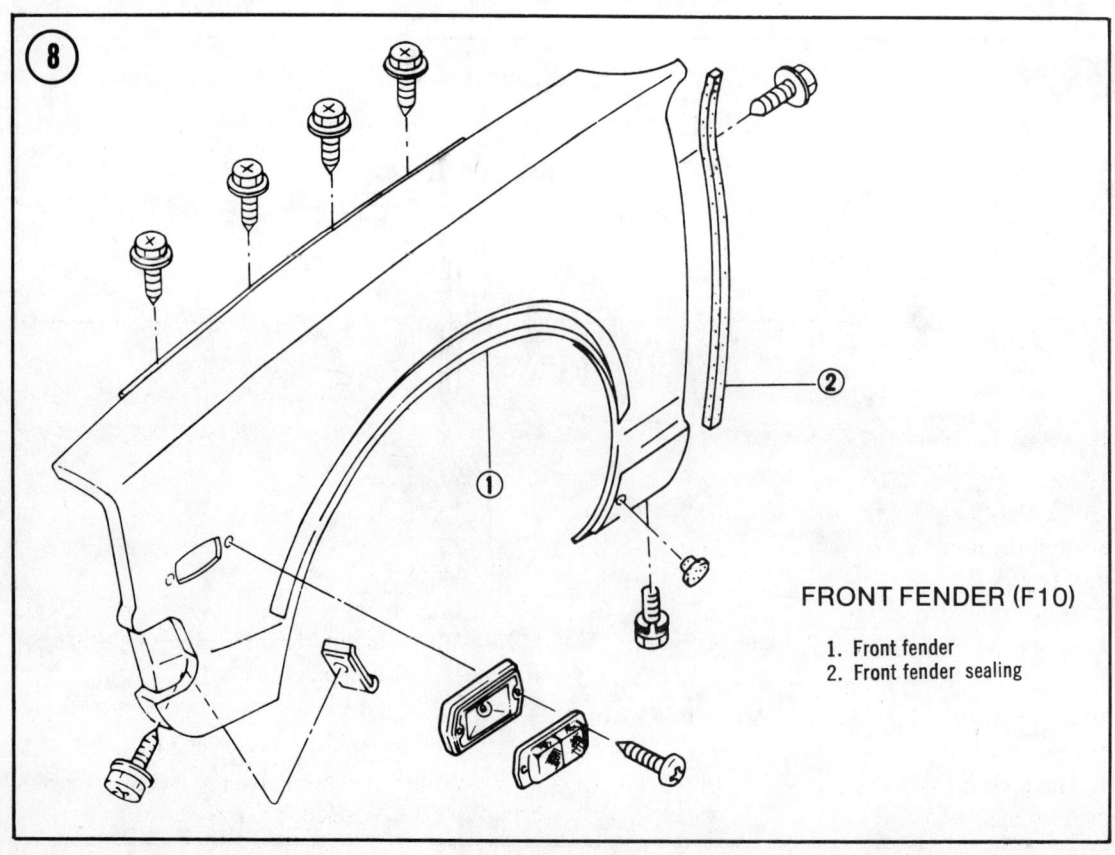

FRONT FENDER (F10)

1. Front fender
2. Front fender sealing

# BODY

**⑨**

FRONT FENDER (310)

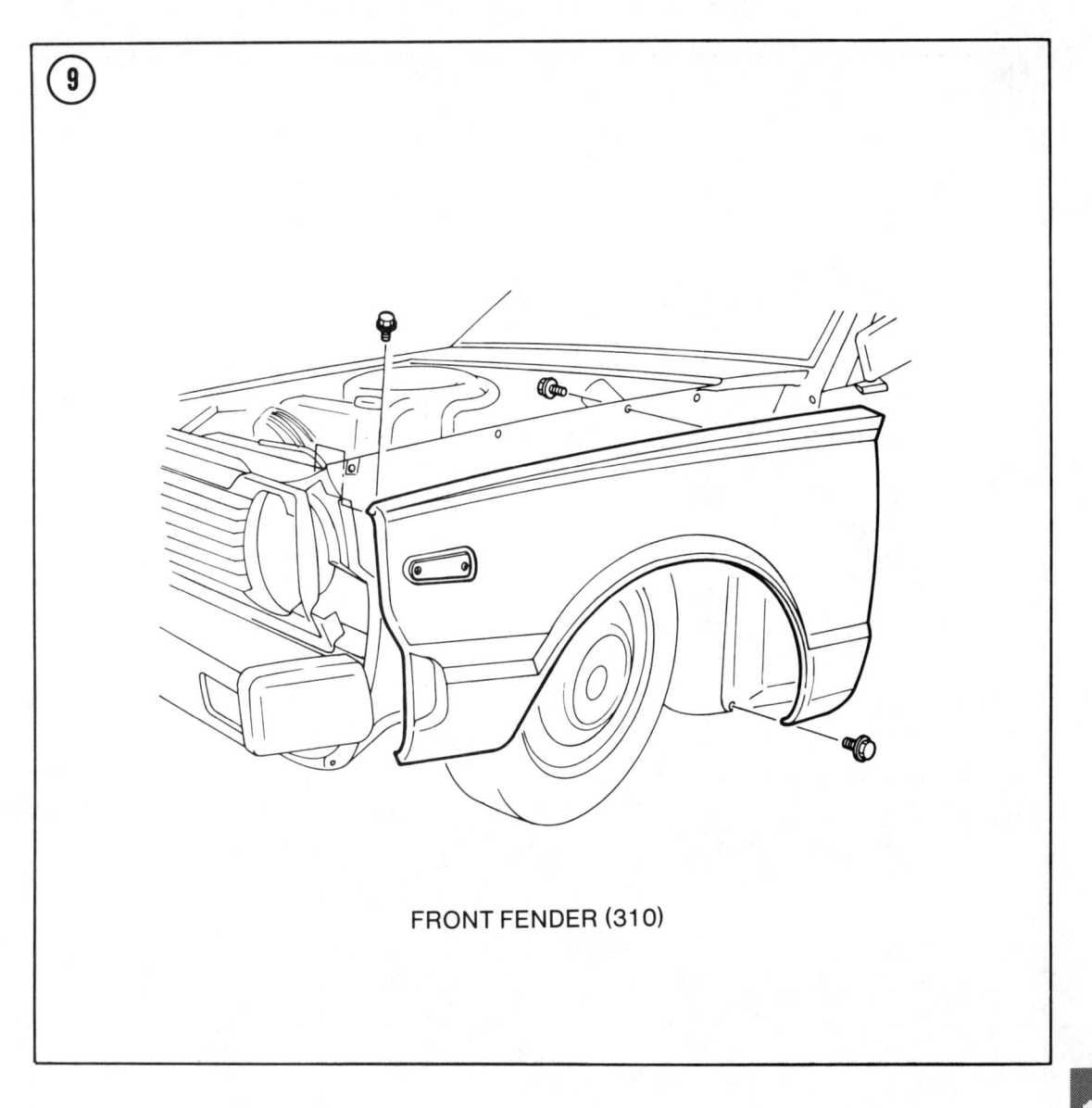

## HOOD

### Removal/Installation

1. Open the hood. Using a soft lead pencil, scribe alignment marks directly onto the hood around the hinges. The marks will ease installation later.

2. Place a thick layer of rags beneath the leading edge of the hood to protect the paint.

3. While an assistant supports one side of the hood, unbolt the hinges. See **Figure 10** (F10) or **Figure 11** (310).

4. Have your assistant unbolt the other hinge, then lift the hood off.

5. Installation is the reverse of removal. Align the hinges with the marks made before removal.

### Adjustment

1. To adjust the leading edge of the hood, loosen the hinge-to-hood bolts or hinge-to-body bolts. See **Figures 12 and 13**.

2. To adjust the trailing edge of the hood, rotate the hood bumpers (**Figure 14**).

3. To adjust the hood latch, loosen its bolts and move it as needed. See **Figure 15**.

250  CHAPTER TWELVE

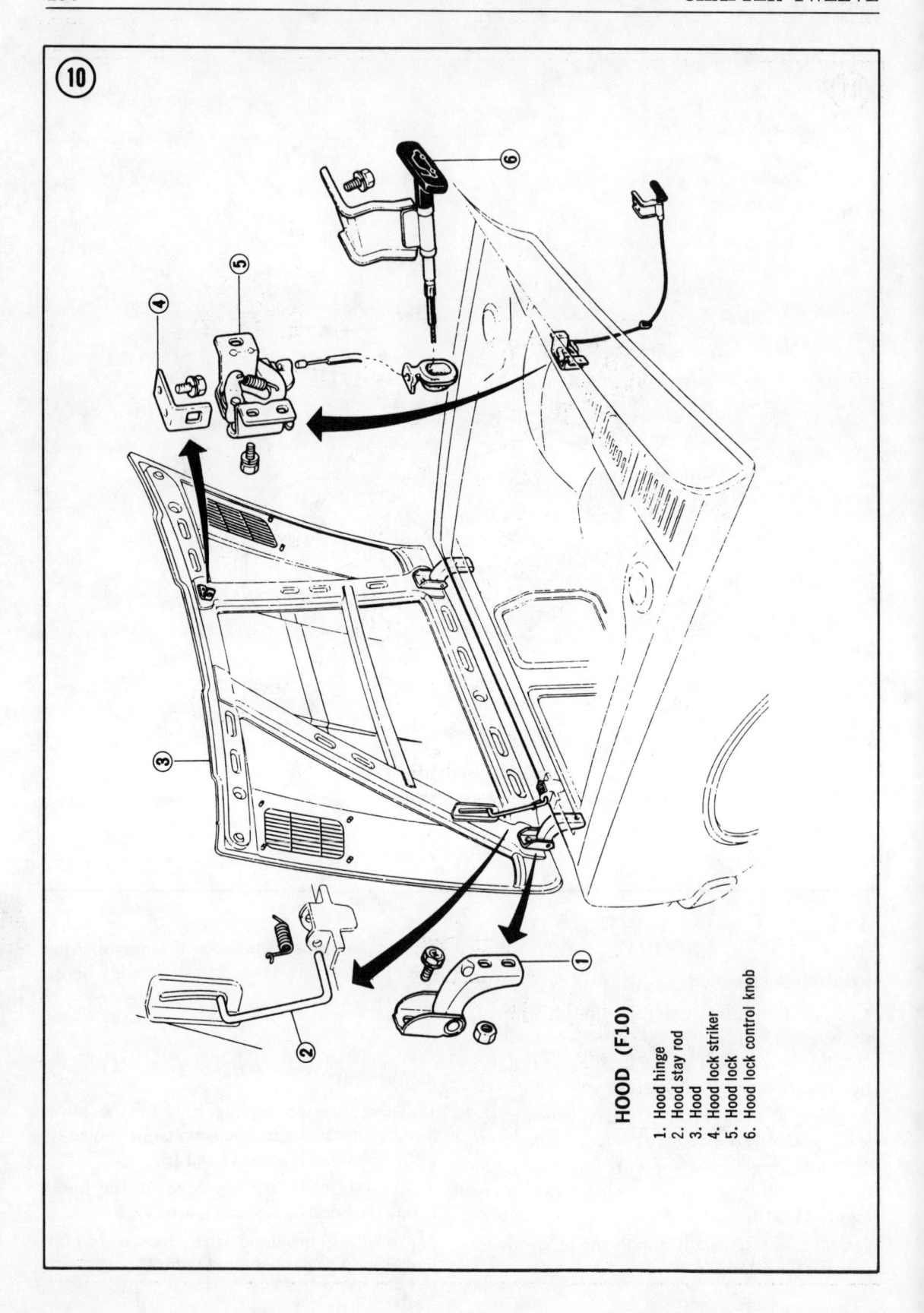

**HOOD (F10)**

1. Hood hinge
2. Hood stay rod
3. Hood
4. Hood lock striker
5. Hood lock
6. Hood lock control knob

**BODY** 251

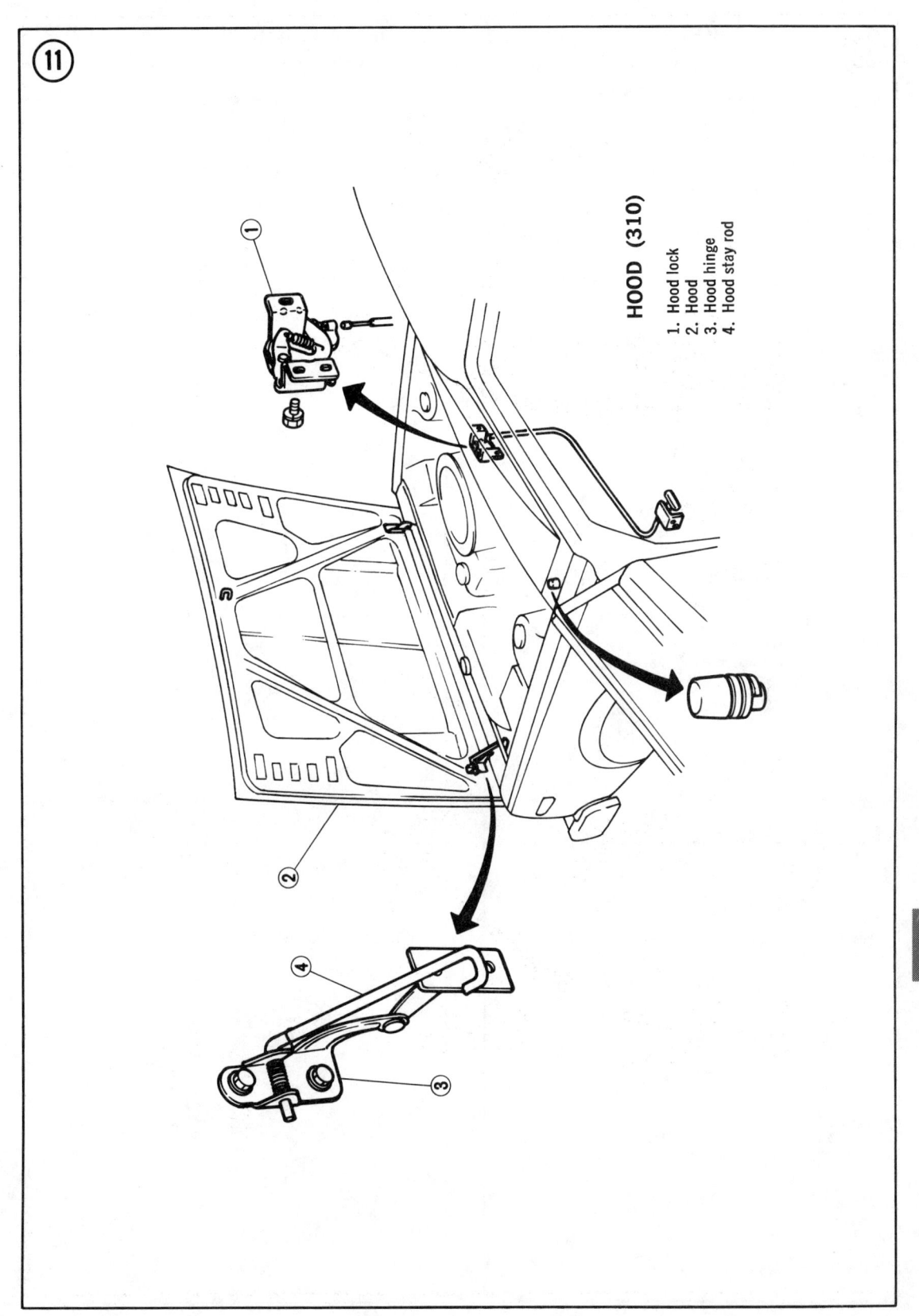

⑪

**HOOD (310)**

1. Hood lock
2. Hood
3. Hood hinge
4. Hood stay rod

12

## DOORS

**Figure 16** shows an F10 front door; **Figure 17** an F10 rear door; and **Figure 18** a 310 door.

### Removal

1. Open the door. Place a jack beneath the door to support it.

> **CAUTION**
> *Place a rag on top of the jack to protect the door's paint.*

2. Unbolt the door from the hinges **(Figure 19)** while an assistant supports the door.

3. Lift the door away from the car.

### Installation

1. Check the door weatherstripping. Replace if damaged.
2. Remove the door striker **(Figure 20)**.
3. Install the door and tighten the hinge bolts.
4. Close the door several times. Make sure the leading edge of the door does not touch the body. Check alignment along the top, at the rear edge, and at the latch.

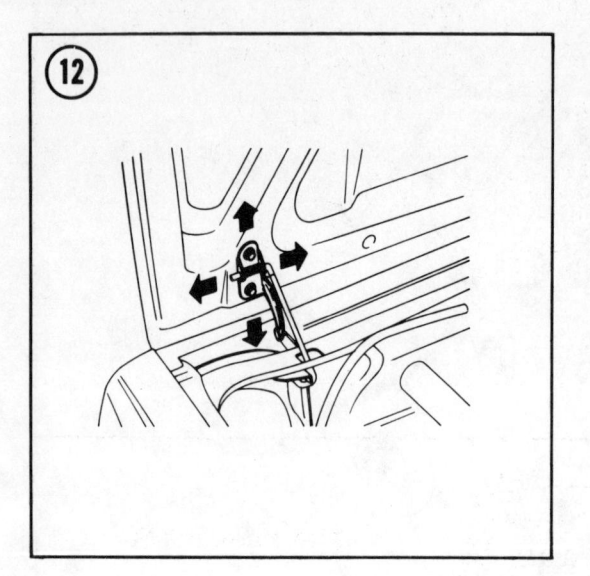

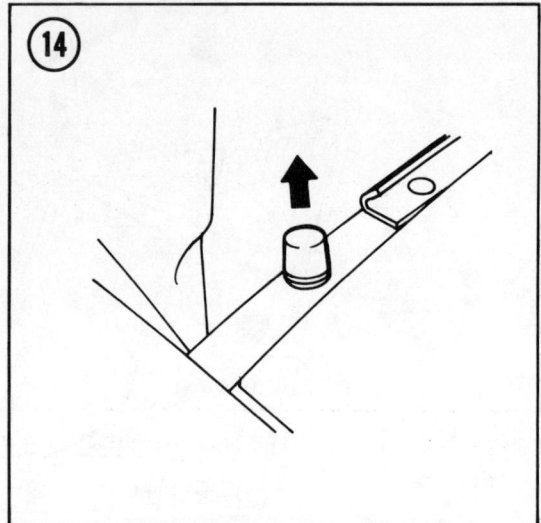

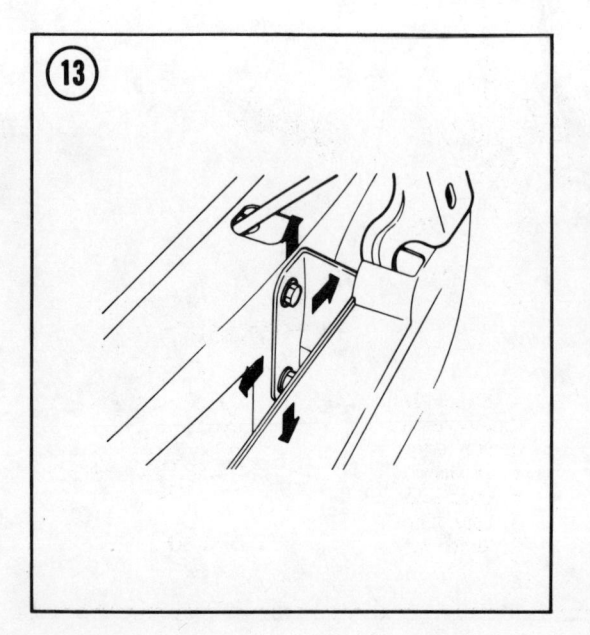

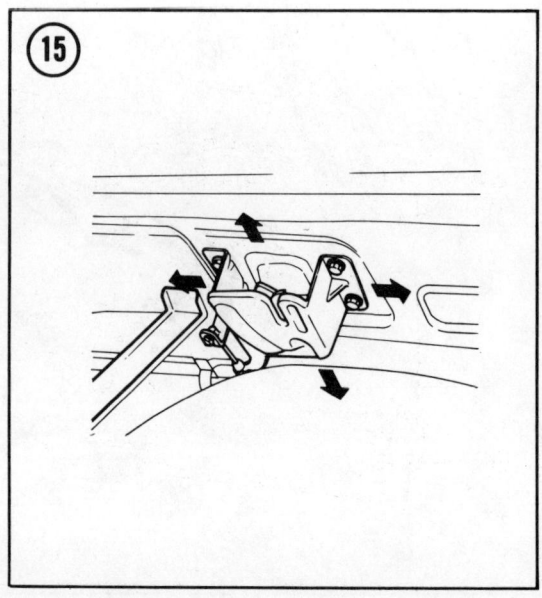

BODY                                                                                               253

⑯                          FRONT DOOR (F10)

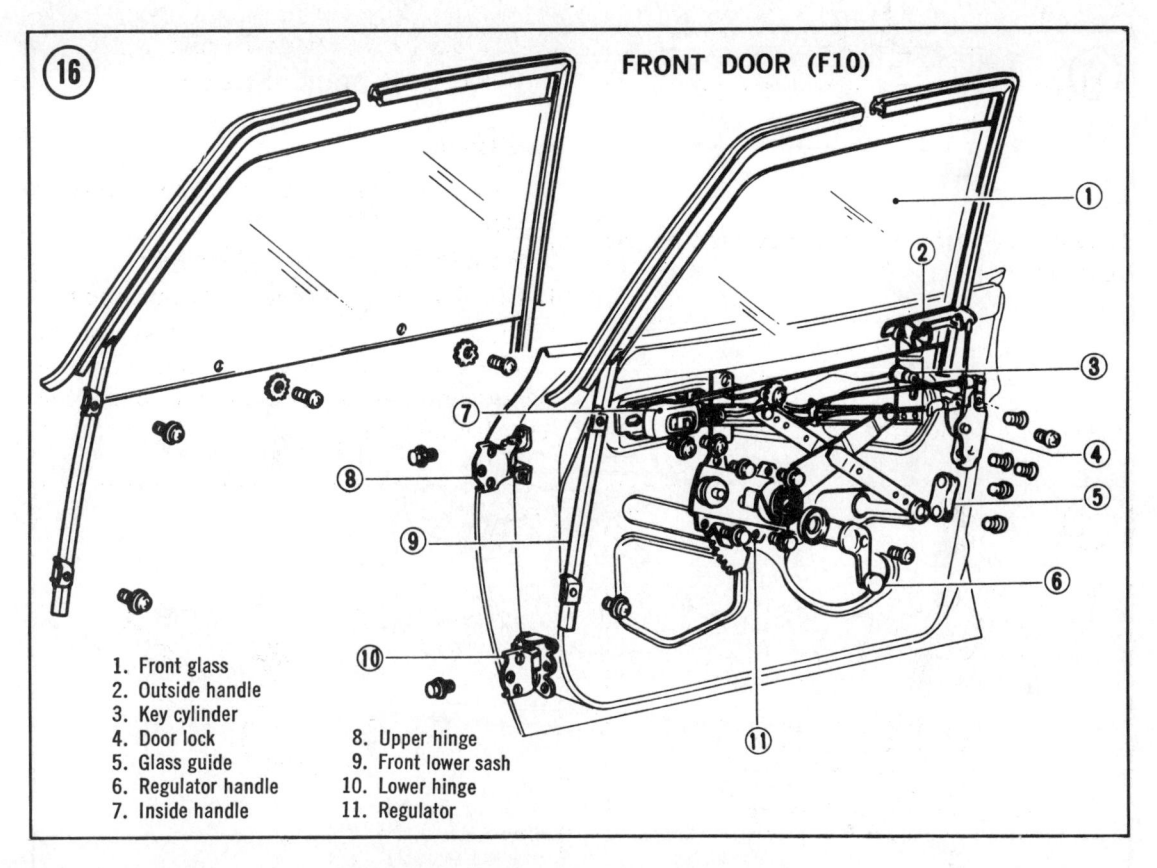

1. Front glass
2. Outside handle
3. Key cylinder
4. Door lock           8. Upper hinge
5. Glass guide         9. Front lower sash
6. Regulator handle   10. Lower hinge
7. Inside handle      11. Regulator

⑰                          REAR DOOR (F10)

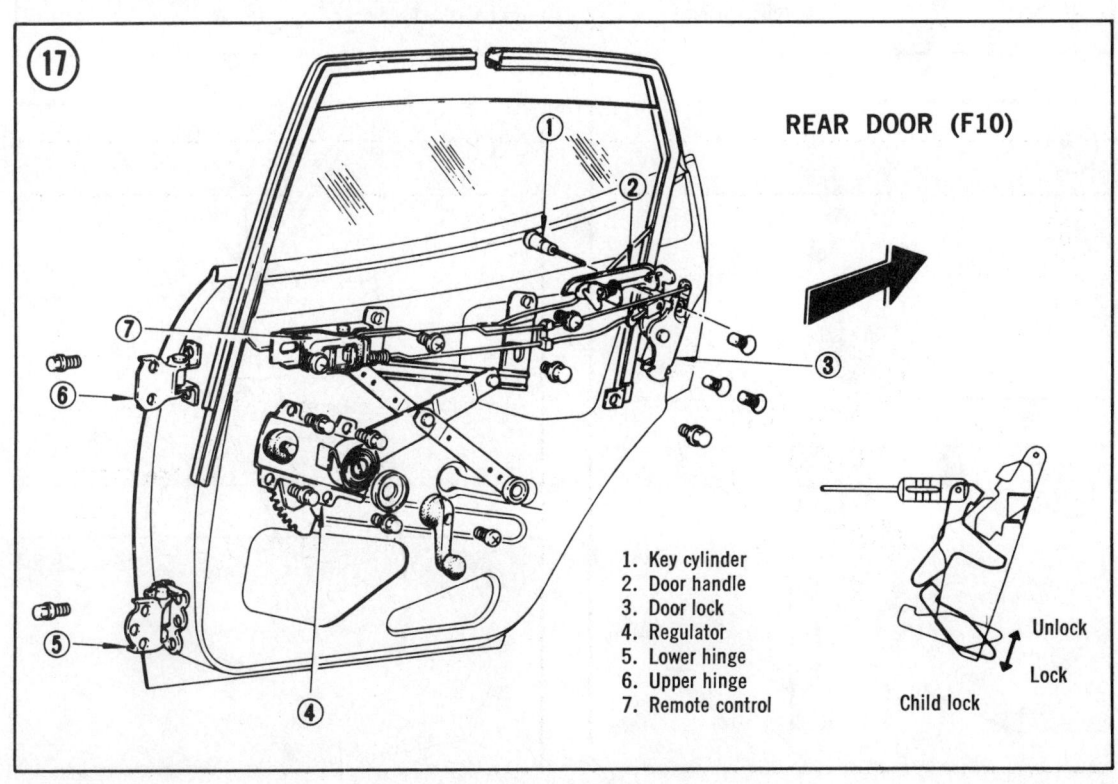

1. Key cylinder
2. Door handle
3. Door lock
4. Regulator
5. Lower hinge
6. Upper hinge
7. Remote control

Unlock
Lock
Child lock

12

CHAPTER TWELVE

**18 DOOR (310)**

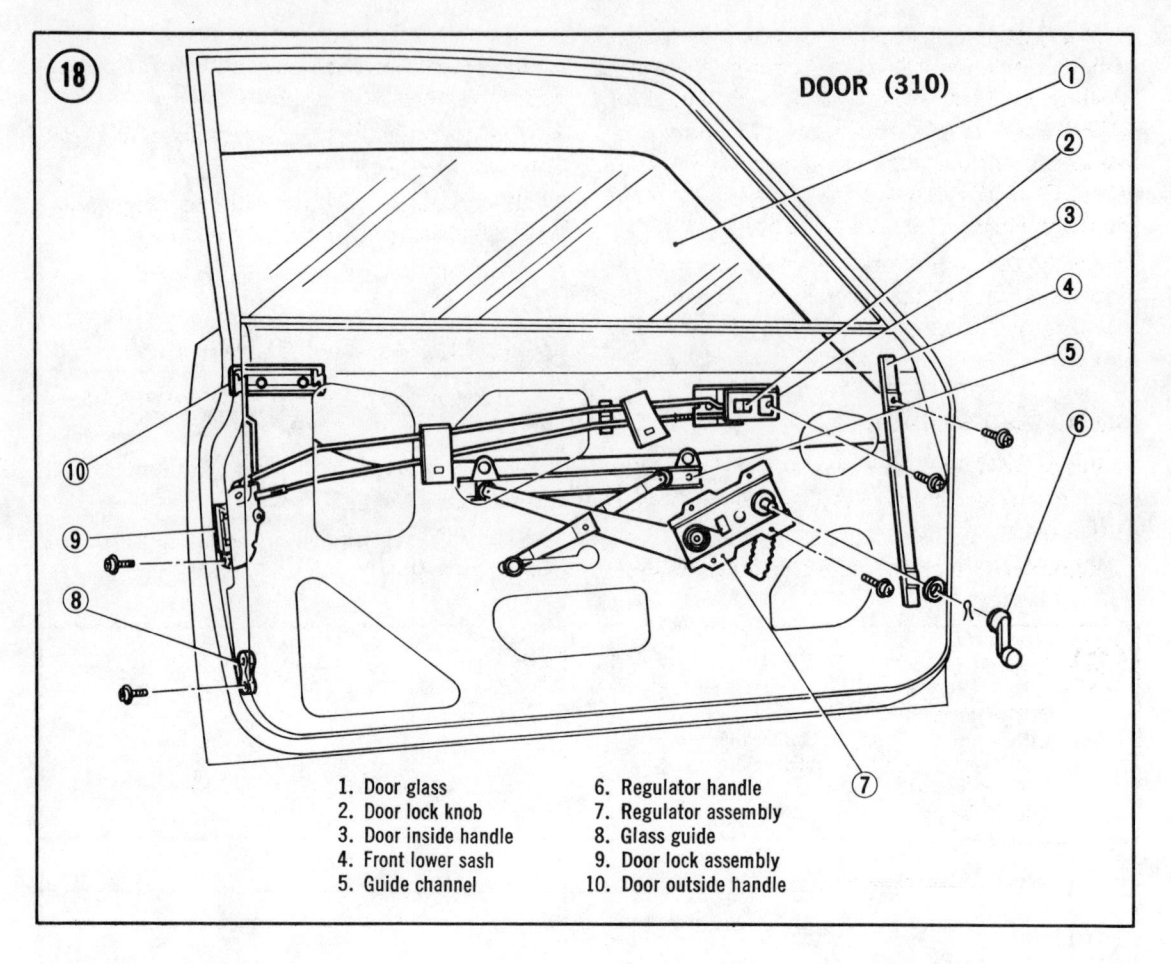

1. Door glass
2. Door lock knob
3. Door inside handle
4. Front lower sash
5. Guide channel
6. Regulator handle
7. Regulator assembly
8. Glass guide
9. Door lock assembly
10. Door outside handle

**19**

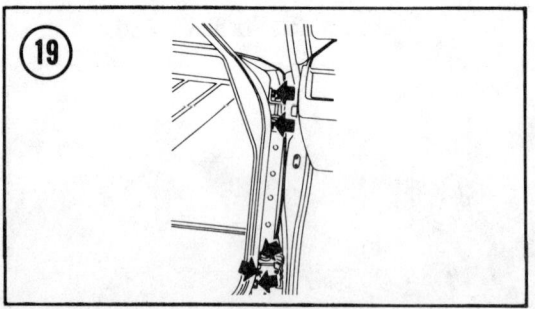

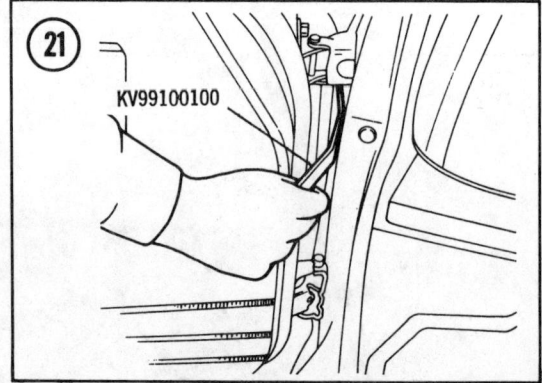

**20**

**21** KV99100100

**22**

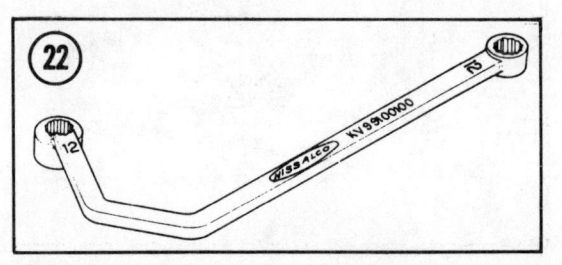

# BODY

5. If adjustment is necessary, loosen the hinge-to-body bolts. On F10's, this requires a specially shaped 12mm wrench such as Datsun tool KV99100100 (Kent-Moore No. J26080). **Figure 21** shows the tool in use; **Figure 22** shows it alone. On 310's, remove the access cover from the front wheel well to reach the bolts.

6. Install the striker and close the door. If it closes too far, not far enough, or if it must be lifted or pushed down to close, reposition the striker.

## Lock Cylinder Replacement

1. Press the door panel against the door. Pull out the window crank clip **(Figure 23)** with a wire hook.

2. Remove the armrest and inside door handle.

3. Insert a wide-bladed screwdriver or similar tool between the door panel and door. Carefully pry the panel away from the door.

4. Peel the plastic sealing screen away from the door. Be careful not to tear it.

5. Remove lock cylinder clip. See **Figure 24**. Take the lock cylinder out of the door.

6. Installation is the reverse of removal.

## Window Replacement

1. Perform Steps 1-3, *Lock Cylinder Replacement*.

2. Temporarily install the window crank. Lower the window.

3. Pry off the outer weatherstripping. See **Figure 25**.

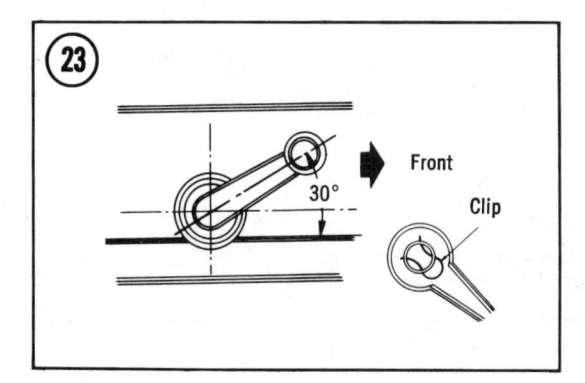

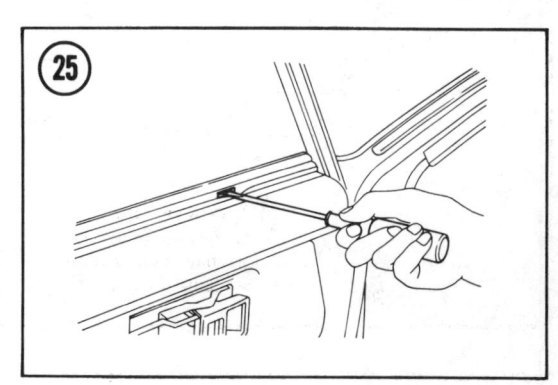

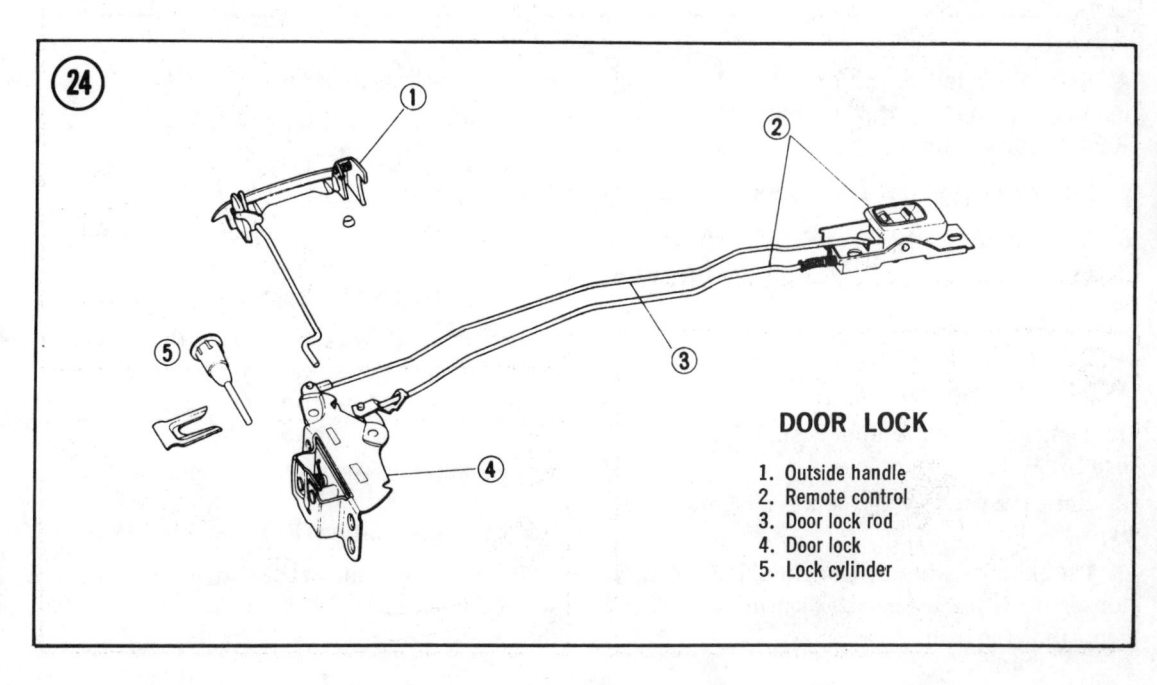

**23**
30° Front
Clip

**25**

**24**
1.
2.
3.
4.
5.

**DOOR LOCK**

1. Outside handle
2. Remote control
3. Door lock rod
4. Door lock
5. Lock cylinder

**26**

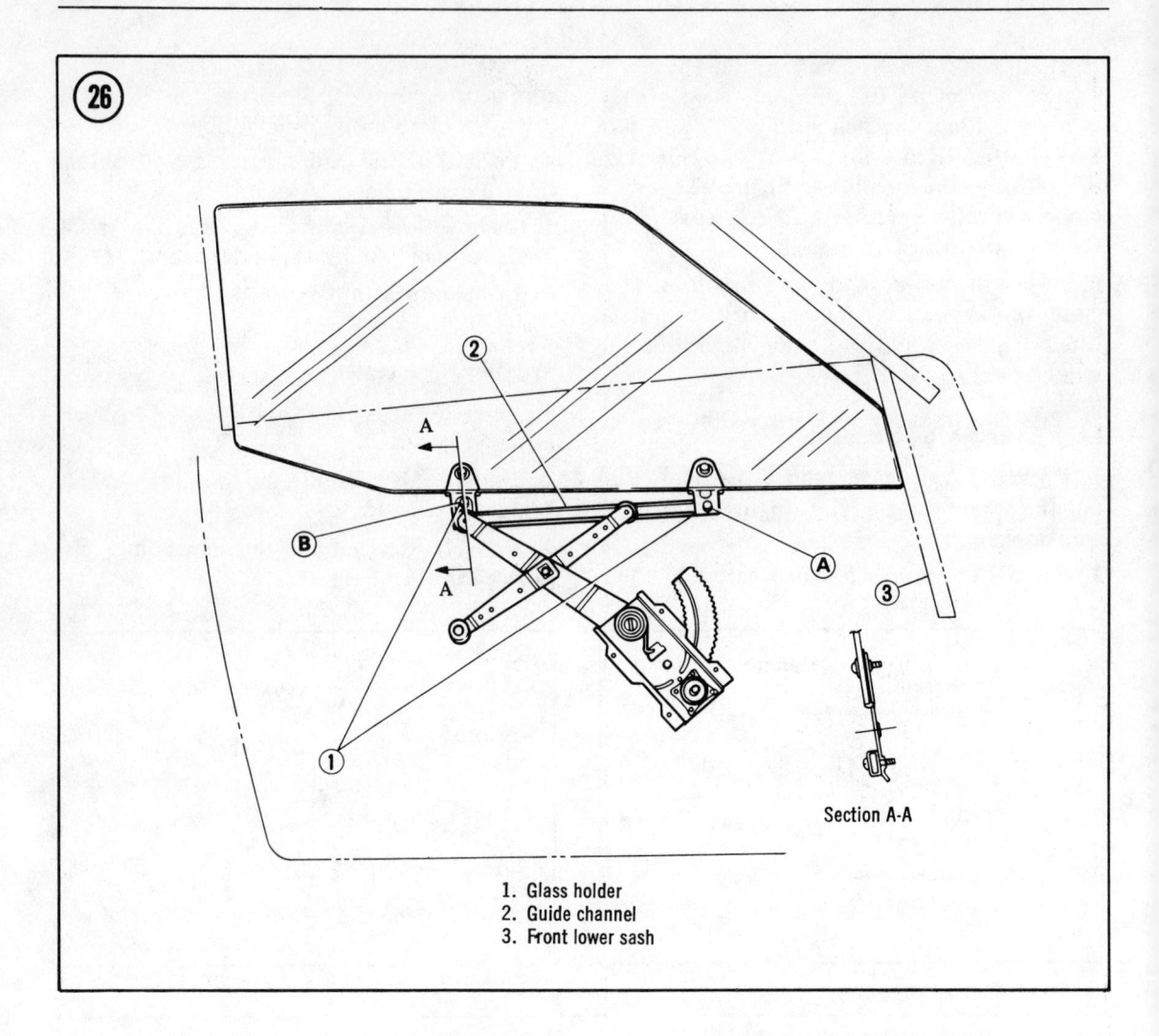

1. Glass holder
2. Guide channel
3. Front lower sash

Section A-A

4. Raise the window until the screws securing its bottom edge appear in the access holes. Remove the screws.

5. Lift the window out of the door.

6. Installation is the reverse of removal. If necessary, adjust as described in the following section.

## Window Adjustment

1. Tighten screws A and B slightly. See **Figure 26**.

2. Raise the window all the way. Tighten screw B.

3. Check door window position. If necessary, loosen the front lower sash securing screws and reposition the sash.

## REAR QUARTER WINDOWS

### Removal/Installation (F10 Hatchback)

1. Carefully pry the interior panel away from the body.

2. Peel back the plastic screen. Be careful not to tear it.

3. Remove the lower sash bolts (**Figure 27**).

4. Lower the window until the glass-to-regulator screws appear in the access hole. Detach the window from the regulator, then take it out through the access hole.

5. Installation is the reverse of removal.

### Removal/Installation (F10 Sedan and Wagon)

To remove, undo the hinge and handle screws (**Figure 28**). Take the glass out. Install in the reverse order.

# BODY

### 27

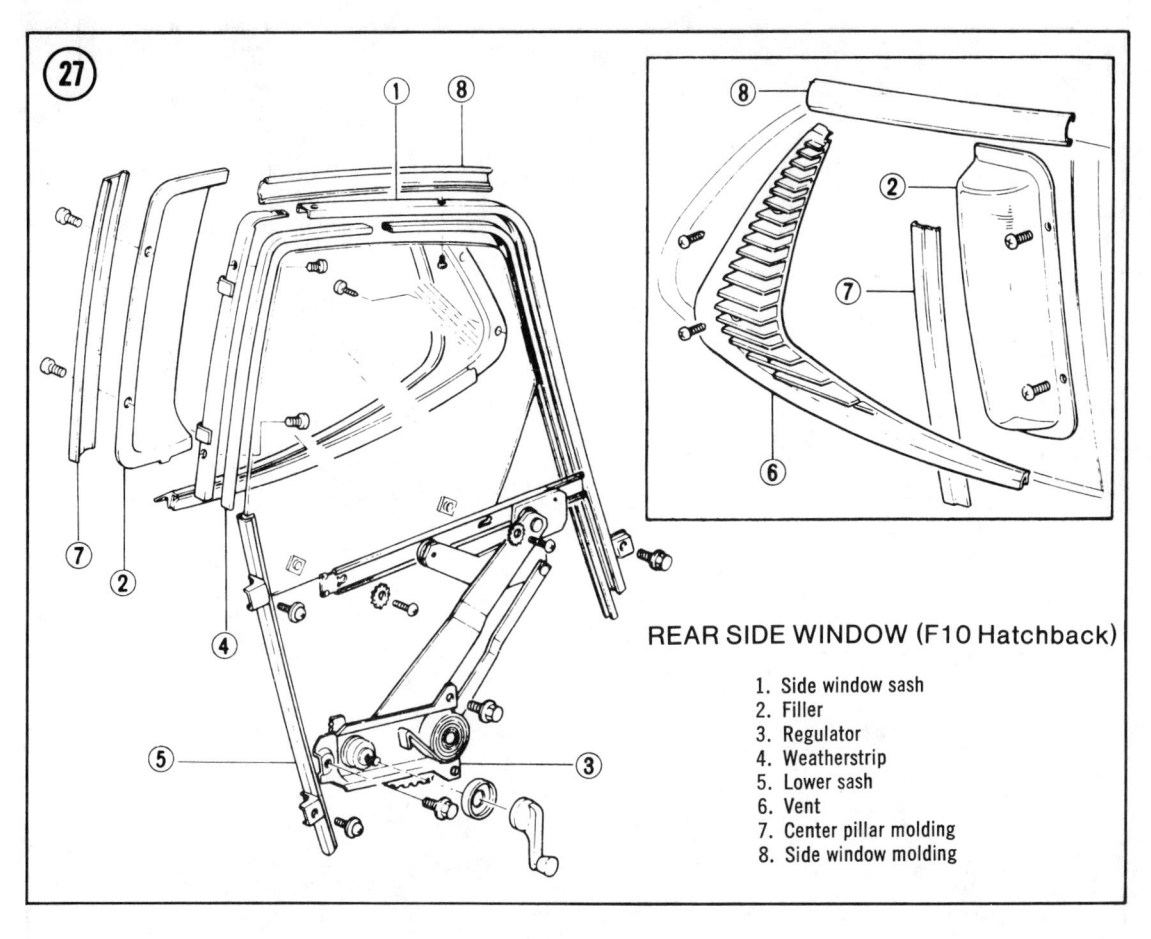

**REAR SIDE WINDOW (F10 Hatchback)**

1. Side window sash
2. Filler
3. Regulator
4. Weatherstrip
5. Lower sash
6. Vent
7. Center pillar molding
8. Side window molding

### 28

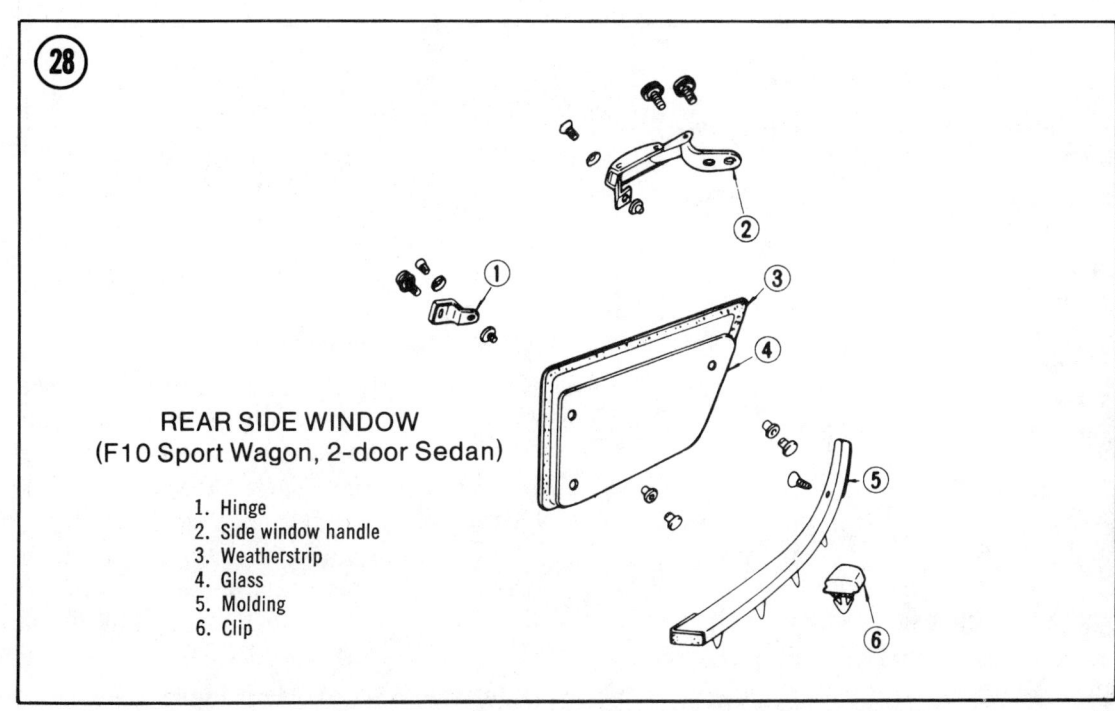

**REAR SIDE WINDOW  
(F10 Sport Wagon, 2-door Sedan)**

1. Hinge
2. Side window handle
3. Weatherstrip
4. Glass
5. Molding
6. Clip

12

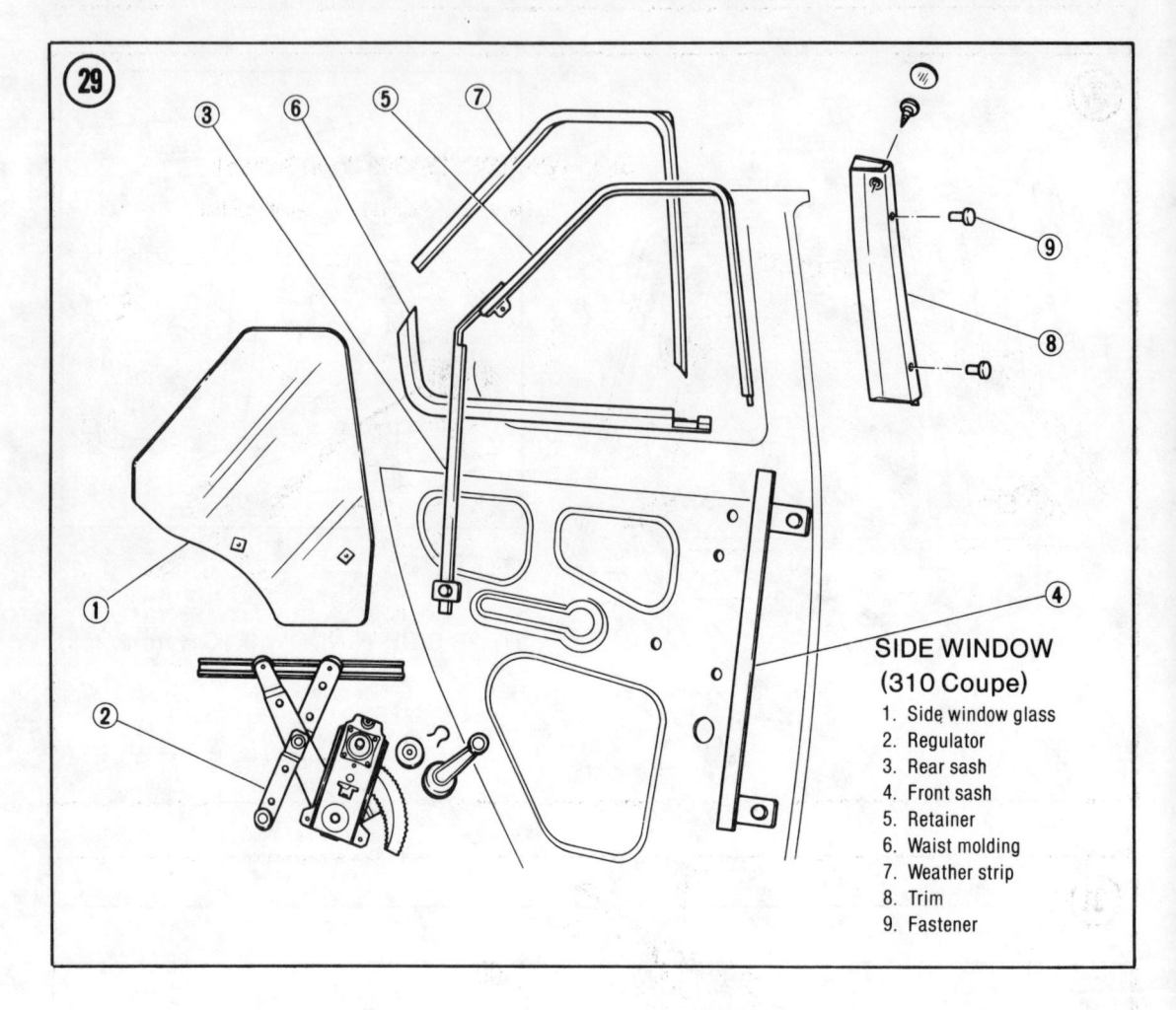

**SIDE WINDOW (310 Coupe)**
1. Side window glass
2. Regulator
3. Rear sash
4. Front sash
5. Retainer
6. Waist molding
7. Weather strip
8. Trim
9. Fastener

### Removal/Installation (310 Coupe)

1. Gently pry the interior panel away from the body.

2. Peel back the plastic sealing screen. Be careful not to tear it.

3. Remove waist molding and trim (6 and 8, **Figure 29**).

4. Lower the window until the bottom edge appears in the access hole.

5. Separate the regulator rollers from the glass channel. Lift the window out.

6. Installation is the reverse of removal.

### Removal/Installation (310 Sedan)

**Figure 30** shows the window with and without the optional remote control.

1. If equipped with remote control, disconnect the cable from the window.

2. Remove the hinge and handle screws. Lift the window out.

3. Installation is the reverse of removal.

### TRUNK LID (F10 SEDAN)

**Figure 31** shows the trunk lid and related parts.

### Removal/Installation

1. Using a soft lead pencil, scribe alignment marks around the trunk hinges onto the trunk lid. The marks will ease installation later.

2. Unbolt the trunk lid from the hinges and lift them off.

3. If hinge removal is necessary, pry the torsion bars out of their brackets with a piece of pipe. Then remove the hinge nuts and take the hinges off.

4. Installation is the reverse of removal.

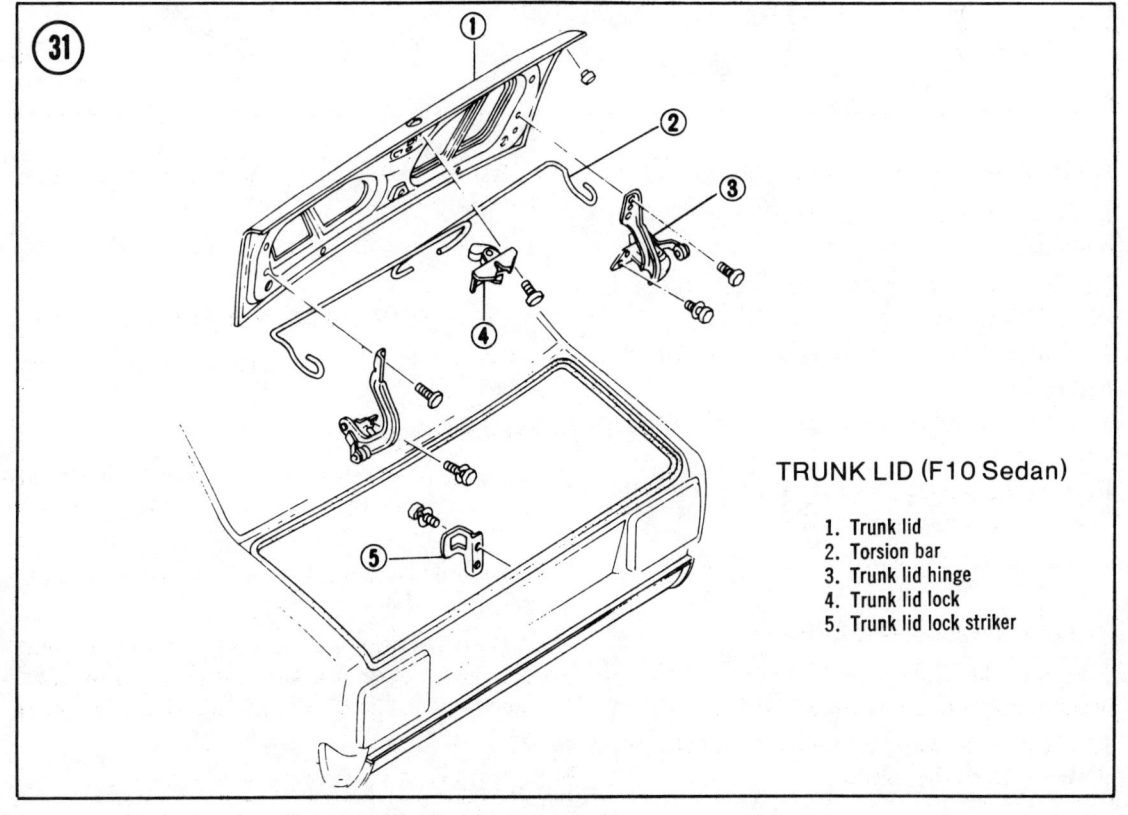

# BODY

### ㉚ SIDE WINDOW (310 2-door Sedan)

For models equipped with remote control

### ㉛ TRUNK LID (F10 Sedan)

1. Trunk lid
2. Torsion bar
3. Trunk lid hinge
4. Trunk lid lock
5. Trunk lid lock striker

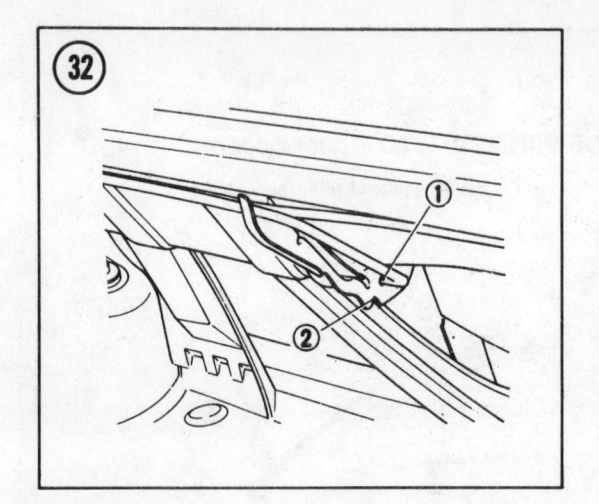

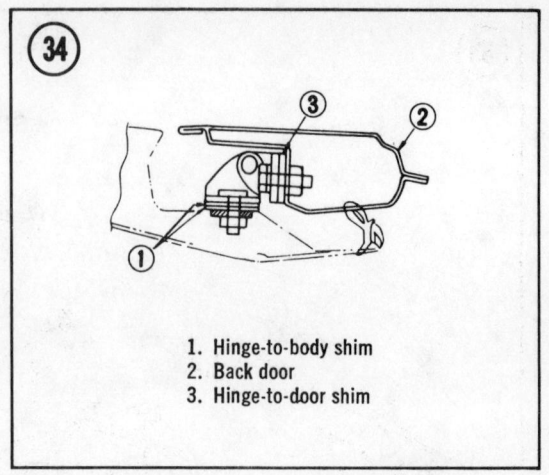

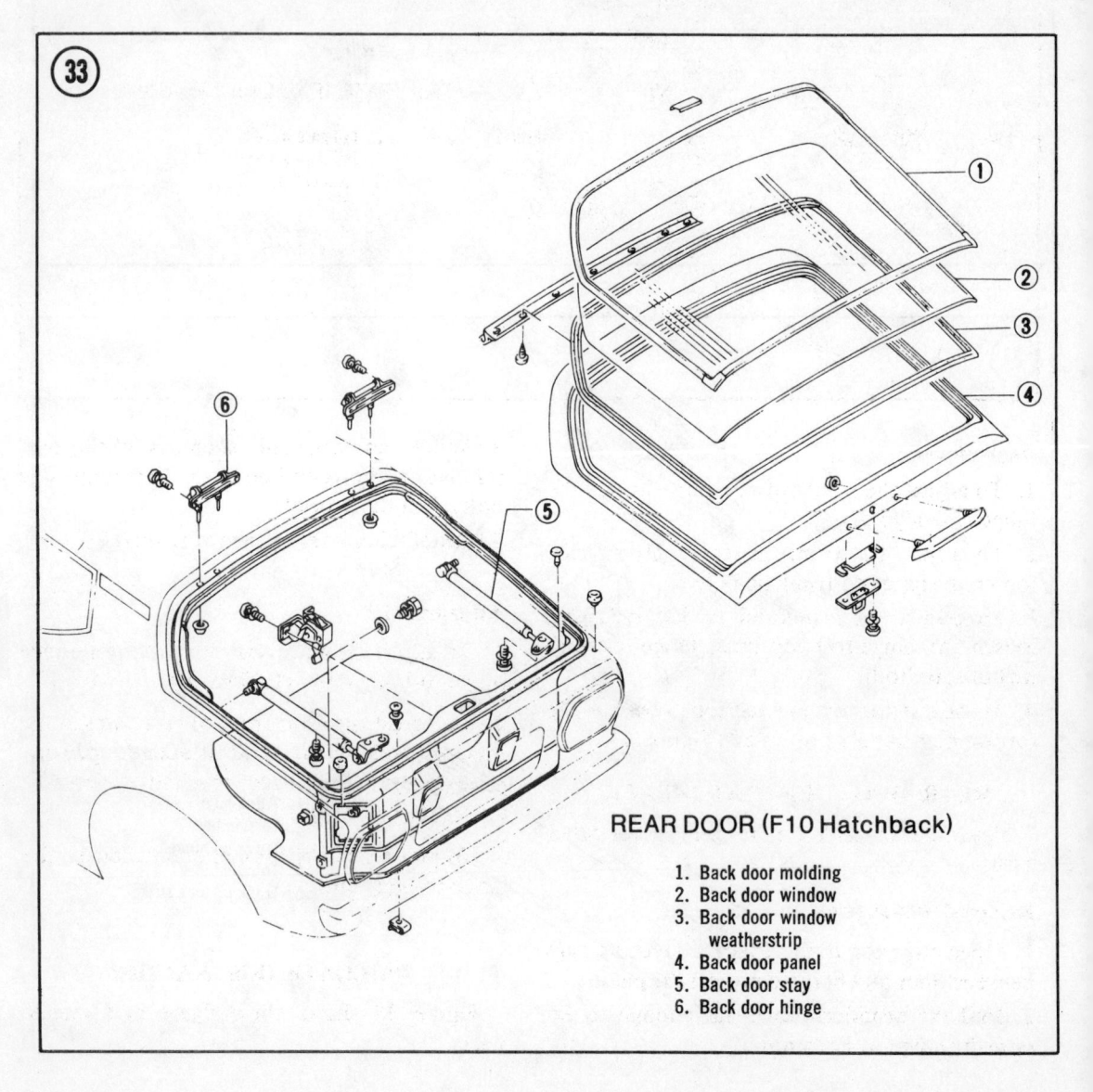

1. Hinge-to-body shim
2. Back door
3. Hinge-to-door shim

**REAR DOOR (F10 Hatchback)**

1. Back door molding
2. Back door window
3. Back door window weatherstrip
4. Back door panel
5. Back door stay
6. Back door hinge

**TAILGATE (F10 Sport Wagon)**

1. Tail gate panel
2. Tail gate window
3. Torsion bar
4. Hinge
5. Tail gate striker
6. Tail gate lock
7. Wedge
8. Back door hinge shim
9. Back door lock shim

### Adjustment

1. To adjust the trunk lid vertically, loosen the hinge-to-body nuts.

2. To adjust the trunk lid to front or rear, loosen the hinge-to-trunk lid bolts.

3. To adjust the trunk lid to left or right, loosen the hinge-to-body nuts, hinge-to-trunk lid bolts, or both.

4. To adjust torsion bar tension, position the torsion bars in hole 1 or hole 2 (**Figure 32**).

## REAR DOOR (F10 HATCHBACK)

**Figure 33** shows the rear door and related parts.

### Removal/Installation

1. Open the rear door. Place a layer of rags between door and body to protect the paint.

2. Pull the headliner down far enough to expose the hinge-to-body nuts.

3. While an assistant supports the door, remove the hinge-to-body nuts and door stay bolts. Lift the door off.

4. Installation is the reverse of removal.

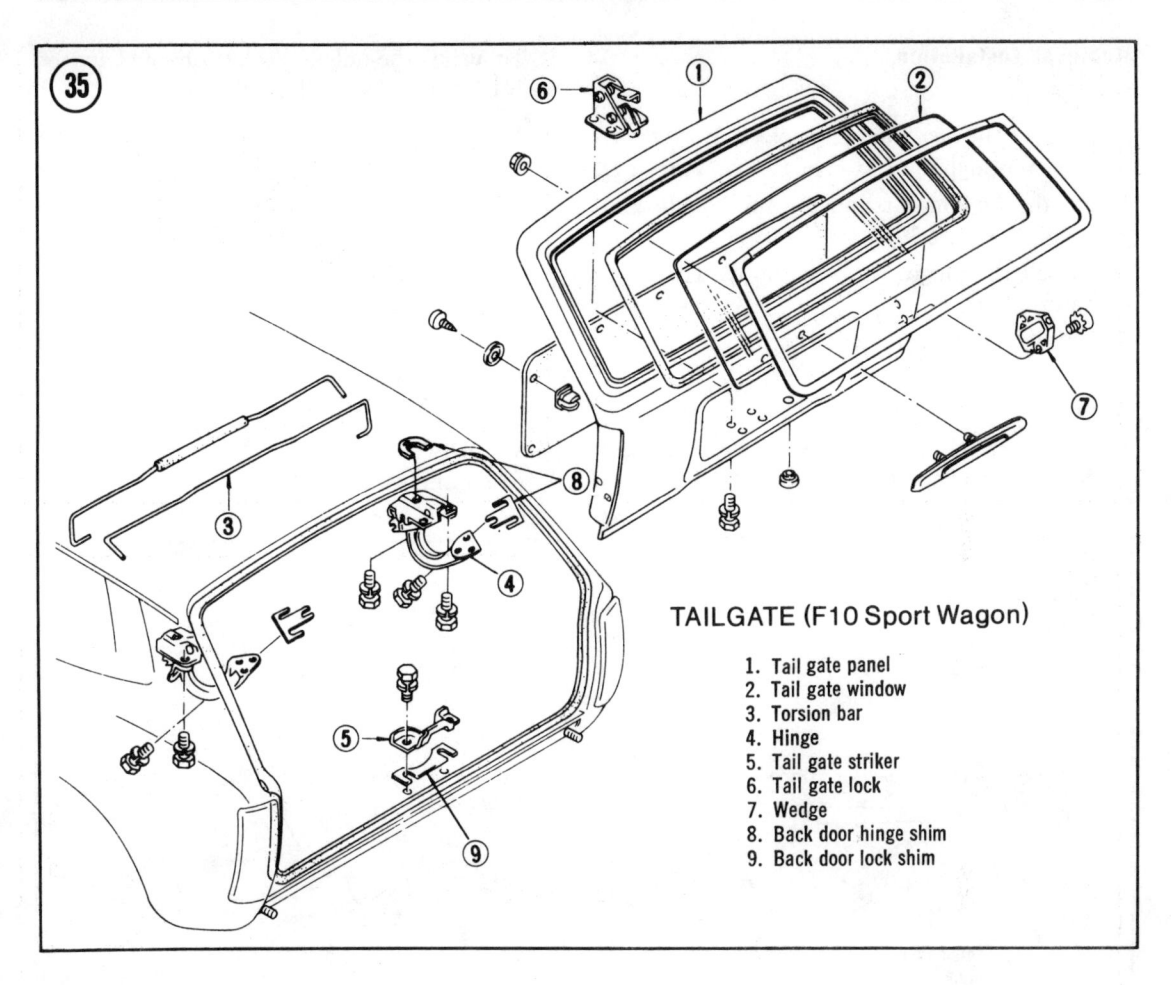

### Adjustment

1. To adjust the door vertically, add or remove hinge-to-body shims (1, **Figure 34**).

2. To adjust the door front to rear, loosen the hinge-to-body nuts. If this doesn't provide enough adjustment, add or remove hinge-to-door shims (3, **Figure 34**).

3. To move the door to right or left, loosen the hinge-to-body nuts.

## TAILGATE (F10 WAGON)

**Figure 35** shows the tailgate and related parts.

## Removal/Installation

1. Remove the hinge covers.

2. Remove the headliner edge piece. Lower the headliner enough to expose the torsion bars.

3. Pry the torsion bars loose from their hooks with a pipe **(Figure 36)**.

4. Unbolt the hinges from the body. Take the tailgate out.

5. Installation is the reverse of removal.

## Adjustment

1. To adjust the tailgate front to rear, loosen the hinge-to-body bolts.

2. To adjust the tailgate to left or right, loosen the hinge-to-tailgate bolts.

3. To raise the door, add shims between the door and hinges.

4. To lower the door, add shims between the body and hinges.

## REAR DOOR (310)

### Removal/Installation

1. Remove the luggage compartment trim panel. Disconnect the harnesses for rear defroster and windshield wiper.

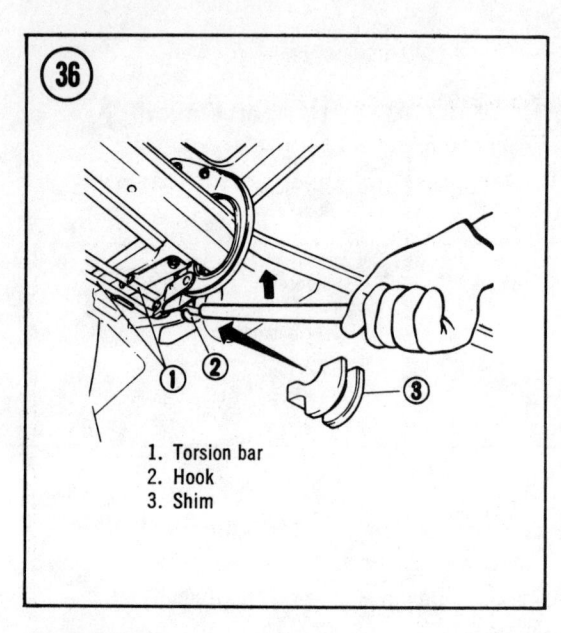

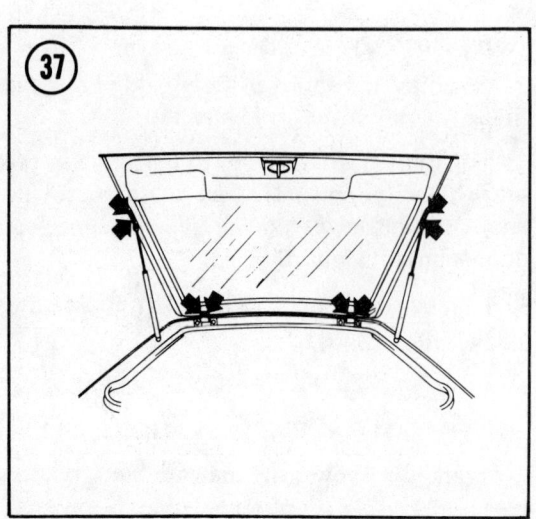

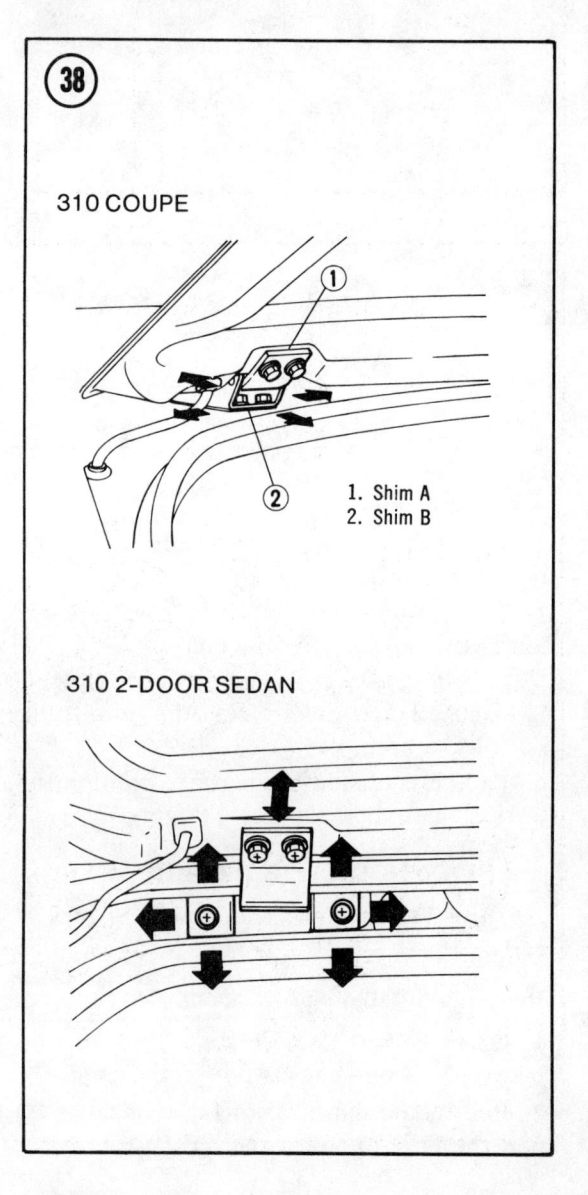

**36**

1. Torsion bar
2. Hook
3. Shim

**38**

310 COUPE

1. Shim A
2. Shim B

**37**

310 2-DOOR SEDAN

# BODY

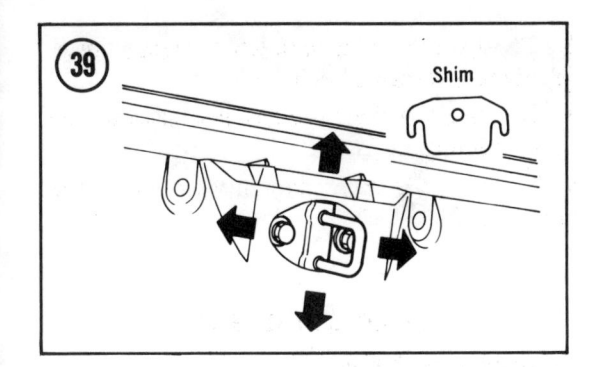

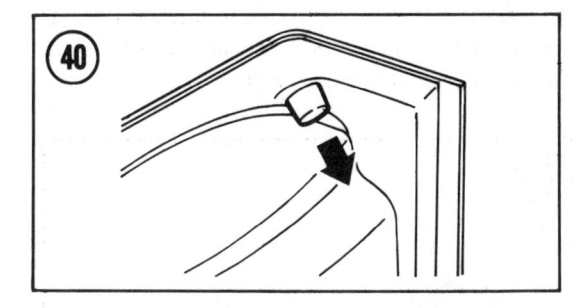

2. While an assistant supports the door, unbolt the door stays and hinges. See **Figure 37**. Lift the door off.

3. Installation is the reverse of removal.

**Adjustment**

1. To align the upper edge and sides, loosen the hinge bolts (**Figure 38**). On coupes, shims may be added or removed from beneath the hinges.

2. If the door shifts vertically or to one side when closed, reposition the striker (**Figure 39**).

3. If the door is loose when closed or must be slammed, reposition the rubber bumpers (**Figure 40**).

## INSTRUMENT PANEL

**Removal/Installation (F10)**

Refer to **Figure 41** for this procedure.

1. Disconnect negative cable from battery.

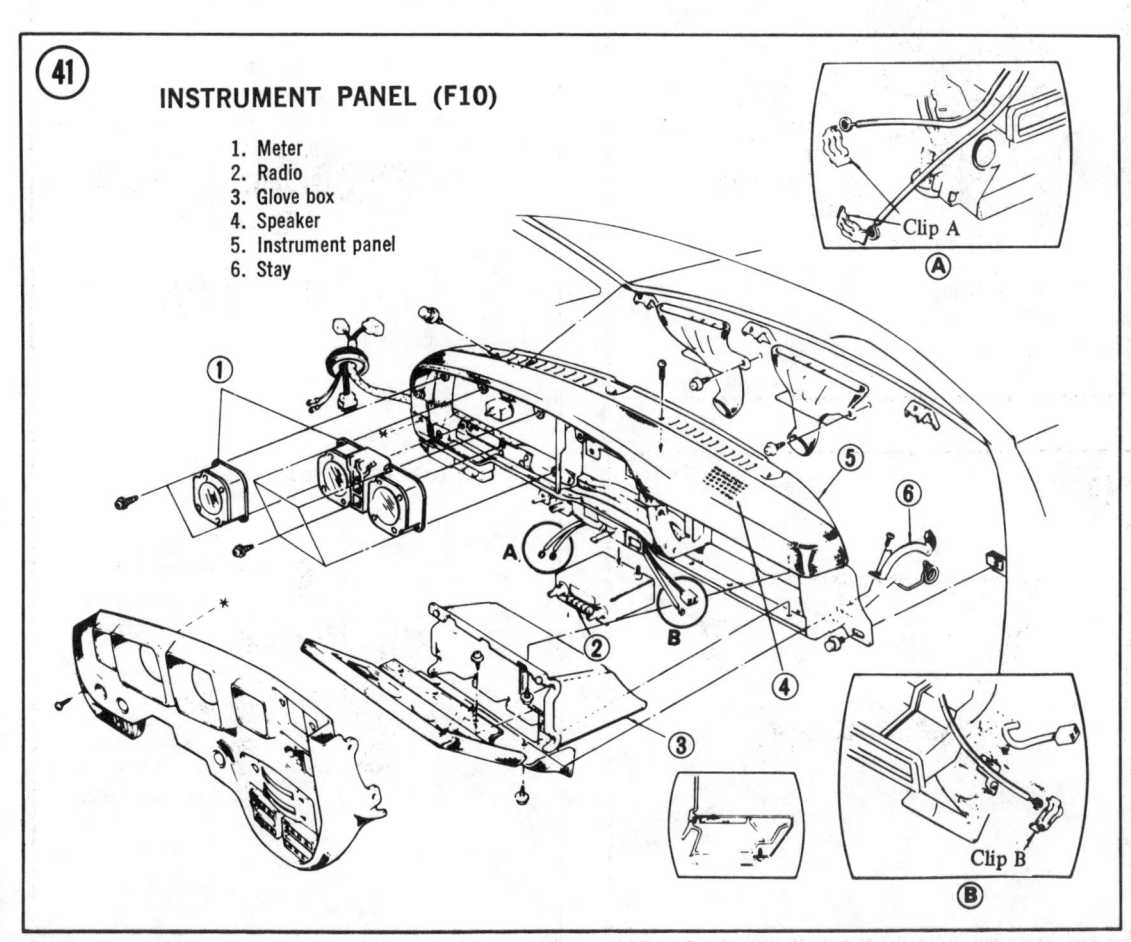

**INSTRUMENT PANEL (F10)**

1. Meter
2. Radio
3. Glove box
4. Speaker
5. Instrument panel
6. Stay

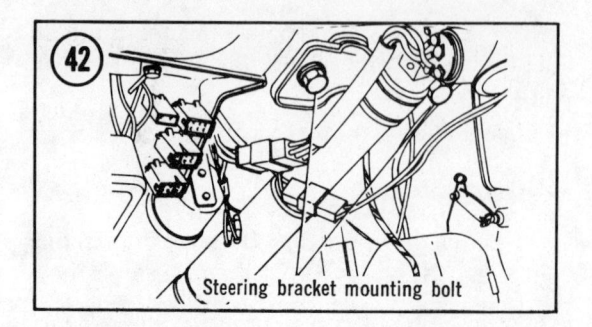

CHAPTER TWELVE

**42**
Steering bracket mounting bolt

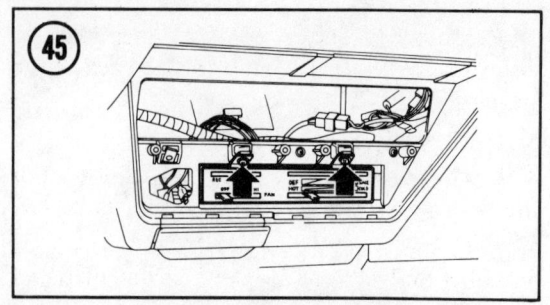

**45**

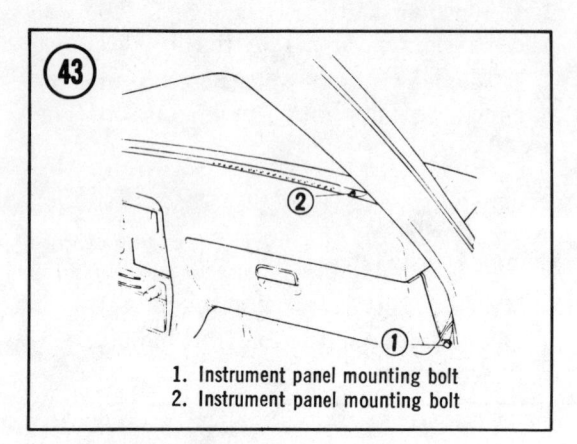

**43**
1. Instrument panel mounting bolt
2. Instrument panel mounting bolt

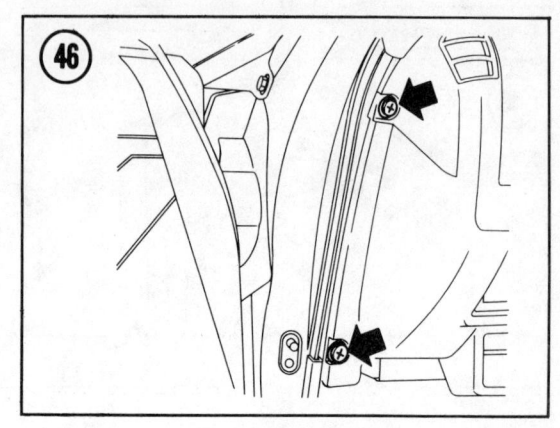

**46**

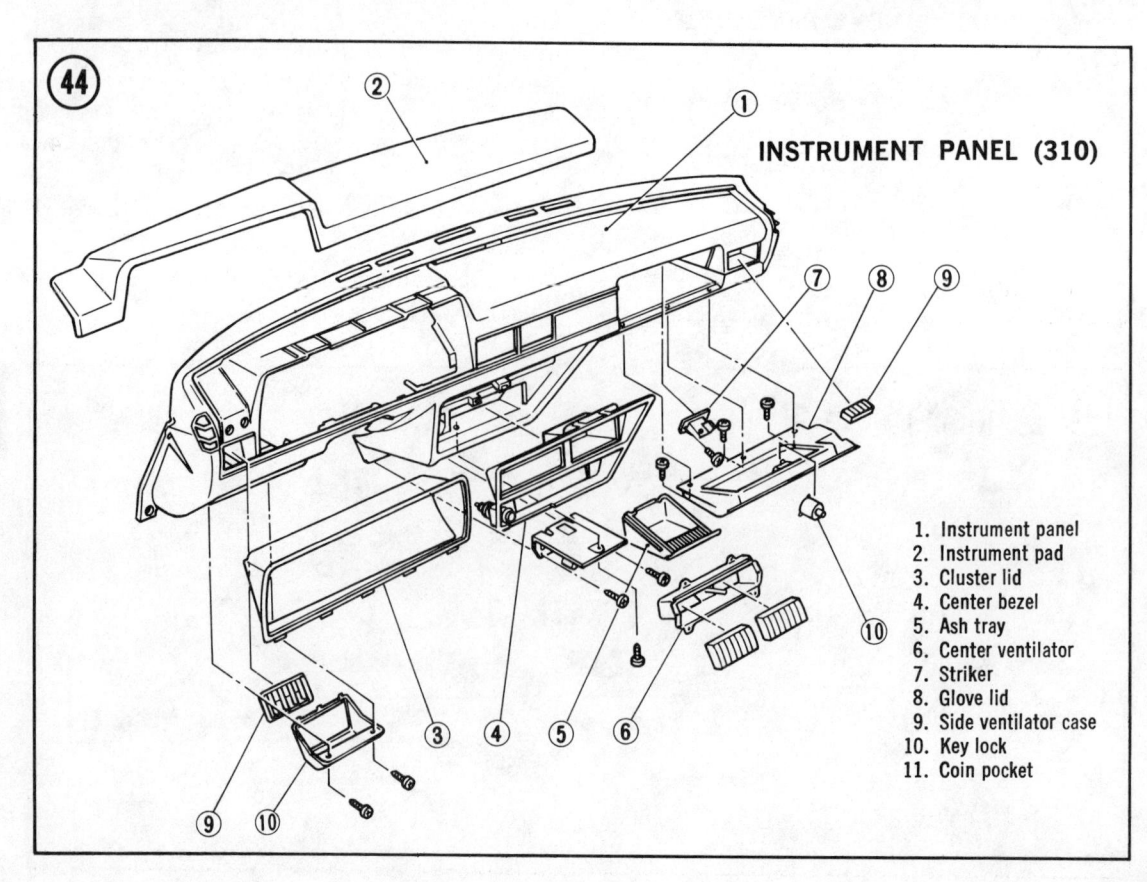

**44** INSTRUMENT PANEL (310)

1. Instrument panel
2. Instrument pad
3. Cluster lid
4. Center bezel
5. Ash tray
6. Center ventilator
7. Striker
8. Glove lid
9. Side ventilator case
10. Key lock
11. Coin pocket

# BODY

2. Reach behind the instrument panel. Disconnect the speedometer cable from the back of the speedometer.

3. Remove the package tray.

4. Disconnect the heater control cables from the heater. See *Heater Removal/Installation*, Chapter Six.

5. Label and disconnect the instrument panel wiring harnesses.

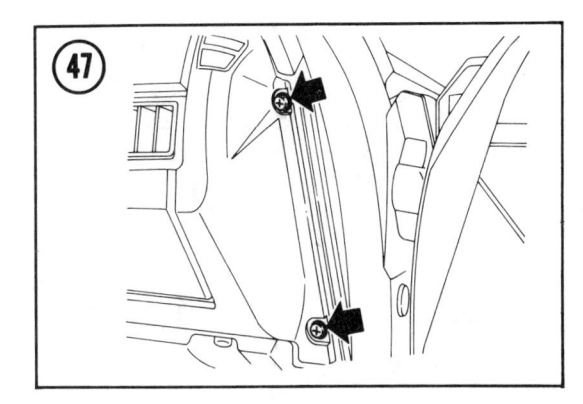

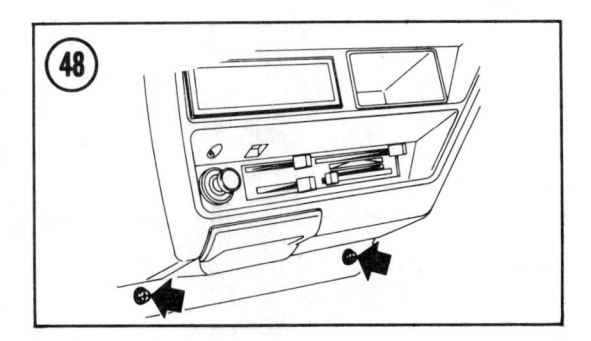

6. Remove the steering column bracket bolts **(Figure 42)**.

7. Remove the screws securing the upper edge of the instrument panel.

8. Remove the mounting bolts at the sides of the instrument panel. See **Figure 43**.

9. Unbolt the instrument panel from the pedal bracket.

10. Remove the panel into the passenger compartment.

11. Installation is the reverse of removal.

**Removal/Installation (310)**

Refer to **Figure 44** for this procedure.

1. Disconnect negative cable from battery.

2. Remove the steering wheel. See *Steering Wheel Removal/Installation*, Chapter Eleven.

3. Remove the steering column shell.

4. Reach behind the speedometer. Disconnect the cable.

5. Disconnect the instrument panel wiring connectors.

6. Remove the heater control knobs. Remove the control panel mounting screws **(Figure 45)**.

7. Remove the panel's left mounting screws, then the right mounting screws. See **Figures 46 and 47**.

8. Remove the lower center mounting screws **(Figure 48)**.

9. Make a wire hook and pull back the instrument mask **(Figure 49)**.

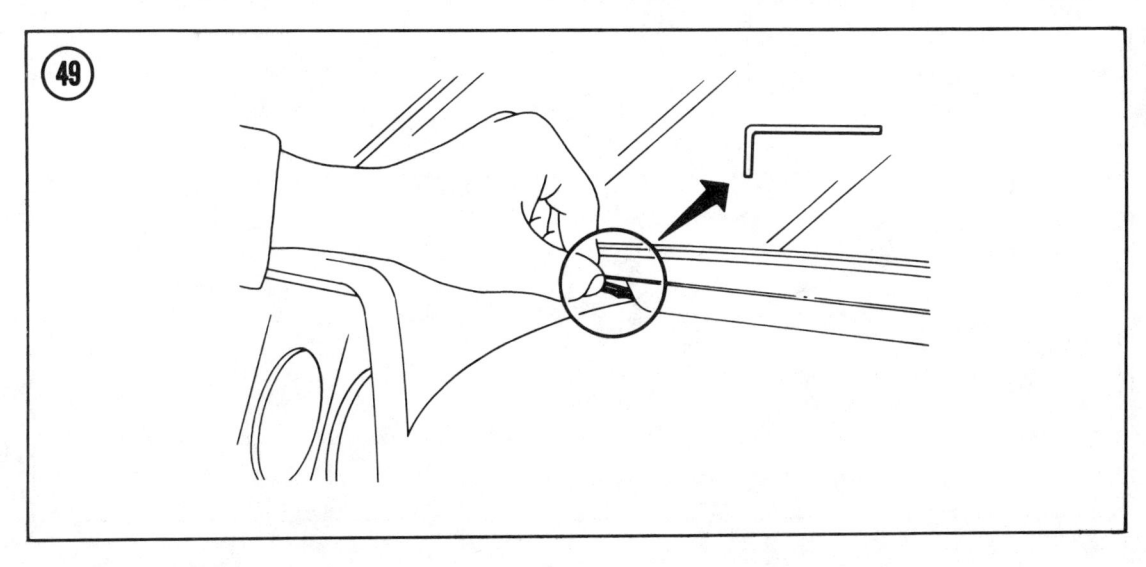

CHAPTER TWELVE

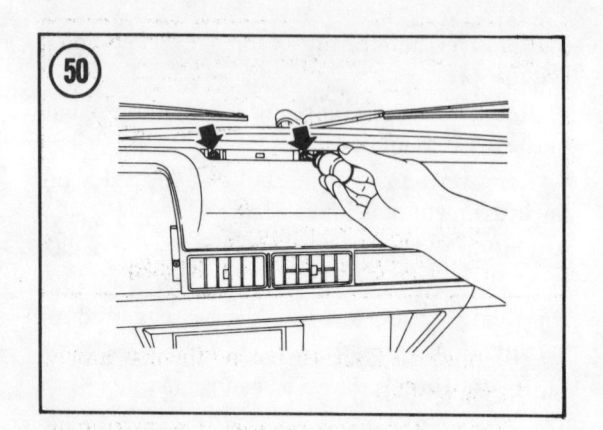

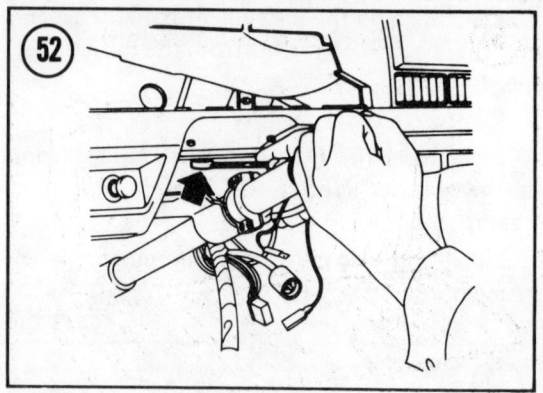

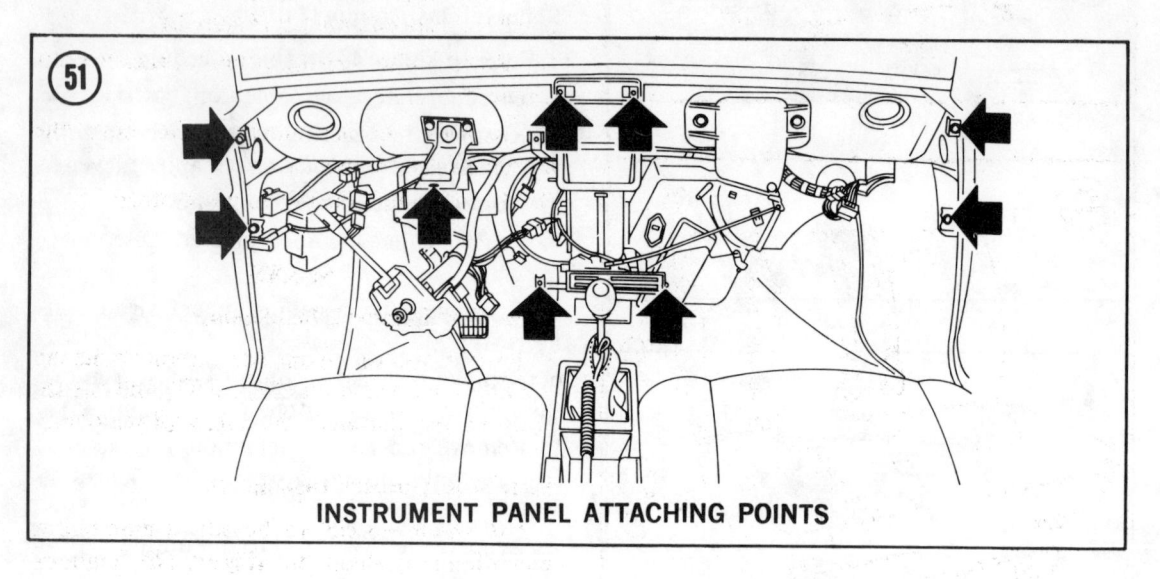

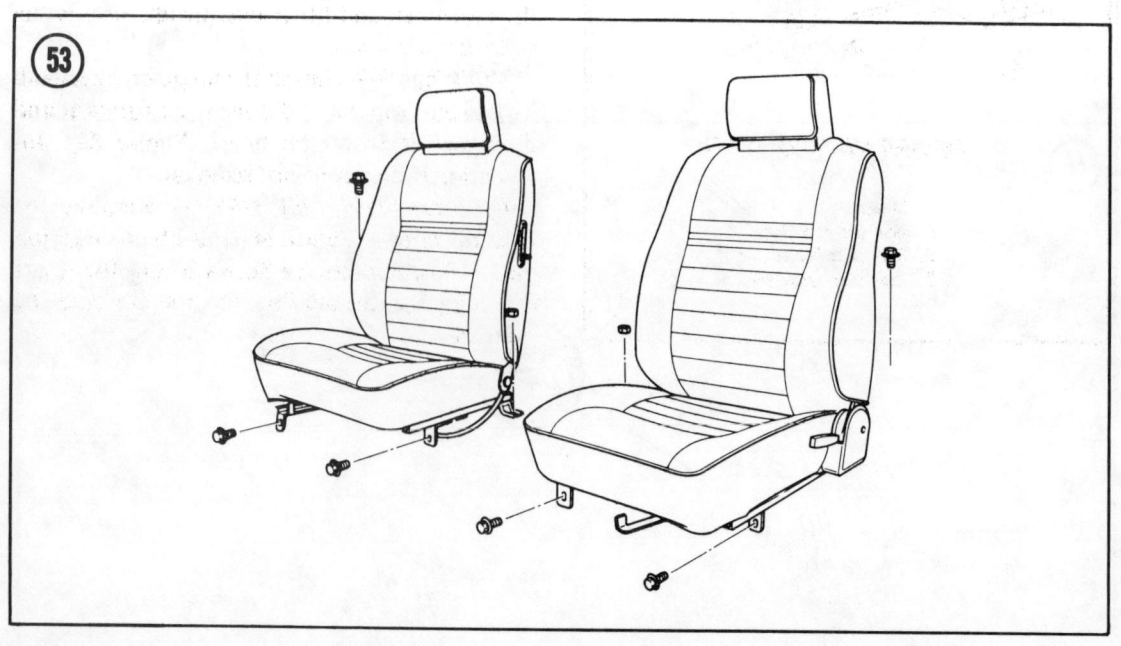

INSTRUMENT PANEL ATTACHING POINTS

# BODY

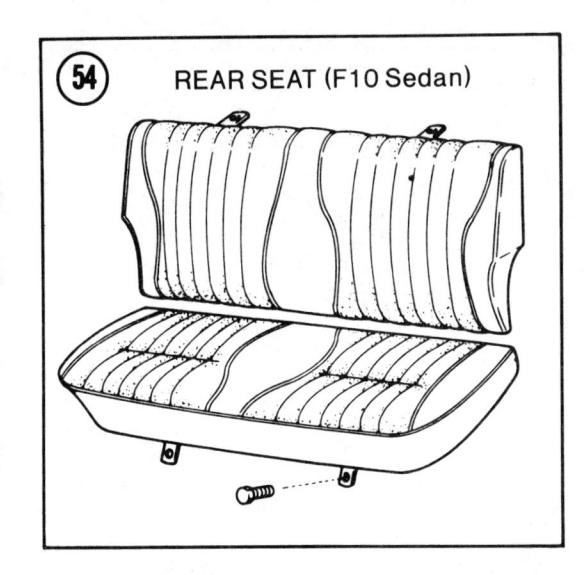

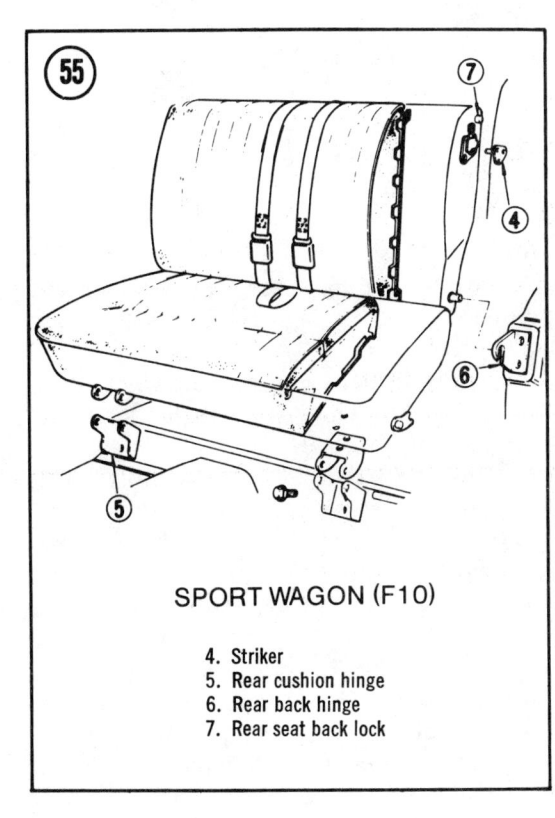

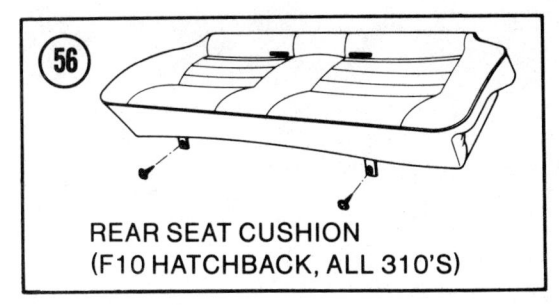

**Figure 54** REAR SEAT (F10 Sedan)

**Figure 56** REAR SEAT CUSHION (F10 HATCHBACK, ALL 310'S)

10. Remove the 2 screws behind the instrument mask (**Figure 50**).

11. Remove the screw securing the instrument panel to the pedal bracket (**Figure 51**).

12. Remove the steering column bracket bolts (**Figure 52**).

13. Remove the instrument panel into the passenger compartment.

14. Installation is the reverse of removal.

## SEATS

### Front Seat Removal/Installation

To remove the front seat, remove its attaching nuts and bolts (**Figure 53**) and lift the seat out. Installation is the reverse of removal.

### Rear Seat Removal/Installation

*F10 Sedan* — Remove the seat cushion screws and lift the cushion out (**Figure 54**). Unhook the seat back and lift it out. Installation is the reverse of removal.

*F10 Wagon* — Unbolt the hinge brackets and lift the cushion out. Fold the back forward and disengage it from the hinge (**Figure 55**). Installation is the reverse of removal.

*F10 Hatchback, all 310's* — Remove the cushion screws (**Figure 56**) and lift the cushion out. Fold the seatback forward and lift it off the hinges. Installation is the reverse of removal.

**Figure 55** SPORT WAGON (F10)

4. Striker
5. Rear cushion hinge
6. Rear back hinge
7. Rear seat back lock

# CHAPTER THIRTEEN

# AIR CONDITIONING

This chapter covers the maintenance and minor repairs that can prevent or correct most air conditioning problems. Major repairs require special training and tools and should be left to a Datsun dealer or air conditioning shop.

## SYSTEM OPERATION

A schematic of the air conditioning system is shown in **Figure 1**. These 5 basic components are common to all air conditioning systems:
  a. Compressor
  b. Condenser
  c. Receiver/drier
  d. Expansion valve
  e. Evaporator

> *WARNING*
> *The components, connected with high-pressure hoses and tubes, form a closed loop. The refrigerant in the system is under very high pressure. It can cause frostbite if it touches skin and blindness if it touches the eyes. If discharged near a flame, the refrigerant creates poisonous gas. If the refrigerant can is hooked up wrong, it can explode. For these reasons,* **read this entire section** *before working on the system.*

For practical purposes, the cycle begins at the compressor. The refrigerant, in a warm, low-pressure vapor state, enters the low-pressure side of the compressor. It is compressed to a high-pressure hot vapor and pumped out of the high-pressure side to the condenser.

Air flow through the condenser removes heat from the refrigerant and transfers the heat to the outside air. As the heat is removed, the refrigerant condenses to a warm, high-pressure liquid.

The refrigerant then flows to the receiver/drier where moisture is removed and impurities are filtered out. The refrigerant is stored in the receiver/drier until it is needed. The receiver/drier incorporates a sight glass that permits visual monitoring of the condition of the refrigerant as it flows. From the receiver/drier, the refrigerant then flows to the expansion valve. The expansion valve is thermostatically controlled and meters refrigerant to the evaporator. As the refrigerant leaves the expansion valve it changes from a warm, high-pressure liquid to a cold, low-pressure liquid.

In the evaporator, the refrigerant removes heat from the passenger compartment air that is blown across the evaporator's fins and

# AIR CONDITIONING

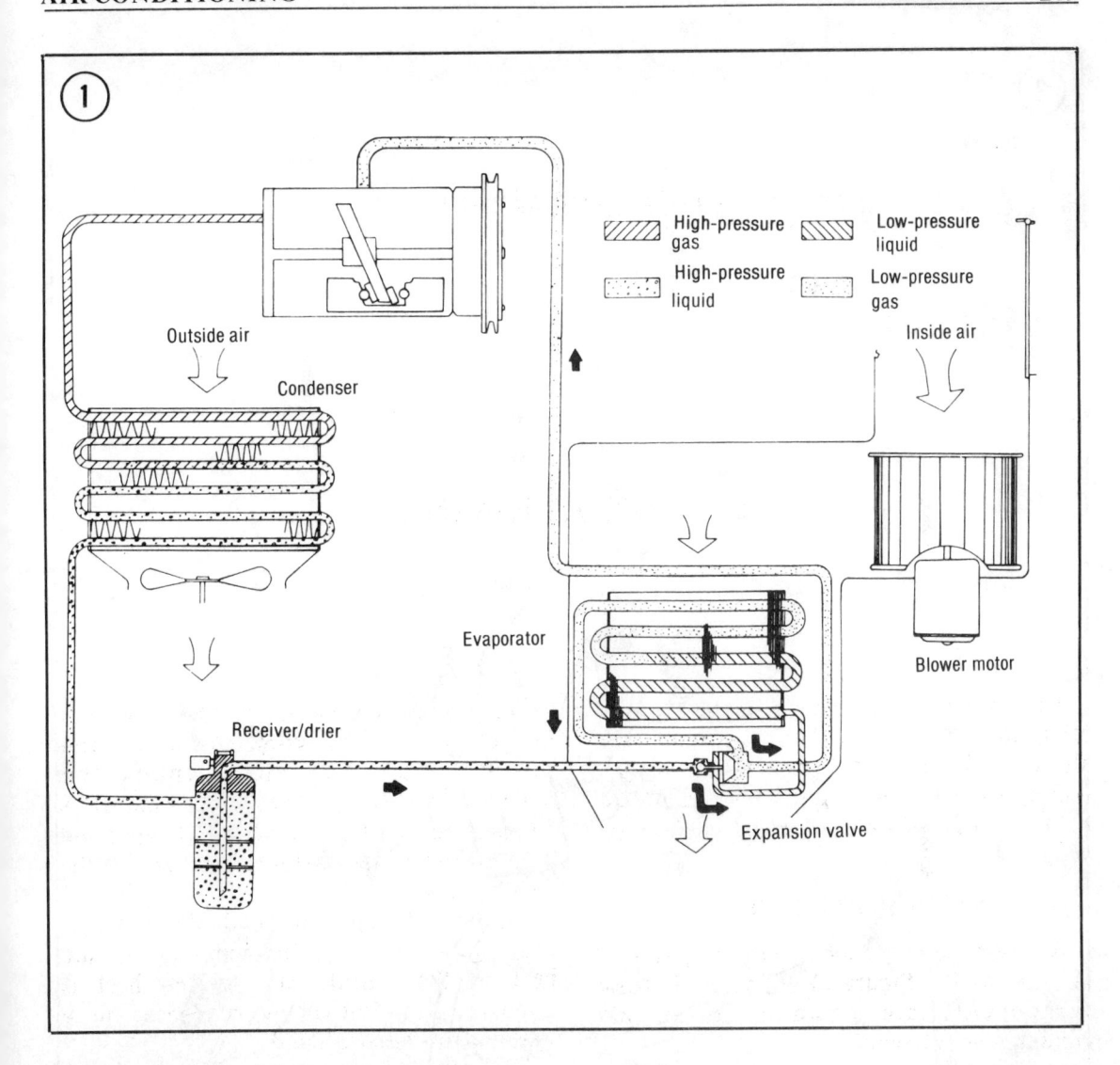

tubes. In the process, the refrigerant changes from a cold, low-pressure liquid to a warm, high-pressure vapor. The vapor flows back to the compressor, where the cycle begins again.

## GET TO KNOW YOUR VEHICLE'S SYSTEM

Locate each of the following components in turn:
a. Compressor
b. Condenser
c. Receiver/drier
d. Expansion valve
e. Evaporator

## Compressor

The compressor (**Figure 2**) is mounted on the fan end of the engine, like the alternator, and is driven by a V-belt. The large pulley on the front of the compressor contains an electromagnetic clutch. This activates and operates the compressor when the air conditioning is switched on.

## Condenser

The condenser (**Figure 2**) is mounted next to the radiator. Air passing through the fins and tubes removes heat from the refrigerant in the same way as it removes heat from the engine coolant as it passes through the radiator.

**CHAPTER THIRTEEN**

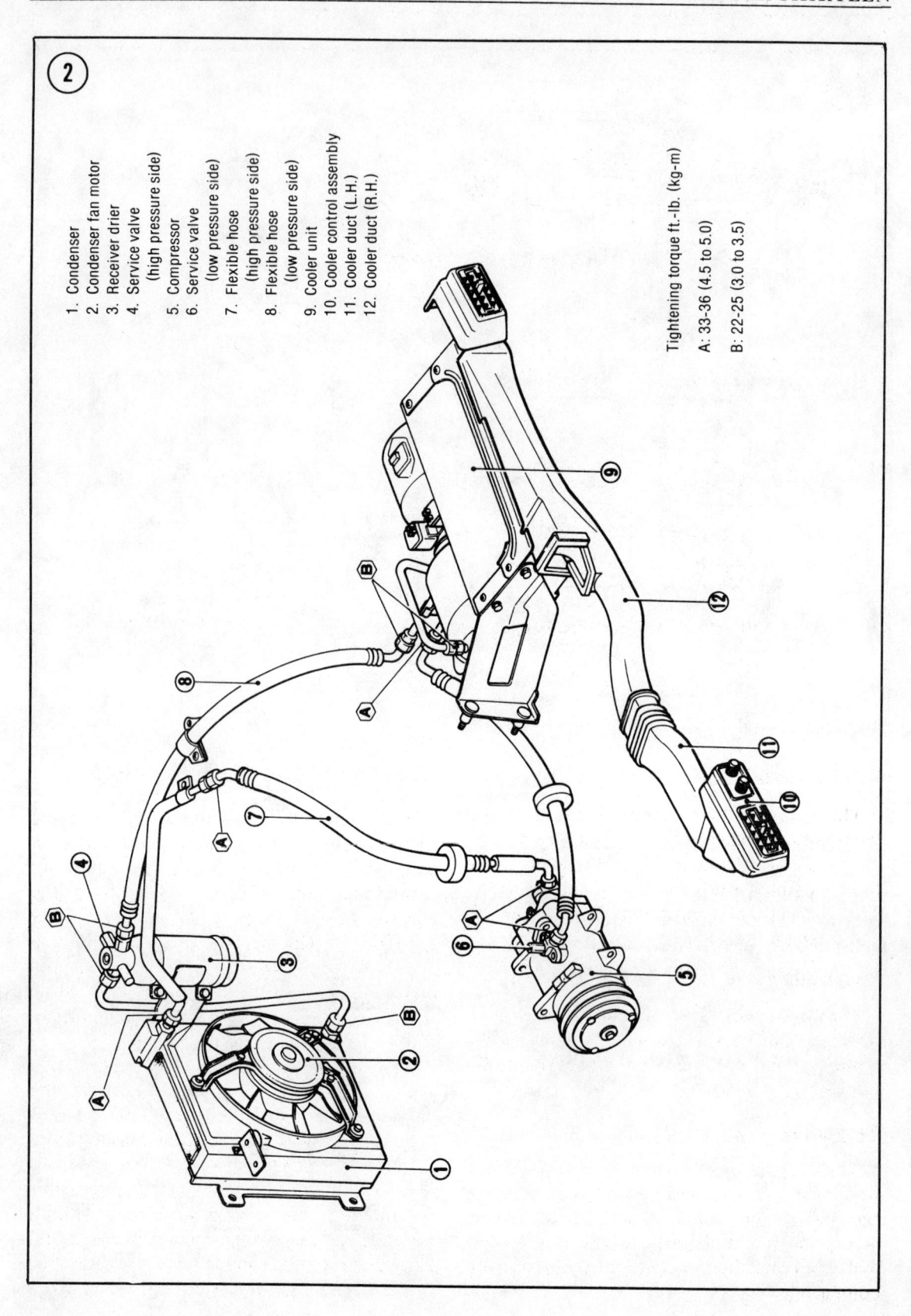

②

1. Condenser
2. Condenser fan motor
3. Receiver drier
4. Service valve
 (high pressure side)
5. Compressor
6. Service valve
 (low pressure side)
7. Flexible hose
 (high pressure side)
8. Flexible hose
 (low pressure side)
9. Cooler unit
10. Cooler control assembly
11. Cooler duct (L.H.)
12. Cooler duct (R.H.)

Tightening torque ft.-lb. (kg-m)
A: 33-36 (4.5 to 5.0)
B: 22-25 (3.0 to 3.5)

# AIR CONDITIONING

1. Sight glass
2. Service valve (high-pressure side)
3. Pressure safety valve
4. Strainer
5. Desiccant

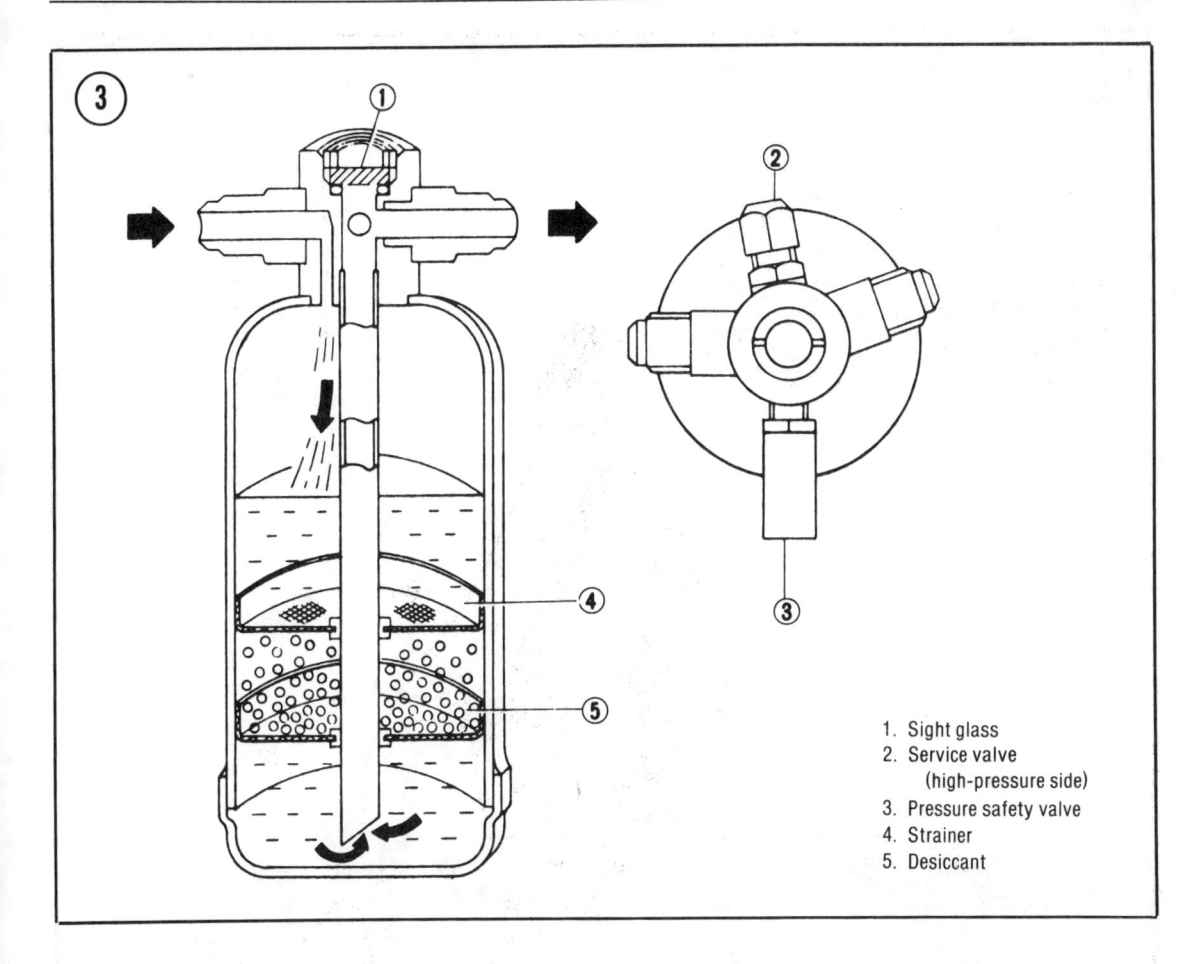

## Receiver/Drier

The receiver/drier (**Figure 3**) is a small tank-like unit, usually mounted to one of the wheel wells. It incorporates a sight glass through which refrigerant flow can be seen. The refrigerant's appearance is used to troubleshoot the system.

## Expansion Valve

The expansion valve (**Figure 4**) is located between the receiver/drier and the evaporator. It is mounted on the evaporator housing.

## Evaporator

The evaporator (**Figure 4**) is located in the passenger compartment cooling unit, beneath the instrument panel. Warm air is blown across the fins and tubes, where it is cooled and dried and then ducted into the passenger compartment.

## ROUTINE MAINTENANCE

Basic maintenance of the air conditioning system is easy; at least once a month, even in cold weather, start your engine, turn on the air conditioner and operate it at each of the control settings. Operate the air conditioner for about 10 minutes, with the engine running at 1,500 rpm. This will ensure that the compressor seal will not deform from sitting in the same position for a long period of time. If this occurs, the seal is likely to leak.

The efficiency of the air conditioning system also depends in great part on the efficiency of the cooling system. This is because the heat from the condenser passes through the radiator. If the cooling system is dirty or low on coolant, it may be impossible to operate the air conditioner without overheating. Inspect the coolant. If necessary, flush and refill the cooling system as described under *Cooling System Flushing* in Chapter Six.

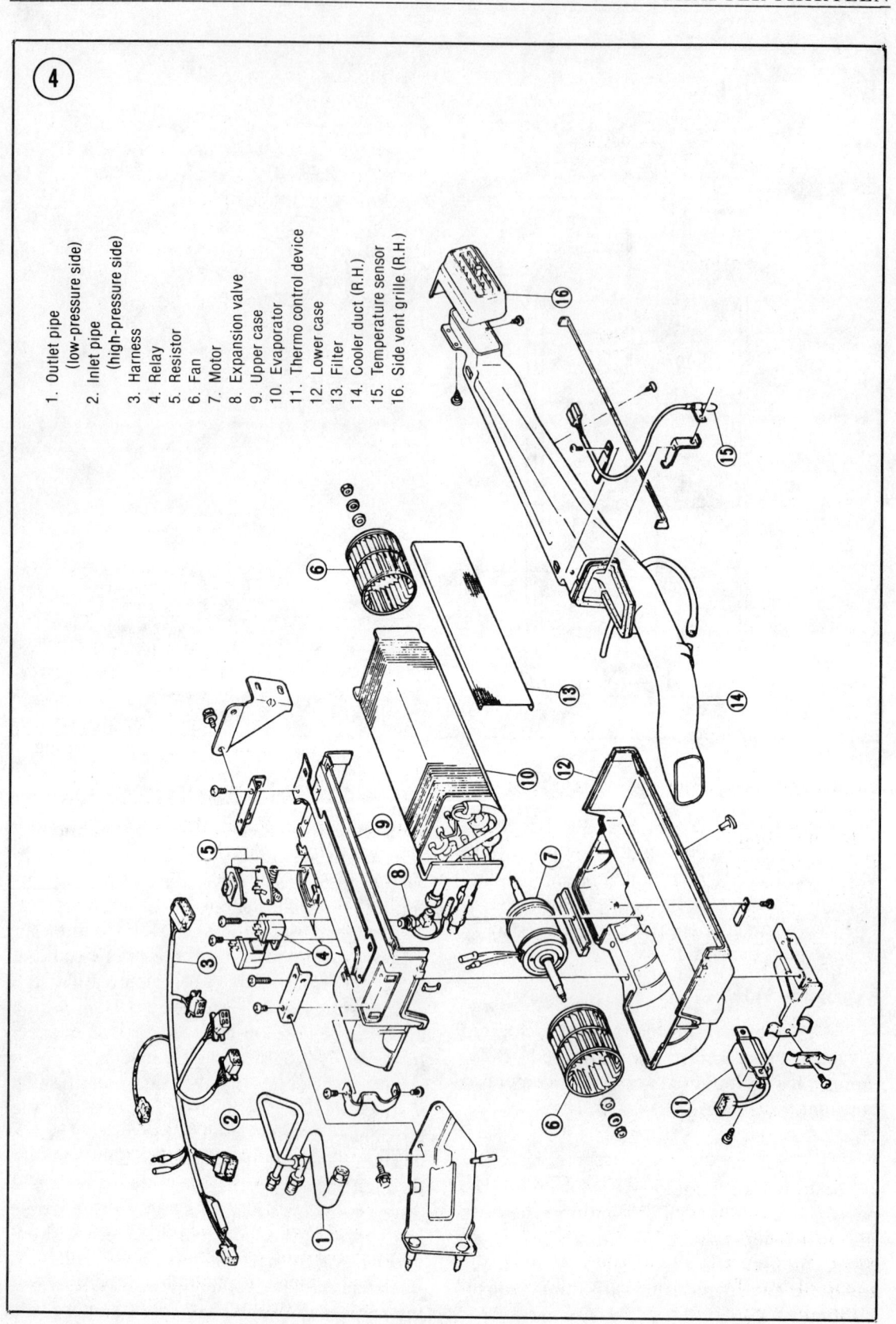

# CHAPTER THIRTEEN

④

1. Outlet pipe
   (low-pressure side)
2. Inlet pipe
   (high-pressure side)
3. Harness
4. Relay
5. Resistor
6. Fan
7. Motor
8. Expansion valve
9. Upper case
10. Evaporator
11. Thermo control device
12. Lower case
13. Filter
14. Cooler duct (R.H.)
15. Temperature sensor
16. Side vent grille (R.H.)

# AIR CONDITIONING

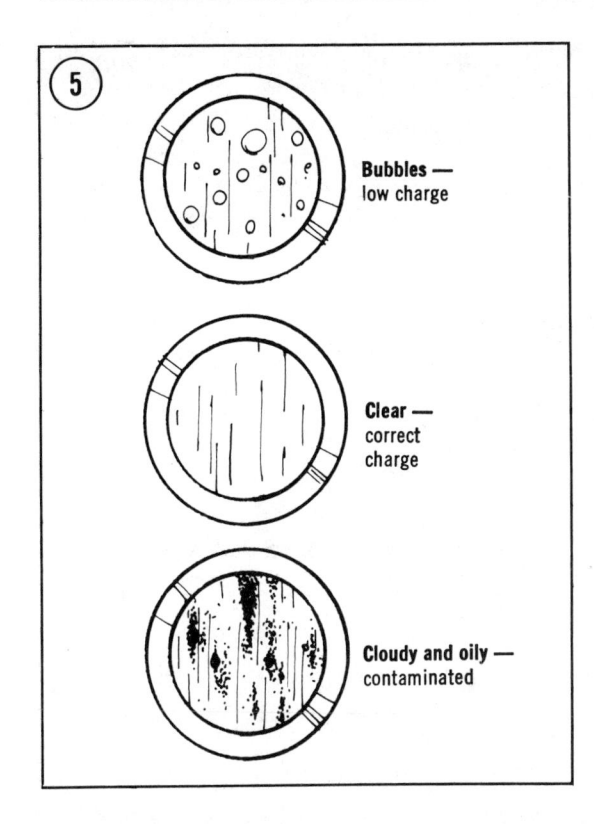

**(5)**
Bubbles — low charge
Clear — correct charge
Cloudy and oily — contaminated

With an air hose and a soft brush, clean the radiator and condenser fins and tubes to remove bugs, leaves and other imbedded debris.

Check drive belt tension as described under *Drive Belts*, Chapter Three.

If the condition of the cooling system thermostat is in doubt, test it as described under *Thermostat* in Chapter Six.

Once you are sure the cooling system is in good condition, the air conditioning system can be inspected.

## Inspection

1. Clean all lines, fittings and system components with solvent and a clean rag. Pay particular attention to the fittings; oily dirt around connections almost certainly indicates a leak. Oil from the compressor will migrate through the system to the leak. Carefully tighten the connection, but don't overtighten and strip the threads. If the leak persists, it will soon be apparent once again as oily dirt accumulates. Clean the sight glass with a clean, dry cloth.

2. Clean the condenser fins and tubes with a soft brush and an air hose or with a high-pressure stream of water from a garden hose. Remove bugs, leaves and other imbedded debris. Carefully straighten any bent fins with a screwdriver, taking care not to puncture or dent the tubes.

3. Start the engine and check the operation of the blower motor, condenser fan and compressor clutch by turning the controls on and off. If any of these fails to operate, shut off the engine and check the fuses. If they are blown, replace them. If not, remove them and clean the fuse holder contacts. Then check operation again. If this doesn't solve the problem, have the air conditioning system tested further by a Datsun dealer or air conditioning shop.

## Testing

1. Place the transmission in NEUTRAL. Set the handbrake.
2. Start the engine and run it at a fast idle.
3. Set the temperature control to its coldest setting and the blower to its highest speed. Allow the system to operate for 10 minutes with the doors and windows open. Then shut them and set the blower on its lowest setting.
4. Check air temperature at the outlet. It should be noticeably colder than the surrounding air. If not, the refrigerant level is probably low. Check the sight glass as described in the following step.
5. Run the engine at a fast idle and switch on the air conditioning. Look at the sight glass (**Figure 5**) and check for the following:
   a. Bubbles—the refrigerant level is low.
   b. Oily or cloudy—the system is contaminated. Have it serviced by a dealer or air conditioning shop.
   c. Clear glass—either there is enough refrigerant, too much or the system is so close to empty it can't make bubbles. If there is no difference between the inlet and outlet air temperatures, the system is probably near empty. If the system does blow cold air, it either has the right amount of refrigerant or too much. To tell which, turn off the air conditioner while watching the sight glass. If the

refrigerant foams, then clears up, the amount is correct. If it doesn't foam, but stays clear, there is too much.

## REFRIGERANT

The air conditioning system uses a refrigerant called dichlorodifluoromethane, or R-12.

*WARNING*
*R-12 creates freezing temperatures when it evaporates. This can cause frostbite if it touches skin and blindness if it touches the eyes. If discharged near an open flame, R-12 creates poisonous gas. If the refrigerant can is hooked up to the pressure side of the compressor, it may explode. Always wear safety goggles when working with R-12.*

### Charging

This section applies to partially discharged or empty air conditioning systems. If a hose has been disconnected or any internal part of the system exposed to air, the system should be evacuated and recharged by a dealer or air conditioning shop. Recharge kits are available from auto parts stores. Be sure the kit includes a gauge set.

1. Carefully read and understand the gauge manufacturer's instructions before charging the system.
2. Place the refrigerant can in a pan of *warm* water, *not hot*.

*WARNING*
*Water temperature must not exceed 104° F (40°C). If it does, the can may explode.*

4. Turn the handle of the refrigerant can tap valve all the way counterclockwise to retract the needle.
5. Turn the disc on the can tap valve all the way counterclockwise. See **Figure 6**. Install the valve on the can.
6. Connect the center hose to the can tap valve.
7. Make sure the gauge valves are closed.
8. Turn the can tap valve clockwise to make a hole in the can. See **Figure 6**.
9. Turn the handle all the way counterclockwise to fill the center hose with air.
10. Slowly loosen the nut connecting the center hose to the gauge set, until hissing can be heard. See **Figure 6**. Let this continue for a few seconds to purge air from the hose, then tighten the nut.

*CAUTION*
*During the next steps, the refrigerant can must remain upright. If it is turned upside down, refrigerant will enter the system as a liquid, which may damage the compressor.*

11. Open the low-pressure valve. Adjust the valve so the gauge reads no more than 2.8 kg/cm$^2$ (40 psi).

*CAUTION*
*Leave the high-pressure valve closed at all times.*

12. Run the engine at a fast idle and turn on the air conditioner. Let the system charge until the sight glass is free of air bubbles. See **Figure 5**.

*NOTE*
*If the system is nearly empty, another can of refrigerant will be needed. Attach it as described in the following steps.*

13. Close the low-pressure valve.
14. Remove the can tap valve and attach a new can. Don't make a hole in the new can yet.
15. Slightly loosen the can tap valve disc. Barely open the low-pressure valve for a few seconds to purge air from the hose. Close the low-pressure valve, then tighten the can tap valve disc.
16. Turn the can tap valve handle clockwise to make a hole in the can. Let the system charge until the sight glass is free of air bubbles.
17. Once the system is fully charged, close the low-pressure valve.
18. Close the can tap valve. Very slowly loosen the charge line to allow any remaining refrigerant to escape.

# AIR CONDITIONING

**6**

(Clockwise) (Counterclockwise)

Purging air

R-12

1. Can tap handle
2. Charging hose
3. Needle
4. Refrigerant can

> *WARNING*
> *Wear gloves and safety goggles to prevent frostbite and blindness. Do not allow any open flame near the refrigerant or poisonous gas may be formed.*

19. Turn off the engine. Cover the compressor service valve fittings with a shop rag, then quickly disconnect them.

20. Install the caps on the service valves.

## TROUBLESHOOTING

If the air conditioner fails to blow cold air, the following steps will help locate the problem.

1. First, stop the car and look at the control settings. One of the most common air conditioning problems occurs when the temperature is set for maximum cold and the blower is set on its lowest speed. This promotes ice buildup on the evaporator fins and tubes, particularly in humid weather. Eventually, the evaporator will ice over completely and restrict air flow. Turn the blower on its highest speed and place a hand over an air outlet. If the blower is running but there is little or no air flowing through the outlet, the evaporator is probably iced up. Leave the blower on and turn the temperature control off or to its warmest setting and wait. It will take 10-15 minutes for the ice to start melting.

2. If the blower is not running, the fuse may be blown, there may be a loose wiring connection or the motor may be burned out. First, check the fuse block for a blown or incorrectly seated fuse. Then check the wiring for loose connections.

3. Shut off the engine and inspect the compressor drive belt. If loose or worn, tighten or replace. See *Drive Belts*, Chapter Three.

4. Start the engine. Check the compressor clutch by turning the air conditioner on and off. If the clutch does not activate, its fuse may be blown or the evaporator temperature-limiting switches may be defective. If the fuse is defective, replace it. If the fuse is not the problem, have the system checked by a Datsun dealer or air conditioning shop.

5. If the system checks out okay to this point, start the engine, turn on the air conditioner and watch the refrigerant through the sight glass. If it fills with bubbles after a few seconds, the refrigerant level is low. If the sight glass is oily or cloudy, the system is contaminated and should be serviced by a shop as soon as possible. Corrosion and deterioration occur very quickly and if not taken care of at once will result in a very expensive repair job.

6. If the system still appears to be operating as it should but air flow into the passenger compartment is not cold, check the condenser for debris that could block air flow. Make sure the condenser fan operates.

7. If the preceding steps have not solved the problem, take the car to a dealer or air conditioning shop for service.

# SUPPLEMENT

# 1980 AND LATER SERVICE INFORMATION

> This supplement contains basic service and maintenance information for the 1980-1981 Datsun 310. This information supplements the procedures in the main body (Chapters One through Thirteen) of the book, referred to in this supplement as the "basic book."
>
> The chapter headings and titles in this supplement correspond to those in the basic book. If a chapter is not included in the supplement, there are no changes affecting 1980 and later models. Use the information given for 1979 models.
>
> If your vehicle is covered by this supplement, carefully read the text then read the appropriate chapter in the basic book before beginning any work.

# CHAPTER THREE

# LUBRICATION, MAINTENANCE, AND TUNE-UP

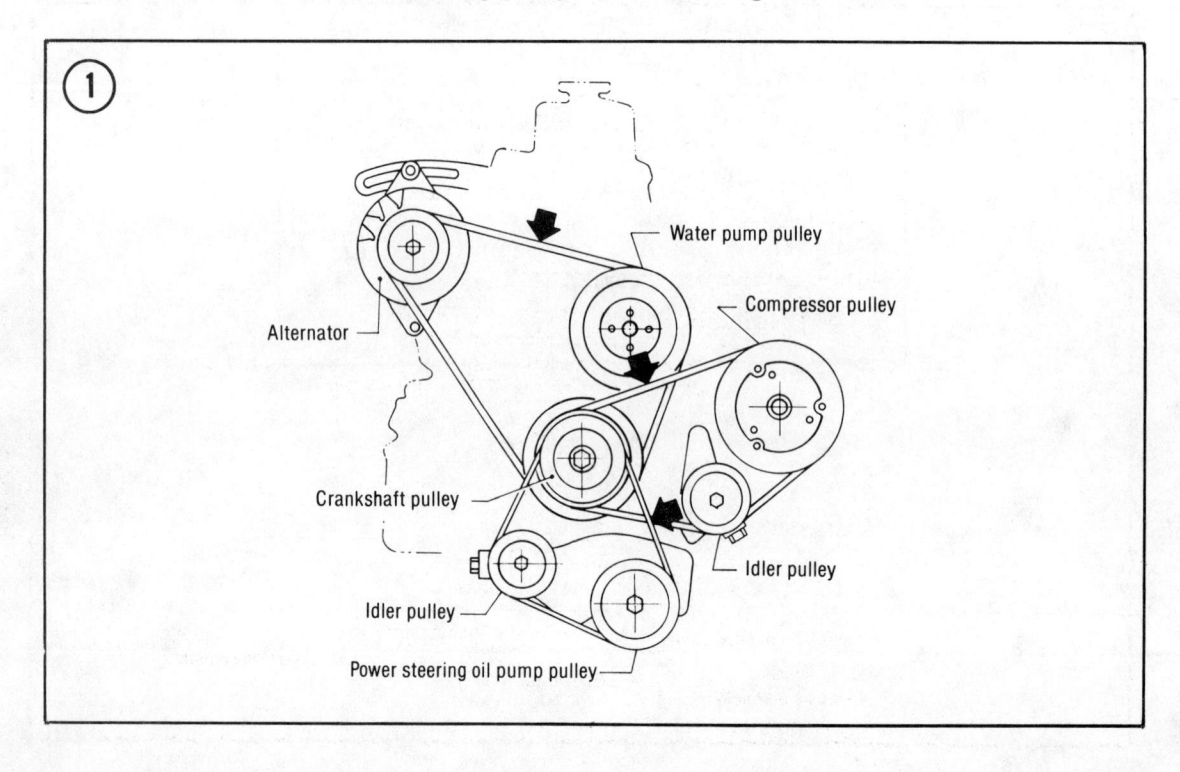

## SCHEDULED MAINTENANCE

The 1980 and 1981 maintenance schedules differ from earlier years. See **Table 1** (1980) or **Table 2** (1981).

### Drive Belts

Drive belt adjustment for 1980 is the same as for earlier models. However, the air pump belt is used only on California cars.

The 1981 models do not have an air pump, and may be equipped with power steering. **Figure 1** shows the drive belts. The alternator and air conditioning compressor belts should deflect 1/3-1/2 in. (8-12mm) when pushed at the points shown. The power steering pump belt should deflect 1/4-1/3 in. (6-8mm). Adjust as follows.

1. To adjust the alternator belt, loosen the alternator mounting and adjusting bolts. Pull or pry the alternator away from the engine to tighten the belt, then tighten the bolts.
2. To adjust the power steering or air conditioning compressor belt, loosen the idler pulley locknut. Turn the adjusting bolt to set belt tension, then tighten the locknut.

### Vacuum Lines

Periodic vacuum line inspection is required only on Canadian models. See **Figure 2** (1980) or **Figure 3** (1981).

# 1980 AND LATER SERVICE INFORMATION

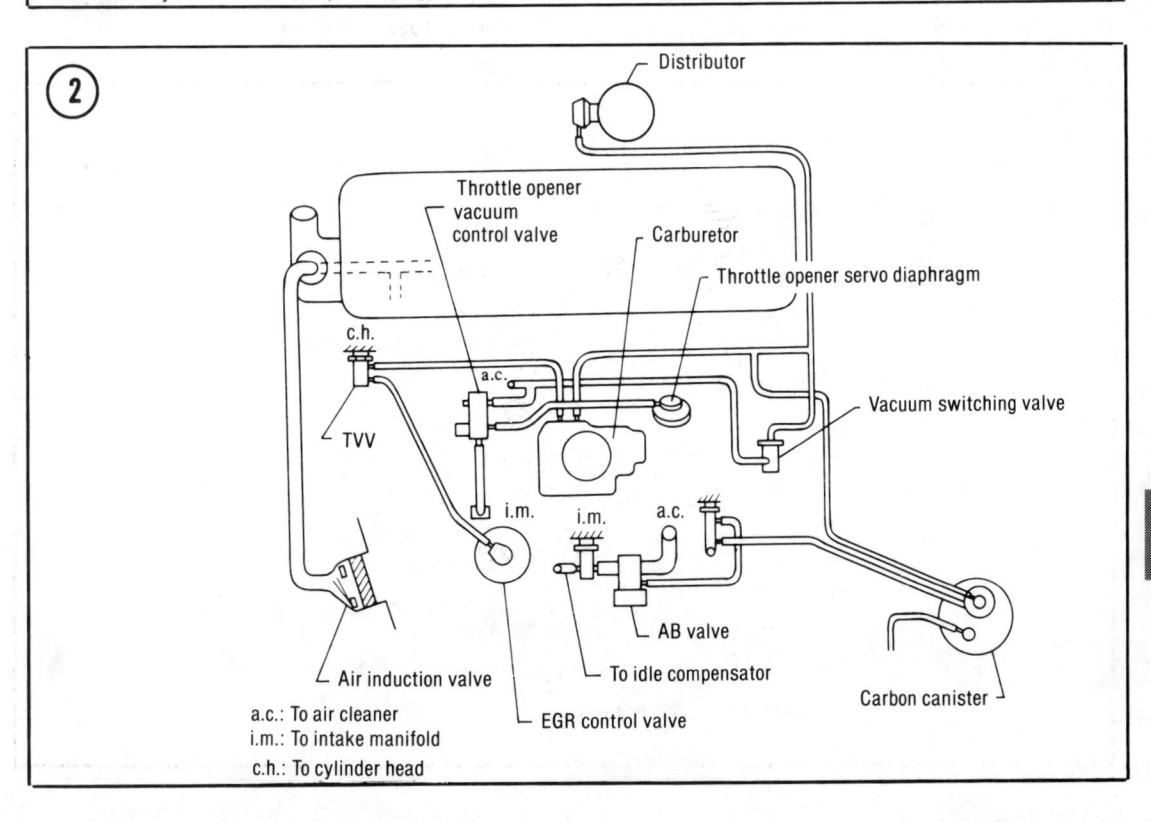

Table 1  SCHEDULED MAINTENANCE, 1980

| | |
|---|---|
| Every 7,500 miles<br>(6 months) | Engine oil and filter<br>Transmission oil check<br>Brakes<br>Hinges, latches, locks<br>Leak inspection |
| Every 15,000 miles<br>(12 months) | Drive belts<br>Coolant hoses and connections*<br>Vacuum lines<br>ATC air cleaner test*<br>Choke plate and linkage*<br>Brake fluid<br>Steering and suspension<br>Wheels and tires<br>Brakes<br>Tune-up |
| Every 30,000 miles | Coolant<br>Air filters<br>Air induction valve filter<br>Fuel filter**<br>Carbon canister filter*<br>PCV filter**<br>Evaporative emission control lines<br>Wheel bearings<br>Transmission oil change |

*Canada only
**Canada only. On U.S. models, replace if clogged.

②

Labels: Distributor; Throttle opener vacuum control valve; Carburetor; Throttle opener servo diaphragm; Vacuum switching valve; c.h.; a.c.; TVV; i.m.; i.m.; a.c.; AB valve; Air induction valve; To idle compensator; Carbon canister; EGR control valve

a.c.: To air cleaner
i.m.: To intake manifold
c.h.: To cylinder head

# SUPPLEMENT

Table 2  SCHEDULED MAINTENANCE, 1981

| Every 7,500 miles (6 months) | Engine oil and filter |
|---|---|
| Every 15,000 miles (12 months) | Transmission oil check<br>Coolant hoses and connections<br>Hinges, latches, locks<br>Leak inspection<br>Drive belts*<br>ATC air cleaner<br>Choke plate and linkage*<br>ATC air cleaner test*<br>Brake fluid<br>Power steering fluid and lines<br>Steering and suspension<br>Wheels and tires<br>Brakes<br>Tune-up |
| Every 30,000 miles (24 months) | Drive belts**<br>Coolant<br>Air cleaner element<br>Air induction valve filter<br>Choke plate and linkage**<br>PCV filter***<br>Evaporative emission control lines<br>Carbon canister filter*<br>Wheel bearings |

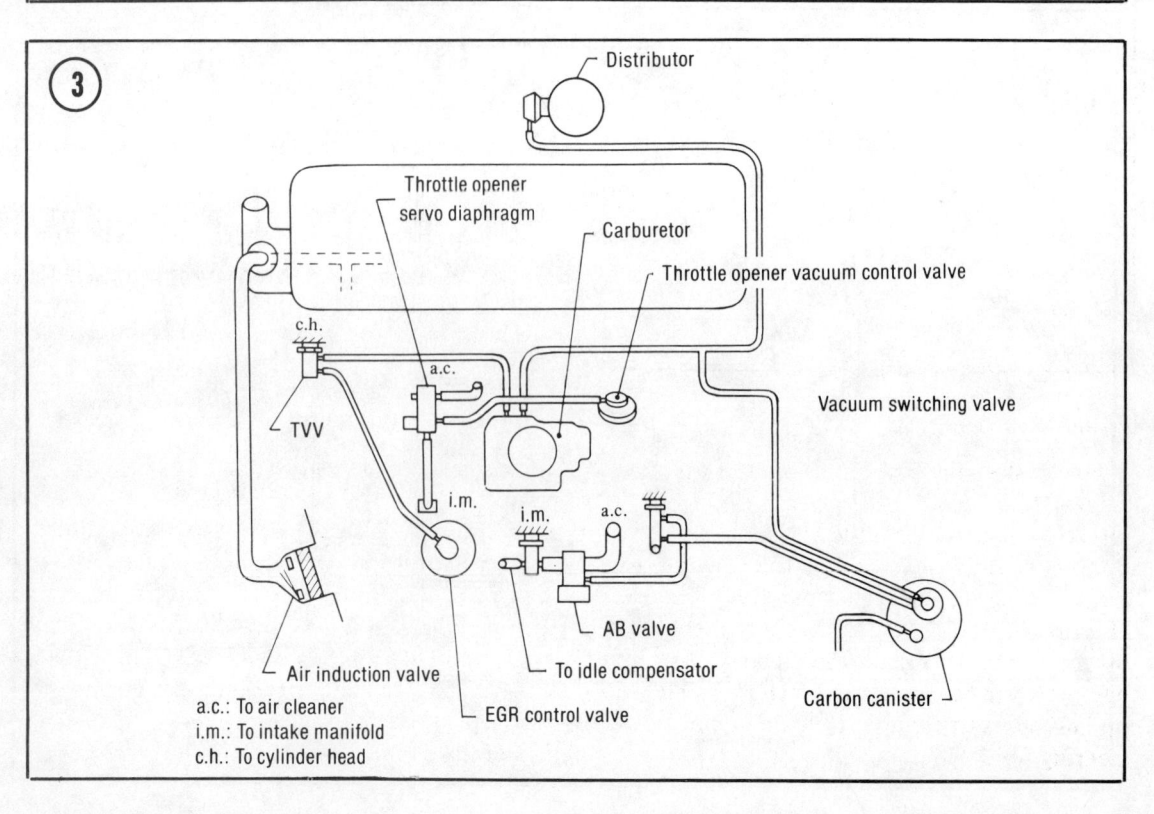

*Canada only  **U.S. only  ***Replace only if clogged.

③

a.c.: To air cleaner
i.m.: To intake manifold
c.h.: To cylinder head

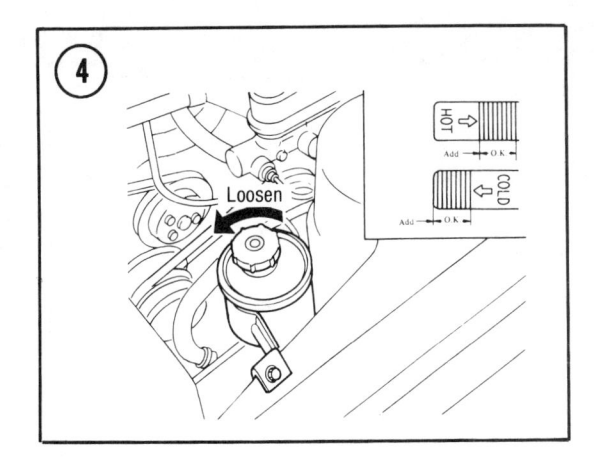

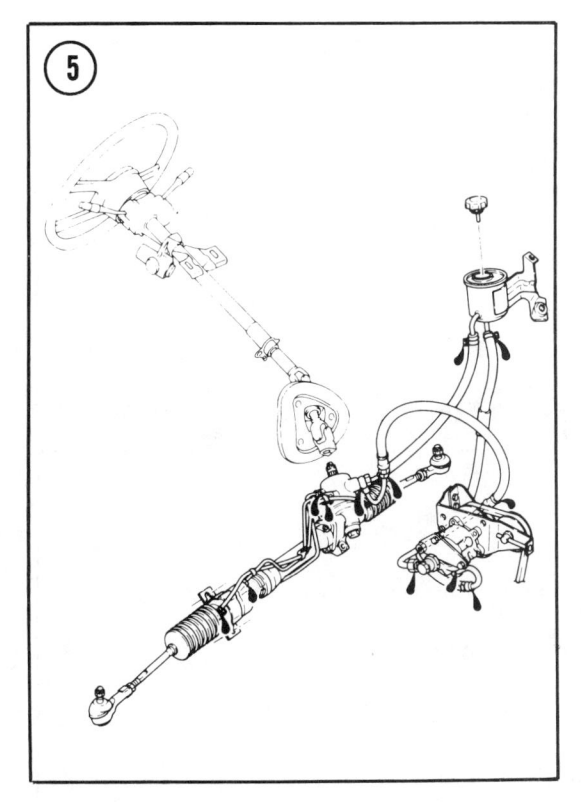

# 1980 AND LATER SERVICE INFORMATION

*CAUTION*
*Do not overfill the reservoir. Overfilling can cause the fluid to foam, resulting in wear or damage.*

2. Check fluid lines for leaks at the points shown in **Figure 5**. Tighten loose connections and replace damaged lines.

## Air Filters

Replacement procedures are the same as for ealier models. An air pump air filter is used only on 1980 California models.

## Air Induction Valve Filter

An air induction valve filter is used on 1980 non-California cars and all 1981 models. At specified intervals (**Table 1** or **Table 2**), replace the filter as described under *Air Induction System*, Chapter Five, main body of book.

## TUNE-UP

Some tune-up specifications differ from earlier models. See **Table 3**.

Valve clearances on all models should be adjusted every 15,000 miles or 12 months.

Spark plugs on U.S. models should be replaced every 30,000 miles or 24 months. On Canadian cars, replace spark plugs every 15,000 miles or 12 months.

Distributor inspection is the same as for earlier models. However, periodic ignition timing checks are not required on U.S. cars. On Canadian cars, check ignition timing every 15,000 miles or 12 months.

Idle mixture adjustment is not required on 1980 California models or 1981 U.S. cars.

## Power Steering

Power steering is optional on 1981 models. Check fluid level and check for leaks at intervals specified in **Table 2**.

1. Check fluid level with the engine off. Remove the dipstick from the fluid reservoir (**Figure 4**). With the engine cold, check level on the "COLD" side of the dipstick. With the engine warm, check on the "HOT" side. Top up if necessary with Dexron type power steering fluid. Do not use any other type.

## Idle Speed Adjustment
## (1980 California Models;
## 1981 U.S. Models)

1. Warm the engine until the temperature needle points to the middle of the gauge.
2. Connect a tune-up tachometer to the engine. Make sure engine speed is less than 1,000 rpm. If it is too high, check the choke linkage for damage or binding. Repair or lubricate as needed.
3. Let the engine idle for 2 minutes.

Table 3  TUNE-UP SPECIFICATIONS

| | |
|---|---|
| Spark plug type (U.S.) | |
|   Standard non-resistor | NGK BP5ES-11 |
|   Hot type non-resistor | NGK BP4ES-11 |
|   Cold type non-resistor | NGK BP6ES-11 or BP7ES-11 |
|   Standard resistor | NGK BPR5ES-11 |
|   Hot type resistor | NGK BPR4ES-11 |
|   Cold type resistor | NGK BPR6ES-11 or BPR7ES-11 |
| Ignition timing | |
|   1980 U.S. | 8 +/- 2° BTDC |
|   1980 Canada | 10 +/- 2° BTDC |
|   1981 | 5 +/- 2° BTDC |
| Idle speed | 750 +/-50 rpm |

4. Rev the engine to 2,000-3,000 rpm 2 or 3 times, then let it idle.
5. Check idle speed with the tachometer. It should be 750 +/- 50 rpm. If not, adjust by turning the idle speed screw.

## Idle Speed and Mixture Adjustment (1980 Non-California Models; 1981 Canadian Models)

A CO meter is required for this procedure. If you don't have one, have idle speed and mixture adjusted by a Datsun dealer or other competent shop.
1. Warm the engine until the temperature needle points to the center of the gauge.
2. Connect a tune-up tachometer to the engine. Make sure idle speed is below 1,000 rpm. If it isn't, check the choke linkage for damage or binding. Repair or lubricate as needed.
3. Disconnect the air induction hose from the pipe. See **Figure 6** (U.S. models) or **Figure 7** (Canadian models). Cap the pipe.
4. Let the engine idle for 2 minutes.
5. On 1980 U.S. models, disconnect and plug the distributor vacuum hose.
6. Race the engine engine at 2,000-3,000 rpm 2 or 3 times, then let it idle.
7. Check ignition timing as described under *Ignition Timing*, Chapter Three, basic book. Adjust as needed.
8. On 1980 U.S. models, unplug the distributor vacuum hose and reconnect the hose to the distributor.

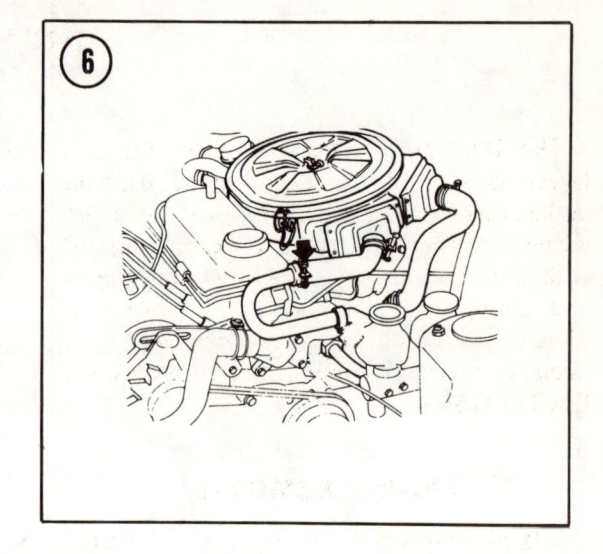

# 1980 AND LATER SERVICE INFORMATION

9. Race the engine at 2,000-3,000 rpm 2 or 3 times, then let it idle.
10. Check idle speed. It should be 750+/-50 rpm. Adjust if necessary by turning the idle speed screw.
11. Race the engine at 2,000-3,000 rpm 2 or 3 times, then let it idle.
12. Check CO percentage with the CO meter. It should be 2 +/- 1%. If not, adjust by turning the idle mixture screw.
13. Turn off the engine. Uncap the air induction pipe, then reconnect the hose to the pipe.
14. Start the engine and race it at 2,000-3,000 rpm 2 or 3 times.
15. Recheck idle speed and adjust as needed.

## CHAPTER FOUR

## ENGINE

The 1981 models use the A15 engine, a larger version of the A14 engine used in earlier cars. Service procedures are the same, except for engine removal on cars equipped with power steering. Valve seat dimensions and some specifications differ. Valve seat dimensions are shown under *Valves and Valve Seats* in this supplement. Specifications are listed in **Table 4**.

### ENGINE REMOVAL

Engine removal is the same as for earlier models, except on cars equipped with power steering. On these cars, detach the power steering pump from the engine as described under *Power Steering* in this supplement. Tie the pump back out of the way. It is not necessary to disconnect the pump lines.

### VALVES AND VALVE SEATS

Service procedures are the same as for 1979 models. For valve seat dimensions, refer to the following illustrations:

**Figure 8**—1980 intake (standard)
**Figure 9**—1980 intake (oversize)
**Figure 10**—1980 exhaust (standard)
**Figure 11**—1980 exhaust (oversize)
**Figure 12**—1981 U.S. intake (standard)
**Figure 13**—1981 U.S. intake (oversize)
**Figure 14**—1981 Canada intake (standard)
**Figure 15**—1981 Canada intake (oversize)
**Figure 16**—1981 exhaust (standard)
**Figure 17**—1981 exhaust (oversize)

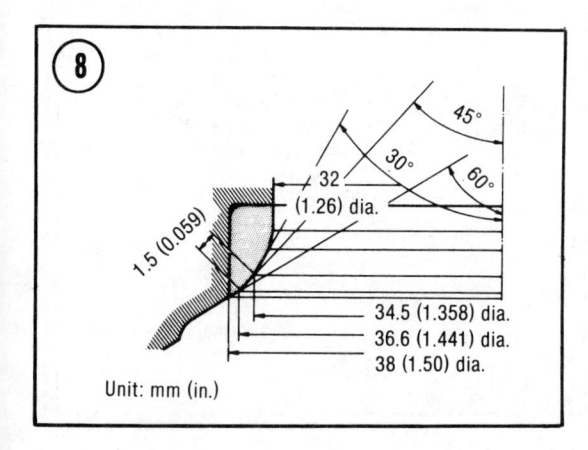

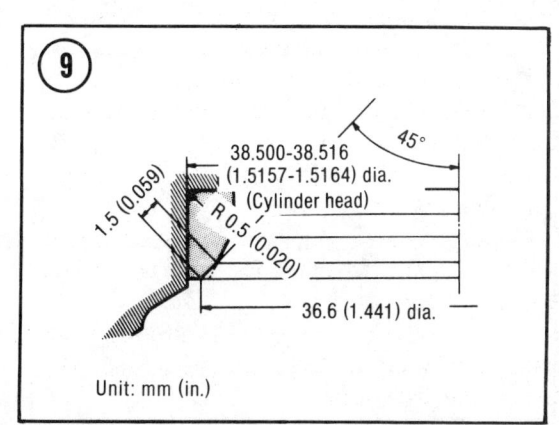

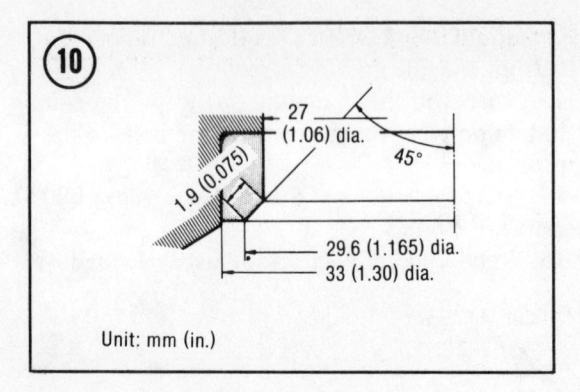

# SUPPLEMENT

### 10
27 (1.06) dia.
45°
1.9 (0.075)
29.6 (1.165) dia.
33 (1.30) dia.
Unit: mm (in.)

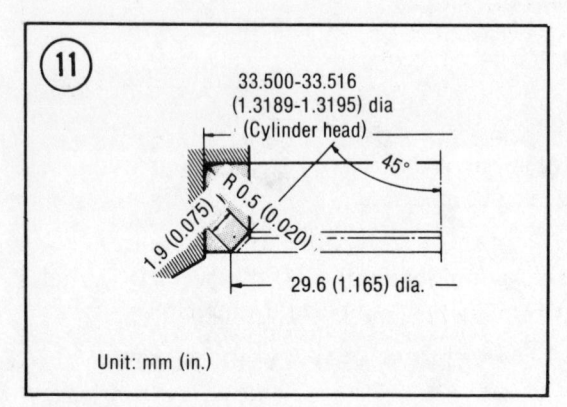

### 11
33.500-33.516 (1.3189-1.3195) dia (Cylinder head)
45°
R 0.5 (0.020)
1.9 (0.075)
29.6 (1.165) dia.
Unit: mm (in.)

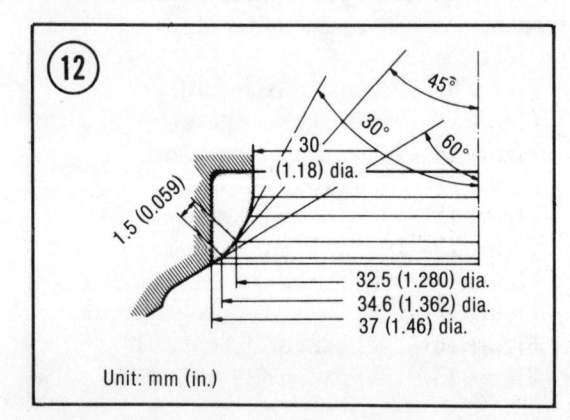

### 12
45°
30°
60°
30 (1.18) dia.
1.5 (0.059)
32.5 (1.280) dia.
34.6 (1.362) dia.
37 (1.46) dia.
Unit: mm (in.)

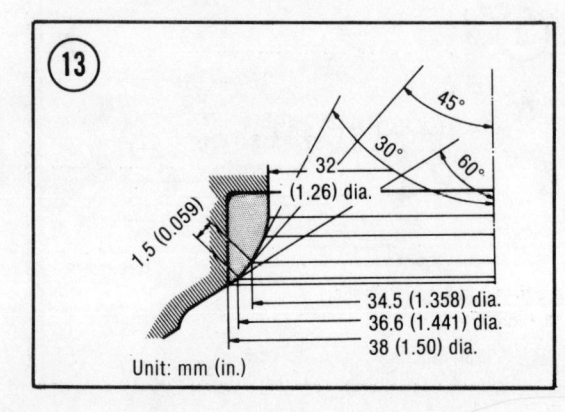

### 13
45°
30°
60°
32 (1.26) dia.
1.5 (0.059)
34.5 (1.358) dia.
36.6 (1.441) dia.
38 (1.50) dia.
Unit: mm (in.)

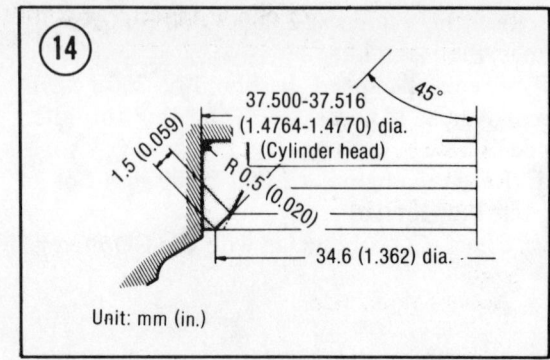

### 14
37.500-37.516 (1.4764-1.4770) dia. (Cylinder head)
45°
1.5 (0.059)
R 0.5 (0.020)
34.6 (1.362) dia.
Unit: mm (in.)

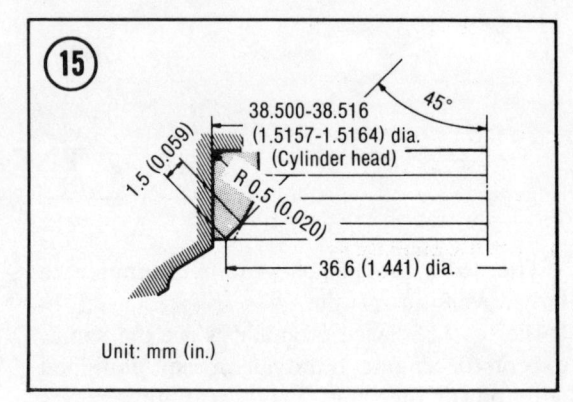

### 15
38.500-38.516 (1.5157-1.5164) dia. (Cylinder head)
45°
1.5 (0.059)
R 0.5 (0.020)
36.6 (1.441) dia.
Unit: mm (in.)

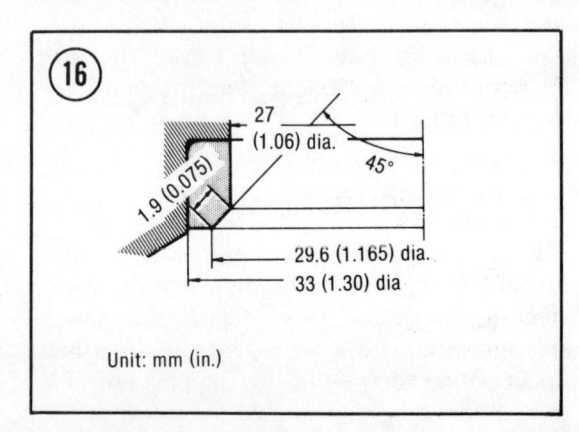

### 16
27 (1.06) dia.
45°
1.9 (0.075)
29.6 (1.165) dia.
33 (1.30) dia
Unit: mm (in.)

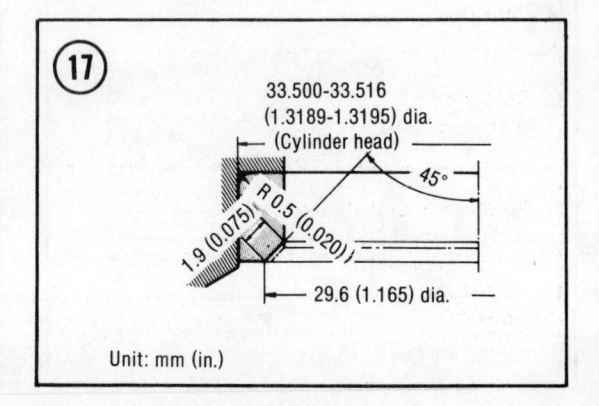

### 17
33.500-33.516 (1.3189-1.3195) dia. (Cylinder head)
45°
R 0.5 (0.020)
1.9 (0.075)
29.6 (1.165) dia.
Unit: mm (in.)

## 1980 AND LATER SERVICE INFORMATION

Table 4 ENGINE SPECIFICATIONS

| | |
|---|---|
| **VALVES** | |
| Intake head diameter, 1981 U.S. | 1.378 in. (35mm) |
| Intake stem diameter, 1981 | 0.3134-0.3140 in. (7.960-7.975mm) |
| Length | |
|   All 1980, 1981 Canada | 4.0748-4.0945 in. (103.5-104.0mm) |
|   1981 U.S. | 3.9173-3.9370 in. (99.5-100.0mm) |
| Valve spring loaded length | 1.189 in. @ 49.0-56.4 lb. (30.2 mm @ 22.2-25.6 kg) |
| **CRANKSHAFT** | |
| Maximum bend | 0.004 in. (0.1mm) |
| **CYLINDER BLOCK** | |
| Maximum out-of-round and taper | 0.0008 in. (0.02mm) |
| **CAMSHAFT** | |
| Intake lobe height, 1980 | |
|   Standard | 1.4252-1.4272 in. (36.200-36.250mm) |
|   Minimum | 1.4055 in. (35.7mm) |
| Exhaust lobe height, 1980 | |
|   Standard | 1.4539-1.4559 in. (36.930-36.980mm) |
|   Minimum | 1.4342 in. (36.430mm) |
| Intake lobe height, 1981 | |
|   Standard, U.S. | 1.4154-1.4165 in. (35.950-35.980mm) |
|   Standard, Canada | 1.4254-1.4272 in. (36.200-36.250mm) |
|   Minimum, U.S. | 1.3957 in. (35.450mm) |
|   Minimum, Canada | 1.4055 in. (35.7mm) |
| Exhaust lobe height, 1981 | |
|   Standard | 1.4146-1.4165 in. (35.930-35.980mm) |
|   Minimum | 1.3949 in. (35.430mm) |
| **OIL PUMP** | |
| Inner to outer rotor clearance | 0.001-0.002 in. (0.03-0.06mm) |
| **PISTONS AND CONNECTING RODS** | |
| Bend or twist (per 3.94 in. or 100mm of connecting rod length) | |
|   Standard | Less than 0.001 in. (0.025mm) |
|   Maximum | 0.002 in (0.05mm) |

# CHAPTER FIVE

# FUEL, EXHAUST, AND EMISSION CONTROL SYSTEMS

## AIR CLEANER

The 1980 air cleaners are the same as for 1979. The 1981 air cleaners differ slightly. See **Figure 18** (U.S.) or **Figure 19** (Canada).

Service procedures for the automatic temperature control system and idle compensator are the same as for earlier models. The altitude compensator used on 1981 California cars is an automatic type, rather than the manual design used earlier.

### Altitude Compensator

The altitude compensator, used on 1981 California cars, compensates for the thin air at high altitudes by admitting additional air to the carburetor.

If the compensator operates when it shouldn't (below 1,641 ft. or 500 meters), it may cause the following problems:
 a. Hesitation and stumbling when the engine is started
 b. Surging at approximately 50 mph
 c. Stumbling during acceleration in the 50-70 mph range
 d. Poor full-throttle acceleration

If the compensator doesn't operate at high altitudes, it may cause the following problems:
 a. Hesitation and stumbling when the engine is started
 b. Smoother running at low altitudes than at high altitudes
 c. Poor throttle reponse in neutral
 d. Poor full-throttle acceleration

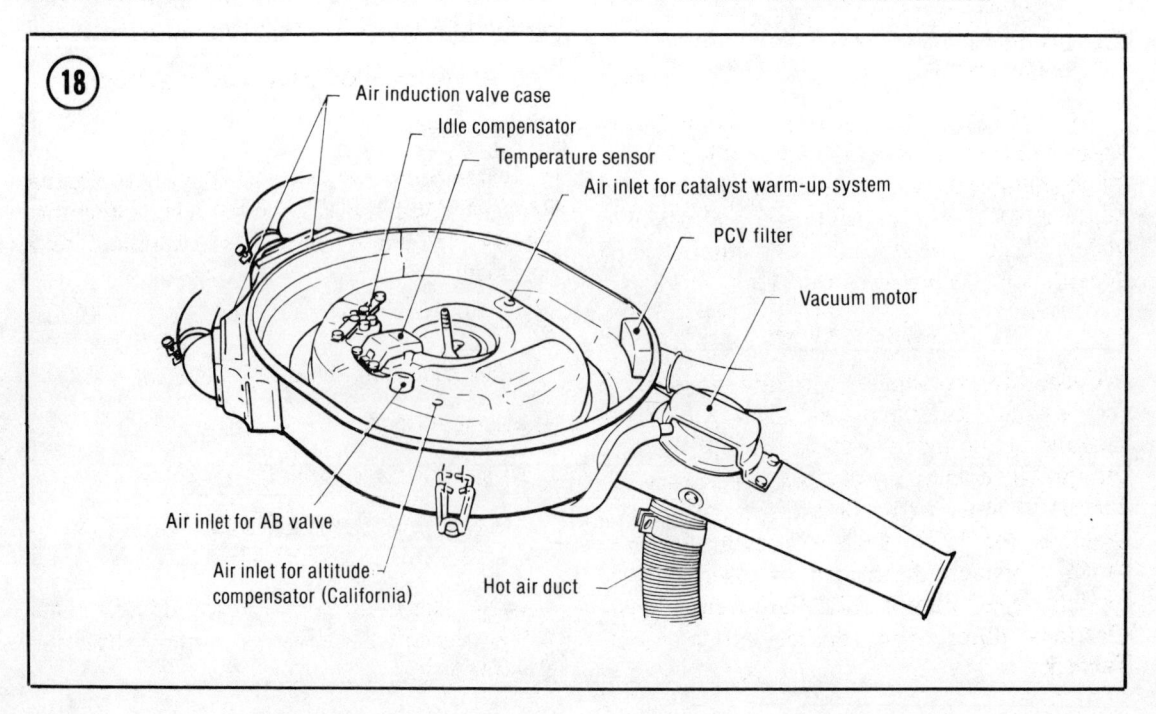

# 1980 AND LATER SERVICE INFORMATION

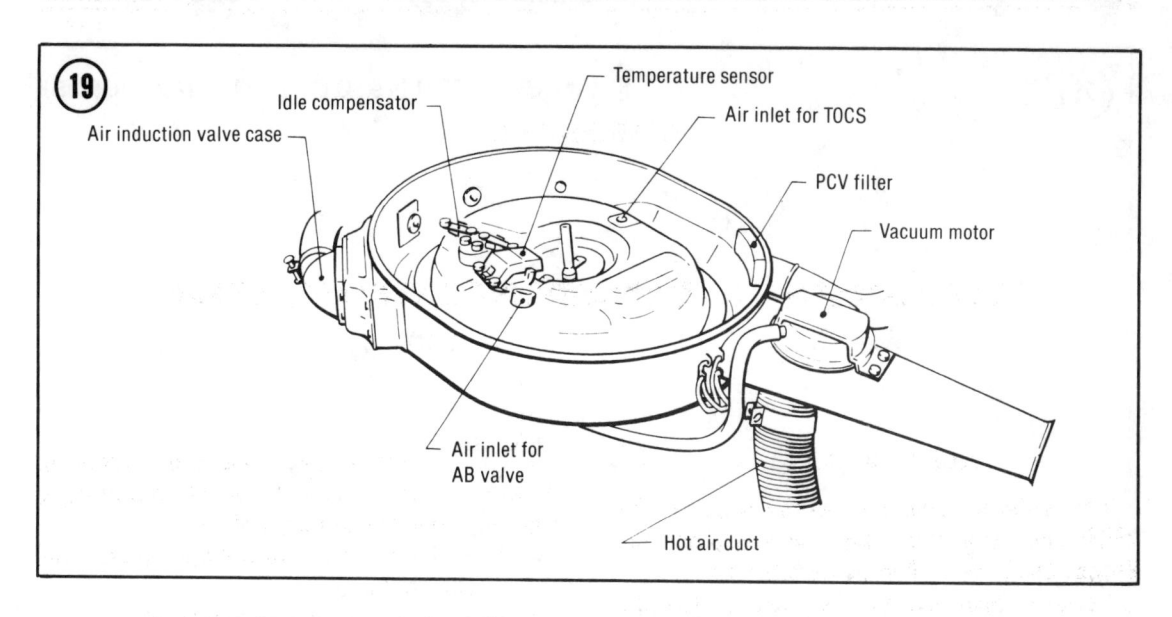

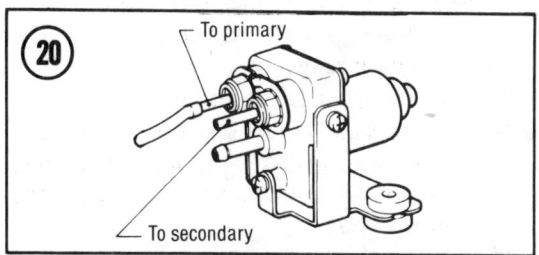

**(19)** Air induction valve case — Idle compensator — Temperature sensor — Air inlet for TOCS — PCV filter — Vacuum motor — Air inlet for AB valve — Hot air duct

**(20)** To primary — To secondary

*NOTE*
*Before testing the compensator, check for color code marks on the compensator and its hoses. If these marks aren't visible, make your own.*

To test the compensator, disconnect its hoses and try to blow air into them. See **Figure 20**. It should be impossible to blow air into the hoses below 1,641 ft. (500 m), and possible above that altitude. If the compensator doesn't perform properly, replace it.

## CARBURETOR

Carburetor overhaul is basically the same as for earlier models. On 1980 California carburetors, a throttle valve switch replaces the throttle opener diaphragm. See **Figure 21**. The 1980 non-California carburetors are the same as for 1979. On 1981 carburetors, a throttle opener diaphragm is used on all models. See **Figure 22**. Carburetor specifications differ from earlier models. See **Table 5**.

### Dashpot

Dashpot adjustment for 1980 cars is the same as for 1979 models. A dashpot is not used on 1981 U.S. models.

On 1981 Canadian cars, adjust as follows.
1. Warm the engine to normal operating temperature.
2. Move the throttle lever by hand until it just touches the dashpot. See **Figure 23**. Engine speed at this point should be 2,300-2,500 rpm. If not, loosen the locknut and turn the dashpot to adjust. Then tighten the locknut.

## AUTOMATIC CHOKE CIRCUIT

### Circuit Test

This is the same as for 1979 models. **Figure 24** shows the 1980 function test connector. **Figure 25** shows the 1981 function test connector.

### Relay Test

This is the same as for 1979 models.

### Heater Coil Test

This is the same as for 1979 models.

## EXHAUST GAS RECIRCULATION SYSTEM

This system differs in some details from 1979 models. Refer to the following illustrations:

# SUPPLEMENT

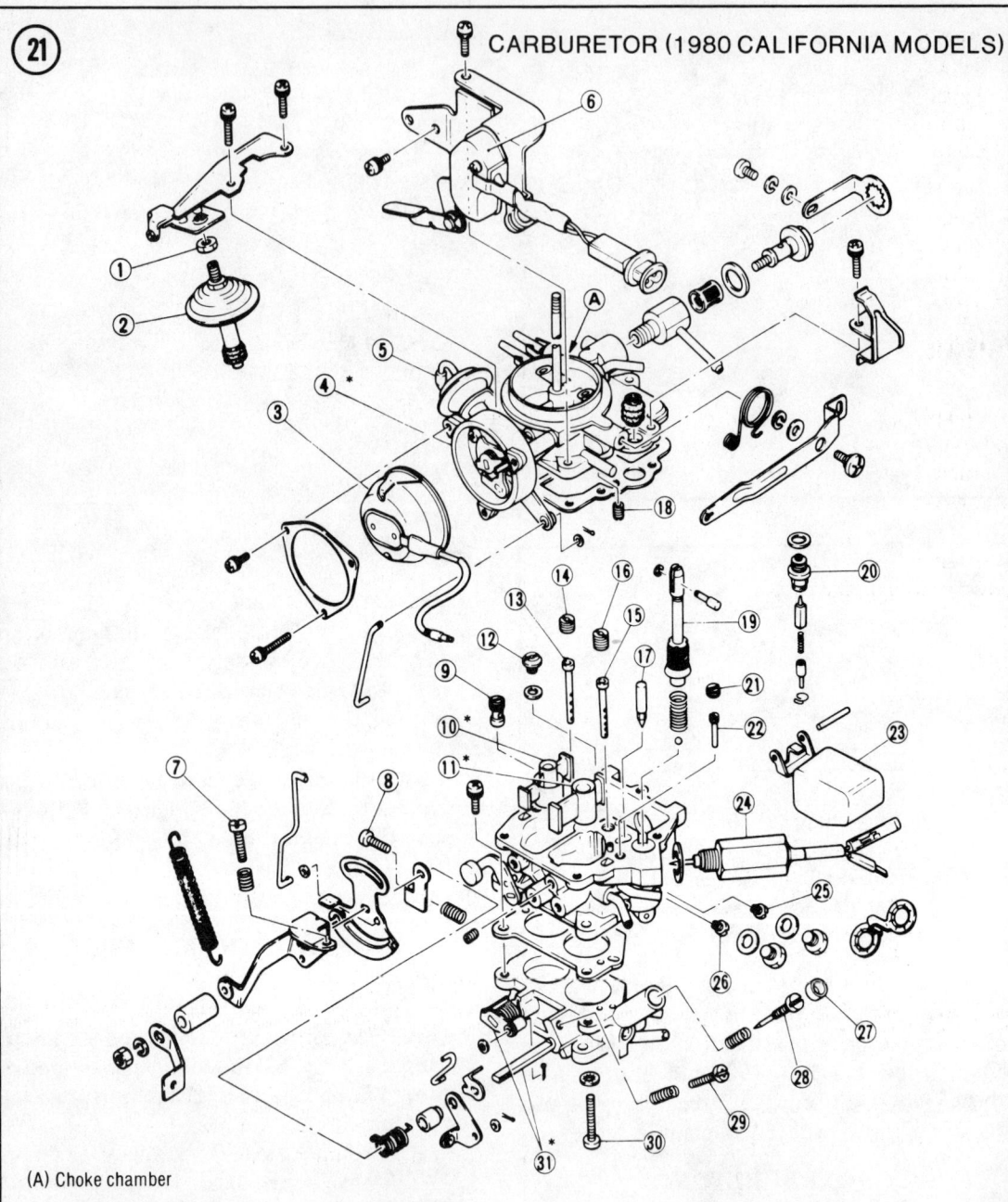

## ㉑ CARBURETOR (1980 CALIFORNIA MODELS)

(A) Choke chamber

1. Dash pot adjusting nut
2. Dash pot
3. Automatic choke cover
4.* Automatic choke body
5. Automatic choke break diaphragm
6. Throttle valve switch assembly
7. Throttle valve switch adjust screw
8. Fast idle adjust screw
9. Secondary slow jet
10.* Secondary small venturi
11.* Primary small venturi
12. Power valve
13. Secondary main air bleed
14. Plug
15. Primary main air bleed
16. Plug
17. Injector weight
18. Primary slow air bleed
19. Accelerating pump
20. Needle valve
21. Plug
22. Primary slow jet
23. Float
24. Anti-dieseling solenoid valve
25. Secondary main jet
26. Primary main jet
27. Idle adjusting screw blind plug
28. Idle adjust screw
29. Throttle adjust screw
30. Vacuum screw
31.* Primary and secondary throttle shaft

Note: Do not remove the parts marked with an asterisk (*).

# 1980 AND LATER SERVICE INFORMATION

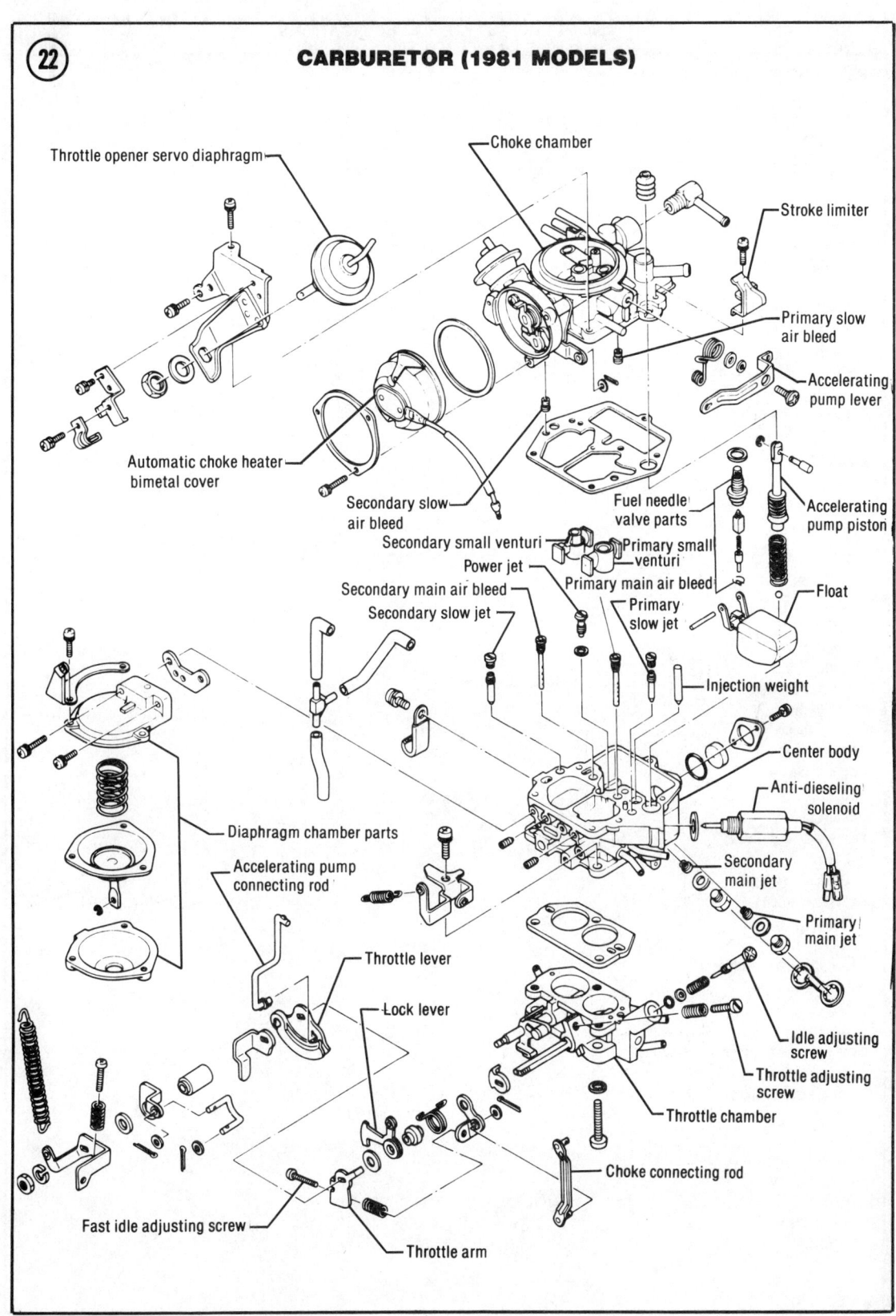

## ㉒ CARBURETOR (1981 MODELS)

Table 5  FUEL SYSTEM SPECIFICATIONS

| | |
|---|---|
| **JETS AND AIR BLEEDS, 1980** | |
| Primary main jet | |
|   U.S. | #107 |
|   Canada | #105 |
| Primary main air bleed | |
|   California | #80 |
|   49-states | #65 |
|   Canada | #95 |
| Primary slow jet | #45 |
| Secondary main jet | |
|   California | #145 |
|   49-states | #143 |
|   Canada | #145 |
| Secondary main air bleed | |
|   California | #80 |
|   49-states | #60 |
|   Canada | #80 |
| Secondary slow jet | #50 |
| Power jet | |
|   California | #38 |
|   49-states | #43 |
|   Canada | #40 |
| **JETS AND AIR BLEEDS, 1981** | |
| Primary main jet | |
|   California | #113 |
|   49-states | #114 |
|   Canada | #100 |
| Primary main air bleed | |
|   California | #60 |
|   49-states | #80 |
|   Canada | #70 |
| Primary slow jet | |
|   U.S. | #45 |
|   Canada | #43 |
| Primary slow air bleed | |
|   U.S. | #190 |
|   Canada | #170 |
| Secondary main jet | |
|   U.S. | #125 |
|   Canada | #145 |
| Secondary main air bleed | #80 |
| Seondary slow jet | |
|   U.S. | #50 |
|   Canada | #70 |
| Secondary slow air bleed | |
|   California | #80 |
|   49-states, Canada | #100 |
| Power jet | |
|   U.S. | #35 |
|   Canada | #40 |
| Fast idle speed | |
|   California | 2,300-3,100 rpm |
|   49-states | 2,400-3,200 |
|   Canada | 1,900-2,700 |

(continued)

# 1980 AND LATER SERVICE INFORMATION

Table 5 FUEL SYSTEM SPECIFICATIONS (continued)

| | |
|---|---|
| EGR operating temperature | |
| 1980 California | 140-203° F |
| | (60-95° C) |
| All others | Above 140° F |
| | (60° C) |
| Throttle opener operating speed | |
| 1980 California | Not used |
| 1980 49-states | 1,900-2,100 |
| 1980 Canada | 1,650-1,850 |
| 1981 (all) | 1,650-1,850 |

**Figure 26** — 1980 California
**Figure 27** — 1980 49 states
**Figure 28** — 1980 Canada
**Figure 29** — 1981 U.S.
**Figure 30** — 1981 Canada

System inspection is the same as for 1979. The venturi vacuum transducer used on 1980 California models and 1981 U.S. models is removed in the same manner as the back pressure transducer used on 1979 models.

## AIR INDUCTION SYSTEM

This system is used on 1980 non-California cars and all 1981 models. Service procedures are the same as for 1979 Canadian cars.

*(23) Locknut, Throttle lever, Dashpot*

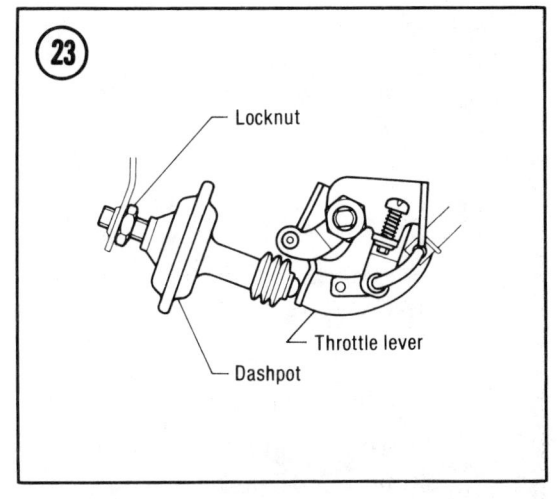

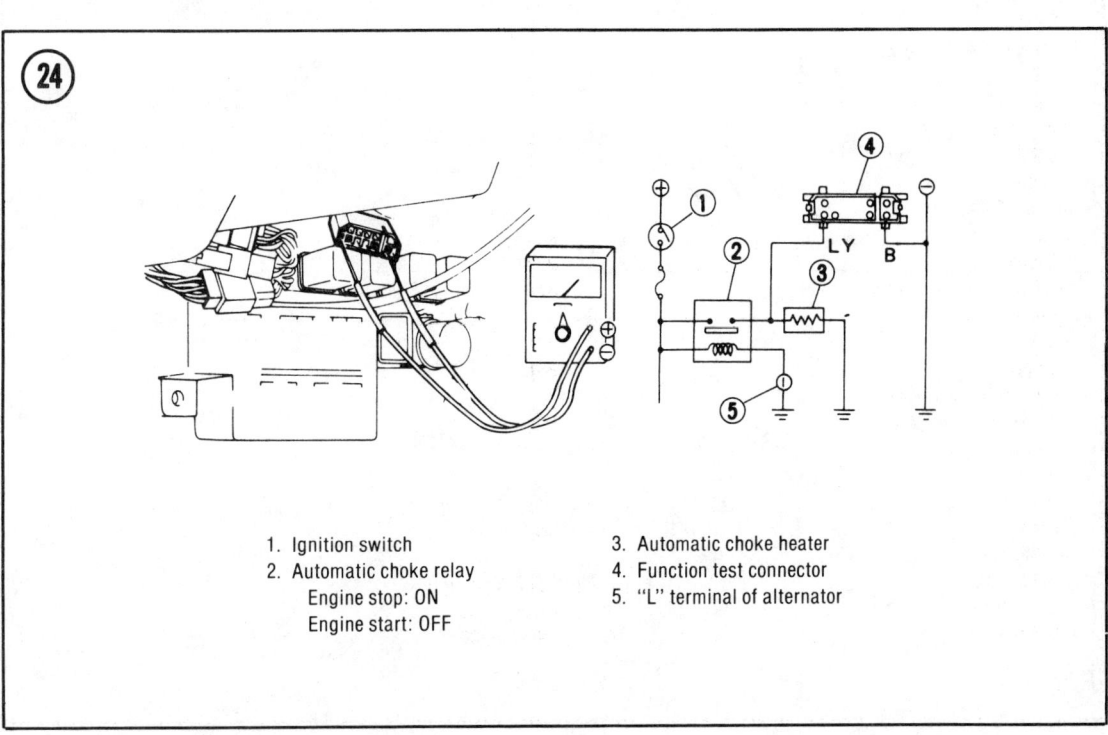

*(24)*
1. Ignition switch
2. Automatic choke relay
   Engine stop: ON
   Engine start: OFF
3. Automatic choke heater
4. Function test connector
5. "L" terminal of alternator

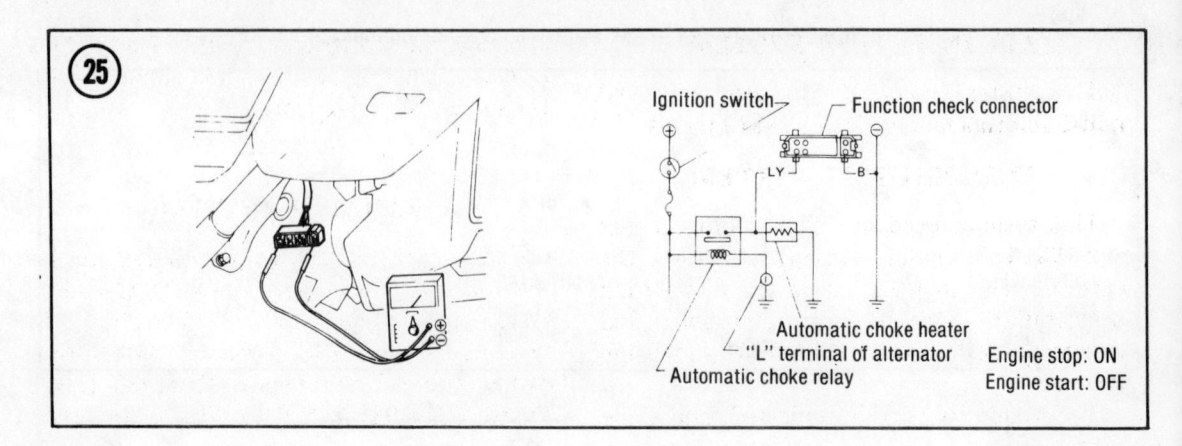

292　　　　　　　　　　　　　　　　　　　　　　　　　　　　　　　　　　　SUPPLEMENT

**25**

Ignition switch — Function check connector
Automatic choke heater
"L" terminal of alternator
Automatic choke relay

Engine stop: ON
Engine start: OFF

**26**

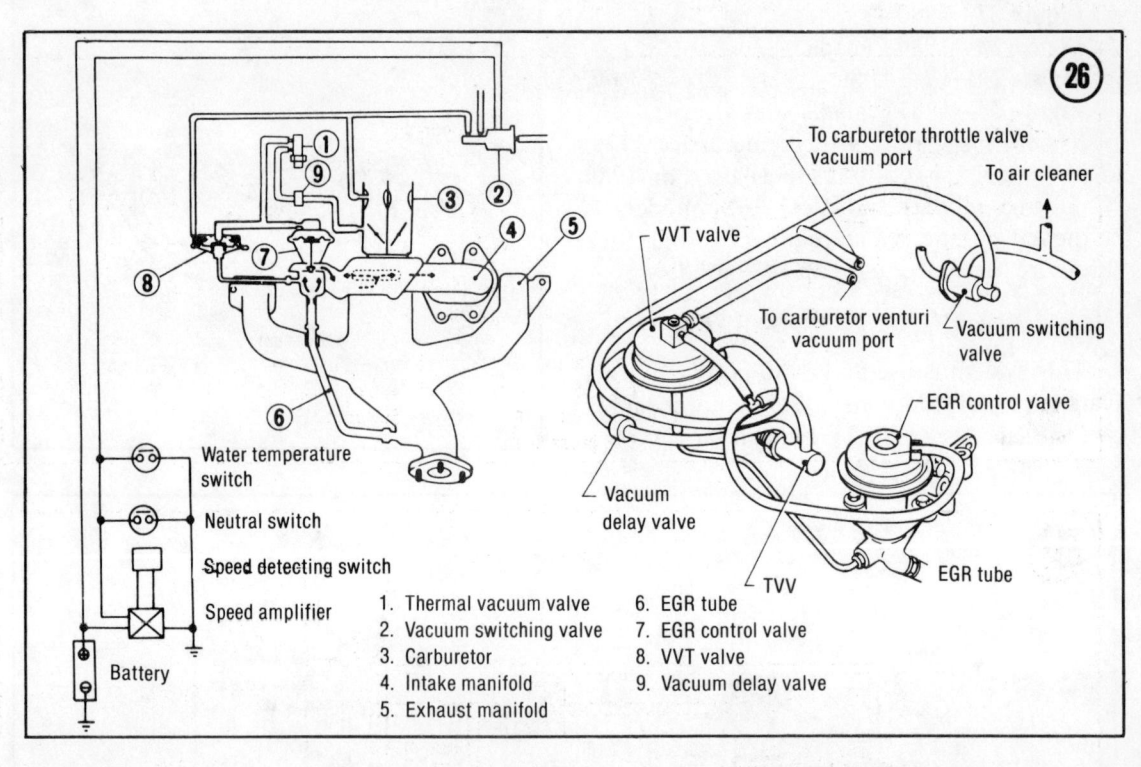

To carburetor throttle valve vacuum port
To air cleaner
VVT valve
To carburetor venturi vacuum port
Vacuum switching valve
EGR control valve
Vacuum delay valve
TVV
EGR tube

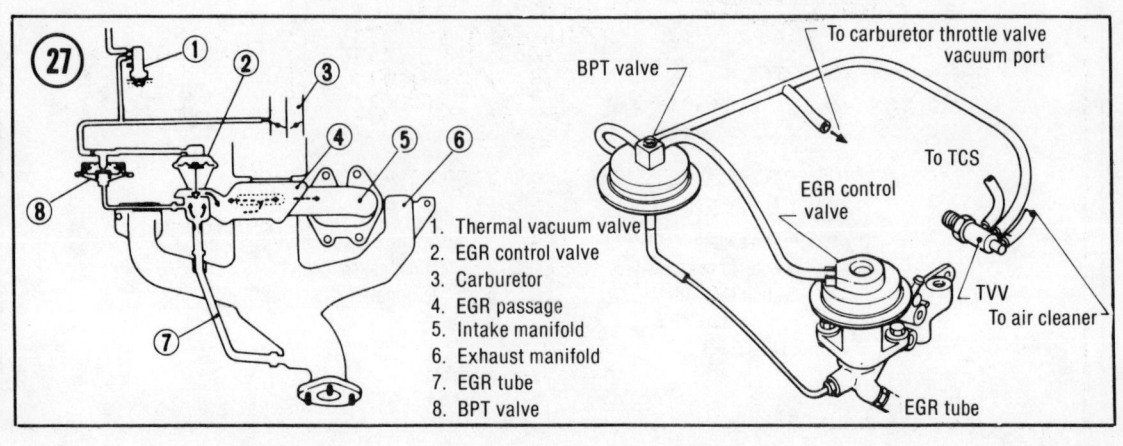

Water temperature switch
Neutral switch
Speed detecting switch
Speed amplifier
Battery

1. Thermal vacuum valve
2. Vacuum switching valve
3. Carburetor
4. Intake manifold
5. Exhaust manifold
6. EGR tube
7. EGR control valve
8. VVT valve
9. Vacuum delay valve

**27**

BPT valve
To carburetor throttle valve vacuum port
To TCS
EGR control valve
TVV
To air cleaner
EGR tube

1. Thermal vacuum valve
2. EGR control valve
3. Carburetor
4. EGR passage
5. Intake manifold
6. Exhaust manifold
7. EGR tube
8. BPT valve

# 1980 AND LATER SERVICE INFORMATION

## AIR INJECTION SYSTEM

This system is no longer used on 1980 non-California models or any 1981 models.

## FUEL SHUTOFF SYSTEM

This system, used on 1980 California cars and 1981 U.S. models, cuts off fuel flow in the carburetor's primary slow circuit during deceleration. This reduces emissions and improves fuel economy. **Figure 31** shows the system.

The system is very complicated, and some components are also parts of other emission control systems. Testing should be done by a Datsun dealer or a mechanic familiar with Datsun emission controls.

## EXHAUST SYSTEM

Exhaust system service procedures are the same as for 1979 models. The 1980 Canadian system is the same as for 1979. The 1980 U.S. systems and all 1981 systems differ from earlier models. See **Figure 32** (1980 U.S.), **Figure 33** (1981 U.S.), or **Figure 34** (1981 Canada).

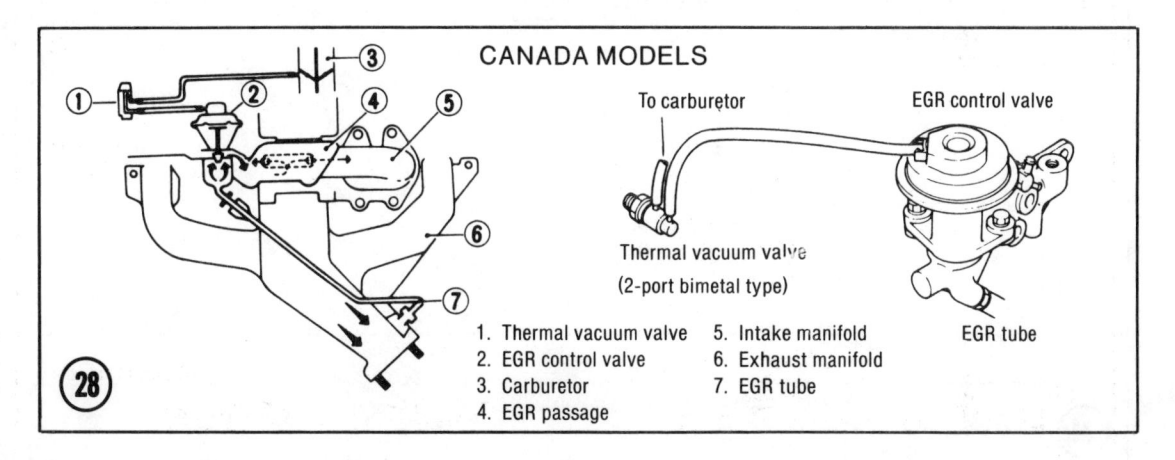

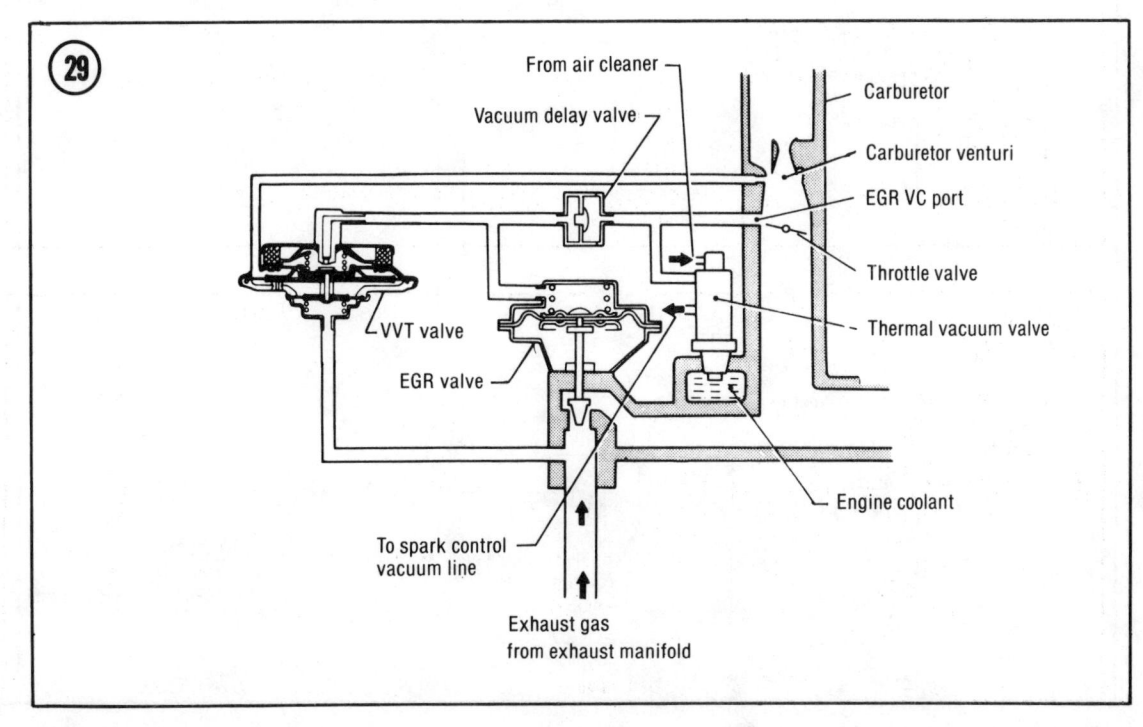

**294**                                                                  **SUPPLEMENT**

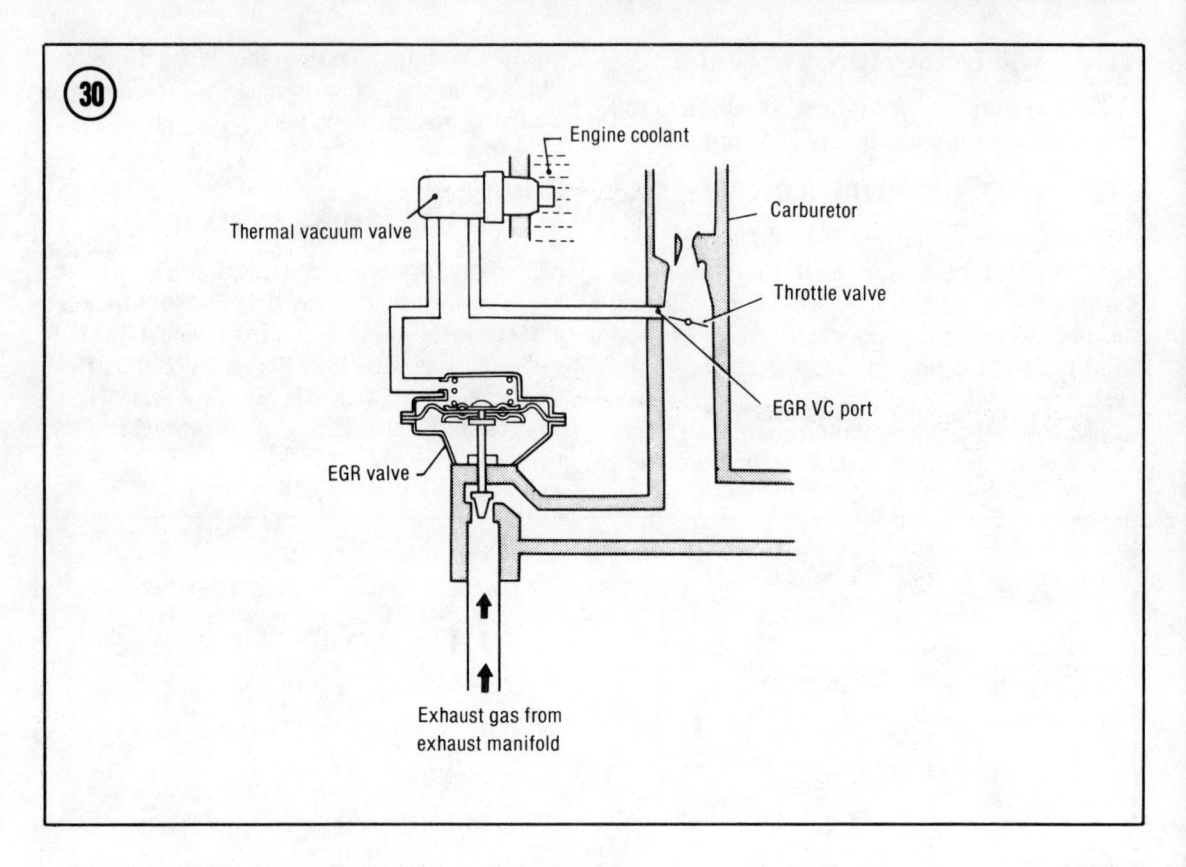

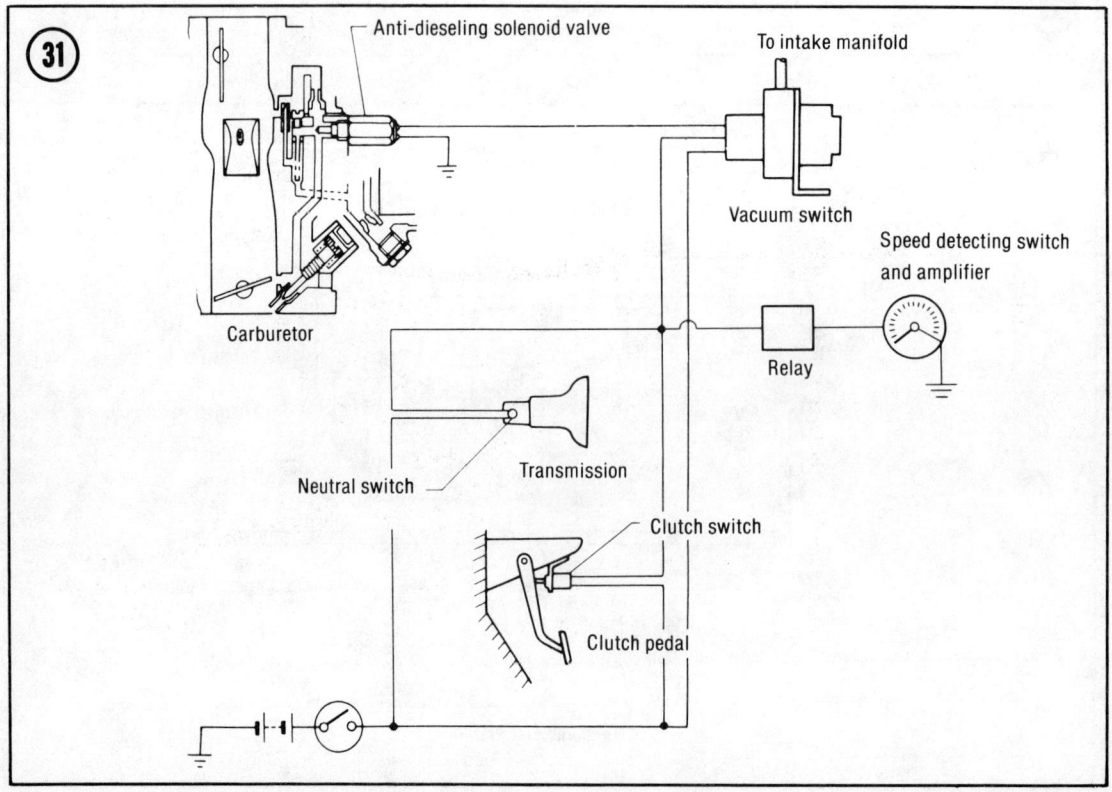

# 1980 AND LATER SERVICE INFORMATION

296  SUPPLEMENT

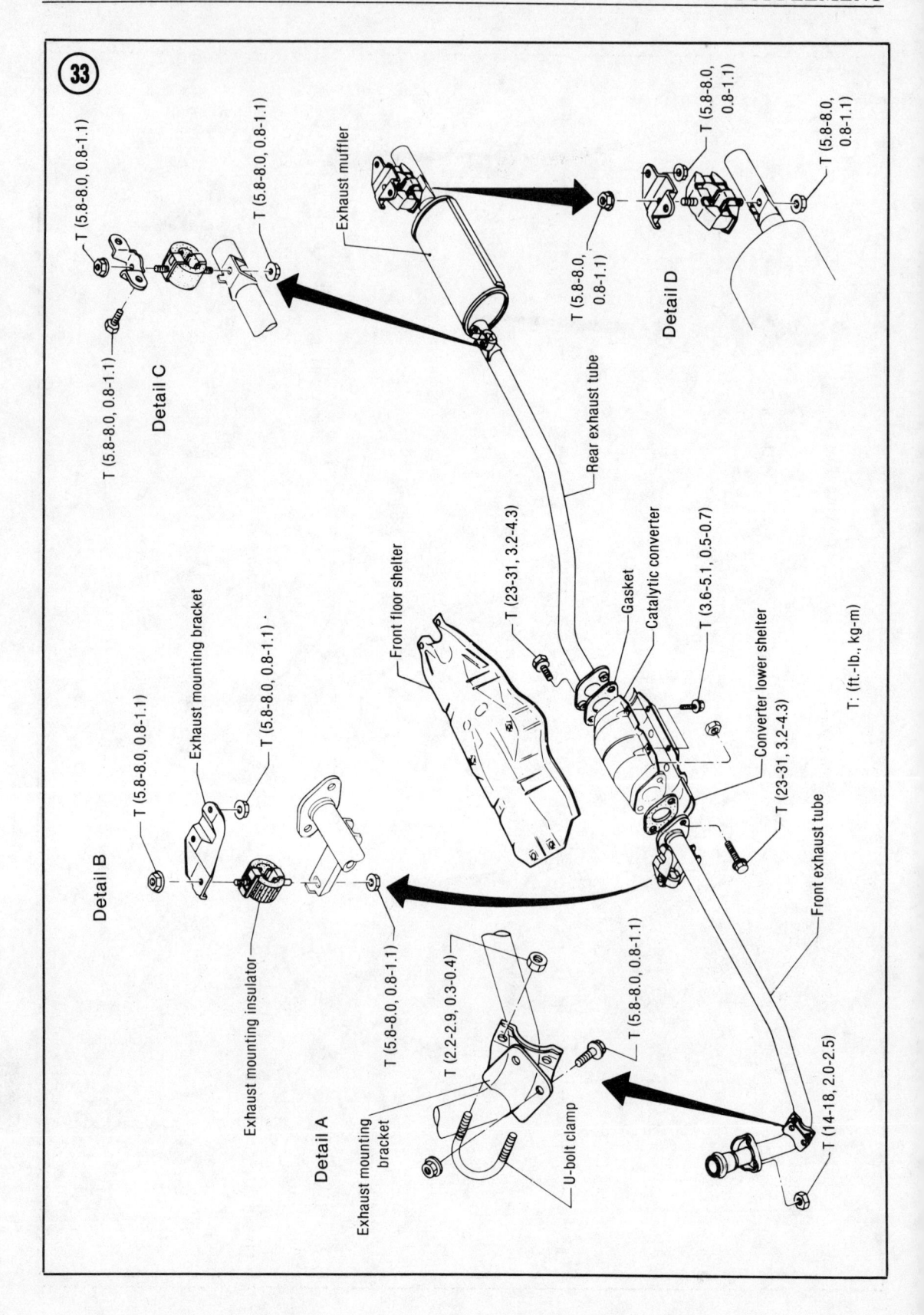

**1980 AND LATER SERVICE INFORMATION** 297

③④

- Rear exhaust tube
- T (5.8-8.0, 0.8-1.1)
- Exhaust mounting bracket
- T (5.8-8.0, 0.8-1.1)
- T (5.8-8.0, 0.8-1.1)
- Detail C
- Exhaust muffler
- T (2.2-2.9, 0.3-0.4)
- Front exhaust tube
- Exhaust mounting bracket
- T (5.8-8.0, 0.8-1.1)
- T (5.8-8.0, 0.8-1.1)
- Exhaust mounting bracket
- Exhaust mounting insulator
- Exhaust mounting bracket
- T (5.8-8.0, 0.8-1.1)
- U-bolt clamp
- Detail B
- Detail A
- T (14-18, 2.0-2.5)
- T: (ft.-lb., kg-m)

14

# CHAPTER SEVEN

# ELECTRICAL SYSTEM

## CHARGING SYSTEM

### Alternator Output Test

The alternator output test is the same as for 1979 models.

### Alternator Overhaul

The 1980 Canadian models use the same 50-amp alternator as 1979 models. Overhaul procedures are the same.

The 1980 U.S. models and all 1981 models use a 60-amp alternator (**Figure 35**). Service procedures are the same except for differences in diodes. To unsolder diodes, see **Figure 36**. To remove the diode and brush holder, see **Figure 37**.

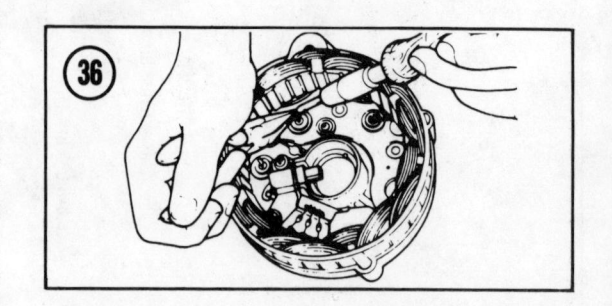

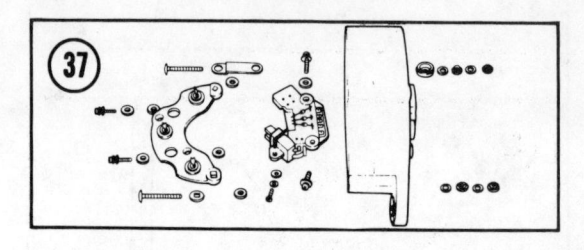

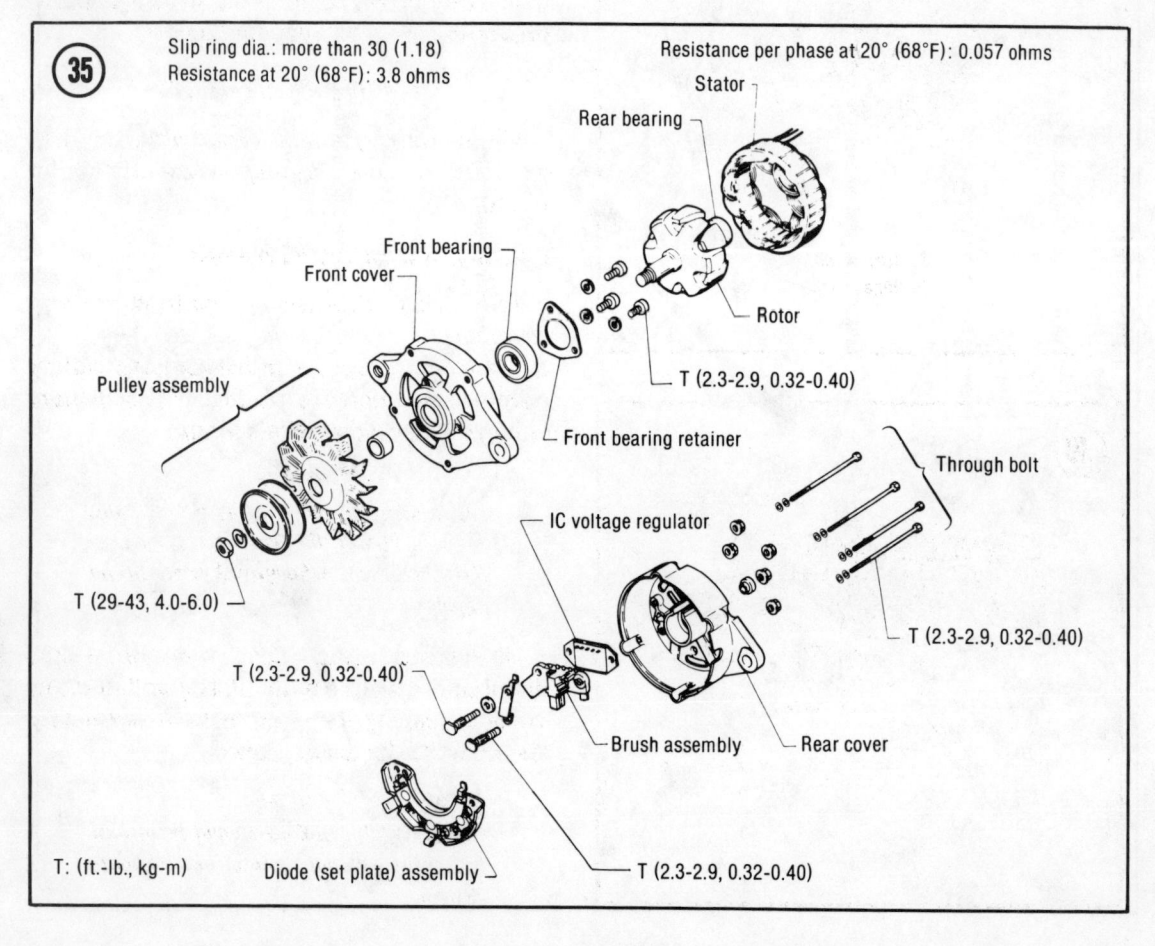

# 1980 AND LATER SERVICE INFORMATION

**Figure 38**
1. SR holder
2. Positive diode

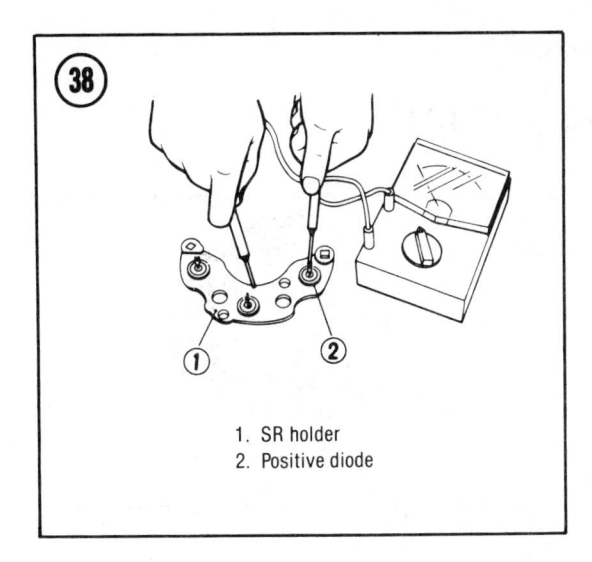

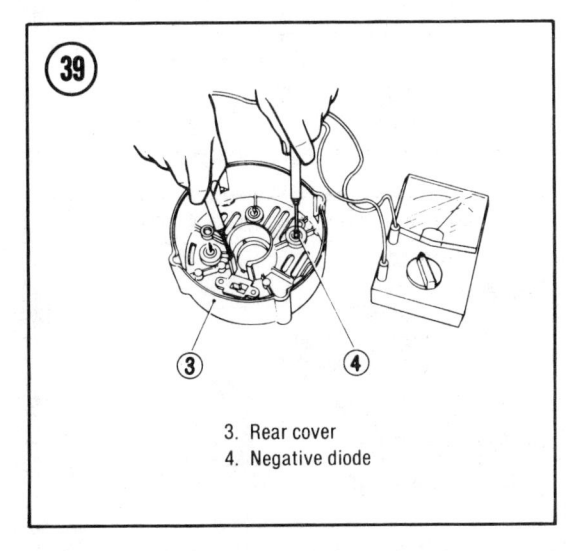

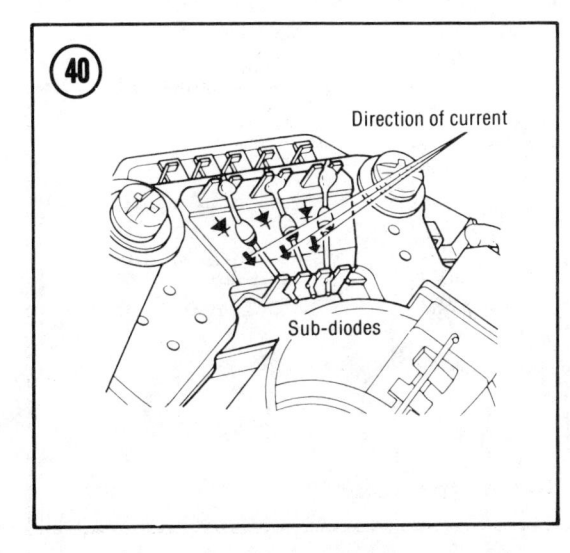

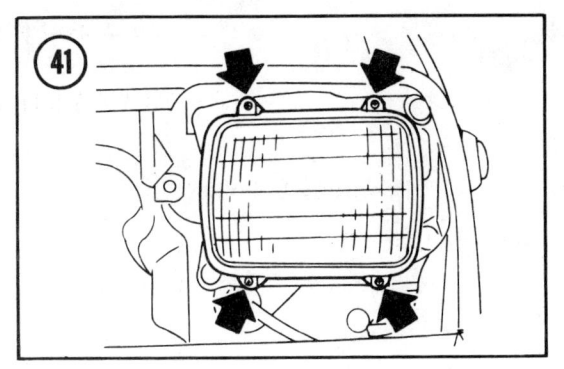

**Figure 39**
3. Rear cover
4. Negative diode

**Figure 40**
Direction of current
Sub-diodes

**Figure 41**

To test positive diodes, see **Figure 38**. To test negative diodes, see **Figure 39**. To test sub-diodes, see **Figure 40**.

## LIGHTING SYSTEM

### Headlight Replacement

On 1980 models, this is the same as for 1979. The 1981 procedure is basically the same, except that the 1981 models use square bulbs (65/55 watts; no SAE trade number). **Figure 41** shows the retaining screws.

## IGNITION SYSTEM

The ignition system is basically the same as on 1979 models. Testing procedures differ slightly.

### Ignition System Test (1980-on)

This test requires a voltmeter and ohmmeter.

1. Connect the voltmeter between the battery terminals and note the reading. This is battery voltage. Write it down for later use.

*NOTE*
*The reading should be at least 11.5 volts. If not, inspect the charging system as described under **Charging System** in this supplement.*

2. Disconnect the thick wire from the distributor cap. Crank the starter and note the voltage reading between battery terminals. Again, write the reading down.

*NOTE*
*Voltage while cranking should be at least 9.6 volts. If not, inspect the charging system.*

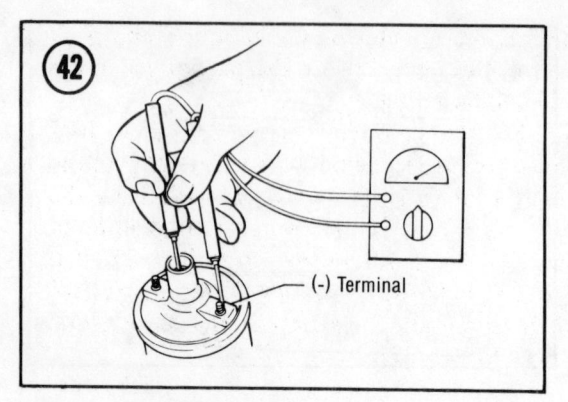

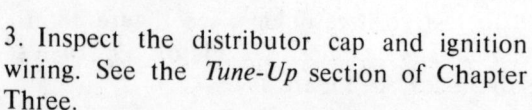

3. Inspect the distributor cap and ignition wiring. See the *Tune-Up* section of Chapter Three.

4. Measure ignition coil secondary resistance with the ohmmeter. See **Figure 42**. It should be 8,200-12,400 ohms. If resistance is not within the specified range, replace the ignition coil.

5. Test the power supply circuit. Connect a voltmeter as shown in **Figure 43**. Turn the key to ON, but don't start the engine. The voltmeter should indicate 11.5-12.5 volts. If it is less than this, check the wiring from ignition switch to integrated circuit (IC) unit on the side of the distributor. Wiring diagrams are at the end of the book.

6. Disconnect the thick wire from the distributor cap. Ground the wire by connecting it to bare metal. Connect the voltmeter as shown in **Figure 44** and crank the starter. The voltage reading should be within 1 volt of battery cranking voltage (written down in Step 2). It should be at least 8.6 volts. If not, check the ignition switch and the wiring to the integrated circuit unit.

7. Connect the voltmeter as shown in **Figure 45**. Turn the key to ON, but don't start the engine. Voltage should be 11.5-12.5 volts. If so, skip the next step. If less than 11.5 volts, perform the next step.

8. Test the coil primary circuit. Connect the ohmmeter as shown in **Figure 46**. Resistance should be 0.84-1.02 ohms. If not within this range, replace the ignition coil. If resistance is correct, check the ignition switch. See *Switches* in this chapter. Also check the wiring from ignition switch to the integrated circuit (IC) unit on the side of the distributor.

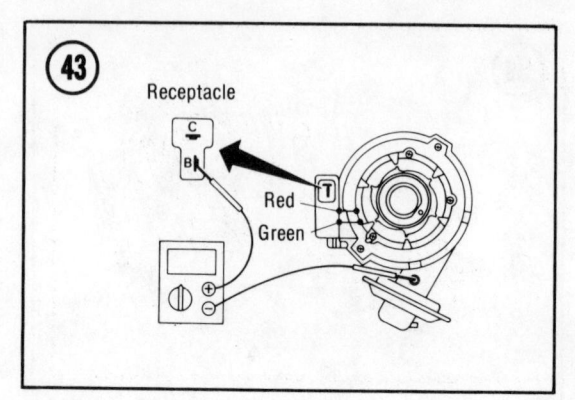

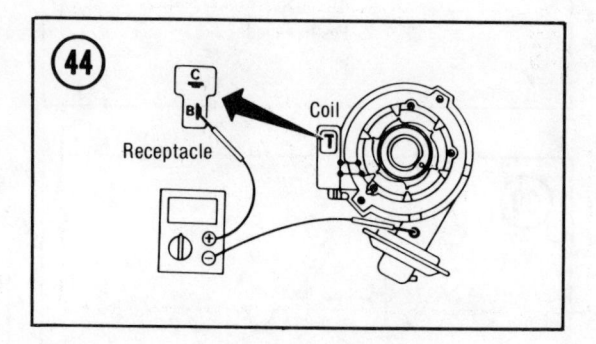

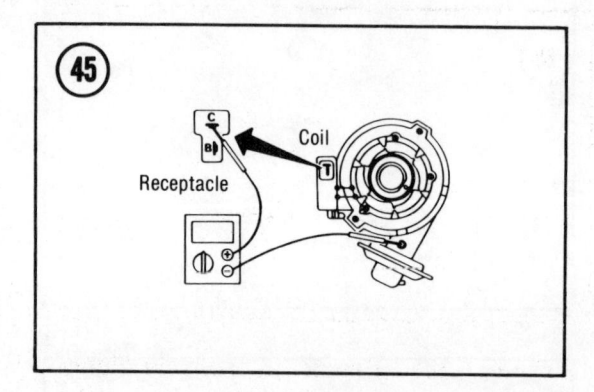

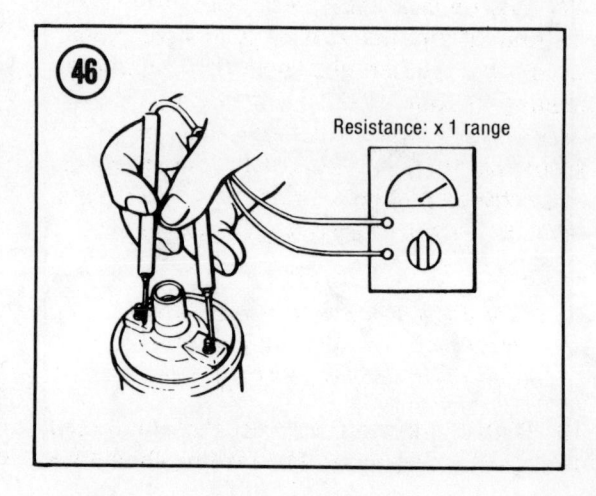

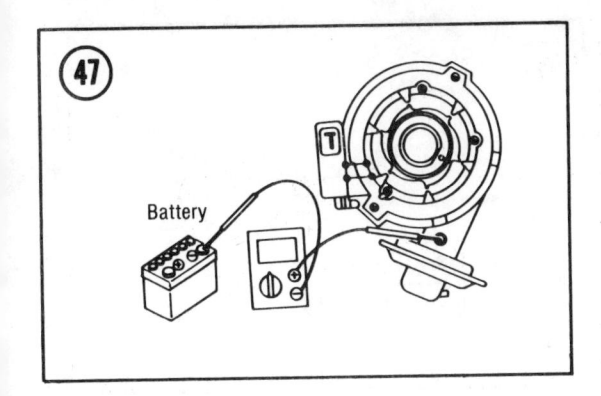

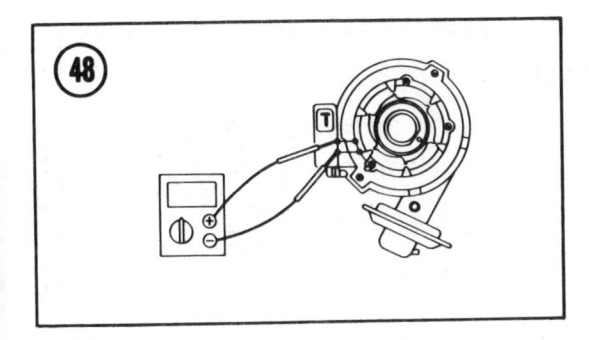

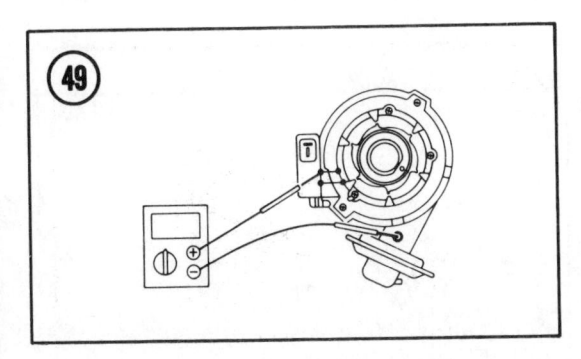

**47**

Battery

**48**

**49**

9. Connect the voltmeter as shown in **Figure 47**. Disconnect the thick wire from the distributor cap and ground it to bare metal. Crank the starter and note the voltmeter reading. It should be 0.5 volts or less. If not, make sure the distributor is properly grounded to the engine. Check the wiring from chassis ground to the battery. Make sure the battery connections are clean and tight.

*NOTE*
*The engine should be at normal operating temperature for Steps 10 and 11. The thick coil wire should remain grounded.*

10. With the key off, connect the ohmmeter as shown in **Figure 48**. The reading should be approximately 400 ohms. If much above or below 400 ohms, check the pickup coil (Step 11) and its wiring.

11. Set the voltmeter to the low scale (0-5 volts preferred). Connect the voltmeter to the distributor as shown in **Figure 49**. Turn the key to START and watch the voltmeter needle. If it holds steady, go to Step 12. If it wavers, hold the disconnected wire about 1/4 inch from bare metal, crank the starter, and check for spark.

*WARNING*
*Hold the wire with a heavily insulated tool. The wire can cause a painful shock, even if the insulation is in perfect condition and you don't touch the end of the wire.*

If there is no spark, replace the integrated circuit unit on the side of the distributor.

12. Remove the distributor cap. Check the pickup coil and reluctor for damage. Replace the distributor if these are found. If the pickup coil and reluctor are good, check the wiring from pickup coil to IC ignition unit. Tighten or replace as needed.

## SPARK TIMING CONTROL SYSTEM

This system is used on all 1980 models and on 1981 U.S. cars. Refer to the following illustrations:
**Figure 50** – 1980 California
**Figure 51** – 1980 49 states
**Figure 52** – 1980 Canada
**Figure 53** – 1981

### System Test (1980 Non-California)

Test procedures for 1980 non-California cars are the same as for 1979 models.

### System Test (1980 California; All 1981)

Start this test with the engine cold.
1. Connect a timing light to the engine, following manufacturer's instructions.
2. Start the engine and note ignition timing.
3. Let the engine warm up. Ignition timing should advance as the engine warms. If not, have the system tested further by a dealer or mechanic familiar with Datsun emission controls.

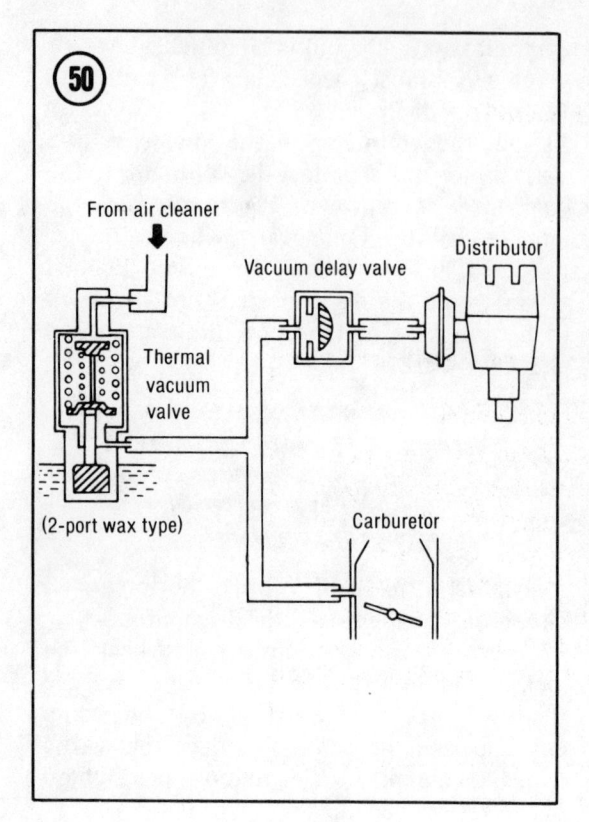

SUPPLEMENT

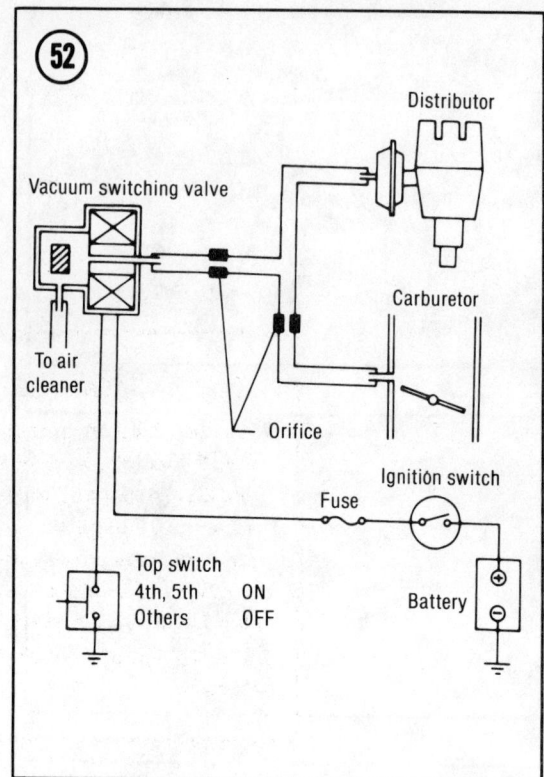

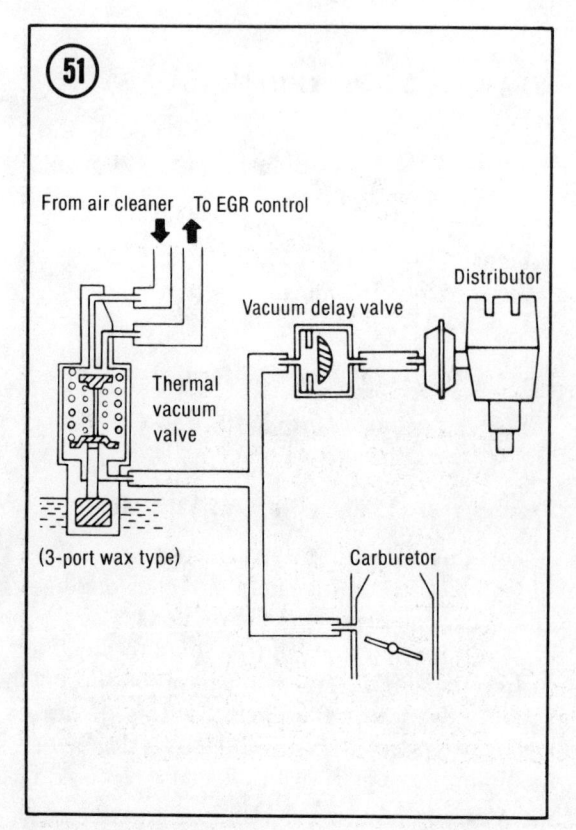

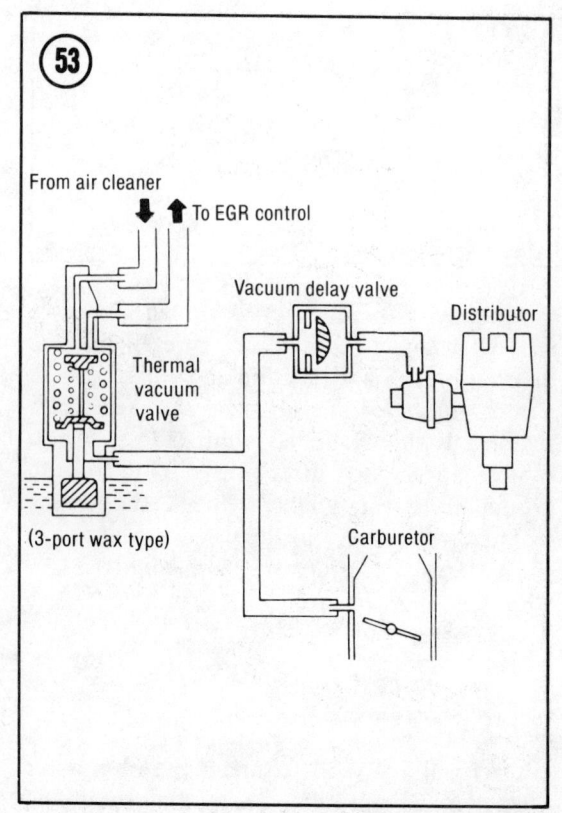

### 50
(2-port wax type)
- From air cleaner
- Thermal vacuum valve
- Vacuum delay valve
- Distributor
- Carburetor

### 51
(3-port wax type)
- From air cleaner
- To EGR control
- Thermal vacuum valve
- Vacuum delay valve
- Distributor
- Carburetor

### 52
- Vacuum switching valve
- To air cleaner
- Distributor
- Carburetor
- Orifice
- Fuse
- Ignition switch
- Battery
- Top switch: 4th, 5th ON / Others OFF

### 53
(3-port wax type)
- From air cleaner
- To EGR control
- Thermal vacuum valve
- Vacuum delay valve
- Distributor
- Carburetor

# CHAPTER NINE

# BRAKES

## MASTER CYLINDER

The 1980 master cylinder is the same as on 1979 models. The 1981 master cylinder uses a single reservoir, rather than the 2 separate reservoirs used on earlier cars. See **Figure 54**. Service procedures are the same as for earlier models. The check valve plugs are tightened to 33-40 ft.-lb. (4.5-5.5 mkg). The stopper bolt is tightened to 1.1-2.2 ft.-lb. (0.15-0.30 mkg).

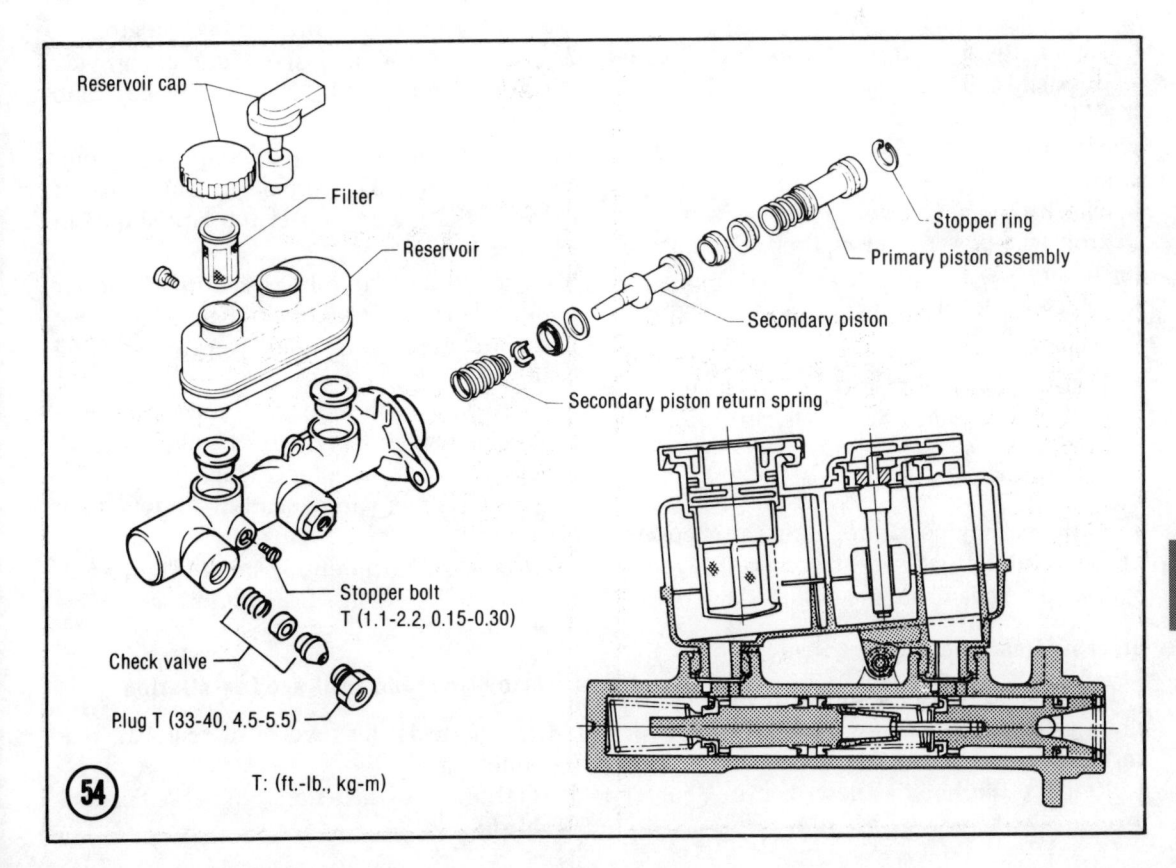

**Figure 54** — Master cylinder components: Reservoir cap, Filter, Reservoir, Stopper ring, Primary piston assembly, Secondary piston, Secondary piston return spring, Stopper bolt T (1.1-2.2, 0.15-0.30), Check valve, Plug T (33-40, 4.5-5.5). T: (ft.-lb., kg-m)

# CHAPTER ELEVEN

# FRONT SUSPENSION AND STEERING

## WHEEL ALIGNMENT

Service procedures are the same as for 1979 models. On 1981 models equipped with power steering, the steering lock angle for outside wheels is 28 1/2° to 31 1/2°.

## POWER STEERING

Power steering is optional on 1981 models. **Figure 55** shows the system.

### System Bleeding

Bleeding is necessary whenever air enters the system.
1. Securely block both rear wheels so the car will not roll in either direction.
2. Jack up the front end of the car and place it on jackstands.
3. With the engine off, turn the steering wheel to full left lock and full right lock 10 times.
4. Check fluid level. Top up as needed with Dexron type power steering fluid. Do not use any other type.
5. Once fluid level stabilizes, start the engine.

*CAUTION*
*During the next step, do not turn the steering wheel hard against the stops. Do not hold the steering wheel at full lock position for more than 15 seconds.*

6. With the engine idling, turn the steering wheel to full left lock and full right lock until bubbles disappear from the fluid.

### Steering Gear and Linkage Removal

The engine must be hoisted and moved slightly forward to remove the steering gear and linkage.
1. Remove the hood as described in Chapter Twelve.
2. Securely block both rear wheels so the car will not roll in either direction.
3. Jack up the front end of the car and place it on jackstands.
4. Remove the upper buffer rod and install an engine sling bracket in its place.
5. Attach a hoist to the engine sling bracket and alternator adjusting bar. See **Figure 56**.
6. Refer to *Engine Removal*, Chapter Four, basic book and disconnect the following:
   a. Transmission shift and select rods from transmission
   b. Exhaust pipe from manifold
   c. Exhaust pipe bracket from transmission.
   d. Lower buffer rod
   e. Front and rear motor mount insulators
7. Hoist the engine just enough to provide removal clearance for the steering gear and linkage.
8. Remove the steering column lower joint cover (**Figure 57**). Loosen, but do not remove, the lower joint clamp bolt shown in **Figure 57**.
9. Remove the clamp bolt securing the lower joint to the steering gear. See **Figure 58**. Disconnect the lower joint from the steering gear.
10. Disconnect the tie rod ends from the steering knuckle arms. Use a puller such as Datsun tool HT72520000 (**Figure 59**).
11. Place a drain pan beneath the hoses, then disconnect them. See **Figure 60**.
12. Remove the mounting bolts (**Figure 61**). Remove the steering gear and linkage to one side.

### Steering Gear and Linkage Installation

Installation is the reverse of removal, plus the following.
1. Tighten all fasteners to specifications (**Table 6**).

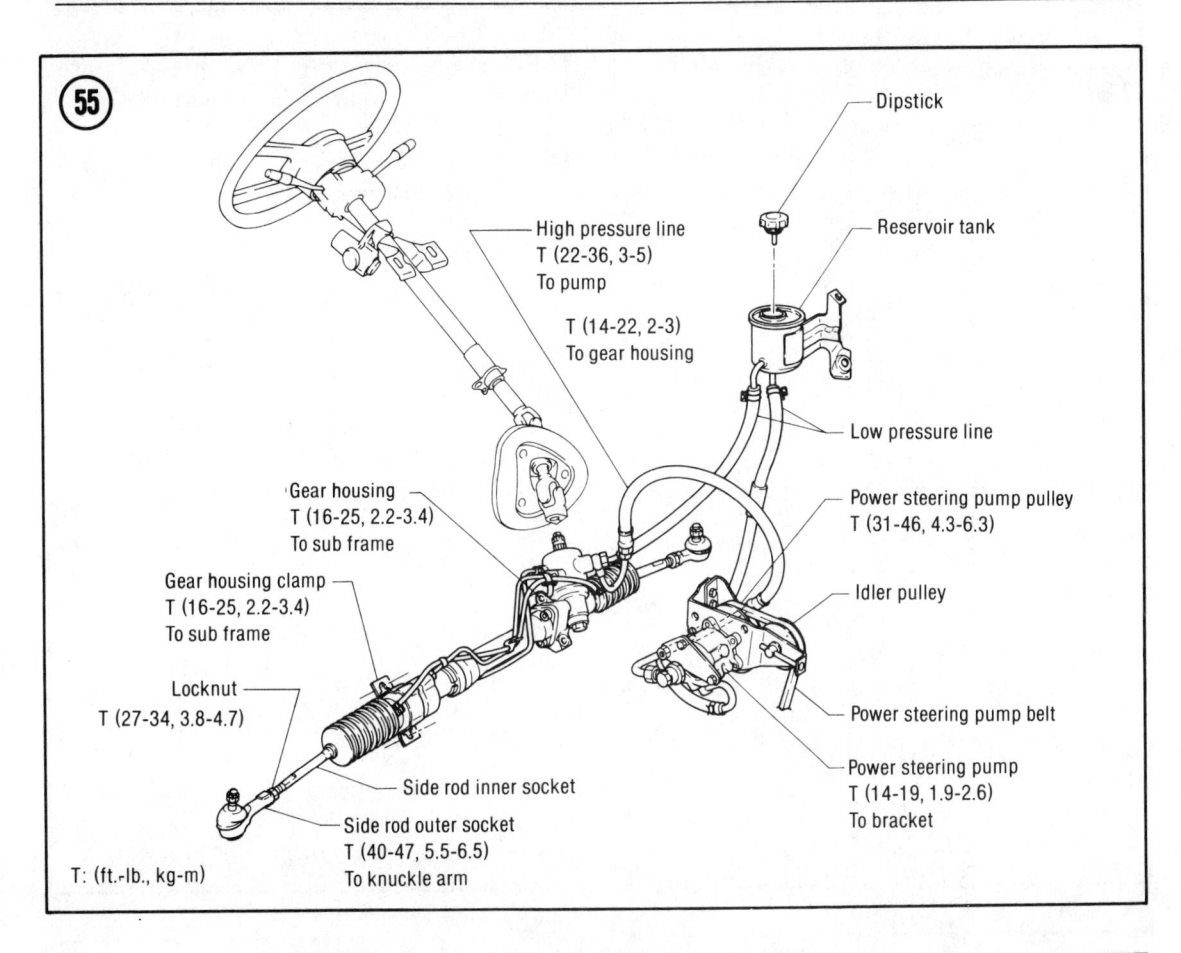

# 1980 AND LATER SERVICE INFORMATION

**55**

- Dipstick
- Reservoir tank
- High pressure line T (22-36, 3-5) To pump
- T (14-22, 2-3) To gear housing
- Low pressure line
- Gear housing T (16-25, 2.2-3.4) To sub frame
- Power steering pump pulley T (31-46, 4.3-6.3)
- Gear housing clamp T (16-25, 2.2-3.4) To sub frame
- Idler pulley
- Locknut T (27-34, 3.8-4.7)
- Power steering pump belt
- Side rod inner socket
- Power steering pump T (14-19, 1.9-2.6) To bracket
- Side rod outer socket T (40-47, 5.5-6.5) To knuckle arm

T: (ft.-lb., kg-m)

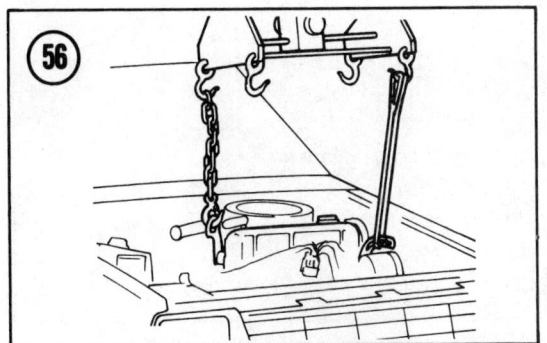

**56**

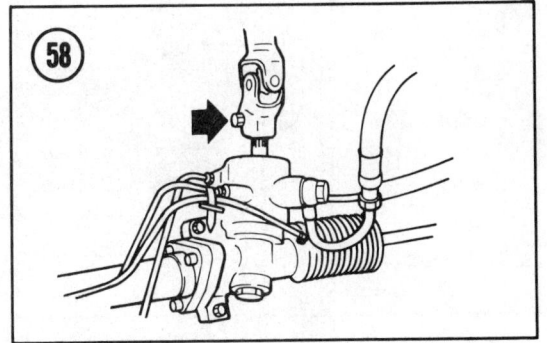

**58**

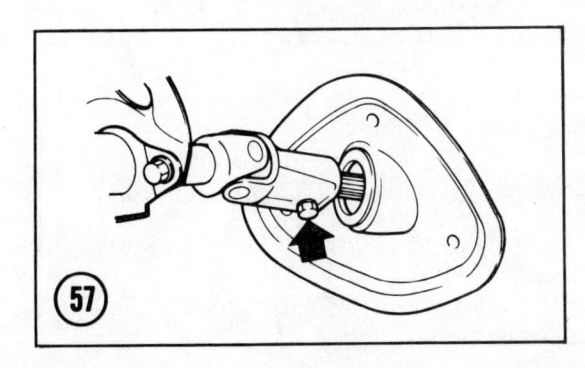

**57**

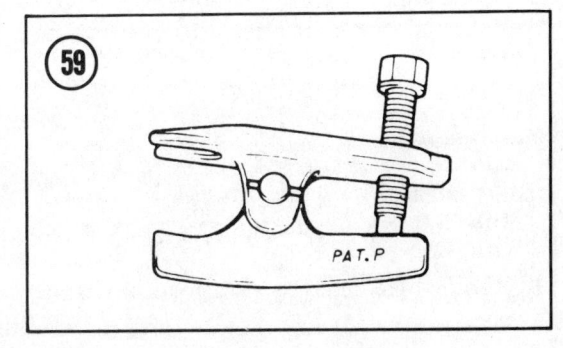

**59**

PAT. P

# SUPPLEMENT

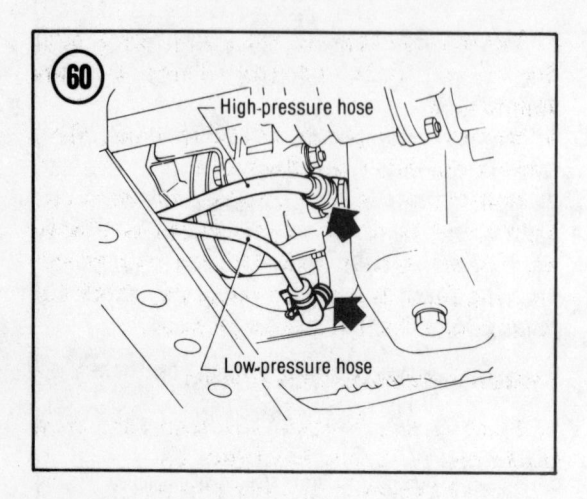

2. Fill the system with Dexron type power steering fluid. Capacity is approximately 7/8 pt. (0.8 liter).

3. Bleed the system and check for leaks as described in this supplement.

**Fluid Leak Check**

1. Warm the engine until power steering fluid temperature is between 140 and 176° F (60-80° C). Check fluid temperature with a thermometer.

2. With the engine running at 1,000 rpm, turn the steering wheel from full left to full right lock several times.

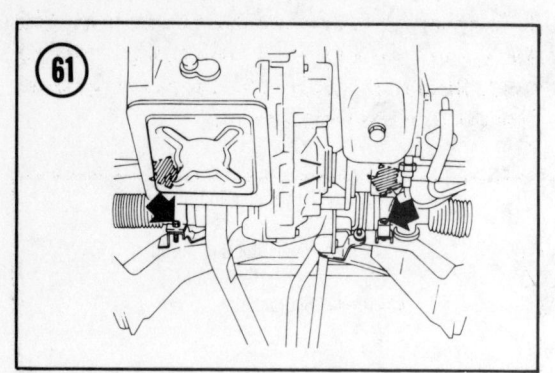

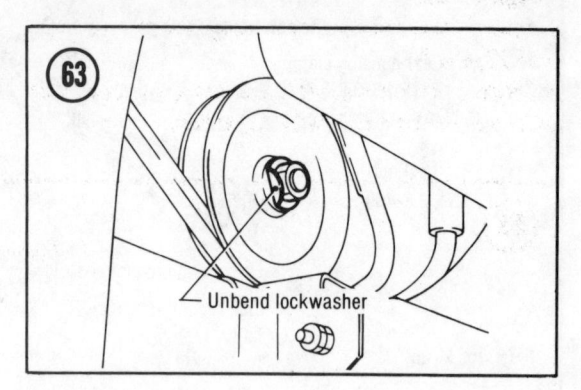

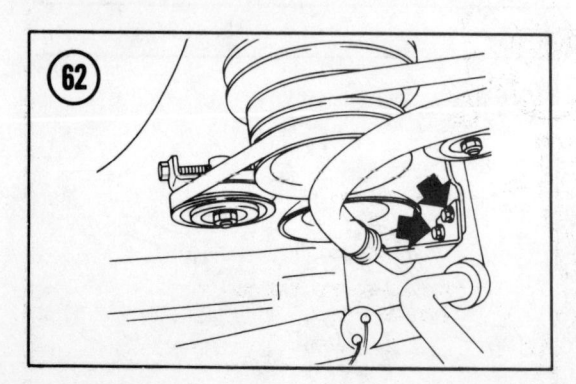

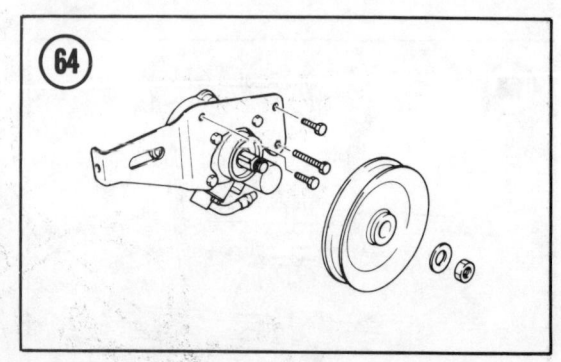

Table 6  POWER STEERING TORQUES

| Fastener | ft.-lb. | Mkg |
| --- | --- | --- |
| Tie rod to knuckle arm | 27-34 | 5.5-6.5 |
| Gear housing clamp | 16-25 | 2.2-3.4 |
| Gear housing mounting bolts | 16-25 | 2.2-3.4 |
| Pump mounting bolts | 14-19 | 1.9-2.6 |
| Pump pulley locknut | 31-46 | 4.3-6.3 |
| High pressure hose to pump | 22-36 | 3-5 |
| High pressure hose to gear | 14-22 | 2-3 |
| Low pressure hose to gear | 14-22 | 2-3 |

*CAUTION*
*During the next step, do not hold the steering wheel in lock position for more than 15 seconds.*

3. Hold the steering wheel in each lock position for 5 seconds. Check carefully for fluid leaks.

**Pump Removal/Installation**

1. Place a drain pan beneath the hoses, then disconnect them. See **Figure 60**.
2. Remove the high pressure hose bracket (**Figure 62**).
3. Bend the lockwasher away from the nut. See **Figure 63**.
4. Remove the nut. If necessary, squeeze the pump drive belt to hold the pulley.
5. Loosen and remove the pump drive belt. See *Drive Belts*, Chapter Three in this supplement.
6. Remove the pump mounting bolts, then remove the pump. See **Figure 64**.
7. Installation is the reverse of removal. Tighten all fasteners to specifications (**Table 6**). Fill the system with Dexron type power steering fluid. Bleed the system and check for leaks as described in this supplement.

**Reservoir Removal/Installation**

1. Place a pan beneath the reservoir, then disconnect the hoses. See **Figure 65**.
2. Detach the reservoir, then take it out.
3. Installation is the reverse of removal. Fill the system with Dexron type power steering fluid. Bleed the system and check for leaks as described in this supplement.

**65**

Dipstick
Reservoir tank
High pressure line
T (22-36, 3-5) To pump
T (14-22, 2-3) To gear housing
Low pressure line
Power steering pump pulley T (31-46, 4.3-6.3)
Idler pulley
Power steering pump belt
Power steering pump T (14-19, 1.9-2.6) To bracket
T: (ft.-lb., kg-m)

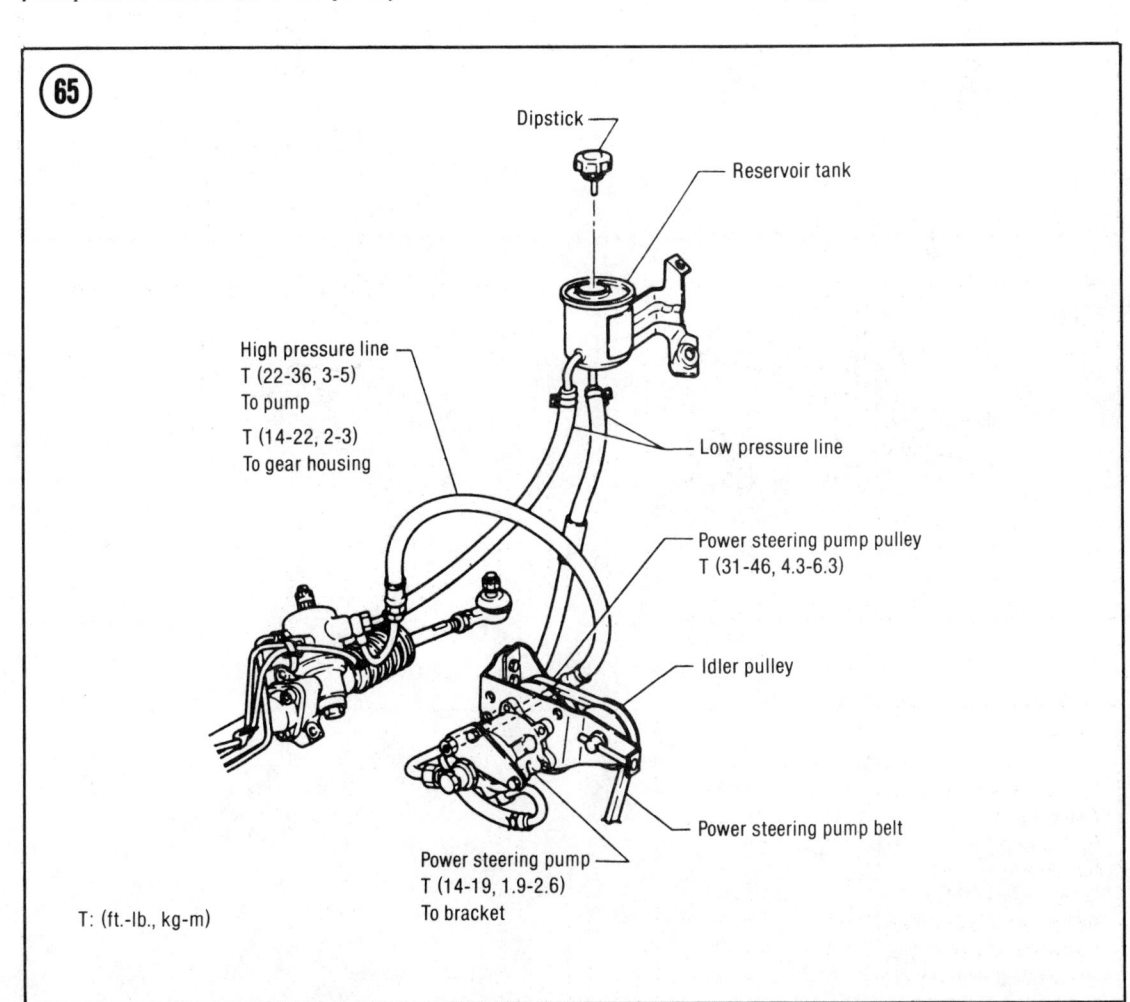

# INDEX

## A

Air cleaner ............................... 45, 83-87, 286-287
Air conditioning
    Compressor ............................................. 269
    Condenser .............................................. 269
    Evaporator ............................................. 271
    Expansion valve ....................................... 271
    Receiver/drier ......................................... 271
    Refrigerant ............................................. 274
    Routine maintenance ..................... 271, 273-274
    System operation ............................... 268-269
    Troubleshooting ...................................... 276
Air filter ............................................... 45, 281
Air induction system .......................... 109-110, 291
Air induction valve filter ............................ 45, 281
Air injection system ................. 22-23, 107-109, 293
Alternator
    Assembly ....................................... 139-141
    Disassembly (1976-1977) ...................... 135-136
    Disassembly (1978 and later) ...................... 136
    Inspection and repair ........................... 136-139
    Overhaul ............................................. 298
    Testing .......................................... 133, 298
Automatic choke circuit ........................ 93-96, 287
Axle shafts ........................................ 238-240

## B

Back-up lights .......................................... 26
Ball-joints .................................... 46, 236-237
Battery ..................................... 9-13, 130-131
Body
    Bumpers ............................................. 244
    Doors ............................................ 252-255
    Front fenders ....................................... 244
    Grille ................................................. 244
    Hood ................................................. 249
    Instrument panel .............................. 263-267
    Locks ................................................. 255
    Rear door (F10) ..................................... 261
    Rear door (310) ............................... 262-263
    Seats ................................................. 267
    Tailgate (F10) ................................. 261—262
    Trunk lid ...................................... 258-261
    Windows ...................................... 255-258
Brake lights ............................................. 26
Brakes
    Adjustments ........................................ 219
    Brake bleeding ................................ 217-218
    Brake booster ................................. 211-217
    Caliper ......................................... 203-205
    Disc .................................................. 206
    Fluid .................................................. 45
    Front ........................................... 201-206
    Lights ................................................. 26
    Maintenance ......................................... 38
    Master cylinder ......................... 208-211, 303
    Pad replacement ............................... 201-203
    Proportioning ....................................... 217
    Rear ............................................ 207-218
    Stop light switch .................................... 26
    Troubleshooting ................................. 28-31
Bumpers ............................................... 244

## C

Caliper ........................................... 203-205
Camshaft .......................................... 71-73
Carburetor ................................ 89, 91-93, 287
Carburetor cooling fan ............................ 96-97
Charging system .......................... 9-14, 298-299
Choke plate and linkage ............................. 45
Clutch
    Bleeding ....................................... 169-171
    Inspection ..................................... 175-176
    Installation ......................................... 176

## INDEX

Master cylinder ........................... 173-174
Operating cylinder ...................... 171-173
Pedal adjustment .............................. 169
Release bearing replacement ......... 176
Removal .................................... 174-175
Troubleshooting ............................ 27-28
Coil springs ................................... 221-223
Compression test ............................. 47-48
Connecting rods and pistons ............ 73-76
Coolant ............................................... 45
Coolant hoses and connections ............ 40
Cooling system
    Fan ............................................ 126
    Flushing ................................... 123
    Radiator ................................... 124
    Refilling ................................... 124
    Thermostat, removal/installation ........ 124-125
    Troubleshooting ........................ 26-27
    Water pump ............................. 127
Crankshaft ....................................... 76-78
Cylinder block inspection ................... 78-79
Cylinder head ................................... 65-66

### D

Directional signals .............................. 26
Distributor ...................................... 50-53
**Doors** ............................. 252-255, 261-263
Drive belts (F10 and 310) ............... 39, 278

### E

Early fuel evaporative system ......... 105-106
Electrical system
    Alternator .......... 9-13, 131-141, 298-299
    Battery ........................ 9-13, 130-131
    Heater ......................................... 26
    Horn ..................................... 155-156
    Ignition ............ 13-14, 52-53, 160-168, 299-301
    Instruments ........................... 151-155
    Lighting system ........ 25-26, 145-147, 299
    Starter ................................... 141-145
    Switches ............................... 147-151
    Wiring diagrams .................. end of book
Engine
    Camshaft .................................. 71-73
    Connecting rods and pistons ........ 73-76
    Crankshaft ............................... 76-78
    Cylinder block inspection ........... 78-79
    Cylinder head ........................... 65-66
    Disassembly sequences ............. 64-65
    Flywheel ..................................... 79
    Front cover, timing chain
      and sprockets ....................... 68-70
    Oil and filter ............................. 36-38
    Oil pressure ................................. 18
    Oil pump ..................................... 71

Removal/installation (F10) .............. 54-60
Removal/installation (310) ....... 61-64, 283
Rocker assembly ............................. 65
Specifications .............................. 80-81
Tightening torques ........................... 82
Troubleshooting ............. 14-17, 24-25
Valve and valve seats ........... 66-68, 283
Evaporative emission
    control system .............. 22, 24, 45, 110
Exhaust gas recirculation
    system (EGR) ......... 24, 100-105, 287, 291
Exhaust system ..................... 110-119, 293

### F

Fan ................................................ 126
Fenders ........................................... 244
Flywheel .......................................... 79
Front cover .................................... 68-70
Fuel evaporation
    control system ............................ 22, 24
Fuel filter ........................................... 45
Fuel pump ...................................... 97-99
Fuel stop checks ..................... 33, 35-36
Fuel system
    Air cleaner ................. 83-88, 286-287
    Air induction system ......... 109-110, 291
    Air injection system ......... 107-109, 293
    Automatic choke circuit ....... 93-96, 287
    Carburetor ........... 89, 91-93, 96-97, 287
    Early fuel
      evaporative system ............ 105-106
    Evaporative emission
      control system ............... 22, 24, 110
    Exhaust gas recirculation
      system ................. 100-105, 287, 291
    Fuel pump ................. 19-20, 97-99
    Fuel tank and lines ................. 120-122
    Manifolds, intake and exhaust .......... 99
    Shutoff system ........................... 293
    Throttle linkage ..................... 119-120
    Throttle opener ............................ 99
    Troubleshooting ........................ 18-20
Fuel tank and lines ..................... 120-122
Fuses and fusible links ................. 159-160

### G

General information ........................... 1-8
Grille ............................................. 244

### H

Headlights ............................. 25-26, 299
Heater .................................. 26, 127-129
Hinges ............................................. 38
Hood .............................................. 249
Horn ........................................ 155-156

# INDEX

## I

Ignition system
    Coil resistance test .................................. 163
    Distributor ................................................ 165
    Ignition coil replacement ....................... 164
    Ignition unit removal/installation ........ 164
    Power supply test ..................................... 163
    Primary circuit test .................................. 163
    Spark plug timing
        control system ................................. 165-168
    Testing ....................... 160-161, 163, 299-301
    Timing ................................................... 52-53
    Troubleshooting .................................... 13-14
Instrument panel ............................... 263-267
Instruments ........................................ 151-155

## L

Latches ......................................................... 38
Leaf springs .......................................... 223-224
Leak inspection ...................................... 38-39
Lighting system ............. 25-26, 145-147, 299
Locks ..................................................... 38, 255
Lubrication (see Maintenance)

## M

Maintenance (also see Tune-up)
    Air cleaner ................................................ 45
    Air filter .............................................. 45, 281
    Air induction valve filter ................. 45, 281
    Ball-joints ................................................. 46
    Brake adjustment .................................. 219
    Brake fluid ................................................ 45
    Brakes ....................................................... 38
    Choke plate and linkage ........................ 45
    Compression test ............................... 47-48
    Coolant hoses and connections ........... 40
    Drive belts (F10) ...................................... 39
    Drive belts (310) ............................... 39, 278
    Engine oil and filter ........................... 36-38
    Evaporative emission
        control system ..................................... 45
    Front and rear wheel bearings ............. 46
    Fuel filter .................................................. 45
    Fuel stop checks ........................... 33, 35-36
    Hinges, latches, locks ............................. 38
    Leak inspection ................................... 38-39
    Manual transmission oil .................. 38, 46
    PCV system ............................................. 46
    Schedule ................................................... 34
    Steering and suspension .................. 45, 281
    Vacuum lines ............................... 40-44, 278
    Wheels and tires ..................................... 45
Manifold, intake and exhaust .................... 99

Manual transmission oil ............................. 46
Master cylinder (brakes) ............... 208-211, 303

## O

Oil pressure light ......................................... 18
Oil pump ..................................................... 71

## P

Positive crankcase
    ventilation (PCV) ......................... 21-22, 46
Power steering .................................... 304-307

## R

Radiator, removal/installation ................ 125
Rocker assembly ......................................... 65

## S

Safety precautions .................................... 3-4
Seats ......................................................... 267
Service hints ............................................. 2-3
Shock absorbers ....................................... 233
Spark plugs ............................................. 48-50
Spark timing control system ....... 165-168, 301
Specifications
    Clutch .................................................... 171
    Engine ................................................ 80-81
    Fuel system ............................................ 90
    Steering and suspension .................... 242
    Tune-up .................................................. 49
Springs ..................................................... 233
Sprockets .............................................. 68-70
Starter ............................................ 9-10, 141-145
Steering and suspension, front
    Axis inclination .................................... 232
    Column ................................................. 241
    Gear and linkage ........................... 241-242
    Lock angles ........................................... 232
    Maintenance ................................... 45, 281
    Power steering .............................. 304-307
    Specifications ....................................... 242
    Torques ................................................. 243
    Troubleshooting .................................... 29
    Wheel ............................................. 240-241
Suspension, front
    Axle shafts ..................................... 238-240
    Ball-joints ...................................... 236-237
    Caster and camber .............................. 232
    Toe-in .................................................... 232
    Shock absorber .................................... 233
    Spring replacement ............................. 233
    Sway bar ........................................ 233-234
    Transverse links ........................... 234-236
    Wheel alignment .......................... 230, 304
    Wheel bearings .................................... 237

# INDEX

Suspension, rear
    Axle tube .................................. 225-227
    Coil springs .................................... 221
    Leaf springs ............................... 223-224
    Rear wheel bearings ................... 227-229
    Suspension arm ........................... 221-223
    Tightening torques ....................... 20, 225
    Troubleshooting .............................. 29
Sway bar ........................................ 233-234
Switches ............................. 25-26, 147-151

## T

Thermostat
    Removal/testing ......................... 124-125
    Installation ..................................... 125
Throttle linkage ............................... 119-120
Throttle opener ..................................... 99
Tightening torques
    Clutch ............................................ 172
    Engine ............................................ 82
    Suspension, front ............................. 243
    Suspension, rear ......................... 20, 225
Timing chain ........................................ 69-70
Tires ................................................... 29, 32
Tools ..................................................... 4-8
Transmission, automatic
    Adjustment ..................................... 200
    Removal/installation ................... 176-177
    Troubleshooting .............................. 28
Transmission, manual
    Five-speed
        Adjustment .................................. 199
        Assembly ............................... 194-196
        Disassembly .......................... 190-194
        Inspection .................................... 194
        Shift linkage .......................... 196-198
    Four-speed
        Adjustment .................................. 199
        Assembly ............................... 186-190
        Disassembly .......................... 177-185
        Inspection ............................. 184-186
        Removal/installation ............. 176-177
        Shift linkage .......................... 196-198
        Troubleshooting ........................... 28
Transverse links ............................... 234-236

Troubleshooting
    Air conditioning ............................... 276
    Brakes ......................................... 28-31
    Charging system ...................... 9, 11-14
    Clutch .......................................... 27-28
    Cooling system ............................ 26-27
    Electrical accessories .................. 25-26
    Emission control system ............. 20-24
    Engine ............................... 14-17, 24-25
    Engine oil pressure light .................. 18
    Equipment ....................................... 5-8
    Fuel system ................................. 18-20
    Ignition system ............................ 13-14
    Starting system ............................. 9-10
    Steering ............................................ 29
    Suspension ...................................... 29
    Tires and wheels ......................... 29, 32
    Transmission ................................... 28
Trunk lid ........................................... 258-261
Tune-up
    Description ...................................... 46
    Distributor ................................... 50-53
    Equipment ....................................... 5-8
    Idle speed adjustment .............. 281-283
    Ignition timing ............................. 51-52
    Specifications ............................ 49, 281
    Spark plugs ................................. 48-50
    Valve adjustment ............................. 48

## V

Vacuum lines ............................. 40-44, 278
Valve adjustment ................................... 48
Valves and valve seats ............... 66-68, 283
Voltage regulator ............................... 9-13

## W

Water pump ......................................... 127
Wheels
    Alignment ............................... 230, 304
    Balancing ........................................ 32
    Bearings ............................. 46, 227-229
Wheels and tires ................................... 45
Windows ...................................... 255-258
Windshield wipers and washers ..... 26, 156-158
Wiring diagrams ......................... end of book

# 1976 F10 WIRING DIAGRAM — PART I

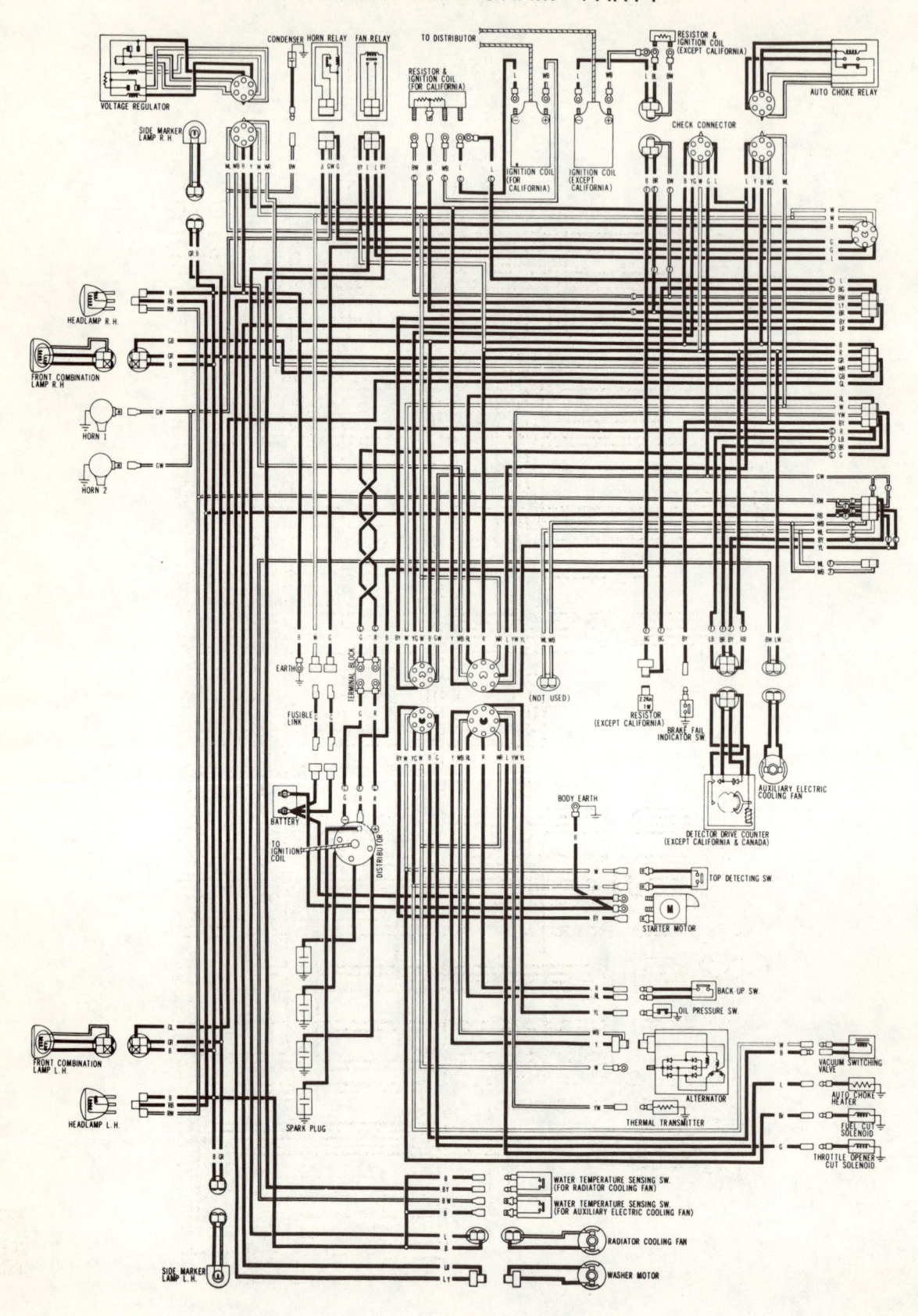

# 1976 F10 WIRING DIAGRAM—PART II

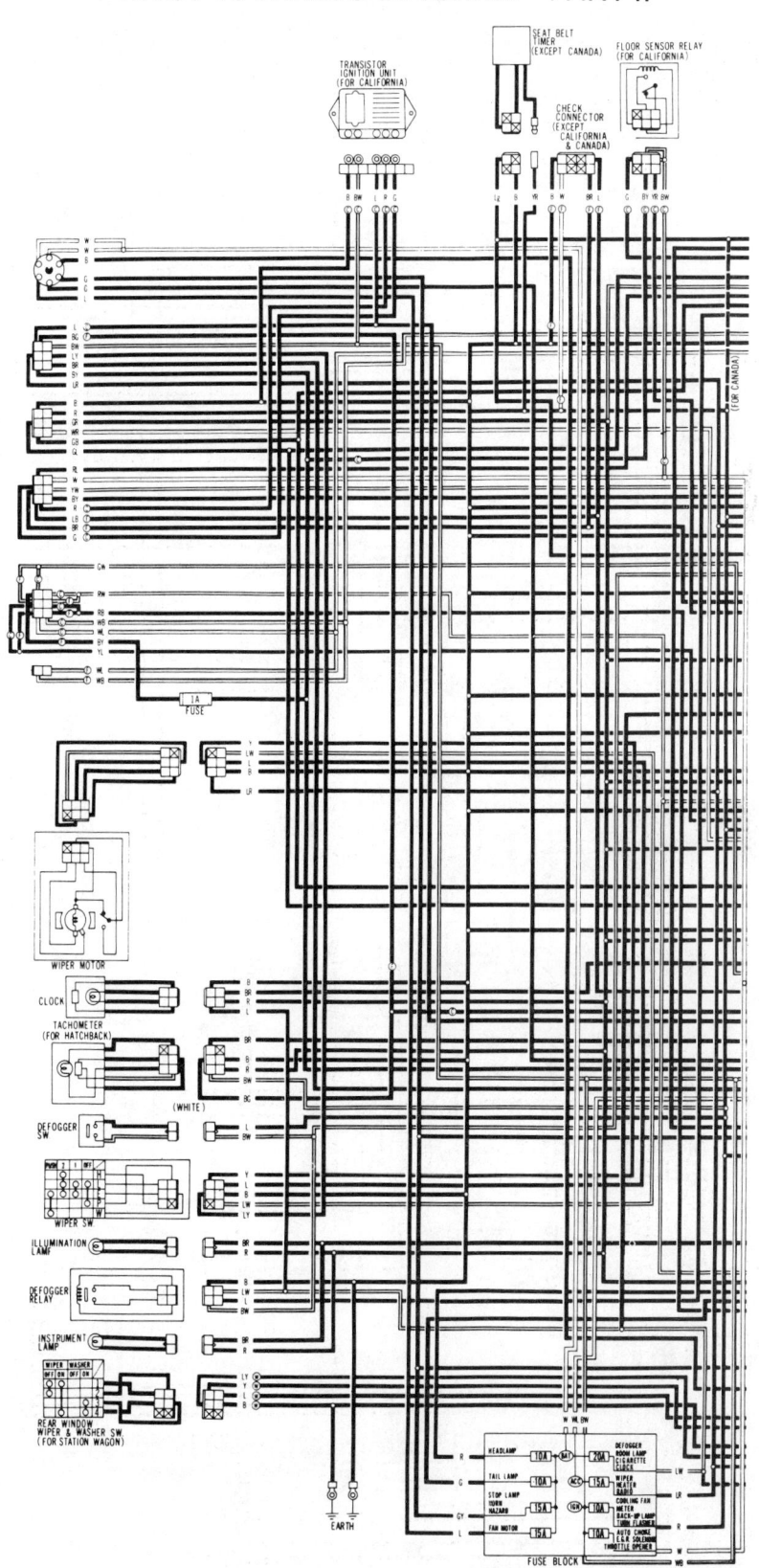

# 1976 F10 WIRING DIAGRAM — PART III

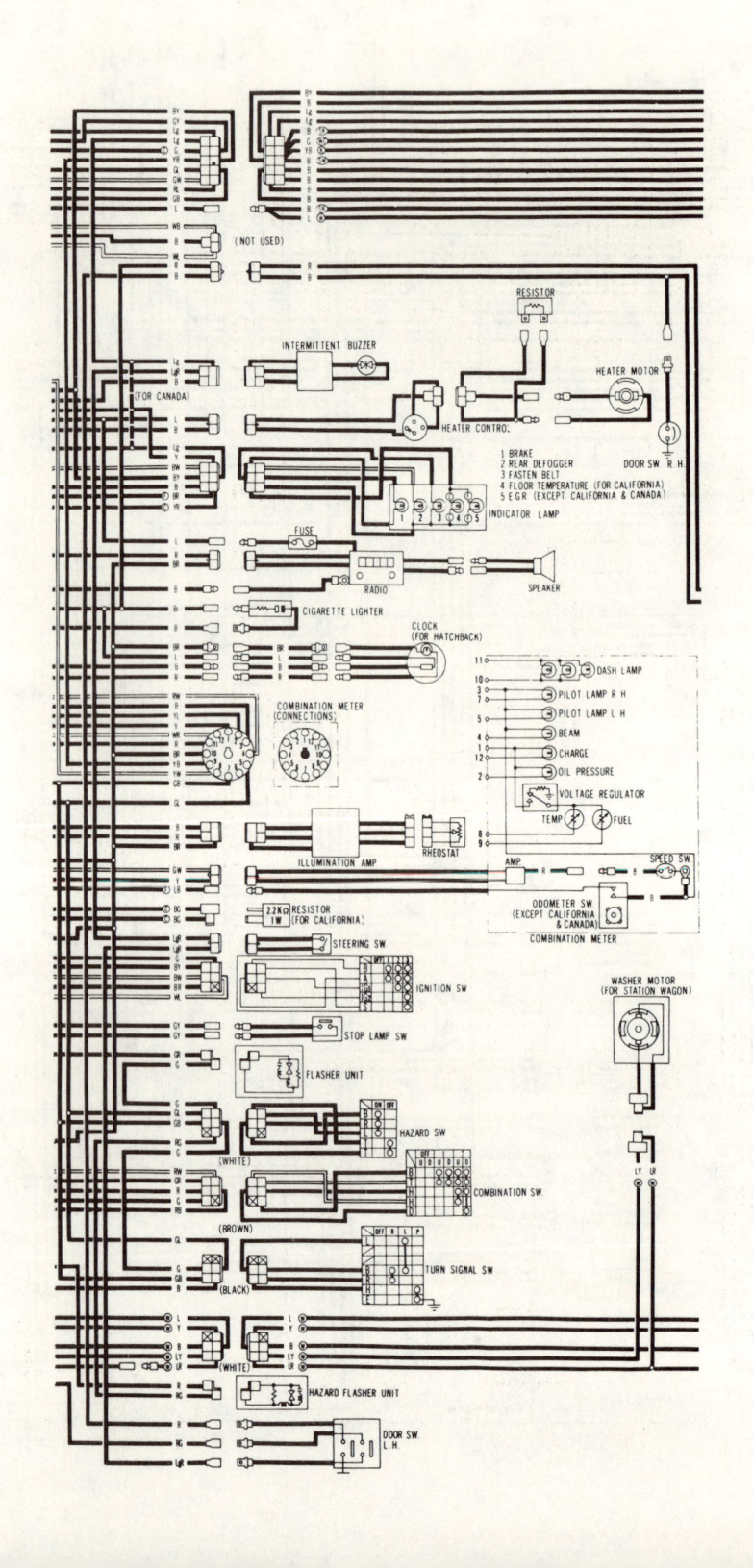

# 1976 F10 WIRING DIAGRAM—PART IV

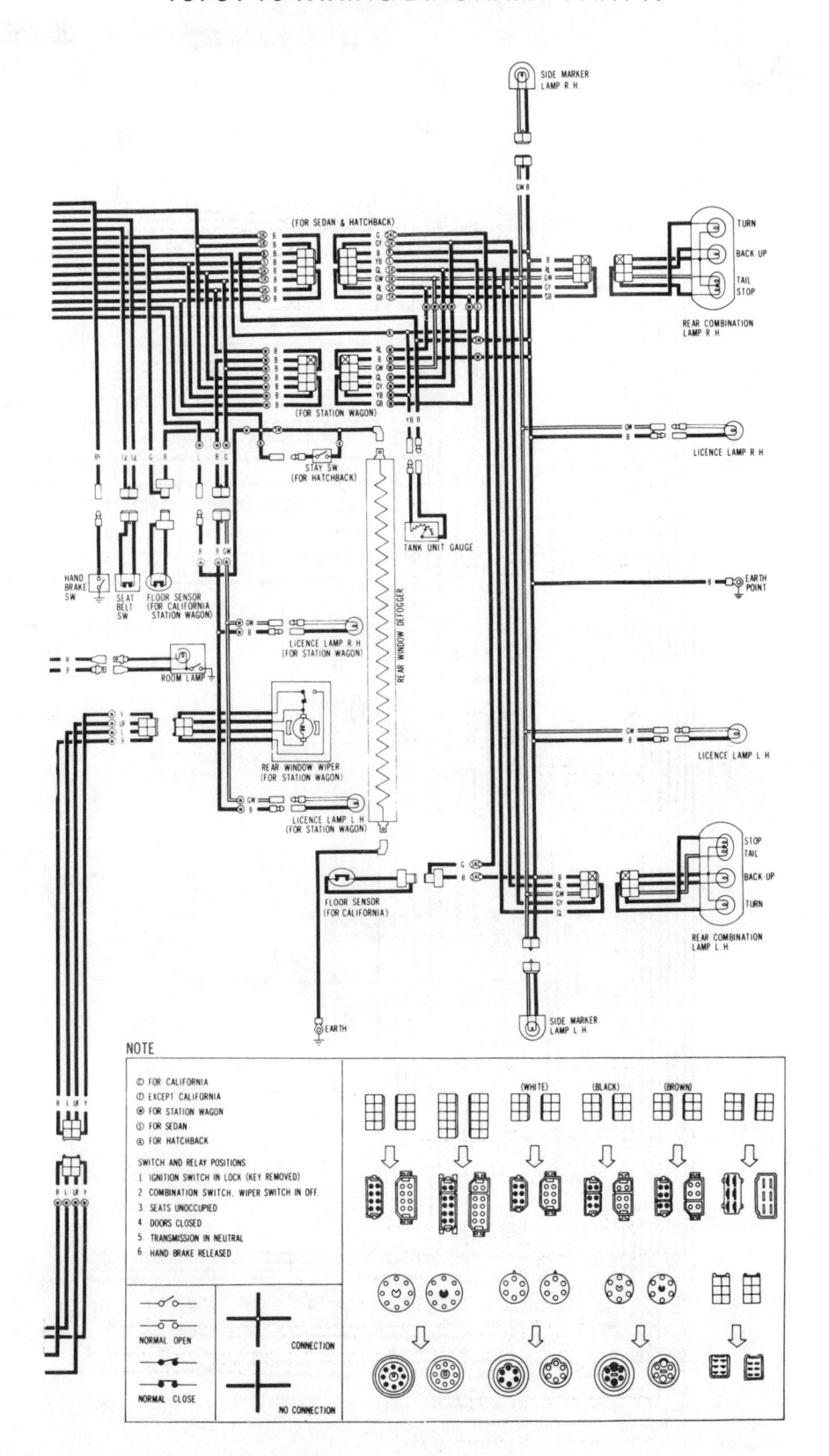

1977 F10 WIRING DIAGRAM—PART I

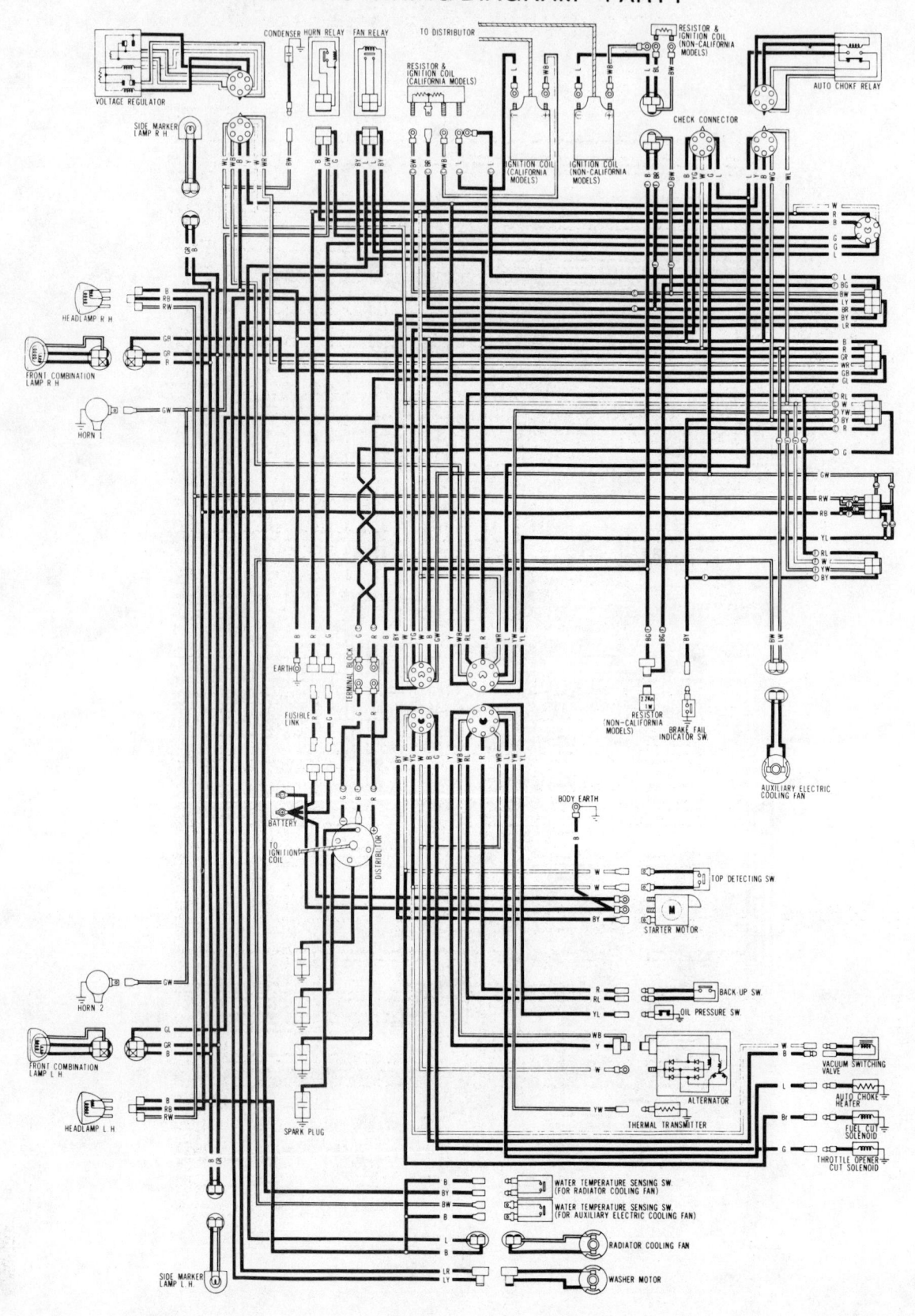

## 1977 F10 WIRING DIAGRAM — PART II

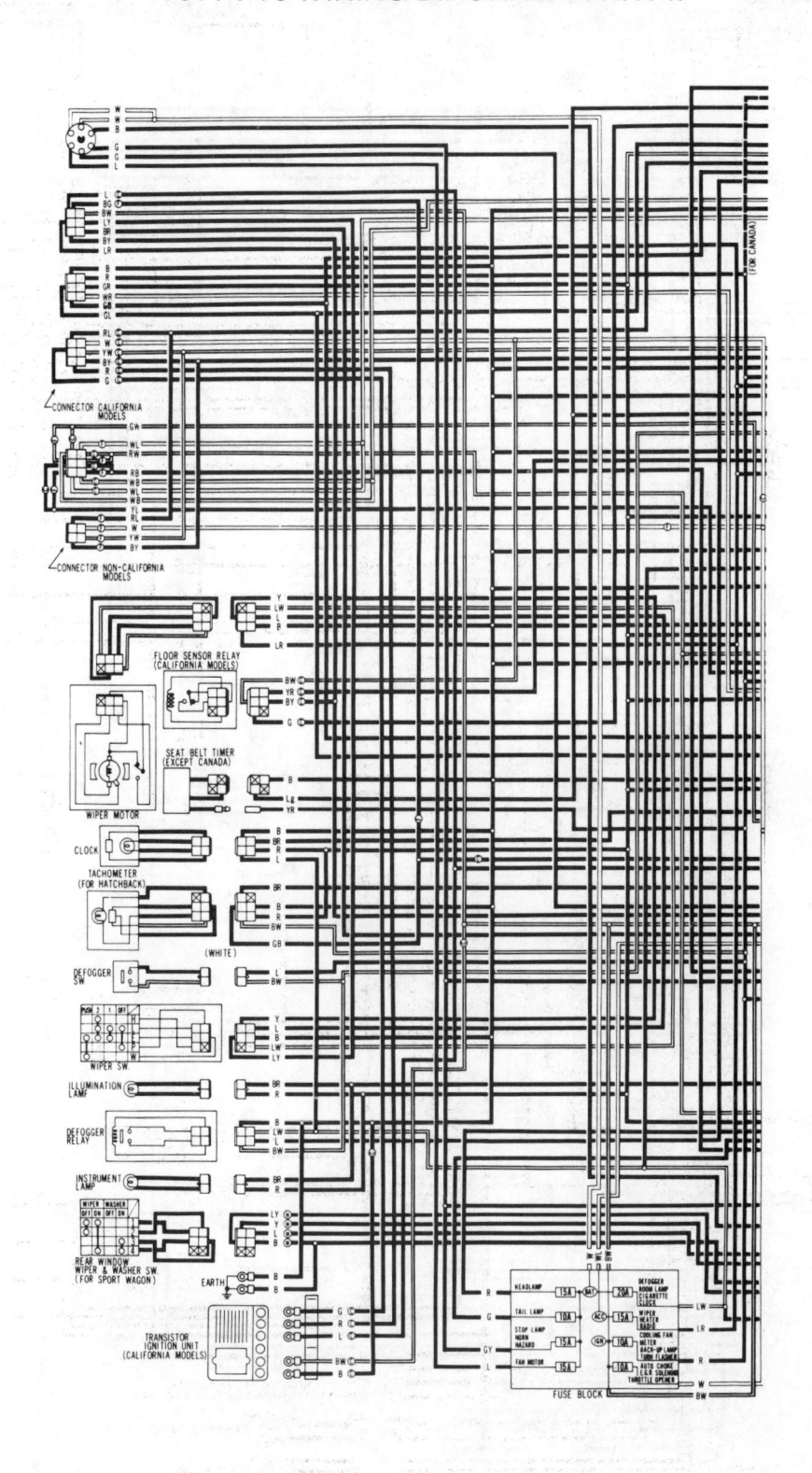

## 1977 F10 WIRING DIAGRAM—PART III

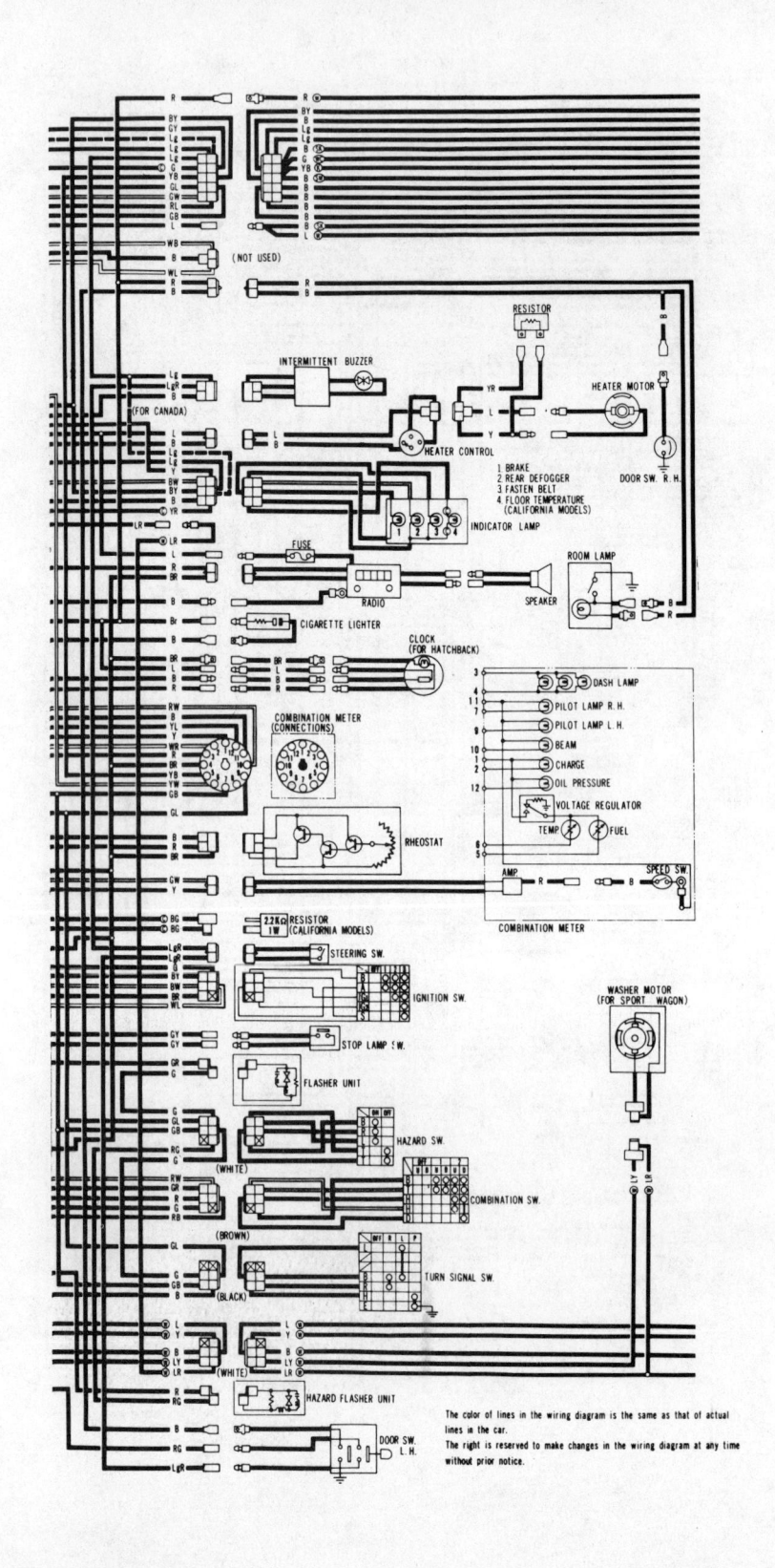

# 1977 F10 WIRING DIAGRAM—PART IV

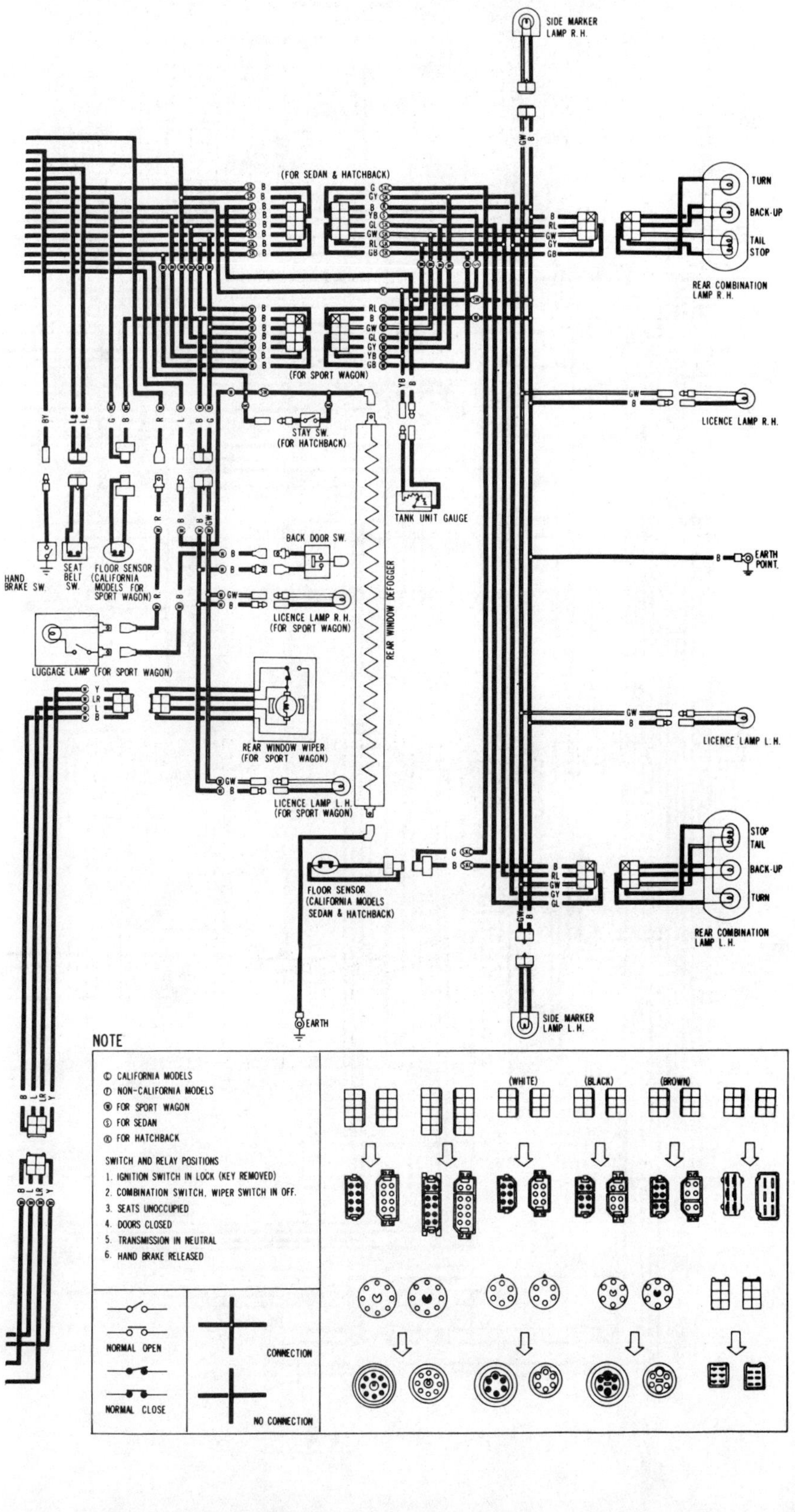

# 1978 F10 WIRING DIAGRAM—PART I

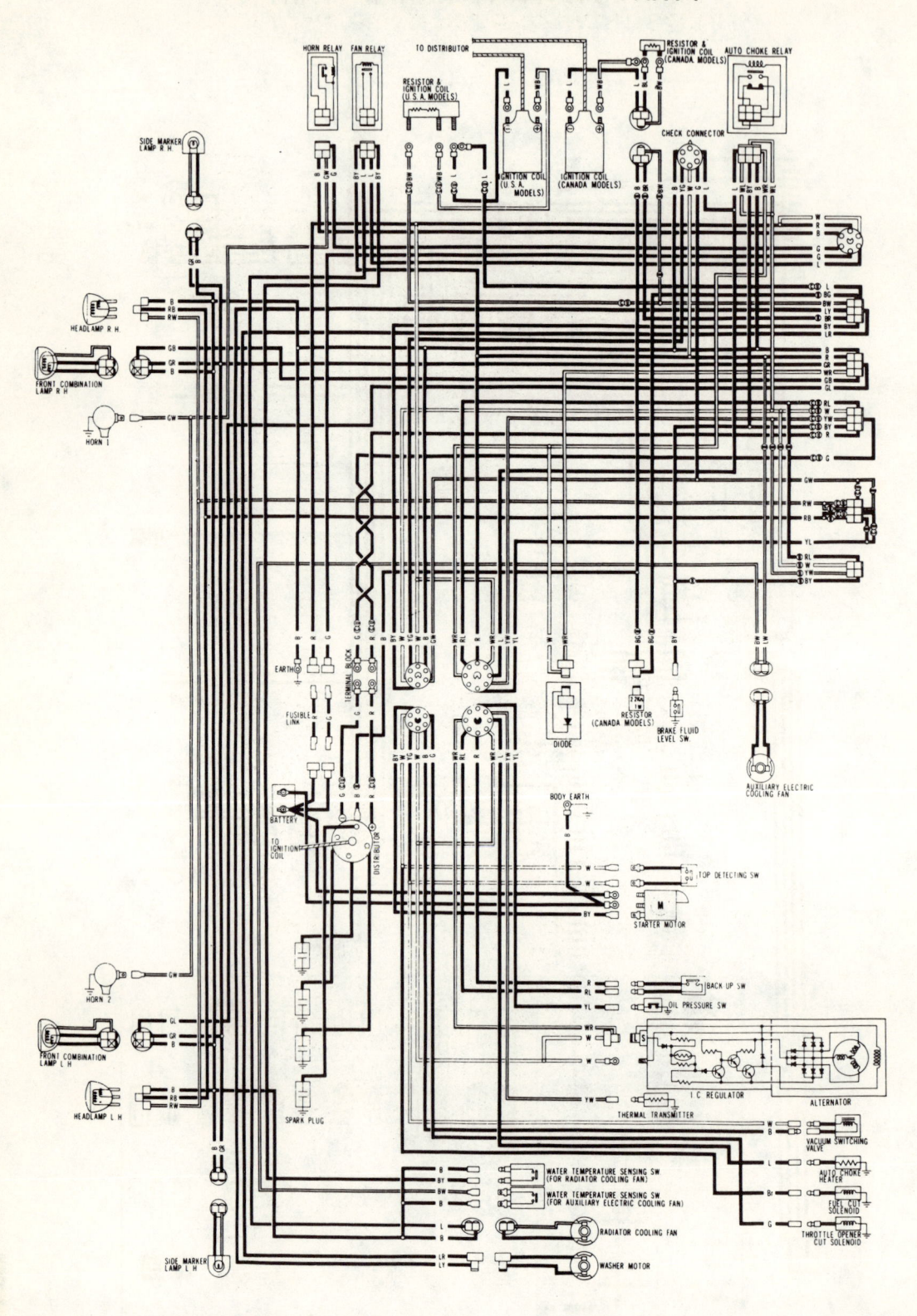

# 1978 F10 WIRING DIAGRAM — PART II

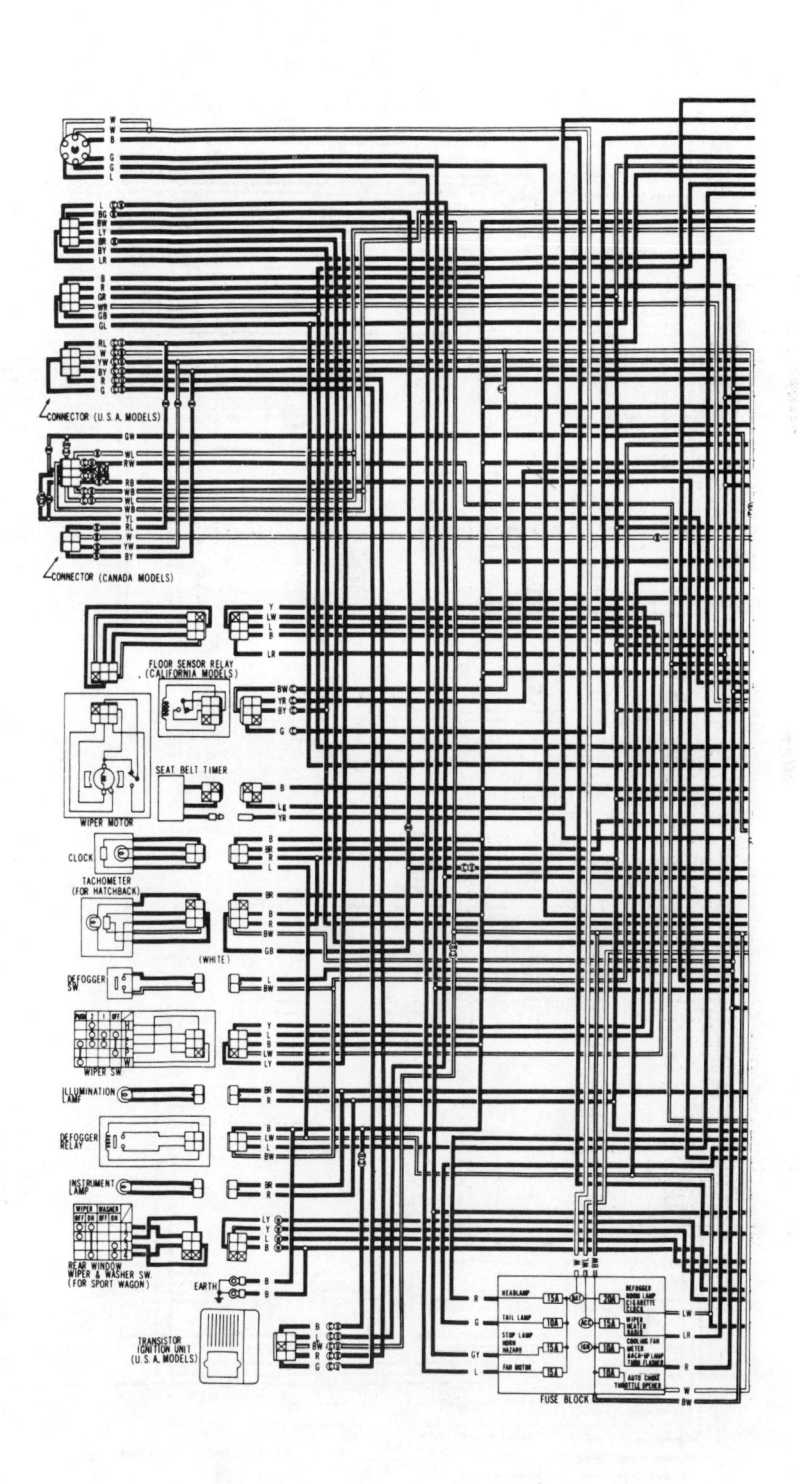

# 1978 F10 WIRING DIAGRAM—PART III

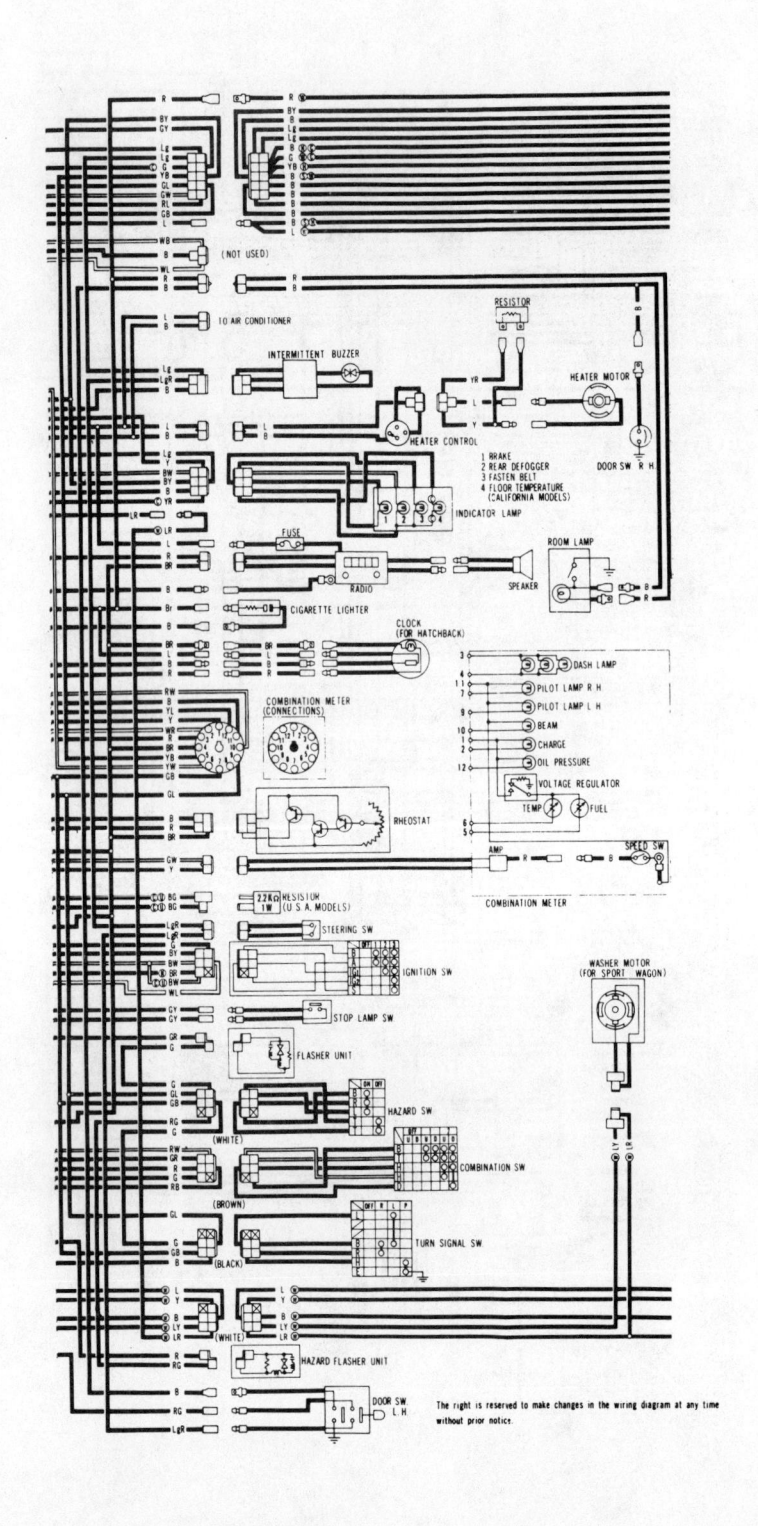

# 1978 F10 WIRING DIAGRAM — PART IV

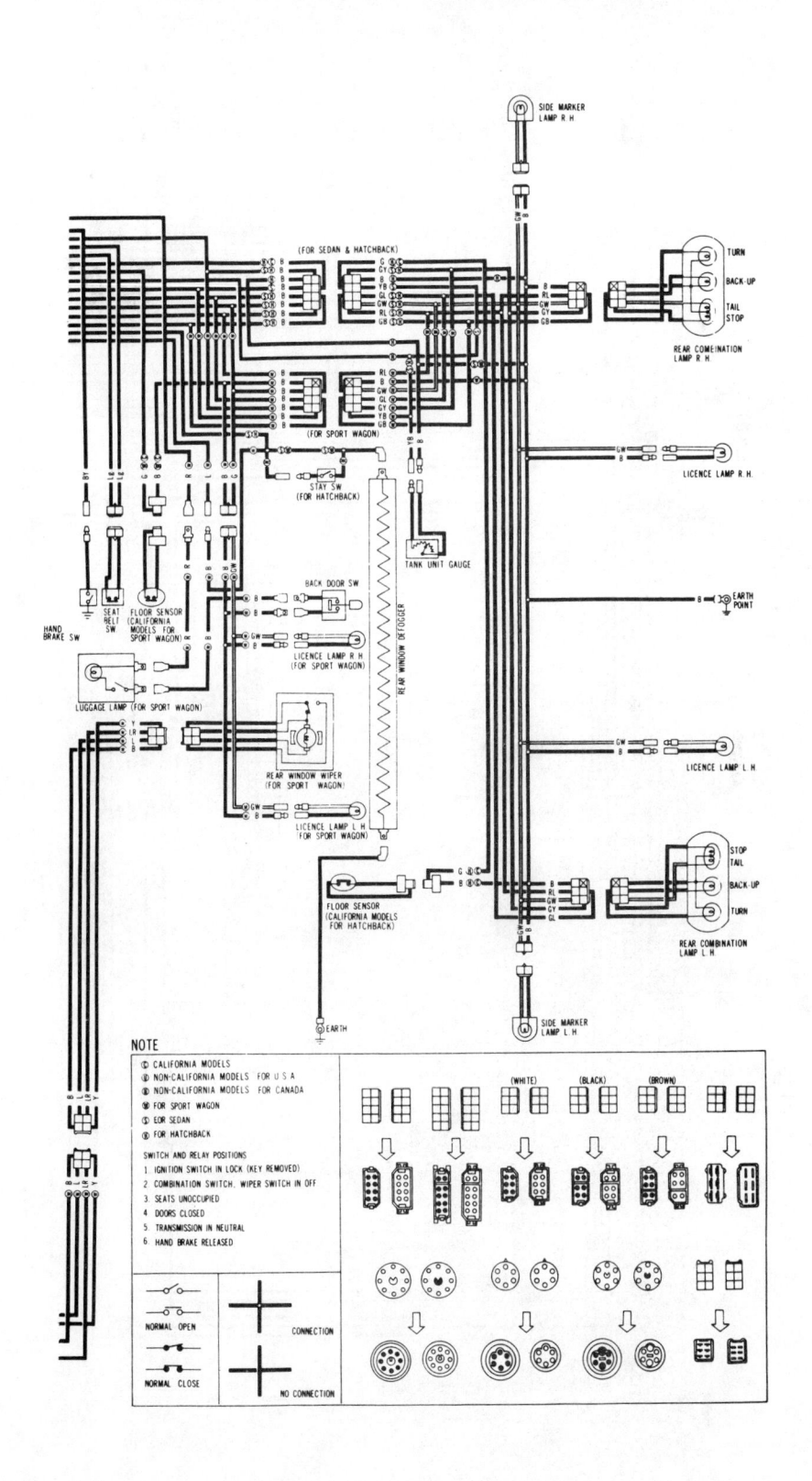

1979 310 WIRING DIAGRAM — PART I

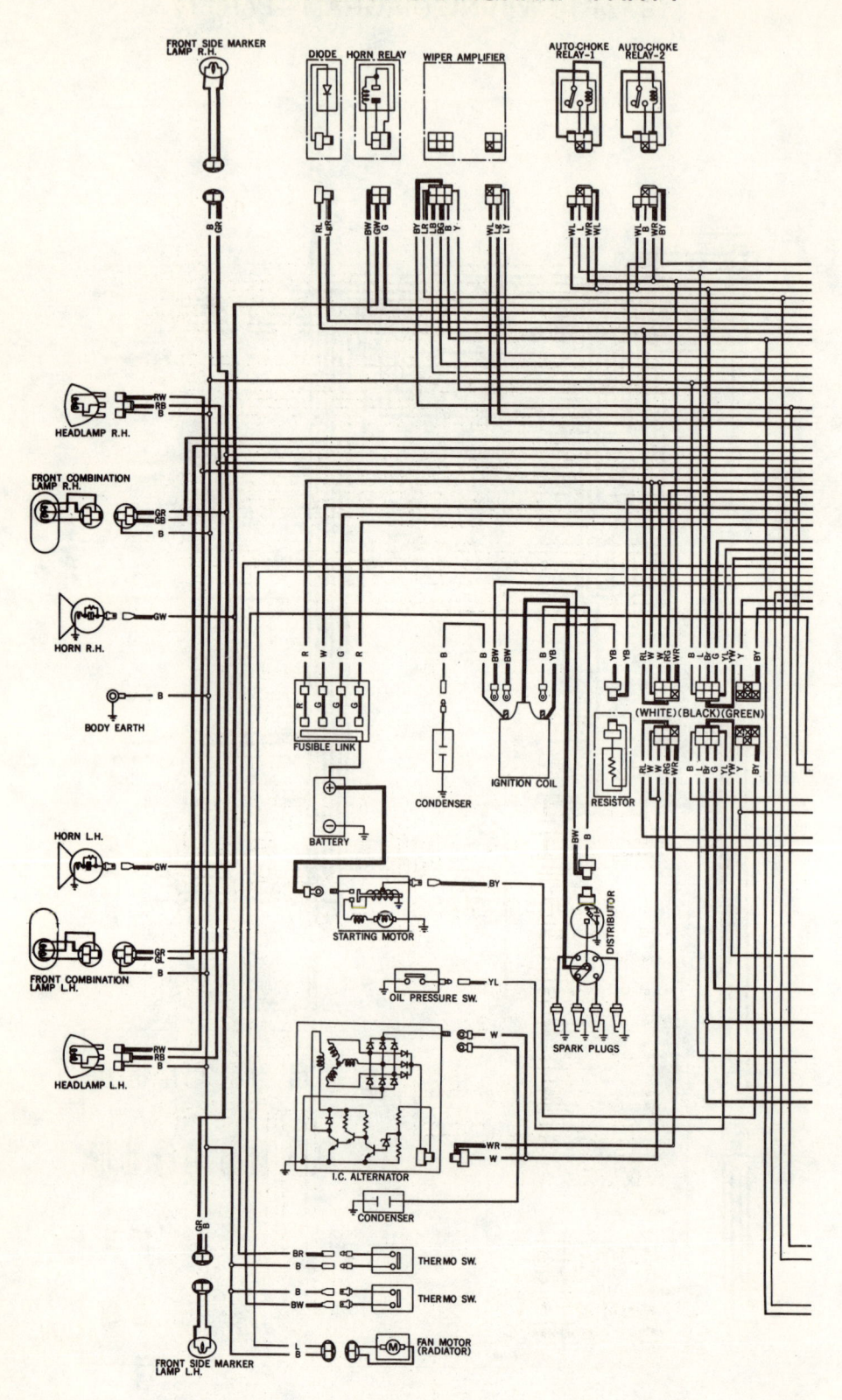

## 1979 310 WIRING DIAGRAM — PART II

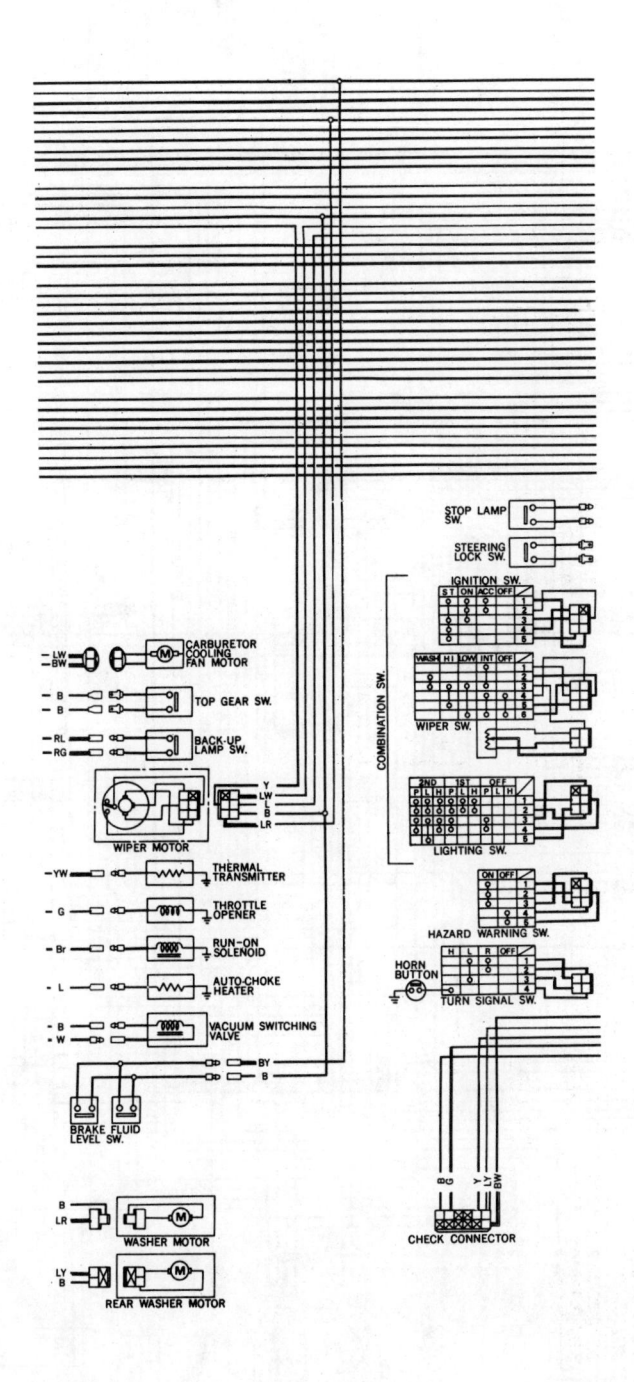

# 1979 310 WIRING DIAGRAM — PART III

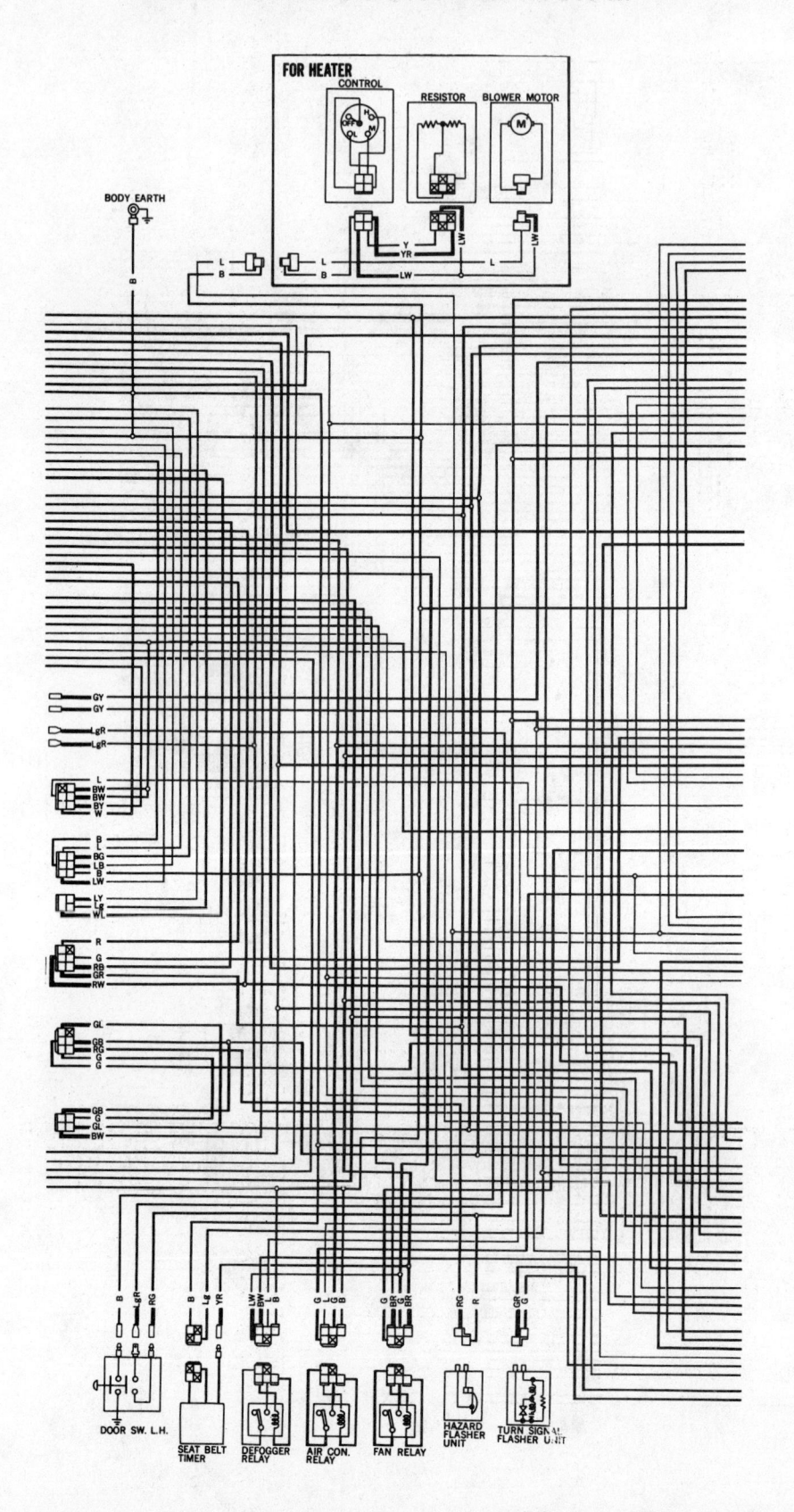

# 1979 310 WIRING DIAGRAM—PART IV

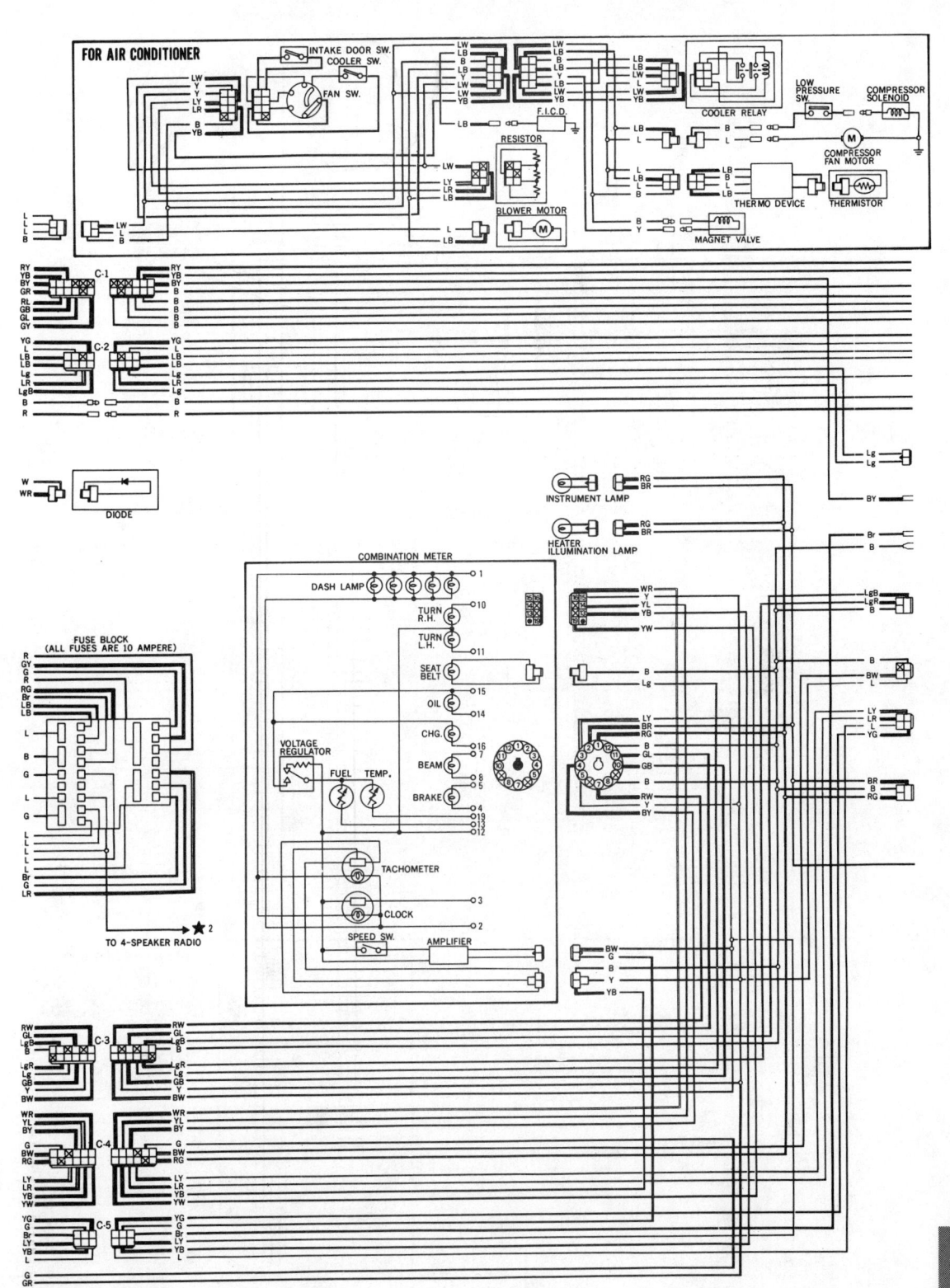

# 1979 310 WIRING DIAGRAM—PART V

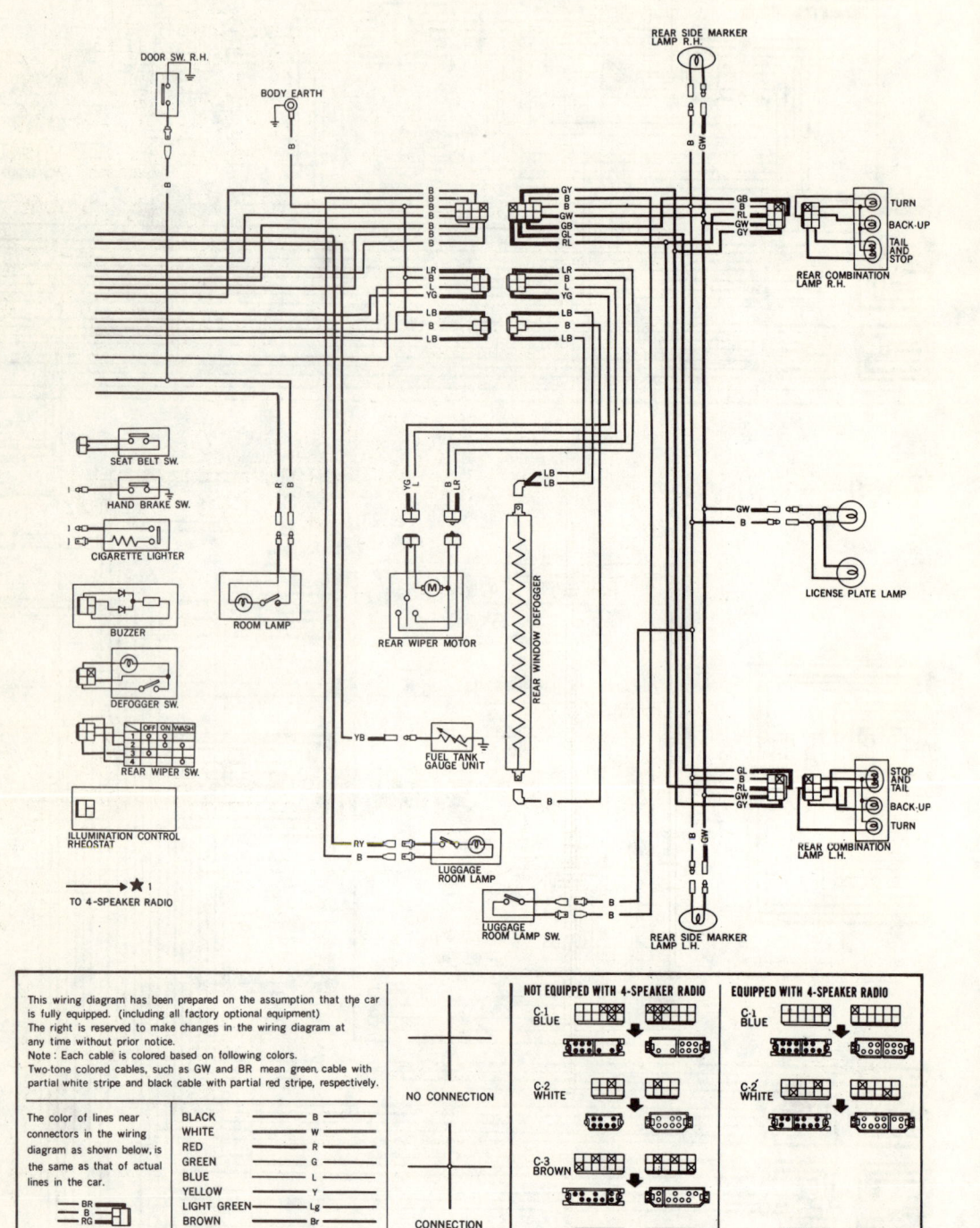

**NISSAN MOTOR CO., LTD.**

# 1980 310 WIRING DIAGRAM — PART I

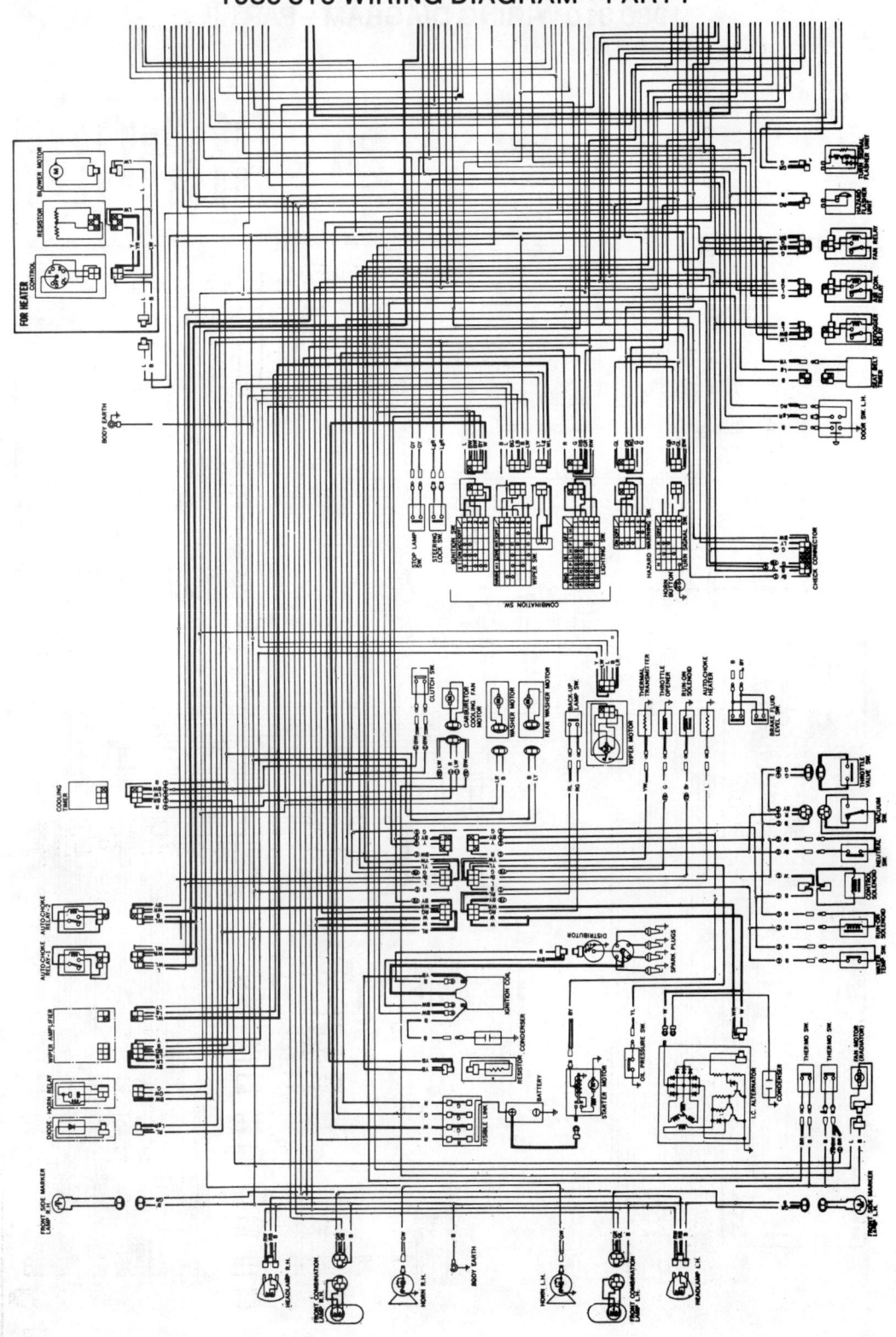

# 1980 310 WIRING DIAGRAM—PART II

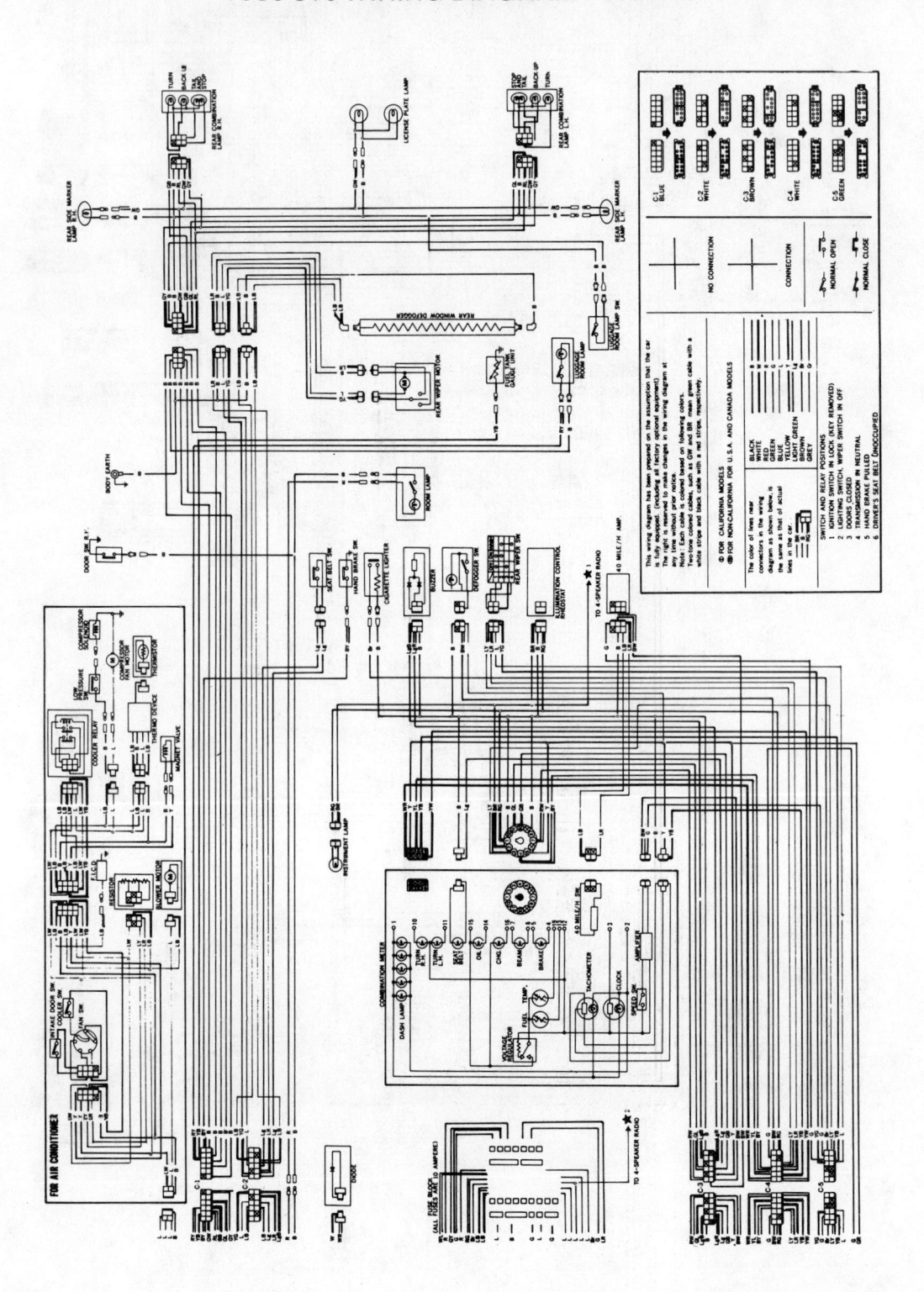

# 1980 310 WIRING DIAGRAM — PART III

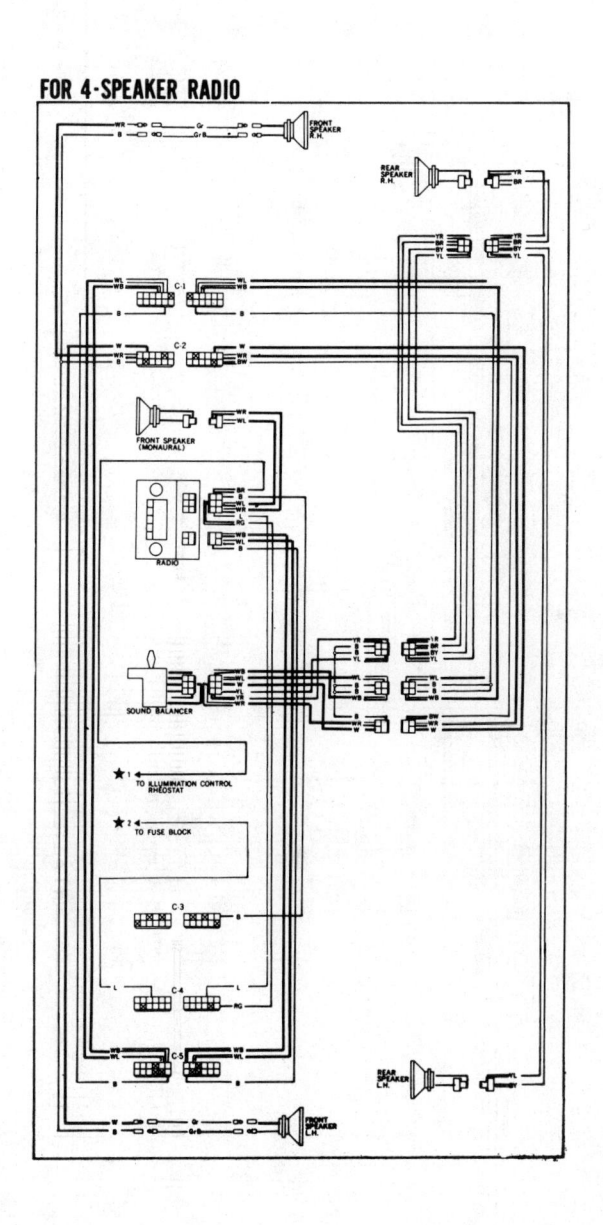

# 1981 310 WIRING DIAGRAM—PART I

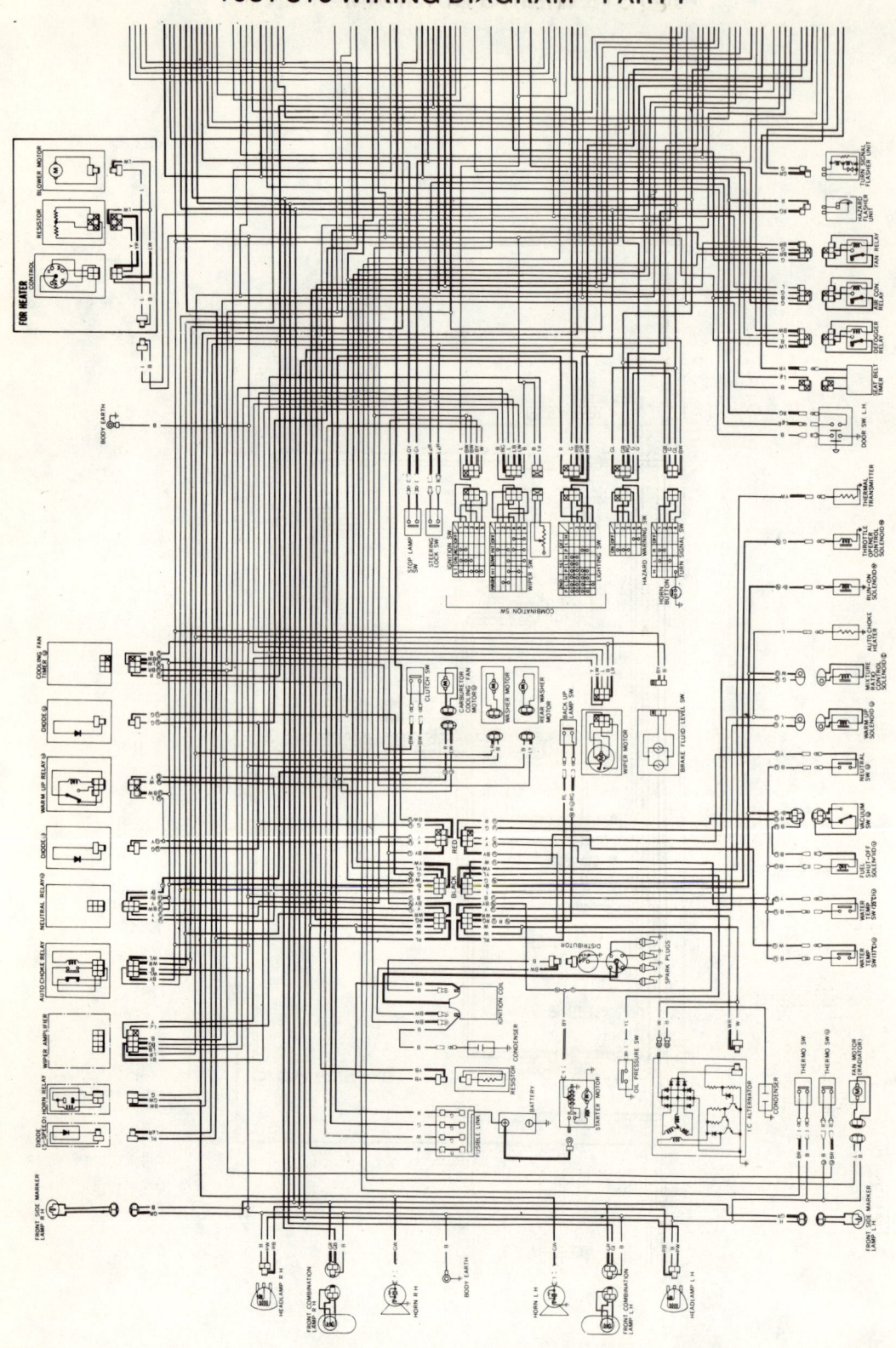

## 1981 310 WIRING DIAGRAM—PART II

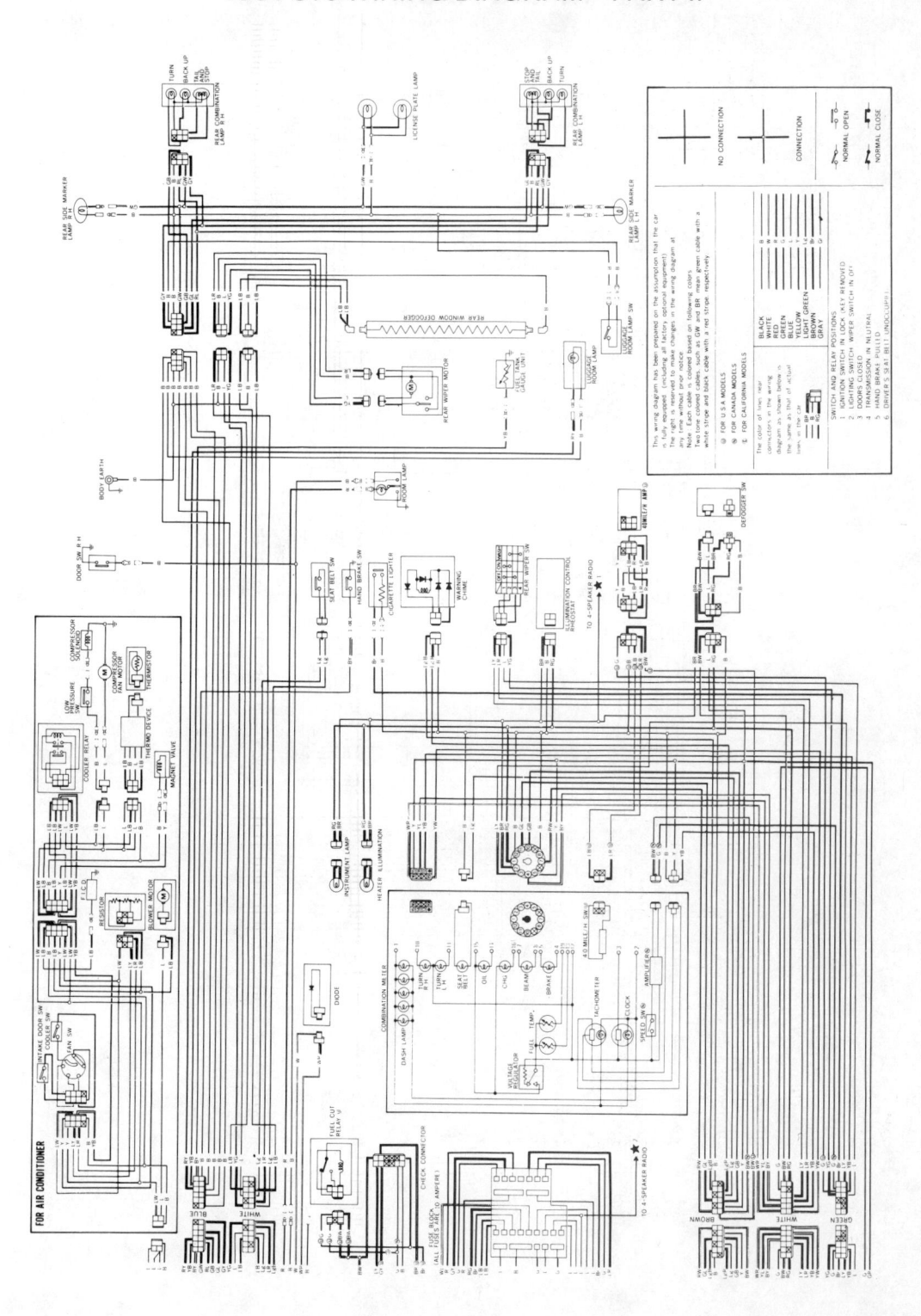

# 1981 310 WIRING DIAGRAM—PART III

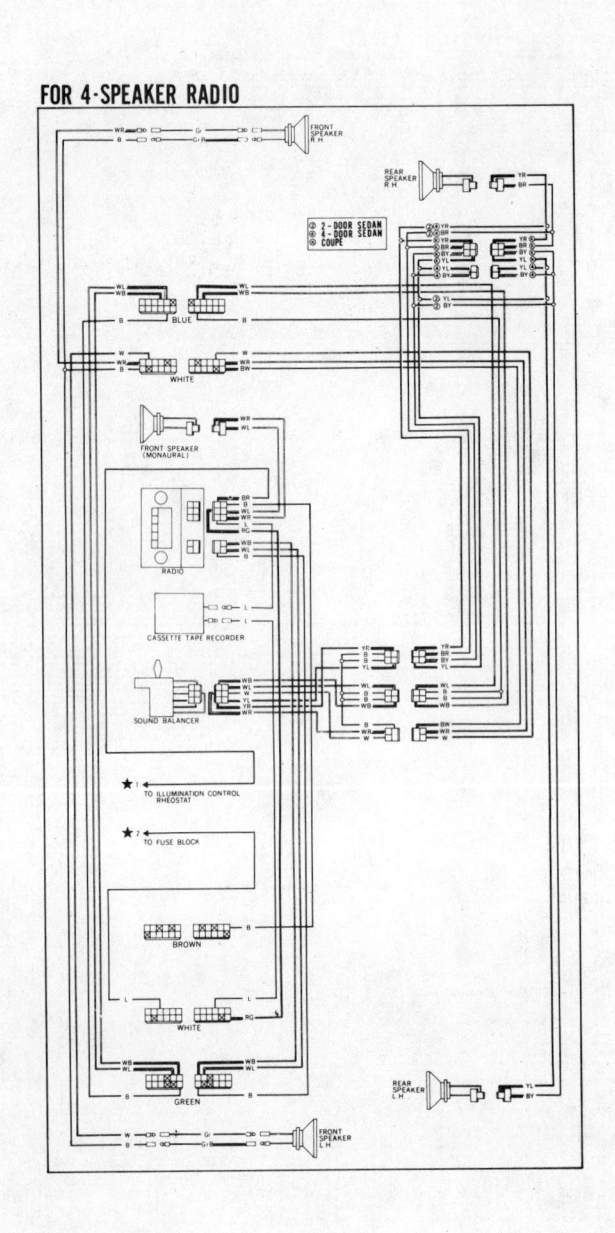